INTRODUCTION TO Marine Biology

INTRODUCTION TO Marine Biology

George Karleskint, Jr.
St. Louis Community College–Meramec

Saunders College Publishing

Harcourt Brace College Publishers

Philadelphia   Fort Worth   Chicago   San Francisco
Montreal   Toronto   London   Sydney   Tokyo

Requests for permission to make copies of any part of the work
should be mailed to: Permissions Department, Harcourt Brace &
Company, 6277 Sea Harbor Drive, Orlando, Florida 32887-6777.

Publisher: Emily Barrosse

Executive Editor: Edith Beard Brady

Marketing Manager: Erik Fahlgren

Senior Development Editor: Gabrielle Goodman

Project Editor: Frank Messina

Director of Production: Joanne Cassetti

Art Director and Text Designer: Carol Bleistine

Photo Researcher: Jane Sanders

Cover Designer and Icon Artist: Ruth Hoover

Cover Photo: Butterfish within the stinging tentacles of a jellyfish
(© Norbert Wu).

Printed in the United States of America

Introduction to Marine Biology

ISBN: 0-03-074191-2

Library of Congress Catalog Card Number: 97-67377

7890123456   032   10 987654321

Dedication

This book is dedicated to

My father, who taught me a love of nature

My mother, who taught me that I could achieve anything if I set my mind to it

My wife, who encouraged me and tolerated the time away from the family while I was writing

My children Brad, Kristy, Tim, and Samantha, who share in my enthusiasm for the sea and its creatures and will inherit the ocean from my generation.

Introduction to Marine Biology is intended for undergraduate students majoring in a variety of disciplines. Many such students may already be interested in the field of marine biology. Others, however, are studying marine biology to fulfill a general education requirement, and they may have a fear of science courses. Like many, these students are generally intrigued by the marine environment and especially marine organisms. Having grown up with Jacques Cousteau and television programs dealing with the ocean and marine organisms, they have a natural interest in this subject area. This text strives to use this interest as a starting point for teaching biological science. For the past 17 years, I have taught introductory marine biology to both majors and nonmajors. In my course, I introduce students to the fundamentals of science—such as the scientific method and basic physical science—as a foundation for understanding marine biology. In both my lectures and field courses, I stress the ecological approach to the study of marine organisms, and I have used this same approach in preparing this text.

The main focus of the text is the *ecology of the marine environment*—that is, the ways in which marine organisms interact with each other and with their physical environment. It is my firm belief that there is no better way to teach about the delicate balance of natural systems than within the context of marine ecosystems. This text also strives to educate students about the importance of marine ecosystems to terrestrial ecosystems and to humankind. I hope that by studying marine biology students will be able to make well-informed decisions when they prepare to vote on legislation or engage in activities that have an impact on the natural world, especially with respect to the oceans. An understanding of topics such as marine pollution and overfishing will be important for our future citizens, and the better informed students are about

such issues, the more likely they are to make sound decisions. It is my hope that this text will provide the factual foundation that is necessary for making such important decisions.

LEARNING AIDS

This text contains the following learning aids to help students master the text material and successfully use their knowledge.

1. **Key Concepts.** These appear at the beginning of each chapter and identify important principles that students should learn as they study the chapter's content.
2. **Boldface Terms.** Throughout the text, important terms appear in **boldface** and are defined.

3. **Pronunciation Guide.** New terms in the text that might pose pronunciation problems for beginning students are accompanied by phonetic transcriptions.

4. **Boxed Readings.** Boxed readings are organized into three categories: Ecology and the Marine Environment, Marine Biology and the Human Connection, and Marine Adaptations. The goals of the boxed readings are to provide students with an interesting and engaging focus on three important aspects of marine biology and to allow instructors to structure their boxed reading assignments (by using the three categories) to fit the course emphasis.

5. **Chapter Summary.** This summary is a synopsis of a chapter's main points. Students will find it useful in reviewing the chapter's overall content.

6. **Selected Key Terms.** This section following the Chapter Summary provides students with a convenient review and helps them with the often daunting task of mastering the language of biology. Each chapter's most important boldface terms are listed in order of appearance with page references directing students to the specific locations in the chapter where the terms were defined.

7. **Questions for Review.** There are three levels of these end-of-chapter questions. The first level comprises objective, **Multiple-Choice** questions that deal primarily with vocabulary and terminology. The second level consists of **Short-Answer** essay questions that test students' recall of facts and their ability to synthesize information. The short-answer essay questions are organized according to the traditional learning taxonomy. The last level of question, **Thinking Critically,** involves higher-order thinking skills and problem solving. The answers to the first-level Questions for Review are provided in the Appendix of the text. Suggested responses to the other questions are found in the *Instructors Manual.*

8. **Suggestions for Further Reading.** At the end of each chapter is a short list of articles and books that supplement the chapter's content. Most of the readings are taken from the popular scientific literature, such as *National Geographic, Natural History, Smithsonian, Discover,* and *Scientific American.* These articles and books are written at a level appropriate for students using this text and are therefore more likely to be enjoyed by them.

9. **Separate Glossary.** Located at the end of the text, the Glossary defines all of the boldface terms that appear in the text, as well as many other biological terms used in the text but not boldfaced.

THE ORGANIZATION OF *INTRODUCTION TO MARINE BIOLOGY*

Part 1 The Ocean Environment

Chapter 1 introduces the science of marine biology, presents the scientific method, and orients the student to the rest of the text. Chapter 2 presents the basic principles of ecology. This information is presented early in the text to support the main theme of ecology of the marine environment. Chapters 3 and 4 introduce the physical aspects of the marine environment and their importance to the organisms that live in the ocean.

Part 2 Marine Organisms

Chapter 5 introduces the student to the evolutionary processes that have led to the development of the various marine organisms and presents the scheme of biological classification that is used to identify and organize these organisms. Chapters 6 through 12 survey all of the major groups of marine organisms and examine their interrelationships. These chapters are organized on the basis of feeding relationships, proceeding from organisms that produce their own food to those that rely on other organisms for food. Descriptions of animals are presented in a traditional format, beginning with invertebrates and working upward through the vertebrate classes to mammals. The focus of the individual chapters is the role that each group of organisms plays in the overall web of marine life.

Part 3 Marine Ecosystems

Chapters 13 to 18 examine the major marine ecosystems. Each chapter in this part examines how the interactions of the physical and biological environment make each ecosystem unique and how these factors influence the number and kinds of marine organisms that can inhabit a given area. The survey begins at the coastline with estuaries (Chapter 13) and the intertidal zone (Chapter 14) and then progresses to offshore ecosystems, coral reefs (Chapter 15), and coastal seas (Chapter 16). Chapter 17 deals with the pelagic ecosystem of the open sea, and Chapter 18 addresses life in the ocean's depths.

Part 4 Humans and the Sea

Part 4 of the text examines the impact that humans have had and continue to have on the marine environment. Chapter 19 addresses the consequences of extracting non-living products from the seas and the importance and impact of commercial fisheries. The chapter also deals with

MediaActive™, Version III is a CD-ROM that contains images from *Introduction to Marine Biology,* as well as figures from other Saunders College Publishing 1998 biology titles. MediaActive is available as a presentation tool to be used in conjunction with Saunders' LectureActive™ presentation software (available with MediaActive™) or with commercial packages such as PowerPoint™ and Persuasion™.

Saunders Web Site provides additional interactive resources for *Introduction to Marine Biology.* Contact our Web site at: http://www.hbcollege.com

how the utilization of these natural resources has changed over the last half century, and it discusses the effects of this change on the marine environment. Chapter 20 examines marine pollution and habitat destruction and presents ways in which interested students can act to stem and possibly reverse environmental damage.

SUPPLEMENTS

Introduction to Marine Biology is accompanied by a supplement package that has been designed to aid students in learning and instructors in teaching. It includes the following[1]:

Instructor's Manual/Test Bank was written by Abel Rajab and Anthony Huntley of Saddleback Community College. It includes chapter outlines and summaries of the major topics addressed in *Introduction to Marine Biology,* as well as 30 to 45 review questions for each chapter—in three different formats—that address the text's learning objectives at various cognitive levels.

Transmasters provide templates for acetate transparencies for *every* piece of line art appearing in the text, with each figure labeled with large type for easy classroom viewing.

Computerized Test Bank is available in three formats—IBM 3.5, Macintosh, and Windows—and allows instructors to revise, add, and delete items appearing in the printed *Text Bank.*

---

[1]Saunders College Publishing may provide complimentary instructional aids and supplements or supplement packages to those adopters qualified under our adoption policy. Please contact your sales representative for more information. If as an adopter or potential user you receive supplements you do not need, please return them to your sales representative or send them to:

Attn: Returns Department
Troy Warehouse
465 South Lincoln Drive
Troy, MO 63379

ACKNOWLEDGMENTS

The production of a text such as this one involves the collaborative efforts and creative talents of many individuals. Without the help of editors, reviewers, and other professionals, this book would never have been published. I would like to acknowledge the highly professional and supportive staff at Saunders College Publishing: I am especially grateful to Publisher Emily Barrosse for her help and support in this project; Executive Editor Edith Beard Brady, who rescued this project from oblivion and enabled it to come to completion; Art Development Editor Ray Tschoepe, who did a fabulous job of transforming my stick figures into impressive art; Photo Editor Jane Sanders, who continued to amaze me by finding the perfect photographs and by searching tirelessly for even the most obscure of images that I requested; Project Editor Frank Messina, who carefully shepherded the project through the production process; and Developmental Editor Jennifer Bortel, who guided me through the initial stages of manuscript development. Jennifer was succeeded by Developmental Editor Gabrielle Goodman, and I am especially indebted to Gabrielle for working tirelessly with me in putting this book together and always being there when I needed encouragement, support, and editorial expertise. I greatly appreciate the wonderful design work of Art Director Carol Bleistine and the efforts of Erik Fahlgren in marketing the book. All of these individuals and a host of other dedicated professionals—including copy editors, proofreaders, and editorial assistants at Saunders—made important contributions to this project, and I thank them all.

I would also like to express my gratitude to my colleagues at St. Louis Community College at Meramec for their support and input and to my students, who have provided me with valuable feedback.

Reviewers

I very much appreciate the diligent work done by the following individuals, who took time from their busy schedules to read and make suggestions concerning the manuscript. Their help and input was invaluable in creating the final text.

Chuck Baxter, Professor Emeritus,
Hopkins Marine Station

Richard Beckwitt,
Framingham State University

Paul Billeter,
Charles County Community College

Larry Brand,
University of Miami

David Bridges,
Unity College of Maine

Richard Carlton,
University of Notre Dame

Jerry Carpenter,
Northern Kentucky University

Joe Conner,
Pasadena City College

Elzbet Diaz de Leon,
Ventura College

Megan Dethier,
University of Washington

Phil Duston,
College of Charleston

Nicholas J. Ehringer,
Hillsborough Community College

Paul Fell,
Connecticut College

Drew Ferrier,
St. Mary's College

Richard Ford,
San Diego State University

Peter Frank,
University of Oregon

Robert Galbraith,
Crafton Hills College

Kathy Griffis,
University of California, Los Angeles

Kathy Griffiths,
University of California, Davis

Charles Holliday,
Lafayette College

Becky Houck,
University of Portland

Anthony Huntley,
Saddleback College

Dennis Kelly,
Orange Coast College

Bruce Kenney,
Duke Marine Laboratory

Robert Klose,
University of Maine

Lester Knapp,
Palomar College

Robert Knowlton,
George Washington University

Larry Liddle,
Long Island University

Dan Lindstrom,
Gordon College

Lewis Lutton,
Mercyhurst College

Paul Lutz,
University of North Carolina at Greensboro

Richard Mariscal,
Florida State University

Richard Matthews,
San Diego Miramar College

Karla McDermid,
University of Hawaii

David McKee,
Corpus Christi State University

Robin Newbold,
Saddleback Community College

Joel Ostroff,
Brevard Community College

David Richard,
Rollins College

Wendy Ryan,
Kutztown University

Erik Scully,
Towson State University

James Small,
Rollins College

Donald Thomson,
University of Arizona

James Thorp,
University of Louisville

Richard Turner,
Florida Institute of Technology

Jacqueline Webb,
Villanova University

Quinton White,
Jacksonville University

Mary Wicksten,
Texas A&M University

Melvin Zucker,
Skyline College

Finally, although every effort was made to produce a product free of errors and inaccuracies, it is inevitable that some will occur. Any errors or oversights are my own and not those of the reviewers or the editorial staff at Saunders. Comments or suggestions for how I might improve this text in future editions are appreciated and can be sent to me in care of the publisher.

George Karleskint, Jr.

July 1997

Contents Overview

Contents

INTRODUCTION TO Marine Biology

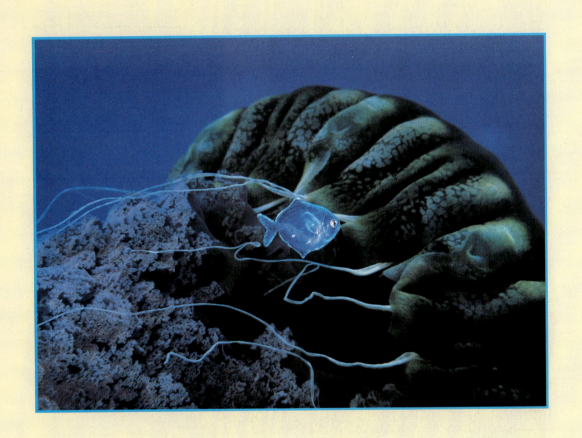

Wind, currents, and waves are a few of the important environmental factors that affect the number and kinds of marine organisms that can inhabit a particular area. (H. Richard Johnston/Tony Stane Images)

Chapter 1

Science and Marine Biology

There is perhaps no better way to learn about the delicate balance of natural systems than to study marine organisms and the communities they form. Not only is this pursuit fascinating in itself but it also teaches us how important the ocean and its inhabitants are to humans and why we should try to preserve and conserve this important resource. The diversity of marine organisms has attracted naturalists and scientists from the earliest days of human history, and the rich history of marine biology has been an important factor in shaping the modern science we know today. This chapter will introduce you to the importance of oceans and their inhabitants, and the developments that have formed the modern science of marine biology.

THE IMPORTANCE OF THE OCEANS AND MARINE ORGANISMS

Oceans are the most important physical feature of our planet. They cover nearly 71% of the earth's surface and represent the last great expanse on this planet to be charted and explored. The physical characteristics of these great bodies of water directly and indirectly affect our everyday lives, and the living organisms that inhabit them are an important source of food and natural products.

Oceans act as enormous solar-powered engines that drive the various weather patterns affecting terrestrial environments (Figure 1–1a). Phenomena such as El Niño Southern Oscillation (ENSO), which can cause droughts in Peru, flooding in Texas, and relatively mild winters in the Midwest, are the result of changes originating in the Pacific Ocean. The action of waves and tides changes the contours of continents and affects the lives of people who inhabit coastal areas.

Ocean productivity—the amount of food marine organisms can produce and the number of organisms the oceans can support—is an important area of research in marine ecology. The sea has provided and still provides a

KEY CONCEPTS

1. Marine and terrestrial environments are interrelated.

2. Oceans are an important source of food and other resources for humans.

3. Marine biology is the study of the sea's diverse inhabitants and their relationships to each other and their environment.

4. The history of marine biology is one of changing perspectives that have shaped the modern science and its applications.

5. Marine laboratories play an important role in education, conservation, and biological research.

6. It is important to study marine biology in order to make informed decisions about how the oceans and their resources should be used and managed.

Figure 1—1

The Importance of Oceans. (a) The exchange of heat energy between the atmosphere and the oceans is responsible for creating the weather patterns that affect terrestrial habitats. The white area in this photo is a tropical storm developing in the Pacific Ocean. (b) The oceans supply a significant amount of food in the form of fish, shellfish, and seaweeds. (a, NASA; b, John Elk/Tony Stone Images)

substantial amount of the world's food supply (Figure 1–1b). The United Nations reports that more than 97 million metric tons (1 metric ton = 1.1 tons) of fish are harvested annually.

Not only are marine organisms an important source of food, but some also provide us with important materials for industry and medicine. The commercial harvest and processing of these organisms and their products provide jobs for millions of people worldwide.

Marine organisms play another important role in scientific research. Biologists have found that many of the species inhabiting the sea and seashores are ideally suited to the study of such varied fields as ecology, physiology, biochemistry, biogeography, behavior, genetics, and evolution. Experiments using marine organisms have provided biologists with important information not only about the organisms studied but also about the workings of biology in general. Discoveries based on experiments with marine organisms have greatly advanced our understanding of biology.

The full extent of what the sea and its inhabitants have to offer has yet to be discovered, and only by learning more about the sea and its creatures will we be able to realize its full potential.

THE STUDY OF THE SEA AND ITS INHABITANTS

Oceanography is the study of the oceans and their phenomena such as waves, currents, and tides. The science of oceanography draws from many different disciplines, in-

Figure 1–2

Marine Biology. The marine organisms in this tidal pool interact with and are influenced by each other (biotic factors) and their physical environment (abiotic factors). The study of marine organisms and these interactions is the science of marine biology. (Larry Ulrich/Tony Stone Images)

cluding chemistry, physics, geology, geography, meteorology, and even biology. The study of the living organisms that inhabit the seas and their interactions with each other and their environment is **marine biology** (Figure 1–2). These two areas of study are not completely distinct from each other, and they frequently overlap. It is necessary to com-

bine elements from both fields in order to form a complete picture of the oceans and their inhabitants.

With this in mind, the theme of this book is the ecology of marine organisms, the interplay and interdependence of organisms with each other and their environment. To set the stage for this study, we will begin with a chapter on the principles of ecology. This chapter will outline the importance of living (**biotic**) and nonliving (**abiotic**) factors in the lives of marine organisms. Since physical and chemical characteristics of the environment play an important role in determining what kinds of organisms can live in a given area, the remainder of Part 1 will introduce you to the physical and chemical characteristics of the marine environment.

In Part 2 we will survey the many groups of marine organisms and examine their relationships to each other. We will organize our study along the lines of feeding relationships among species, proceeding from organisms that produce their own food (**producers**) to those that rely on other organisms for food (**consumers**). Our focus in this part will be the role different groups of organisms play in the web of marine life. There is tremendous diversity among the organisms in the world's oceans, and we know very little about many of them. Some of these organisms may be important new sources of food, sources of useful commercial products, or sources of chemicals that could be used as medicines (Figure 1–3). Millions of visitors are drawn yearly to oceanside resorts to admire the beauty and variety of marine organisms. This is a key factor in industries such as ecotourism and sports such as SCUBA diving. Unless we strive to learn more about these organisms, we will never know all of the important contributions that they make to our lives now and could make to our lives in the future.

In Part 3 of the book, we will examine the different marine habitats and learn how the interactions of the physical and biological environments make each habitat unique. Then, in Part 4, we will examine the effects of human activities on these various habitats.

We hear a great deal today about the impacts that humans are having on the environment, including the sea. There are even some who claim that the sea is doomed and that we cannot reverse the damage that has been done. Although the media play an important role in informing us about human impact on the sea, it also interprets much of this information for us as well. The attitude that people take toward such topics as the dumping of trash, disposal of radioactive and industrial wastes, oil spills, and overfishing is greatly influenced by the media. In order to make intelligent decisions about the ocean, as well as the rest of the environment, responsible citizens need to have a background of facts with which to analyze these complex issues. This is perhaps the most important reason for a concerned citizen to learn more about marine biology.

Figure 1—3

Marine Organisms and Medicine. The cartilage that makes up the skeletons of sharks is an important source of antiangiogenesis factor, a chemical that prevents tissues from establishing a blood supply. This chemical may be useful in the fight against cancer by depriving tumors of blood, thus killing them. (Stuart Westmorland/ Tony Stone Images)

MARINE BIOLOGY: A HISTORY OF CHANGING PERSPECTIVES

Human interest in the sea probably dates back to the time we first set eyes on it, fished its waters, and sailed across it. The great expanse of water with its variety of strange and wonderful creatures has inspired awe, wonder, curiosity, myth, and, at times, fear. This interest in the sea and its creatures was the beginning of the sciences of oceanography and marine biology.

Like living organisms, the science of marine biology has evolved over time. Early investigations of the sea's creatures centered on simple observations of marine organisms that were easily accessible from the shore. As ships and equipment became available, humans set out to conquer the sea and attempted to control its awesome magnitude and power. Improvements in shipbuilding and navigation opened up new frontiers in marine exploration. In their quest for discoveries, explorers (and merchants alike) eagerly set about emptying the sea of its contents, often overexploiting its resources. These same technical improvements also allowed the science of marine biology to expand as a body of knowledge. In more recent times, advances in technology such as submersibles, robotics, and computers have broadened our view of the marine environment even more. This new knowledge has led marine biology to a more global perspective of not only the interrelatedness of ocean habitats to each other but also of interactions between ocean and terrestrial habitats.

Early attempts to study the sea's creatures can be traced back to the ancient Greeks and Romans. The Greek philoso-

pher Aristotle, who was an accomplished naturalist, was one of the first to develop a scheme of classification, which he called the "ladder of life." His writings contain descriptions of over 500 species, of which almost one third are marine. He also studied fish gills, proposing that they functioned in gas exchange, and made detailed observations of the anatomy of the cuttlefish (*Sepia*).

Pliny the Elder was the foremost of the Roman naturalists. His only surviving work is the 37-volume *Natural History,* which contains mostly information about terrestrial animals, but it does include references to marine fishes and bivalves (shellfish such as clams and mussels). During the Middle Ages, the Catholic Church became the primary overseer of scholastic pursuits but gave most of its attention to matters of theology and philosophy. The study of natural history was still dependent on the works of the early Greek and Roman naturalists. Arabian philosophers of the Middle Ages involved themselves with interpreting and explaining the works of the ancient naturalists, rather than engaging in their own studies and observations. It would be the late 18th century before biologists would again conduct studies of the marine environment based on original observations.

During the late 18th and early 19th centuries, biology expanded into several disciplines. This was a time of great discovery, fueled in part by the development of better sailing ships and navigational instruments. The exploration of new lands and new sea routes provided information for the various branches of science, including biology. Scientists, such as the French naturalist Lamarck and the anatomist Cuvier, studied and described many marine organisms during this time. In December 1831, HMS *Beagle* set sail on a five-year voyage of exploration that would take it around the

Figure 1–4

Charles Darwin. Although better known for his theory of evolution by natural selection, Darwin was an accomplished marine biologist. This page is from Darwin's monograph on Cirripedia, a reference work that is still used by marine biologists today. (Copyright Library, Academy of Natural Sciences, Philadelphia)

globe. Among the members of the expedition was Charles Darwin. During the voyage, Darwin, the father of the theory of evolution by natural selection and an early marine biologist, was able to observe marine life firsthand and collect many specimens of marine organisms. Based on his observations of atolls (ring-shaped coral reefs that enclose a lagoon) in the Pacific, he proposed an explanation of coral atoll development that is still widely accepted. It was during this voyage that he began to formulate what would eventually become his theory of evolution. In 1859, Charles Darwin published his landmark work, *On the Origin of Species by Means of Natural Selection.* This work stimulated many scientists to investigate the causes of adaptations observed in marine organisms. It also sparked study of the interrelationships among marine organisms themselves and between marine organisms and their environment. In the years following the voyage of the *Beagle,* Darwin engaged in a detailed study of the barnacles that inhabited the rocky coasts of England and produced a monograph on the subject that is still used today (Figure 1–4).

In the early part of the 19th century, it was generally agreed that living organisms could not survive in the cold and darkness of the ocean depths, an idea proposed by the English naturalist Edward Forbes. Evidence to the contrary was produced when a transatlantic telegraph cable, linking the United States and England, failed shortly after it was laid in 1858 and had to be retrieved. The cable was located at a depth of approximately 1.7 kilometers (1 mile) in the northern Atlantic Ocean. As the repair ship retrieved the cable, it was found to be covered with all sorts of marine organisms that had never been seen before. These organisms apparently flourished in the depths of the ocean, and their discovery sparked investigations by several countries into life on the ocean floor. Subsequent dredging expeditions recovered animals from as deep as 4.42 kilometers (14,500 feet or over 2 3/4 miles).

As a result of increasing interest in the marine environment, the British Admiralty organized the *Challenger* expedition, which lasted three and one-half years. The expedition was named after the research vessel HMS *Challenger,* a ship containing state-of-the-art equipment and research facilities. When the *Challenger* returned to England in May 1876, it had criss-crossed the major oceans of the world and brought back enough information to fill 50 volumes of scientific reports. During this expedition, more than 4,700 new species of marine organisms were collected and described. Many of these new species were dredged from great depths (Figure 1–5).

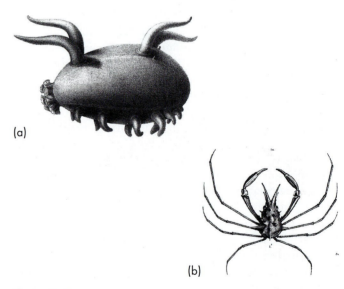

(a)

(b)

Figure 1–5

The *Challenger* Expedition. Drawings from the *Challenger* reports show two organisms, (a) a sea cucumber (*Scotoplanes globosa*) and (b) a crab (*Anamathia pulchra*), that were first discovered by the expedition in dredgings from over 1,000 meters in the Pacific Ocean off the Philippine Islands. Very little is known about the behavior and ecology of deep-sea organisms such as these, and it remains a task for today's marine biologists to discover more about how these animals live. (a and b, Copyright President and Fellows of Harvard College/Museum of Comparative Zoology)

The *Challenger* expedition gave birth to the modern sciences of marine biology and oceanography, and, even today, marine scientists continue to refer to the *Challenger* reports in their research. Later expeditions by Great Britain and other countries would add to and build upon the broad base of information accumulated by this groundbreaking expedition.

Charles Wyville Thomson, the driving force behind the *Challenger* expedition and its chief scientist, collected samples of the microscopic organisms that were floating in the water (Figure 1–6). Previously, biologists had paid little attention to these organisms, but now they were coming under close scrutiny. In 1887, Victor Hensen coined the term **plankton** to describe all of the different organisms that float or drift in the sea's currents. These tiny organisms are at the base of the ocean's complex food webs, and only in the last 50 years have marine biologists started to understand the specific roles that these organisms play.

In 1877 the American naturalist Alexander Agassiz began the first of several expeditions he would direct to investigate the organisms of the sea. Agassiz collected samples of animals from hundreds of locations and dredged animals from depths of 180 to 4,240 meters (600 to 14,000 feet). The amount of material he collected rivaled that which was brought back from the *Challenger* expedition (Figure 1–7).

In addition to collecting and cataloging marine organisms, Agassiz studied coloration in marine animals. He noted that the most brightly colored animals were found in surface waters and that, as one proceeded deeper, brilliant colors gave way to blues and greens and ultimately reds and blacks. He theorized that the colors were related to the absorption of different wavelengths of light at different

Figure 1–7

Alexander Agassiz. Alexander Agassiz was one of the foremost U.S. marine biologists of the 19th century. He is pictured here in his laboratory with jars containing some of the marine specimens that he collected during his many expeditions. (Copyright President and Fellows of Harvard College/Museum of Comparative Zoology)

depths, a theory that later proved to be correct. He also noted that there was a great deal of similarity in the deepwater organisms on the east and west coasts of Central America and hypothesized that the Pacific and the Caribbean were at one time connected. Agassiz spent much of the latter part of his life studying the structure and formation of coral reefs.

Alexander's father, Louis Agassiz, founded the Museum of Comparative Zoology at Harvard University. He also founded the first marine biology laboratory in the United States in July 1873. Originally located on Penikese Island off the Massachusetts coast and called the Anderson Summer School of Natural History, it was founded to help teachers at all levels improve their methods of teaching natural history to their students. Agassiz was a firm believer in learning through observation and in the "hands-on" acquisition of knowledge. It was Agassiz's hope that the Summer School would allow him the opportunity to teach his methods to others using marine organisms as subjects of study. A primary reason for locating his school by the sea was that all the major groups of animals would be within easy reach and could be studied in their natural habitats. Agassiz's goal of teaching biology through direct observation is still a major objective of the courses that are offered at marine laboratories around the world.

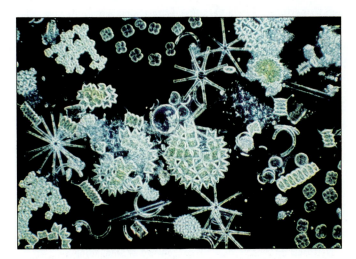

Figure 1–6

Plankton. Some examples of marine plankton, organisms that float or drift in the sea's currents. Charles Wyville Thomson, the chief scientist of the *Challenger* expedition, was one of the first scientists to seriously investigate the role of plankton in marine communities. (Roland Birke/Peter Arnold, Inc.)

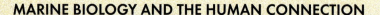

The Process of Science

A particular endeavor or study becomes a science when the principles on which it is based can be presented as **hypotheses,** explanations that can be tested by experiments. A good hypothesis can explain past events and predict the outcome of current or future experiments. All branches of science seek to organize observations so that hypotheses can be formed that suggest relationships between the observations. These hypotheses are then tested by experiments. The data gathered from these experiments are evaluated, and logical conclusions are drawn from the information at hand. This orderly approach to gathering and analyzing information is called the **scientific method.** It should be emphasized that the steps of the scientific method are simply an idealized approach to problem solving in science. Individual researchers tend to bring their own variations to the plan as they construct and conduct their experiments.

Generally, the first step in the process is to gather together observations concerning the subject that is under study. Making observations is a very important aspect of the scientific process, but observations alone are not very helpful without some thread or pattern that ties them together. It is not enough just to note facts. A good researcher is not only a good observer but also a creative thinker. For instance, like many visitors to a rocky shore, a biologist named J. H. Connell noticed that barnacles grew on the rocks in groups forming definite zones. Connell also noticed that two different types of barnacles occupied distinctly different areas on the rocks, and he wondered whether this separation of species might be related to competition of some sort. Through a series of experiments, Connell was able to test his hypothesis and show that the zonation of

barnacles on the rocks was indeed due to competition between the two types (for further details of Connell's work see Chapter 14).

After noticing a recurring pattern or a relationship between observations, a researcher uses inductive reasoning to form a hypothesis. Inductive reasoning involves looking at individual events and proposing a general explanation for them. For instance, researchers have found that night-feeding fishes like moray eels, catfish, and some shark species are unable to locate their food when their nostrils are plugged. These observations have led the researchers to suggest that night-feeding fishes rely on their sense of smell to locate their prey. A hypothesis may be as simple as proposing a cause-and-effect relationship between observed events, or it may be more complex and propose a model for the way in which a particular process may work. Regardless of how the hypothesis is stated, to be valid, it must be testable.

Hypotheses are tested by performing experiments or, sometimes, in fieldwork, by making systematic and detailed observations. A well-designed experiment involves running two trials at the same time. In these two trials only one factor, the experimental variable, is altered. The trial that contains the experimental variable is called the **experimental set.** The trial without the experimental variable is known as the **control set.**

For example, a researcher may observe that a certain species of shrimp exhibits a body color similar to the color of the algae on which it feeds. Based on this observation, the researcher may hypothesize that the color of this shrimp species is due to pigments derived from the algae. To test this hypothesis, two experimental setups would be devised.

Each setup would contain the same species and number of shrimp, and the shrimp would be subjected to the same environmental conditions such as temperature, salinity (the amount of salt dissolved in the water), and the number of hours they are exposed to light. The only factor that would differ between the two sets would be the algae that the shrimp were fed. In this example, the type of food is the experimental variable. The control group would be given the usual diet, and the experimental group would be fed another type of algae that contained different pigments. If the shrimp that ate the new algae changed color to match that of their new food source, the researcher may conclude that the experimental results support the hypothesis.

Before finally accepting the hypothesis, however, the researcher will most likely perform the experiment again several times. Results that cannot be duplicated may have occurred purely by chance and cannot be considered valid. Alternative experiments would also be performed to ensure that other variables had not been overlooked. Whenever possible, the data that are collected in an experiment are quantitative. Numerical data are easier to compile and subject to statistical analysis.

If the results of an experiment do not support the hypothesis, the investigator will then alter or modify the hypothesis in light of the new information and begin a new round of testing. This process may be repeated several times before a researcher arrives at a hypothesis that is consistently supported by experimentation. Over a period of time, some hypotheses are expanded upon and eventually become generally accepted as theories. A scientific **theory** represents a body of observations and experimental support that have stood the test of time.

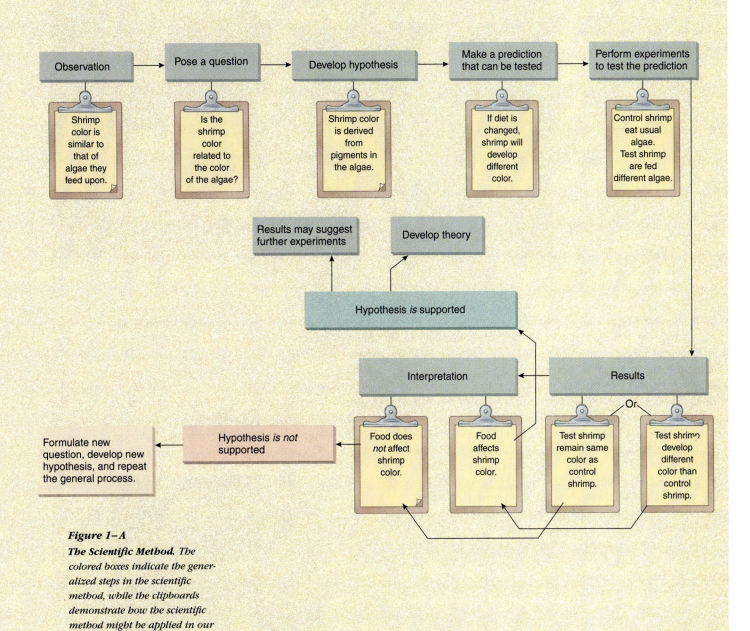

Figure 1–A

The Scientific Method. The colored boxes indicate the generalized steps in the scientific method, while the clipboards demonstrate how the scientific method might be applied in our specific example.

Figure 1–8

Marine Biological Laboratory at Woods Hole. This Massachusetts facility is one of many institutions worldwide whose goal is to learn more about the organisms that inhabit the oceans. (George Whitely/Photo Researchers, Inc.)

Agassiz's school was the inspiration and predecessor of the Marine Biological Laboratory at Woods Hole (Figure 1–8), which was founded in 1888 in a little fishing village on Cape Cod, Massachusetts. In 1922, the Woods Hole Oceanographic Institute was constructed down the street from the Marine Biological Laboratory, and their proximity allowed for a great deal of interchange between the two institutions. The Marine Biological Laboratory is one of the foremost institutions of its kind in the United States and is a major, internationally recognized research institution. Louis Agassiz's influence on the field of marine biology can still be seen at the Marine Biological Laboratory at Woods Hole. One of his favorite sayings, "Study nature, not books," is prominently posted in the library to remind students and researchers alike of this important concept.

Later in this century, other important research institutions were founded, such as the Scripps Institution of Oceanography in California, the University of Miami's Rosenstiel School of Marine and Atmospheric Science in Florida, the Friday Harbor Laboratories of the University of Washington, and Duke University Marine Laboratory in North Carolina, to name a few.

Understanding commercial fishery production was a driving force in the establishment of many marine laboratories and continues to be an important goal. In addition to research on ocean productivity and the interrelationships of marine organisms, modern marine laboratories focus on the use of marine organisms to solve fundamental problems in biology, such as the control of cell division, the process of embryological development, the functioning of the nervous system, and the applications of biotechnology.

Early in this century, expeditions were mounted to study the Arctic and Antarctic seas. Individuals such as the Norwegian Fridtjof Nansen and the Englishman Sir Alistair Hardy (Figure 1–9) led expeditions that collected information and organisms from these two areas. Nansen was interested in reaching the magnetic North Pole as well as charting the waters around the pole. He was not successful in his attempt to reach the pole, but his expedition to gather important information about the polar seas was quite success-

Figure 1–9

Sir Alistair Hardy. The Englishman Sir Alistair Hardy led expeditions to the Antarctic Sea to study whales. (Godfrey Argent Ltd.)

Figure 1-10

The Submersible *Alvin*. Submersibles like *Alvin* allow marine scientists to investigate life in the ocean's deepest recesses. (R. DeGoursey/ Visuals Unlimited)

ful. Hardy was interested in the biology of whales for the purpose of commercial exploitation, and it was this interest that took his expeditions to the Antarctic Sea. In addition to making important observations concerning whales, he also increased the amount of information available concerning the Antarctic Sea. His book *The Open Sea: Its Natural History* remains a classic in the field of marine biology.

Some of today's investigators center their attention not only on the relationships of marine organisms to each other, but also the impact of human activities such as fishing and pollution on the marine environment. The new frontiers of the sea are its depths where deep-sea vessels, such as *Alvin* (Figure 1-10), can take marine scientists to the very floor of the ocean to view and collect organisms that live in its deepest recesses.

Since the latter part of the 19th century, advances in the field of marine biology have literally given us both a broader look and a deeper view into the sea itself. With this new perspective came a broader and deeper understanding about global ecology. For years, our technology for exploiting the sea has outpaced our understanding of marine biology, but advances in the science have led to new attitudes about conservation and resource management.

MARINE BIOLOGY TODAY

We live in the Information Age. Each day, new findings that increase our comprehension of the living world that surrounds us are added to the world's data banks. Knowledge of the inner workings of the sea as well as its inhabitants also is increasing at a breakneck speed, aided by the many advances in technology. Deep-sea submersibles can reach the sea's deepest recesses, and researchers can live in underwater habitats while they directly observe the activities of marine organisms. The advent of computers with connections to the Internet and the "information superhighway" allows scientists and nonscientists around the world to share the enormous amounts of information collected on a variety of subjects, including marine biology. With the aid of these new tools, marine biologists are discovering the multifaceted interrelationships among marine organisms and marine systems and their ties to terrestrial environments.

In the following chapters, you will learn more about the sea and the organisms that inhabit it. You will learn how both physical and biological factors interact to produce the complex ecosystems of the marine environment. You will be introduced to the methods of science as they apply to the study of marine biology, and you will learn what impact humans have had on this truly wondrous realm.

CHAPTER SUMMARY

The world's oceans play an important role in everyday life as they affect weather patterns, provide food, and provide important resources.

The science of oceanography is the study of the oceans and their phenomena. Marine biology is the study of the organisms that inhabit the sea, their interrelationships, and their interactions with their environment. An understanding of marine biology is important so that we can understand how marine organisms relate to us and how human activities affect the marine environment. A basic knowledge

of marine biology is also important so that conscientious citizens can make prudent decisions about activities that involve and affect the sea.

The science of marine biology has changed over the years as new technologies have developed. Human interest in the sea and its creatures can be traced back to the ancient Greeks and Romans. It was not until the 19th century, however, that the foundations for the modern sciences of marine biology and oceanography were formed. The research expedition of HMS *Challenger* in 1876 was a landmark in the study of the sea. Information collected during this expedition and subsequent expeditions laid the foundations for modern marine science. The late 19th century saw the first marine laboratories founded in the United States, beginning with the Marine Biological Laboratory at Woods Hole. These laboratories continue to play a vital role in both marine and basic biological research.

SELECTED KEY TERMS

oceanography, *p. 3*
marine biology, *p.4*
plankton, *p. 7*

hypothesis, *p. 8*
theory, *p. 8*

QUESTIONS FOR REVIEW

MULTIPLE CHOICE

1. One of the first naturalists to systematically study marine organisms was
   a. Darwin
   b. Forbes
   c. Agassiz
   d. Aristotle
   e. Pliny

2. The publication of Darwin's book *On the Origin of Species by Means of Natural Selection* sparked an interest in the study of
   a. physical oceanography
   b. animal and plant adaptations
   c. plankton
   d. barnacles
   e. polar seas

3. Marine biology is the study of
   a. the physical characteristics of the ocean
   b. the organisms that inhabit the sea and their relationships to each other and their environment
   c. marine animals but not marine plants
   d. the organisms found in the open sea but not along the shoreline
   e. marine fishes and mammals only

4. The modern science of marine biology can trace its beginnings to
   a. the invention of steam-powered ships
   b. the founding of the first marine laboratory at Woods Hole
   c. the *Challenger* expedition
   d. the voyage of HMS *Beagle*
   e. the development of modern fishing techniques

5. The term *plankton* refers to
   a. all kinds of marine plants and algae
   b. microscopic animals only
   c. organisms that float or drift in the sea's currents
   d. animals that are active swimmers
   e. all of the marine organisms that can produce their own food

SHORT ANSWER

1. What event proved that some organisms could live in the dark recesses of the ocean's depths?

2. What was Louis Agassiz's goal in founding the first marine laboratory in the United States?

3. Explain the significance of the *Challenger* expedition.

4. Describe how the focus of marine biology has changed from early times to the present day.

5. Describe the kinds of research performed at marine biology laboratories today.

THINKING CRITICALLY

Apply the scientific process to a scenario of your own (real or imagined). Develop a hypothesis to explain something you've observed, and design an experiment to test your hypothesis. Finally, fill in the blank clipboards in the figure on the next page to demonstrate how the specific steps in your process relate to the generalized steps of the scientific method. (You may want to review the figure in this chapter's box, "The Process of Science.")

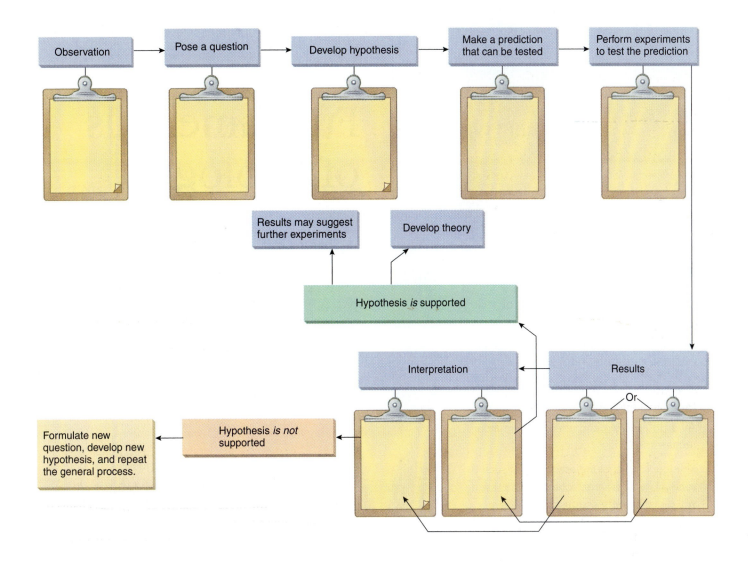

SUGGESTIONS FOR FURTHER READING

Edmunds, P. J. 1996. Ten Days Under the Sea, *Scientific American* 275(4):88–94.

Hartwig, E. O. 1990. Trends in Ocean Science, *Oceanus* 33(4):96–100.

Orange, D. L. 1996. Mysteries of the Deep, *Earth* 5(6):42–45.

Chapter 2

Fundamentals of Ecology

The term *ecology* is derived from the Greek word *oikos* meaning "home," in reference to nature's household and the economy of nature. The science of ecology deals with the interactions of organisms with each other and with their environment and how these interactions affect survival and reproduction. Frequently, the popular press uses the term *ecology* to refer to environmental science: the study of the effects of human activity on the environment. In this text, we will use the term *ecology* in its scientific sense.

The organisms that inhabit the seas are integrated components of a living network that encompasses the globe. Just as cells are parts of living organisms, organisms are parts of **ecosystems,** systems composed of living organisms and their nonliving environment. All of the earth's ecosystems taken together compose the **biosphere.** The structure of the biosphere is determined by the basic principles of life: the capture of energy, the cycling of nutrients, survival and reproduction, and the process of evolution that has shaped the natural world. The world's ecosystems are all interconnected, and what happens to one ultimately affects the others. Each organism and each ecosystem on the planet plays a crucial role in the function of the biosphere. Energy flows through the biosphere, and within the biosphere nutrients are cycled. The oceans play an important role in all of this as they produce food and recycle wastes.

CHARACTERISTICS OF THE ENVIRONMENT

An organism's environment consists of all the external factors acting upon that organism. These factors can be either physical or biological. The physical or **abiotic environment** consists of the nonliving aspects of an organism's

---

KEY CONCEPTS

1. Ecology is the study of relationships among organisms and the interactions of organisms with their environment.

2. An organism's environment consists of biotic and abiotic factors.

3. An organism's habitat is where it lives, and its niche is how it makes its living.

4. All organisms expend energy to maintain homeostasis.

5. Physical factors of the environment such as sunlight, temperature, salinity, and pressure will dictate where organisms can live.

6. Species interactions that influence the distribution of organisms in the marine environment include competition, predator–prey relationships, and symbiosis.

7. Marine ecosystems consist of interacting communities and their physical environments.

8. Energy in ecosystems flows from producers to consumers.

9. The average amount of energy passed from one trophic level to the next is about 10%.

10. With the exception of energy, everything that is required for life is recycled.

surroundings. For marine organisms, the abiotic environment includes temperature, salinity, pH, the amount of sunlight, ocean currents, wave action, and the type and size of sediment particles. The living or **biotic environment** consists of living organisms and the ways in which they interact with each other. For example, some organisms serve as food, some are predators, and others are parasites. Although we speak of the biotic and abiotic environments as two separate aspects of an organism's surroundings, in reality the two are difficult to separate.

HABITAT AND NICHE

The Habitat: Where an Organism Lives

The specific place in the environment where an organism is found is called its **habitat.** Marine habitats are characterized primarily by their abiotic features, the physical and chemical characteristics of the environment. Some of the habitats that we will examine in more detail in later chapters include estuaries (areas where rivers meet the sea), rocky shores, sandy beaches, coral reefs, open ocean, and deep water. Each of these habitats is characterized by its own set of physical and chemical characteristics, and these characteristics dictate what types of organisms can live in that habitat. Each habitat can be divided into smaller subdivisions called **microhabitats.** For instance, the sandy beach habi-

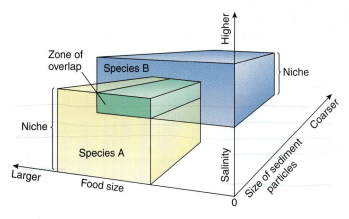

Figure 2–2

A Niche. An organism's niche is determined by a variety of abiotic and biotic factors acting together on the organism. This three-dimensional graph shows how several factors (food size, salinity, and size of sediment particles) interact to form niches for two species of burrowing worms. On the basis of this graph, we can see that species A prefers to burrow in substratum composed of smaller sediment particles where the salinity of the water is low and prefers to feed on medium- to large-sized food items. Species B, on the other hand, prefers coarser sediments where the salinity of the water is higher and prefers smaller food items. The zone of overlap indicates the combination of sediments, salinity, and food that would meet the requirements of both organisms.

tat contains several different microhabitats for microscopic organisms in the spaces between the sand granules. These habitats are characterized by the size of the sand particles, the amount of space between them, and the ability of these spaces to hold water between the tides. As a general rule, the more complex the habitat, the more microhabitats it contains. Coral reefs contain literally thousands of microhabitats for the many organisms that live in this community (Figure 2–1).

The Niche: An Organism's Environmental Role

What an organism does in its environment (in a sense, its occupation) is its **niche.** For example, mussels stick to rocks and filter seawater for food, crabs scavenge, and some worms burrow into the bottom sediments, extracting organic material as they do. A full description of an organism's niche would include the range of environmental and biological factors that affect its ability to survive and reproduce. Because a niche is so complex and involves so many different factors, it is not possible to show a picture of a niche as you could a habitat, but it is possible to examine different aspects of the niche separately in order to see how each of these factors affects an organism (Figure 2–2).

If we were to examine the organisms on a rocky shore, we would notice that their distribution from high tide line

Figure 2–1

The Coral Reef Habitat. Large habitats, such as the coral reef, can contain many smaller microhabitats. Microhabitats in the coral reef include the crevices in the coral, the sediments surrounding the coral stands, and even the tissues of the organisms themselves.

(Jeffrey Rotman)

to low tide line is determined by factors such as available moisture, the length of time exposed to air during low tide, their ability to withstand the force of waves, and the characteristics of the rock itself. This would be an abiotic view of this particular niche.

Biological factors that describe a niche include predator–prey relationships, parasitism, competition for the same resources, and organisms that provide shelter for other organisms. If we return to the example of a rocky shore, we would find that some organisms, like blue mussels (*Mytilus edulis*), are distributed in their particular zone because of the zone's abiotic characteristics. Predators, such as sea stars, are found in an overlapping zone because of the abundance of prey, namely the mussels. The seaweed that grows on the rocks provides food and shelter for a variety of small crustaceans, and the distribution of snails is determined by the amount and distribution of seaweed.

An organism's behavior also plays an important role in defining its niche. Behavioral factors, such as when and where an organism feeds, how it mates, where it bears its young, and social behaviors, influence an organism's niche. For instance, two very similar species that require the same kind of food can coexist if one feeds at night and the other during the day, provided the amount of available food will support both organisms. This is because each species' niche is defined by so many factors.

ENVIRONMENTAL FACTORS THAT AFFECT THE DISTRIBUTION OF MARINE ORGANISMS

Maintaining Homeostasis

One of the greatest challenges faced by living organisms, whether they are microscopic and consist of a single cell or are larger and multicellular, is to maintain a stable internal environment. Factors such as temperature, the amount of water, salts, nutrients, and the levels of waste products all have to be maintained within narrow limits for an organism to survive. When any one of these factors changes, the organism must make the proper adjustments to reestablish a balanced state. This important internal balance is called **homeostasis,** and the means of maintaining homeostasis is vital to the life of all organisms. An organism's ability to maintain homeostasis limits the areas of the world in which it can survive and reproduce.

If there is a range of environmental conditions and animals are able to move freely within the range, the animals will preferentially occupy those areas that offer the best set of conditions. If, however, animals are forced to occupy habitats that have a less than optimal range of environmental factors, they may fail to reproduce or, worse, die. The same is true of organisms that rely on sunlight to provide energy for food production (a process called *photosynthesis*). These organisms will thrive in environments with the proper amounts of sunlight and nutrients, like nitrogen and phosphorus, but will fail to reproduce or will die if the environment is too deficient in these factors.

For every species there is an optimal range for each environmental factor that affects its life. As long as the factors remain within the optimal range, the organism should be able to thrive and reproduce. When one or more of the environmental factors is outside of the optimal range, an organism's chances of survival are decreased (Figure 2–3). Zones of stress are regions above or below the optimal range of an environmental variable. An organism may be able to exist in a stress zone if the stress is not too great. If the stress is high, the organism may have to expend so much energy maintaining homeostasis that it will not have enough left to reproduce. Beyond the stress zones are zones where the environmental variable is so far from the optimum range that the organism cannot survive. These areas are called **zones of intolerance.**

Sunlight

Sunlight plays an important role in the marine environment. It powers the process of photosynthesis that, directly or indirectly, provides energy to nearly all forms of life on earth (this process is described in more detail on page 23). The most important photosynthetic organisms in marine environments are **phytoplankton,** the tiny plantlike organisms and bacteria that float in ocean currents. Phytoplankton, together with seaweeds and plants, are the primary source of food for marine animals. The distribution of these important food producers is determined by the available sunlight. In cloudy coastal waters, as in those of some North Atlantic bays and estuaries, phytoplankton can only survive in the shallowest areas because sunlight penetrates to a depth of less than 1 meter (3.3 feet). In the very clear water of the South Pacific, there may be enough sunlight for photosynthesis at depths of 200 meters (660 feet).

Sunlight is also necessary for vision. Many animals rely on their vision to capture prey, avoid predation, and communicate with each other. The distribution of these animals is affected by the depth that the sunlight can penetrate and allow for accurate vision. Fishes and other animals that live in the dark recesses of the ocean's depths generally have poor eyesight and rely on other senses, such as taste and smell, to find food, avoid predation, and find mates.

Excessive sunlight can be a problem as well. Organisms that live in the harsh environments of the intertidal zone are subject to the intense heat of the sun and the problem of **desiccation** (drying out) that results from overexposure to sunlight (solar energy). Many algae suffer pigment destruction when exposed to intense sunlight, limiting their ability to photosynthesize.

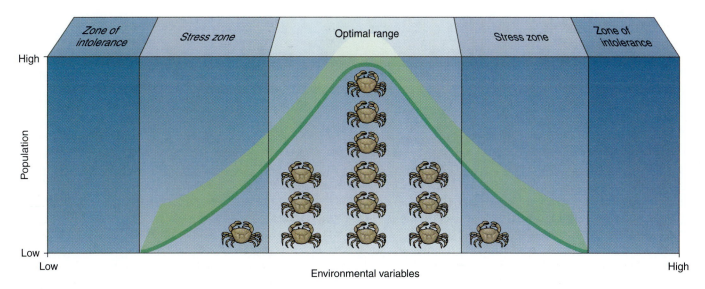

Figure 2–3

Optimal Ranges. An organism survives and reproduces best when environmental factors affecting it fall within an optimum range. Although organisms can live outside of their optimal ranges, they expend more energy maintaining homeostasis, leaving less energy available for reproduction.

Temperature

The majority of marine animals are **ectotherms,** which means they obtain most of their body heat from their surroundings. Ectotherms become sluggish when the temperature drops and more active when temperatures rise. Marine mammals and birds, on the other hand, are **endotherms.** An endotherm can maintain a constant body temperature because its **metabolism** (the chemical reactions within its cells) generates heat internally. Since these animals maintain temperatures that are usually higher than those of their surroundings, they have to be very well insulated so that they do not lose too much heat to the water that surrounds them (see Chapter 12).

Temperature often influences the distribution of organisms in shallow water and in the intertidal zone, the region that is covered at high tide and exposed at low tide. Shallow water, as in tidepools, can change temperature dramatically in a short time period as a result of hot summer sun or freezing nights. Organisms that survive in this habitat must be able to adapt quickly to a wide range of temperatures. The same is true of organisms that live in intertidal zones. The temperatures in these environments change quickly and substantially, and only the hardiest organisms are able to survive in this type of habitat. Large bodies of water, like oceans, do not change temperature rapidly (see Chapter 4 for more details), and, as a result, organisms that live in the open ocean away from the shore experience relatively constant temperatures at a given depth.

Most organisms can tolerate only a specific range of environmental temperatures. Temperatures above or below this critical range disrupt metabolism, resulting in decreased ability to reproduce, injury, or even death.

Salinity

Salinity is a measure of the concentration of dissolved inorganic salts in the water. Substances that are dissolved in water are generally referred to as **solutes.** The membranes of living cells are almost always permeable to water, but they are not always permeable to the substances that are dissolved or suspended in water. All organisms must maintain a proper balance of water and solutes in their bodies (maintain homeostasis) in order to keep their cells alive. When a solute cannot move across the cell membrane to reach a balanced state on both sides (equilibrium), water moves instead to achieve the balance. The movement of water across a membrane in response to differences in solute concentration is called **osmosis** (Figure 2–4). The process of osmosis is vital to the life of a cell. If a cell loses too much water, it will become dehydrated and die. On the other hand, if a cell takes in too much water, it will swell and, in the case of animal cells, possibly burst. Living organisms invest both time and energy in maintaining the proper amount of salts and water in their cells and the fluid that surrounds them. This is a particularly important challenge for many marine organisms, since they live in water that has a high concentration of solutes. Some marine animals, such as the spider crab (*Macrocheira*) and other inhabitants of the open ocean, cannot regulate the salt concentration of their body fluids. Since their body surfaces are permeable to both

Figure 2-4

Osmosis. Water tends to move from areas of lower solute concentrations to areas of higher solute concentrations. (a) An isotonic solution contains the same concentration of solute molecules (green) and water molecules (blue) as a cell. Cells placed in isotonic solutions do not change since there is no net movement of water. (b) A hypertonic solution contains a higher concentration of solute than a cell. A cell placed in a hypertonic solution will shrink as water moves out of the cell to the surrounding solution by osmosis. (c) A hypotonic solution contains a lower solute concentration than a cell. A cell placed in a hypotonic solution will swell and possibly rupture as water moves by osmosis from the environment into the cell.

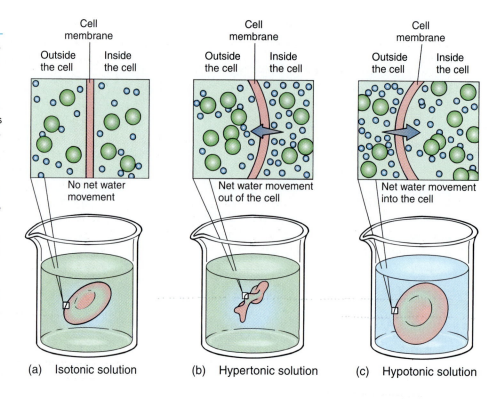

salts and water, the concentration of solutes in their body fluids rises and falls with changes in the concentration of solutes in seawater. This is not a problem for these animals because the open ocean is a very stable environment, and they are seldom exposed to fluctuations in salt and water content. They do, however, have very little ability to withstand changes in salinity, and if they are moved to a less stable environment, they will die.

Along coastal areas, the concentration of salts in seawater can vary greatly, especially in bays, estuaries, and tidepools. In these areas, pockets of water can lose moisture to evaporation, thus concentrating the salt content, or rain and freshwater runoff can dilute the normal salt concentration of the seawater. Animals that thrive in this type of environment, such as the fiddler crab (*Uca pugnax*), must be able to adjust the salt content of their body tissues by regulating salt and water retention (see Chapter 13). As we will see in the following chapters, salinity is very important in determining the distribution and types of organisms in many marine habitats.

Pressure

The pressure at sea level is 760 mm Hg or 1 atmosphere (14.7 lbs per square inch). Since water is so much denser than air, for every 10 meters (33 feet) below sea level in the ocean the pressure increases by 1 atmosphere. For instance,

the pressure at an average ocean depth of 3,700 meters is 370 atmospheres (2.7 tons per square inch). A Styrofoam cup lowered to a depth of 1,000 meters, where the pressure is 100 atmospheres, is compressed to about one third of its original size. As you might imagine, very few surface-dwelling organisms are able to survive at these depths (although pressure may not be the only factor limiting the number of organisms at great depths). Not only does the pressure of the water affect organisms that inhabit the deep regions of the seas, but it poses problems for animals that sometimes frequent these depths, such as whales and some species of deep-sea fishes. These animals possess specialized adaptations that allow them to survive at great depths (see Chapters 10, 12, and 18).

Metabolic Requirements

The availability of nutrients is another important factor that influences the distribution of organisms in the marine environment. The term **nutrient** refers not just to food but also to all of the organic and inorganic materials that an organism needs in order to metabolize, grow, and reproduce. Some of the most important nutrients, like nitrogen and phosphorus, are those needed by phytoplankton, seaweeds, and plants. The chemical composition of seawater supplies many of the mineral needs of marine organisms (see Chapter 4). For instance, the mineral calcium, which is so important for the synthesis of mollusc shells, coral skeletons, and the exoskeleton of crustaceans, is readily available in seawater.

A particularly important requirement for metabolism is oxygen. Oxygen is produced as a by-product of the photosynthesis performed by phytoplankton, seaweeds, and plants. It was probably the planet's first major pollutant. Life first evolved in an environment that lacked free oxygen, and when free oxygen first entered the environment it was toxic to most early life forms. The presence of free oxygen revolutionized life on earth by producing an environment that would allow the evolution of multicellular organisms.

Oxygen dissolves in seawater at the surface from the higher concentration in the atmosphere or from photosynthesis. The ability of water to hold oxygen depends on its temperature. Cooler water of the open sea contains more oxygen than the very warm water found in a tidepool.

Not all organisms require oxygen for life. Some bacteria are **anaerobic** organisms, meaning that they can survive and even thrive in the absence of oxygen. Many of these organisms are found at the ocean's depths, in salt marshes, in sand and mud flats, and in the spaces between sediment particles where the amount of oxygen is very limited. These are the organisms responsible for the familiar odor of "rotten eggs" associated with many of these areas. **Aerobic** organisms require oxygen for their survival and are limited to regions of the ocean that contain sufficient quantities of oxygen.

Metabolic Wastes

All organisms produce waste products when they metabolize. Most living organisms release carbon dioxide as a product of respiration. Animals excrete nitrogen-rich waste products, and plants release oxygen when they photosynthesize. Most of the time, waste products of metabolism are either removed from the environment or broken down and recycled by a variety of organisms, especially bacteria. In some environments, waste products can accumulate to toxic levels and prohibit the growth of all but the hardiest organisms. Certain small tidepools and coastal marsh areas are especially susceptible to the problems of accumulating metabolic wastes because the exchange of waste-laden water with new (uncontaminated) marine water is limited.

SPECIES INTERACTIONS

Competition

When organisms require the same limited resource, such as food, living space, or mates, competition occurs. Competition may occur between similar species (**interspecific competition**) or between members of a single species (**intraspecific competition**), and it prevents two groups of organisms from occupying the same niche. In other words, usually no two groups of organisms can utilize exactly the same resources in exactly the same place at exactly the same time. If this were to occur, chances are good that one group will be more efficient at what it does than the other and will have better success at survival and reproduction. A possible result of competition is the extinction of the less successful competitor, a process that ecologists call **competitive exclusion.**

Those individuals that are successful at using resources not in demand by other individuals experience less competition and usually produce more offspring. For example, several species of angelfish (*Holocanthus*) feed almost exclusively on sponges, a food source utilized by few other species. Many times, two groups of organisms competing for the same resource have evolved anatomical and behavioral specializations that permit them to use the particular resource more efficiently. If a variety of similar yet distinct resources are available, a single niche can be subdivided into two or more smaller niches that have only a minimal amount of overlap. This process that allows organisms to share a resource is called **resource partitioning.** For instance, plankton feeders on coral reefs divide up their food species in a number of different ways. Plankton-feeding damselfishes (Pomacentridae) stay close to the reef. Sea bass (*Anthias*) swarm after plankton in the water above the reef during the day, whereas soldierfish (Holocentridae) hunt for plankton at night.

Many similar species can occupy similar small niches if the anatomy, feeding behavior, and preferred territory of each species are just a little different from another's. Many butterflyfishes (Chaetodontidae) have elongated jaws that allow them to reach deep into cracks and crevices for small invertebrates. Some species, however, have a blunt snout and must feed on the surface of coral heads, where they suck food-laden mucus from the coral animals (Figure 2–5). These different characteristics allow several similar species to thrive in the same coral reef community.

Predator–Prey Relationships

Another important interaction between species is that between predators and their prey. The number of **herbivores** (plant-eating animals) that a given area can support is relative to the amount of vegetation that is available for food. The number of herbivores, in turn, supports a certain number of predators that use the herbivores as food. If the number of herbivores in an area increases so that there is not enough food for all of them, some will starve and the number of individuals will decline. As the number of individuals declines, the vegetation may be able to grow back, and as a

Figure 2–5

Competition. Competition among butterflyfishes is limited by the shape of their mouths, which dictates where they can find food and the type of food they can eat. This saddled butterflyfish (*Chaetodon ulietensis*) has a blunt mouth that restricts it to feeding on the surface of corals. (Hal Beral/Visuals Unlimited)

result of more food, the population of herbivores may increase again.

A similar situation can exist between **carnivores** (meat-eating animals) and their prey. If the number of carnivores increases too quickly, there will not be enough prey to support them, and the carnivores will starve and begin to die off. As the carnivore population declines, the number of prey species may start to increase again as the result of less predation.

In some habitats, the presence and activities of a particular organism prevent one or a few highly aggressive species from multiplying and crowding out others and thus dominating a community. For example, in rocky intertidal communities along the northern Pacific coast of the United

States, the ochre sea star (*Pisaster ochraceus*) is a dominant predator. The ochre sea star feeds on a variety of prey but seems to prefer mussels (*Mytilus*). In experiments performed by Robert Paine in the early 1970s, the ochre sea stars were removed from certain rocky areas, and those areas were kept relatively free of sea stars for five years. In the absence of the sea stars, mussels quickly colonized more of the rocky habitat, crowding out many other species in the process. The predation of mussels by sea stars keeps these highly competitive animals in check, allowing many other species, such as sea anemones, chitons, snails, and seaweeds, to survive in this habitat. Animals such as the ochre sea star whose presence in a community makes it possible for many other species to live there are called **keystone predators** or **keystone species** (Figure 2–6).

If something happens to upset the natural balance of predator–prey relationships, a "boom or bust" cycle frequently occurs. For instance, in the early part of this century, sea otters were nearly hunted to extinction for their furs along the Pacific coast. Sea otters are predators, and one of their favorite foods is the sea urchin, an animal that feeds on a variety of seaweeds. As the number of sea otters decreased, the number of sea urchins increased, and they began to overgraze on the forests of kelp, a giant seaweed (brown alga) that dominates their habitat. In localized areas, the kelp was almost decimated. Since many fish species relied on the kelp for cover, the loss of kelp in these areas led to a reduction in the local fish populations, which in turn depressed local populations of eagles. Fortunately, sea otters became protected by the International Marine Mammal Protection Act, and, as a result, the sea urchin population was again brought under control, although populations of sea urchins still thrive in sewage-polluted areas where sea otters cannot survive. Commercial harvest of sea urchins for

Figure 2–6

Keystone Predator. This ochre sea star limits the size of the mussel population in this community. This prevents the mussels from crowding out other species of rock dwellers. (Adam Jones/Dembinsky Photo Associates)

food in southern California may now threaten the balance of this community once again. This is just one of many examples of human intervention disrupting the delicate interactions between predators and their prey.

Symbiosis: Living Together

Some organisms have developed very close relationships with each other, to the extent that one frequently depends upon the other in order to survive. Any change in environmental factors that affects one partner will invariably affect the other as well. This arrangement is called **symbiosis,** a term that means "living together."

There are three types of symbiotic relationships that are distinguished by the nature of the relationship between the two organisms: mutualism, commensalism, and parasitism. In **mutualism,** both organisms benefit from the relationship. Sometimes, as in the case of coral animals and their zooxanthellae (a type of single-celled, photosynthetic organism; see Chapter 15), the two organisms are so dependent that they appear and function as a single organism. The coral animal provides the zooxanthellae with nitrogen, phosphate, and carbon dioxide—nutrients needed in photosynthesis. In return, the zooxanthellae provide the polyp with food in the form of carbohydrates. Also, the removal of carbon dioxide from the polyp by the photosynthesizing zooxanthellae makes it easier for the polyp to form its stony skeleton.

This type of mutualistic relationship is at one extreme of the spectrum. Not all mutualistic relationships are equally beneficial to both organisms involved. In some, the two organisms can be separated, but one or both may exhibit decreased survival or die as a result. A good example is

the Pacific clownfish (*Amphiprion*) and its mutualistic anemone (Figure 2–7a). The body of the clownfish is coated with a special mucus that protects it from the anemone's toxic stings. The fish acclimates the anemone to its presence by rubbing against it so that the fish's mucous covering picks up the anemone's scent. This way the anemone doesn't recognize the fish as food or foe. The clownfish gains protection by living within the stinging tentacles of the anemone and in nature cannot survive very long without its symbiotic partner. In return, the clownfish defends the anemone from other fishes that might eat the anemone's tentacles, rendering it defenseless and unable to feed. In many instances, the anemone will still survive if the clownfish is removed. Although both organisms in this case derive a mutual benefit from the relationship, the benefit is greater for the clownfish.

In **commensalism,** one organism benefits from the relationship, while the other partner is neither harmed nor benefited. An example of commensalism is the relationship between the remora fish (*Echeneis*) and some species of sharks and rays (Figure 2–7b). Remoras have a flattened suction cup on top of their heads that allows them to attach to the shark's body. In addition to getting a "free ride" on the shark, they eat the scraps of food that remain when a shark feeds. Remoras are also less likely to be attacked by predators since they are attached to the shark. The shark, on the other hand, does not appear to benefit from the relationship, but is not harmed either, although a certain amount of drag presumably results from having the remoras attached.

In **parasitism,** one organism, the parasite, lives off another organism, the host. The parasite benefits from the relationship, while the host is harmed. Many fishes and marine mammals are infected by parasitic worms, such as tapeworms (Figure 2–7c). Tapeworms live in the intestines

(a)

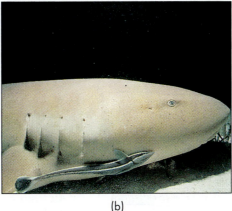
(b)

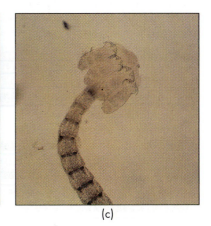
(c)

Figure 2–7

Symbiotic Relationships. (a) Mutualism: a clownfish and sea anemone. (b) Commensalism: a remora fish and shark. (c) Parasitism: a portion of a tapeworm taken from the intestine of a dogfish shark.

(a, Hal Beral/Visuals Unlimited; b, Jeffrey Rotman; c, Roxanna Smolowitz, D. V. M./School of Veterinary Medicine, University of Pennsylvania)

of their hosts and derive their nourishment from them (see Chapter 8). As a result of the tapeworm infestation, the host is weakened and is more vulnerable to disease and predation.

Although in theory each of these symbiotic relationships is clearly defined, determining where one ends and the other begins can be difficult. In nature the distinctions between commensalism and mutualism and commensalism and parasitism are frequently vague and open to interpretation.

POPULATIONS, COMMUNITIES, AND ECOSYSTEMS

Populations and Communities

To a biologist, a **population** is a group of the same species that occupies a specified area. The population, rather than the individual, is the basic unit that many ecologists study. A biological **community** is composed of populations of different species that occupy one habitat at the same time. The species that make up a community are linked together to some degree by competitive relationships, predator–prey relationships, and symbiosis. For instance, we can talk about the populations of barnacles, mussels, seaweeds, sea stars, and snails that inhabit a rocky shore on the Pacific coast. This assemblage of populations would make up a rocky shore community (see Figure 2–6). Communities can be large or small, depending on the area that is being discussed.

Zones of the Marine Environment

Marine communities can be designated by the regions of ocean that they occupy. Ecologists frequently divide the marine environment into two major divisions: the **pelagic division,** composed of the ocean's water (the water column), and the **benthic division,** the ocean bottom. These divisions can be subdivided into zones on the basis of three characteristics: (1) distance from the land, (2) light availability, and (3) depth (Figure 2–8).

The pelagic division is subdivided into the neritic zone and the oceanic zone. The **neritic zone** is composed of the water that overlies the continental shelves. The larger **oceanic zone** consists of the water that covers the deep ocean basins. The pelagic division can also be divided into the **photic zone,** where sunlight is present to support photosynthesis, and the **aphotic zone,** where sunlight is absent. The photic zone not only contains the greatest number of photosynthetic organisms, but it also contains the largest number of animals. As we saw earlier in typical predator–prey relationships, large populations of photosynthesizers can support correspondingly large populations of

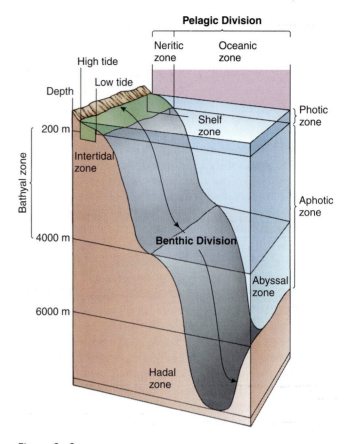

Figure 2–8

Ocean Divisions and Zones. Ecologists frequently divide the ocean into two major divisions: the pelagic division, consisting of the water column, and the benthic division, consisting of the sea bottom. The pelagic division can be subdivided based on the availability of sunlight (photic zone and aphotic zone) or distance from the shore (neritic zone and oceanic zone). The benthic division can be subdivided on the basis of depth (intertidal zone, shelf zone, bathyal zone, abyssal zone, and hadal zone).

herbivores and the predator species that feed on them. The organisms that inhabit the pelagic division consist of **plankton** (drifters) and **nekton** (swimmers).

The benthic division begins at the shore with the **intertidal zone.** This region of ocean bottom is covered with water only during high tide. During low tide it is exposed. The **shelf zone** extends from line of lowest tide to the edge of the continental shelf. From the edge of the continental shelf to a depth of 4,000 meters (13,200 feet) is the **bathyal** (BATH-ee-uhl) **zone.** The **abyssal** (uh-BIS-uhl) **zone** extends from 4,000 to 6,000 meters. At a bottom deeper than 6,000 meters (19,800 feet) lies in the **hadal** (HAYD-uhl) **zone.**

Benthic organisms live primarily in or on the bottom sediments. Benthic animals that live on the bottom are called **epifauna;** those that live in the bottom sediments are called **infauna.**

We will examine the physical characteristics of these zones in more detail in later chapters.

Ecosystems: Basic Units of the Biosphere

Communities are rarely isolated from each other, and, as a result, there are interactions among different communities. An ecosystem consists of both the biological communities and the abiotic environment, which interact to produce a stable (homeostatic) system. Some of the major marine ecosystems are estuaries, salt marshes, mangrove swamps, rocky shores, sandy beaches, kelp forests, coral reefs, and open ocean.

Estuaries occur where partially enclosed areas of the sea receive freshwater runoff to produce an area of mixed salinities. Because estuaries are such changeable environments, they are home to opportunistic species that can cope with environmental variation.

The intertidal zone is that area of the shore that is defined by the high tide and low tide mark. In the northeastern United States and along much of the Pacific coast, this area is rocky shoreline. In the southeastern part of the country, sandy beaches predominate. These areas offer particular challenges to organisms in dealing with the extremes of temperature, moisture, and salinity that occur between the tides.

Kelps are cold-water brown algae that are found off the coasts of North America, Japan, Siberia, South America, Great Britain, Scandinavia, and the Atlantic coast of South Africa. These algae form an amazing undersea forest that provides food and shelter for many important marine organisms.

Coral reefs are complex ecosystems formed by living organisms including coral animals and coralline algae. They are found in subtropical and tropical waters and are home to thousands of species. The great diversity of life on coral reefs reflects the thousands of niches that are available.

As with communities, the different ecosystems are not independent, and there is a great deal of interaction among them. We will study the characteristics of these ecosystems in more detail in Part 3 of this book.

ENERGY FLOW THROUGH ECOSYSTEMS

Producers

All living organisms require energy in order to live, grow, and reproduce. The source of this energy for practically all life on earth is the sun. Some organisms contain special pigment molecules, like chlorophyll, that are able to capture the sun's energy. This energy is then stored in organic molecules that can serve as a source of energy for the organisms that produce them or as food for other organisms. The

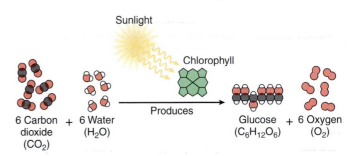

Figure 2–9

Photosynthesis. In the process of photosynthesis, carbon dioxide and water combine to form a sugar called glucose. Oxygen is a by-product of the reaction. The energy for the process is supplied by sunlight. Special molecules, like the green pigment chlorophyll, absorb light energy and make it available to power the photosynthetic process. The glucose produced by photosynthesis can be used by the photosynthetic organism as food or to make other important molecules.

process by which the energy of sunlight is captured and stored in organic molecules is called **photosynthesis** (Figure 2–9). In photosynthesis, light energy is used to combine molecules of carbon dioxide with water to form carbohydrate molecules like the sugar glucose. In advanced photosynthetic organisms, such as plants, oxygen gas is also released in the process. In the marine environment, the primary photosynthetic organisms are phytoplankton, seaweeds, and plants. Since these organisms are able to produce their own food, as well as food for other organisms, they are called **autotrophs** (*auto,* meaning "self," and *troph,* meaning "feed") or **producers.** Over one half of the photosynthesis (measured in kilocalories per year) that occurs on earth occurs in the oceans.

In the marine environment, not all producers are photosynthetic. Some are **chemosynthetic,** using the energy from chemical reactions to form organic molecules from carbon dioxide and other molecules. For example, certain bacteria that inhabit regions of the ocean floor where water heated by the earth's core seeps through (deep-sea vents) use the available chemical energy to produce their food by chemosynthesis, since no light is available at these depths for photosynthesis.

Consumers

Organisms that rely on other organisms for food are collectively called **heterotrophs** (*hetero,* meaning "other," and *troph,* meaning "feed") or **consumers.** First-order consumers (also known as "primary consumers") are those that feed directly on producers. These organisms are called **herbivores.** Second-order consumers (also known as "secondary consumers") are **carnivores** that feed on herbivores. Third-order consumers (also known as "tertiary

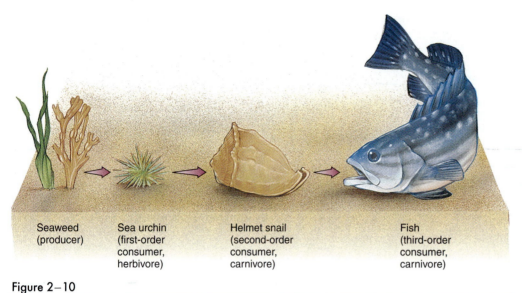

Figure 2–10

A Food Chain. Food chains depict the feeding relationships among a group of organisms as a linear sequence from producers to higher level consumers.

consumers") are carnivores that feed on other carnivores, and so on. **Omnivores** are consumers that feed on both producers and other consumers. **Detritivores** are organisms that feed on **detritus,** organic matter such as wastes and bits of decaying tissue. **Decomposers** are organisms that break down the tissue of dead plants and animals and help to recycle important nutrients. Detritivores and decomposers are also considered consumers since they cannot produce their own food, and they rely on other organisms for their organic nutrients.

In every ecosystem, producers and consumers are linked together by feeding relationships called **food chains** (Figure 2–10). For instance, in tropical regions, sea urchins feed on seagrass and seaweeds and helmet snails feed on the urchins. Humans, fishes, and marine mammals, in turn, prey upon the urchins and helmet snails. This is an example of a food chain. The concept of food chains, although helpful in establishing feeding relationships, is too simplified for what actually occurs in an ecosystem. If we were to study the dynamic feeding relationships in a particular community or ecosystem, we would notice that the relationships are interconnecting and much more complex than the simple food chain implies. These complex feeding networks are referred to as **food webs.** Figure 2–11 shows an example of a simplified marine food web.

Energy Flow Through Trophic Levels

Energy flows from the sun through producers to the various orders of consumer organisms. The energy that producers receive from photosynthesis or that consumers get from eating other organisms is temporarily stored until the organism is consumed by another or decomposed. Each of these energy storage levels is called a **trophic level,** with the producers making up the first trophic level. Theoretically, at least, there is no limit to the number of trophic levels, but in reality there are limits because only a fraction of the energy available on one level is passed along to the next when an animal feeds.

Energy transfer from one trophic level to the next is not efficient. First-order producers capture and store less than 1% of the available solar energy. The percentage of energy that is taken in as food by one trophic level and is passed on as food to the next higher trophic level is called **ecological efficiency.** The ten percent rule of ecology states that, on average, only about ten percent of the energy available at one trophic level is passed along to the next, although in reality the amount can be quite variable. This small amount of transferred energy reflects the amount of energy that is expended on finding, capturing, eating, and digesting food, as well as the heat lost during metabolism. As a result, the higher the trophic level, the lower the amount of available energy. This controls the number of trophic levels possible. Because energy is lost going from one trophic level to the next, most food chains are short (two or three links), although open-water food chains may be five or six links because nutrients are so scarce and the primary producers are very small.

The flow of energy from one trophic level to another can be pictured in the form of an energy pyramid (Figure 2–12a). Not only do energy pyramids indicate that the amount of available energy decreases with each trophic level, but they also indicate that each trophic level supports fewer organisms as a result. It is for this reason that herbivores are generally more numerous in a community than are carnivores.

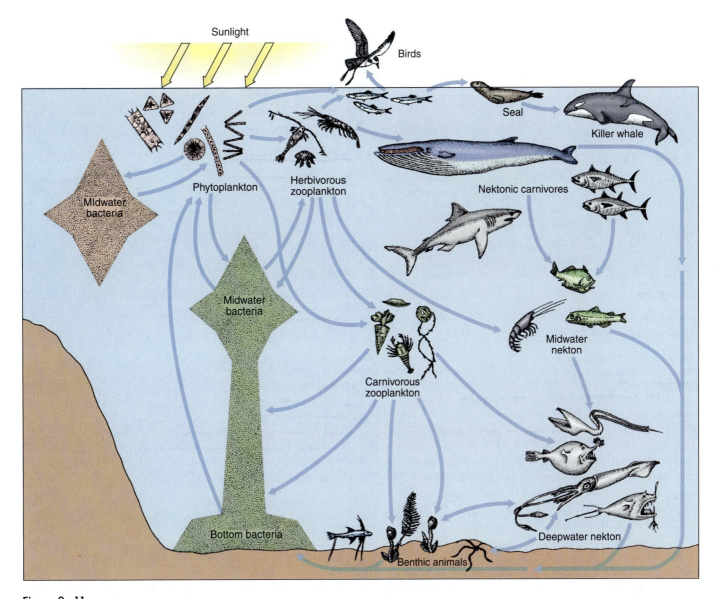

Figure 2–11

A Food Web. Food webs show the complex, interconnecting feeding relationships among members of a community or ecosystem.

These patterns are only generalizations. The actual energy relationships in ecosystems are much more complicated than this. Although pyramids of energy give the best overall picture of community structure, ecologists also use ecological pyramids of biomass and pyramids of numbers to indicate relationships among trophic levels (Figure 2–12b).

BIOGEOCHEMICAL CYCLES

Unlike energy, the nutrients that are necessary for life are available in limited supply and are recycled. The cycling of these various nutrients involves biological, physical, and chemical processes. For this reason they are frequently referred to as **biogeochemical cycles.** A cycle exists for each of the nutrients needed for life, such as water, carbon, nitrogen, and phosphorus. In this section we will examine a few of the more important cycles.

The Hydrologic Cycle

The most abundant compound in all living organisms is water. Obviously the cycling of water is of primary importance to all living things as well as to the marine environment (Figure 2–13). Since the regions of the earth around the equa-

text continues on page 27

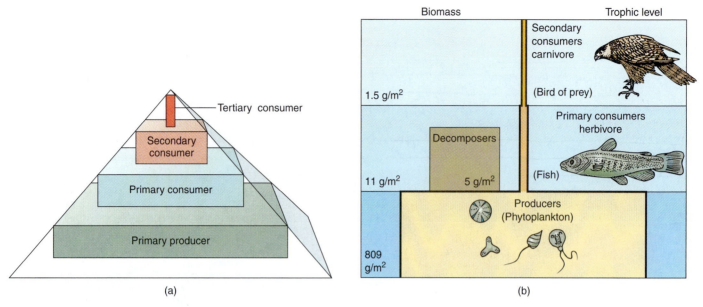

Figure 2—12

Ecological Efficiency. (a) The flow of energy from one trophic level to another can be represented as an energy pyramid, emphasizing the decrease in available energy from one level to the next. This diagram also represents a pyramid of numbers. The size of each box indicates the relative number of organisms that can be supported by each level. (b) A pyramid of biomass indicates the mass (grams) of living organisms that a given area (m^2) can support. Typically a pyramid of biomass resembles a pyramid of numbers.

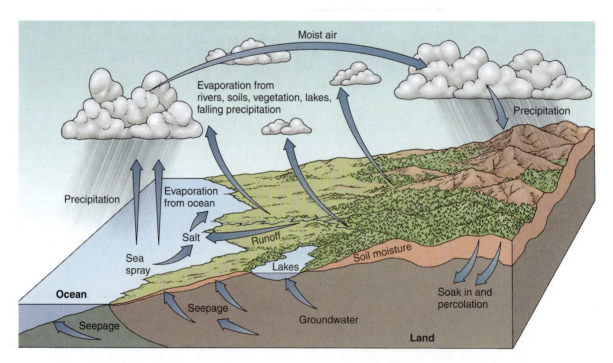

Figure 2—13

The Hydrologic Cycle. Water leaves the oceans by way of evaporation and returns in the form of precipitation. Rivers and streams collect the precipitation that falls on land and return it to the sea.

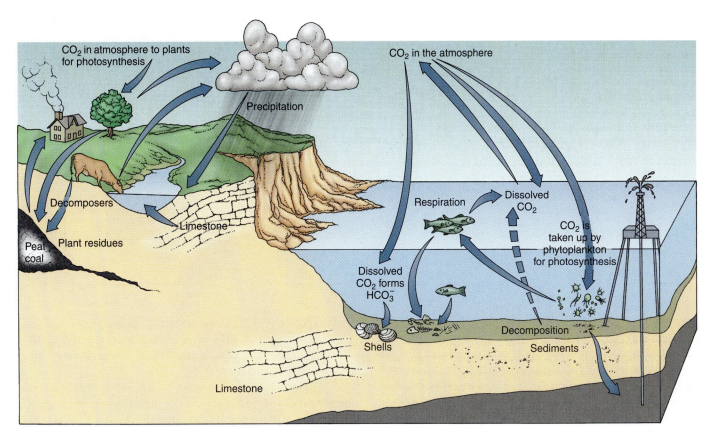

Figure 2—14

The Carbon Cycle. Carbon dioxide from the atmosphere that is dissolved in seawater is used by producers to make food through the process of photosynthesis. When the food is metabolized in respiration, the carbon dioxide is returned to the environment. Some carbon dioxide is converted into bicarbonate ions and incorporated into the shells of marine organisms. When these organisms die, their shells sink to the bottom. Some of the shells are compressed to form limestone, some of which is moved by geological processes to the surface, where erosion will return the carbon to the sea.

tor receive the most heat energy and sunlight, this is the area of the ocean that supplies the greatest amount of water to the atmosphere via evaporation.

Water vapor is carried north and south from the equator and from west to east within each hemisphere. As the air masses rise and cool, the water falls to earth as precipitation. Although most sea salt remains in the ocean, some enters the air as a result of waves crashing on the beach and similar action. Sea salt plays the important role of precipitation nuclei, airborne particulates that attract water droplets. When these get heavy enough, they fall back to earth as precipitation.

Precipitation over the sea returns a great deal of water back to the ocean. Precipitation that falls on land collects in the many rivers and streams that carry water back to the sea. On the way, minerals and organic substances are dissolved in the water. Thus, fresh water that enters the marine environment not only returns water but also large amounts of organic and inorganic nutrients as well.

The Carbon Cycle

Compounds containing the element carbon are essential components of all living organisms. Carbon makes up the backbone of carbohydrates, proteins, lipids, and nucleic acids, the basic molecules of life. Carbon cycling then is very important in the scheme of life. Living organisms produce carbon dioxide when they respire. When an organism dies, a variety of decay organisms start to break down the tissues (see Chapter 6), and a major product of this process is carbon dioxide. Marine producers use the carbon dioxide in photosynthesis to make carbohydrates. The carbohydrates can then be used to make other organic molecules (Figure 2—14).

Carbon dioxide (CO_2) dissolves in seawater and forms bicarbonate ions (HCO_3^-), as shown in the following equation:

$$CO_2 + H_2O \longrightarrow H_2CO_3 \longrightarrow H^+ + HCO_3^-$$

Figure 2–15

The Nitrogen Cycle. Upwellings and runoff from the land bring nitrogen into the photic zone, where producers can incorporate it into amino acids. Nitrogen-fixing bacteria in the photic zone can convert atmospheric nitrogen into forms that can be used by producers. Nitrogen is returned to the environment when organisms die or animals eliminate wastes.

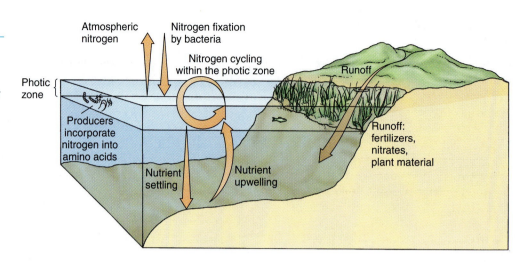

The bicarbonate ions can then be taken up by some marine organisms and combined with calcium to form the calcium carbonate needed for their shells and skeletons. The calcium carbonate from shells, corals, animal skeletons, and coralline algae eventually collects in bottom sediments and becomes limestone. Geological upheavals sometimes bring the limestone to the surface, where it is weathered by wind, rain, and other physical forces. The calcium and carbonate released from the limestone then washes back into the sea by way of rivers and streams. Another source of calcium and carbonate is the recycling of dead shells and skeletons by organisms such as boring sponges.

When some marine plants and animals die, their tissues are trapped in sediments and ultimately decay, forming deposits of fossil fuel. Most of the world's oil reserves were formed in this way. When these fuels are burned, carbon dioxide is released. The carbon dioxide can then be recycled by the process of photosynthesis.

The Nitrogen Cycle

Producers, like plants, seaweeds, and phytoplankton, require nitrogen "fertilizer" for protein synthesis and thus for proper growth and reproduction. The nitrogen they need is usually in a simple form such as ammonia (NH_3), ammonium (NH_4^+), nitrite (NO_2^-), or nitrate (NO_3^-). The producers use energy from photosynthesis to concentrate the nitrogen in their tissues and assemble it into amino acids and then proteins. The nitrogen is passed in the form of amino acids and proteins to consumers when they eat producers or other consumers. Animals process the amino acids and proteins and excrete nitrogen in the form of ammonia, urea, and uric acid. Certain bacteria convert some of the ammonia into nitrites and nitrates that can be used again by the producers to make more amino acids and pro-

teins (Figure 2–15). When an organism dies, decomposers release nitrogen into the environment from the tissue, and various groups of bacteria then convert the nitrogen into forms used by producers.

The atmosphere consists of almost 79% nitrogen and represents a major reservoir for this important nutrient. Electrical discharges during thunderstorms produce nitrates that reach the oceans with precipitation. Some microorganisms are capable of converting atmospheric nitrogen into forms that are useable by producers in a process called **nitrogen fixation.** The major nitrogen-fixing organisms in the marine environment are cyanobacteria, and we will examine the role they play in this process more closely in Chapter 6.

Runoff from the land may contain nitrogen from fertilizers, sewage, and dead plants and animals, as well as from animal wastes. This nitrogen can collect in shallow coastal waters and support a large amount of phytoplankton growth. In the open ocean, dead organisms and the nutrients they contain sink from the photic zone and are unavailable for the producers in those areas. As a result, large quantities of nutrient-laden material end up on the ocean floor. In certain areas, such as the coasts of California and Peru, the combination of winds and ocean currents brings large quantities of this nutrient-laden material from the bottom back into the photic zone, a phenomenon known as **upwelling** (see Chapters 4 and 16). In some areas, upwellings are seasonal, while in others they are more or less a constant event. In the areas where upwelling occurs, phytoplankton thrive and rapidly reproduce. This forms the basis for productive food webs that may include anchovies, herring, lobster, cod, hake, bluefish, tuna, and other important commercial species. As much as 50% of the world's commercial fish catch comes from upwelling areas that amount to only about 0.1% of the ocean's surface.

CHAPTER SUMMARY

An organism's environment consists of all the external factors acting on that organism. The specific place in the environment where an organism is found is called its habitat. Marine habitats are characterized primarily by their physical features. The organism's role in the environment, in a sense its profession, is its niche.

Many environmental factors influence the distribution of marine organisms. Abiotic factors include sunlight, temperature, salinity, pressure, and the levels of nutrients and wastes. Biotic factors include competition, predator–prey relationships, and symbioses.

In ecological terms, a population is a group of the same species that occupies a specific area. A biological community consists of several populations of different species that occupy one habitat at the same time and to some degree depend on each other. An ecosystem consists of both the biotic communities and abiotic environment, which interact to produce a stable system.

The energy for most life on earth comes from sunlight. Producers capture the energy of sunlight in the chemical bonds of organic molecules. Consumer organisms rely on these molecules as a source of food, since they cannot synthesize their own. In every ecosystem, producers and consumers are linked together by feeding relationships called food chains. In reality, the interactions among most living organisms are more complex than simple food chains. These complex feeding networks are known as food webs.

The energy that an organism receives from photosynthesis or from feeding on other organisms is temporarily stored until the organism is consumed by another or decomposed. These energy storage levels are called *trophic levels.* Energy transfer from one trophic level to the next is not efficient. The ten percent rule of ecology states that the average amount of energy passed from one trophic level to the next is about ten percent. This energy flow from one trophic level to another can be pictured in the form of an energy pyramid.

Although energy constantly flows through ecosystems, nutrients necessary for life do not. These nutrients are constantly recycled from one generation to the next through biogeochemical cycles.

SELECTED KEY TERMS

ecosystem, *p. 14*	salinity, *p. 17*	community, *p. 22*	heterotroph, *p. 23*
abiotic environment, *p. 14*	solute, *p. 17*	pelagic division, *p. 22*	consumer, *p. 23*
biotic environment, *p. 15*	osmosis, *p. 17*	benthic division, *p. 22*	herbivore, *p. 23*
habitat, *p. 15*	resource partitioning, *p. 19*	neritic zone, *p. 22*	carnivore, *p. 23*
niche, *p. 15*	keystone predator, *p. 20*	oceanic zone, *p. 22*	omnivore, *p. 24*
homeostasis, *p. 16*	symbiosis, *p. 21*	photic zone, *p. 22*	detritivore, *p. 24*
phytoplankton, *p. 16*	mutualism, *p. 21*	aphotic zone, *p. 22*	detritus, *p. 24*
ectotherm, *p. 17*	commensalism, *p. 21*	nekton, *p. 22*	food chain, *p. 24*
endotherm, *p. 17*	parasitism, *p. 21*	photosynthesis, *p. 23*	food web, *p. 24*
	population, *p. 22*	autotroph, *p. 23*	
		producer, *p. 23*	

QUESTIONS FOR REVIEW

MULTIPLE CHOICE

1. The specific place in the environment where an organism is found is called its
   a. community
   b. habitat
   c. niche
   d. ecosystem
   e. trophic zone

2. The role that an organism plays in its living and nonliving environment defines its
   a. community
   b. habitat
   c. niche
   d. ecosystem
   e. trophic zone

3. The area of the ocean that receives the greatest amount of sunlight for photosynthesis is the
   a. coastal zone
   b. benthic zone
   c. photic zone
   d. nutrient zone
   e. trophic zone

4. Based on Figure 2–2, species B prefers or tolerates _____ better than species A.
   a. higher salinity, coarse sediments, larger food
   b. lower salinity, fine sediments, smaller food
   c. higher salinity, fine sediments, larger food
   d. lower salinity, coarse sediments, smaller food
   e. higher salinity, coarse sediments, larger food

5. The upside-down jellyfish depends on algae in its body for certain nutrients. The algae are protected by the jellyfish and supplied with nutrients. This relationship would be an example of
   a. mutualism
   b. commensalism
   c. parasitism

6. According to the ten percent rule, how many kilograms of phytoplankton would be needed to produce 10 kilograms of fish that were second-order consumers?
   a. 1 kilogram
   b. 10 kilograms
   c. 100 kilograms
   d. 1,000 kilograms
   e. 10,000 kilograms

7. The ultimate source of energy for most life in the ocean is
   a. photosynthesis
   b. the sun
   c. thermal vents
   d. predation
   e. phytoplankton

8. The most important primary producers in marine ecosystems are
   a. seaweeds
   b. plants
   c. phytoplankton
   d. detritivores
   e. filter feeders

9. The most important abiotic factor influencing life in the ocean is
   a. temperature
   b. pressure
   c. salinity
   d. sunlight
   e. predation

10. An organism that feeds on organic wastes and decaying material would be a(n)
    a. herbivore
    b. carnivore
    c. omnivore
    d. detritivore
    e. piscivore

SHORT ANSWER

1. What are six abiotic factors that affect the distribution of organisms in an ecosystem?

2. What would probably happen to the natural balance in a community if the population of predators dramatically decreased or was wiped out?

3. What is the difference between a community and an ecosystem?

4. Describe the three types of symbiotic relationships found in nature.

5. Describe the marine nitrogen cycle.

6. Why are there fewer marine organisms in the ocean's depths?

7. Why are so many marine organisms ectotherms?

8. Why is energy transfer between trophic levels inefficient?

9. How can groups of similar species avoid competition?

10. Does the area of overlap in Figure 2–2 indicate that species A and species B have overlapping niches and will be in direct competition under those conditions? Why or why not?

THINKING CRITICALLY

1. Why does it make good ecological sense for whales to feed on plankton?

2. Although the open ocean receives plenty of radiant energy and has a larger area, it is not nearly as productive as the shallow coastal seas. Why?

3. Why do organisms that live in tidepools have to be more tolerant of changes in salinity than organisms that live in the open sea?

4. Which type of competition would you think would be more intense: interspecific or intraspecific? Why?

5. Would it be possible to have a pyramid of numbers where the trophic level above was larger than the trophic level below? Explain using an example with marine organisms.

SUGGESTIONS FOR FURTHER READING

Brewer, R. 1988. *The Science of Ecology.* Saunders College Publishing, Philadelphia.

Grall, G. 1992. Pillar of Life, *National Geographic Magazine* 182(1):95–114.

Paine, R. T. 1974. Intertidal Community Structure: Experimental Studies on the Relationship Between a Dominant Competitor and Its Principal Predator, *Oecologica* 15:93–120.

Rennie, J. 1992. Living Together, *Scientific American* 266(1):122–133.

Saffo, M. B. 1987. New Light on Seaweeds, *Bioscience* 37(9):654–664.

Geology of the Oceans

The physical characteristics of the environment play an important role in determining the kinds of organisms that can live in a given area and the characteristics that they will exhibit. Before we begin our study of the ocean's inhabitants and their interactions with each other and their environment, we need to gain a basic understanding of the physical characteristics of the ocean itself. In this chapter you will learn how oceans were formed and about the forces that shaped the physical features of the ocean in the past and continue to shape them today. You will be introduced to the methods scientists and navigators use to locate their position when they are at sea with no landmarks for reference points and how this information is used to produce charts that accurately represent the geography of the ocean. We will also examine the role that living organisms play in forming some of the physical features of the ocean environment.

THE FORMATION OF THE OCEANS

Current theories hold that our solar system was formed about 4.6 billion years ago, and along with it, our planet earth. It is generally believed that for the first billion years of its existence, the earth was mainly composed of silicon compounds, iron, magnesium oxide, and small amounts of other elements. Geologists hypothesize that originally the earth was composed of cold matter but that over time several factors, such as energy from space and the decay of radioactive elements, contributed to raising its temperature.

The process of heating continued for several hundred million years until the temperature at the center of the earth was high enough to melt iron and nickel. As these elements melted, they moved to the earth's core, displacing lighter, less dense elements, and eventually raising the core temperature to approximately 2000°C. Molten material from the earth's core moved to the surface and spread out, creating some of the features of the early earth's landscape. It is

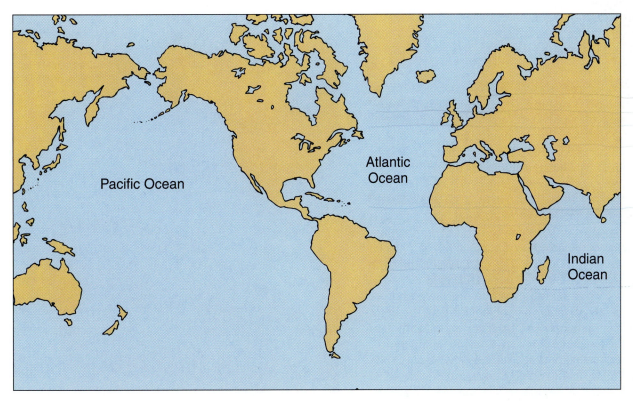

Figure 3-1

The World's Oceans. The three principal oceans of the world are the Atlantic Ocean, Indian Ocean, and Pacific Ocean.

probable that the process of melting and solidifying happened repeatedly, ultimately separating the lighter elements of the earth's crust from the deeper, denser elements. This separation of elements eventually produced the various layers of the planet that will be discussed in the next section.

In these early times, any water present on the planet was probably locked up in the earth's minerals. As the cycles of heating and cooling took place, water, in the form of water vapor, was carried to the surface, where it cooled, condensed, and formed the oceans.

The earth's early atmosphere is thought to have been formed by some of the hot gases escaping from deep within the planet. Initially, the atmosphere contained only reducing compounds such as hydrogen, carbon monoxide, carbon dioxide, and nitrogen. Because oxygen is so chemically active, any free oxygen gas originally present would have combined with other elements to form oxides. Gaseous oxygen did not begin to accumulate in the atmosphere until the evolution of modern photosynthesis (see Chapter 2).

The distance of the earth from the sun, the time for the earth to make one rotation on its axis, and the presence of the atmosphere all contribute to the ability of water to exist as a liquid on the earth's surface. The orbit and rotation of the earth around the sun help to ensure an average global surface temperature of 16°C (61°F). The atmospheric gases also help to maintain the temperature by reducing solar radiation.

Today 1.37 billion cubic kilometers (approximately 362×10^{18} gallons) of water covers 71% of the earth's surface and forms its oceans. There are three principal oceans of the world: the Atlantic, Pacific, and Indian Oceans (Figure 3-1). Of the three, the Pacific Ocean is the largest, the Indian Ocean is the smallest, and the Atlantic Ocean is the shallowest. Each ocean has its own characteristic surface area, volume, and depth, and these are shown in Table 3-1.

Table 3-1
Comparative Characteristics of Major Oceans

Ocean	Surface Area (km^2)	Volume (km^3)	Average Depth (m)
Atlantic	82,400,000	323,600,000	3,926
Pacific	165,200,000	707,600,000	4,282
Indian	73,400,000	291,000,000	3,962

CONTINENTAL DRIFT

The Layers of the Earth

The earth is made up of several layers (Figure 3–2). At the center of the planet is an inner core, which is solid (because of the extremely high pressure), very dense, very hot, and rich in iron and nickel. Surrounding the inner core is an outer core that consists of a transition zone and a thick layer of liquid material. The liquid material has the same composition as the inner core, only it is cooler and under less pressure. The next layer, called the **mantle,** is the thickest layer and contains the greatest mass of material. The mantle is mainly composed of magnesium–iron silicates. Although the outer region of the mantle is thought to be rigid, the inner layer is able to flow slowly. The outermost layer of the earth is the thinnest and coolest of all, the **crust.**

Since the crust is the most easily accessible portion of the earth for study, more is known about it than the other layers. The continental crust is thicker and slightly less dense than the oceanic crust and is mainly composed of granite rock that contains mostly silicate-rich minerals, such as quartz and feldspar. The oceanic crust, by comparison, is primarily composed of basalt-type rock that has a lower silicate content and is higher in iron and magnesium. The layer of the mantle just below the crust is also composed of rigid, basalt-type rock that is fused to the crust. The region of crust and upper mantle is called the **lithosphere** (LITH-uh-sfeer), and the region of mantle below the crust is known as the **asthenosphere** (as-THEN-uh-sfeer). The asthenosphere is thought to be liquid, and it is able to flow under stress.

The Moving Continents

With the advent of more accurate world maps in the later 17th century, several scientists became intrigued with the observation that the continents were shaped like pieces of a jigsaw puzzle (Figure 3–3). For instance, the eastward bulge of South America would fit well with the west coast of the African continent. Also, the comparative study of rock formations, the placement of mountain ranges, and the distribution of fossils all suggested that the continents were at one time connected.

This idea was not new. As early as the 1600s, the English scholar Sir Francis Bacon suggested that the continents may have once been connected to each other. In the latter part of the 19th century, the Austrian geologist Edward Suess even proposed a name, Gondwanaland, for the fusion of southern continents. In 1915, the German meteorologist Alfred Wegener proposed that at one time there was only one supercontinent, which he named Pangaea. He suggested that forces associated with the earth's rotation were responsible for breaking the supercontinent into a northern

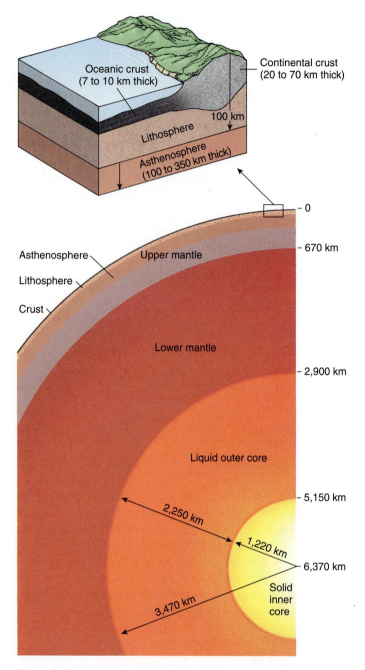

Figure 3–2

Composition of the Earth. The earth is composed of several layers. The surface layer is the crust, which is fused to the lithosphere, the outermost portion of the mantle just beneath it. The asthenosphere lies beneath the lithosphere and is able to flow under stress.

portion, Laurasia, composed of Europe, Asia, and North America, and a southern portion, Gondwanaland, composed of India, Africa, South America, Australia, and Antarctica. Following World War II, there was renewed interest in the theory of moving continents. With more sophisticated equipment, studies were made during the 1950s to deter-

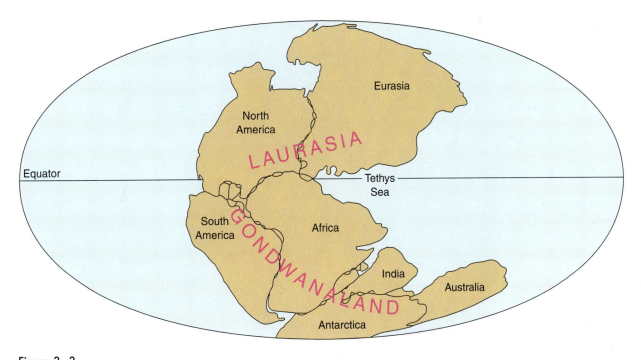

Figure 3—3

The Supercontinent Pangaea. This map of Pangaea shows how today's continents were once connected to form a single supercontinent around 400 million years ago.

mine what forces might be acting to produce the continental movement.

In the early 1960s, H. H. Hess proposed that molten material located deep in the earth's mantle, called **magma,** moved by convection currents to the region below the solid upper mantle and crust (Figure 3–4). The convection currents are driven by the heat of the mantle and the cooling of the magma. The current of molten magma would then flow along laterally under this zone, cooling as it did so, and then sink back toward the core again. Occasionally, when the upward-moving magma breaks through the crust of the ocean floor, volcanos are formed. Over time, a long mountain range called a **midocean ridge** will form along the crack produced by the erupting magma, and the magma that oozes out of these mountain ranges, or ridge systems, will cool and form new crust known as oceanic basaltic crust. In some areas of the range, a rift valley runs along the length of a portion of the mountain crests. These are areas of high volcanic activity. Steep-sided fracture zones, linear regions of unusually irregular ocean bottom, may also occur running perpendicular to the ridges and rises and separating sections of the range.

Since new crust is being formed and the earth is not growing larger, there must be some area where old crust is being removed. The removal of old crust takes place in regions called **subduction zones,** such as the deep recesses of ocean trenches. Old crust at the bottom of the trenches

sinks down and eventually reaches the mantle, where it is liquified and recycled by convection currents into the earth's core (Figure 3–5).

Most of the magma that rises from the deep mantle is turned back by the rigid lithosphere and moves in a lateral direction, ultimately descending along the sides of ascending convection currents and carrying pieces of the lithosphere with it. This activity produces a lateral movement of the crust called **seafloor spreading.** This is the force that causes **continental drift.** Seafloor spreading causes movement of the basaltic crust of the seafloor, and, since the continents rest on this crust, they also move, much in the same way that boxes move on a conveyor belt.

Along with the correct fit of the continental boundaries, other evidence that supports this theory includes observations on the distribution of earthquakes, the temperature of the sea bottom, the age of rock samples from the seafloor, and the analysis of core samples drilled through the ocean sediments. Earthquakes are known to occur around the globe in narrow zones that correspond to areas along ridges and trenches, the most active areas of crustal movement (Figure 3–6a). The highest seafloor temperatures occur in the regions of the ridges, and they decrease as one moves away from the ridge. These observations are consistent with the oozing of heated magma at the ridges and its cooling as it moves away laterally. Dating of the rocks from the ocean floor indicates that they are much younger

text continues on page 39

Figure 3—4

Formation of Oceanic Crust and Mountains. Molten material called *magma* rises from the earth's core to the upper mantle. The hot magma (red arrows) rises because it is less dense than the surrounding material. As magma reaches the mantle it cools (blue arrows), becomes more dense, and sinks back toward the core. This cycling of magma from the core to the mantle and back that results from changes in temperature and density is called *convection*. Occasionally, the heated magma breaks through the earth's crust, forming volcanos. Lines of volcanos form mountain ranges called *midocean ridges*.

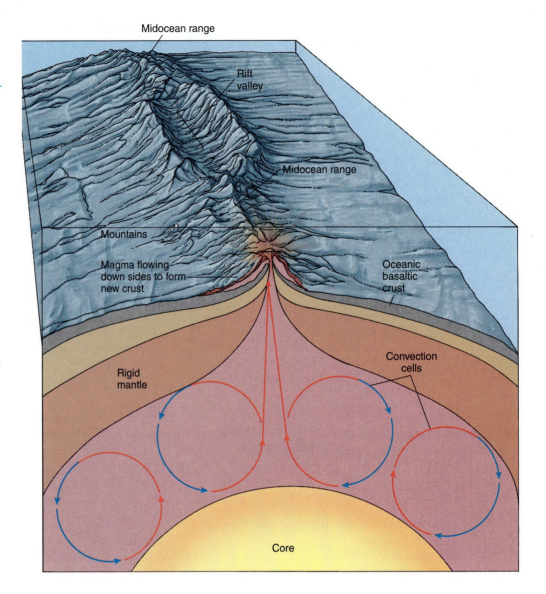

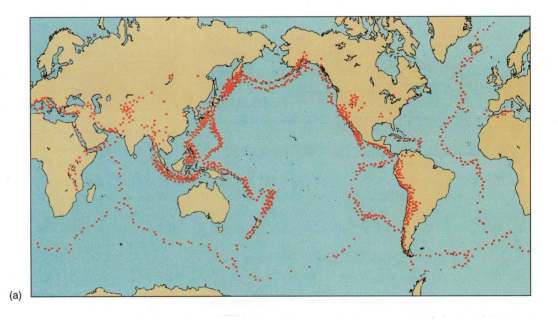

(a)

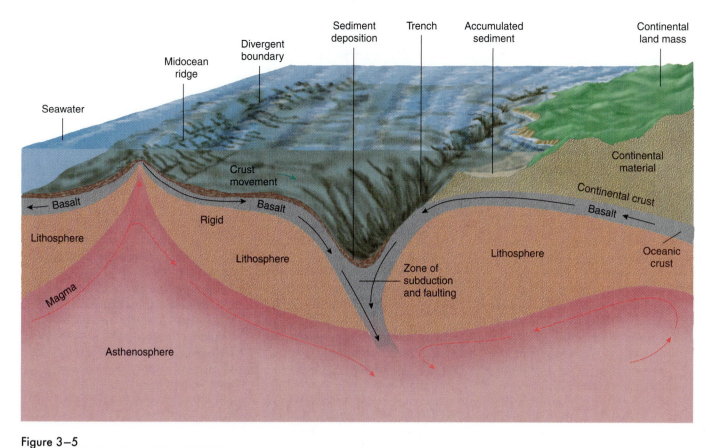

Figure 3–5

Seafloor Spreading and Continental Drift. Rising magma forms new oceanic crust that moves laterally from the midocean ridges. At subduction zones, old crust sinks and is ultimately returned to the mantle, where it melts and forms new magma. Since the continents rest on the basaltic crust, as the crust moves, the continents are carried along.

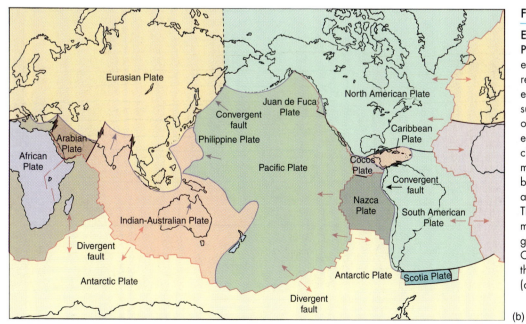

Figure 3–6

Earthquake Zones and Tectonic Plates. (a) This map shows the earth's earthquake zones. Each red dot represents an earthquake epicenter (the area of the earth's surface directly above the origin of an earthquake). Notice that the earthquake zones correspond closely to the location of the major plate boundaries (see part b). The distribution of volcanos also follows along these lines. (b) This map shows the location of the major tectonic plates and their general direction of movement. Compare the plate boundaries to the earthquake zones shown in (a).

(b)

MARINE BIOLOGY AND THE HUMAN CONNECTION

Magnetic Evidence for Continental Drift

Over the past 76 million years, there have been 170 reversals of the earth's magnetic field. No one knows what causes these reversals, but their existence can be demonstrated by measuring the magnetic poles of magnetized rocks that were formed at different periods of time. When particles such as iron are heated in a magnetic field, they become magnetized as they cool, forming miniature compass needles. The north and south magnetic poles of these particles are lined up with the north and south magnetic poles of the earth at the time they form. Examination of the north and south poles of magnetized iron found in rocks of various ages in the earth's crust indicates that reversals in the earth's magnetic field have occurred in the past.

In 1960, oceanographers from the Scripps Institution of Oceanography towed magnetometers (devices that measure magnetic fields) over the ocean floor and recorded a series of band patterns (Figure 3–A) that represented changes in the polarity of magnetic components in the earth's crust. No one was sure of the significance of these patterns until 1963, when F. J. Vine and D. H. Matthews of Cambridge University hypothesized that the band patterns corresponded to changes in the earth's magnetic field that had been frozen in the rock of the seafloor and moved by shifting plates. They suggested that as magma containing iron particles rose to the surface, the iron particles would orient their north and south magnetic poles in the same direction as the north and south poles of the earth at the time they were formed. As the magma reached the crust, cooled, and hardened, the direction of the field was frozen in time in the magnetized iron particles in the rocks. These changes would occur in vertical layers of rock, and as new crust formed, it would carry these magnetized rocks away from the ridge system in a lateral direction. Over time, this process would produce alternate areas of crust in which the polarity of the magnetic field would reverse. This, then, would account for the banding patterns observed by the research team from the Scripps Institution.

Vine and Matthews deduced that if their hypothesis was correct, there would be symmetrical band patterns moving in both directions from a ridge and that the oldest rocks would be found farthest away from the ridge. Furthermore, the age of the rocks and the polarity that they demonstrated should be the same as those found in similarly dated rocks found on land.

To test their hypothesis, they performed experiments to measure the polarity and the age of rock samples taken from various sites relative to ridges on the ocean bottom. The conclusions drawn from these experiments, that new crust was being formed on the seafloor and was spreading, added strong evidence for the theory of plate tectonics.

Figure 3–A

Magnetometer Data. This pattern of bands was produced when a magnetometer was towed across the floor of the ocean. The alternation of dark and light bands indicates reversals in the earth's magnetic field over millions of years. Notice that the older rock is farther away from the site of crust formation and that the newer rock is closer to it.

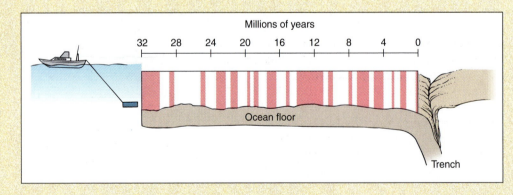

than terrestrial rocks. This can be explained by the constant formation of new rock at the ridges and the removal of old crust at the trenches.

In 1968, a specially constructed drilling ship, the *Glomar Challenger,* set out to take core samples from the ocean's bottom. From 1968 to 1983, the *Glomar Challenger* gathered core samples for analysis, and the results revealed that the crust was thinnest and the rock youngest in the area of the ridges. This observation is consistent with the hypothesis that new crust is being formed in these areas. As one moves away from the ridges, the rocks become older and the amount of sediment overlying the rocks is greater. This is to be expected since new crust moves the older crust away from the ridges, and since this crust was in place longer, it accumulated more sediment falling from above. In no case was any rock found that was older than 180 million years, much younger than the age of terrestrial rocks. This observation supports the constant formation of new crust and removal of old crust from the ocean floor. All of these data support the theory of continental drift.

The Theory of Plate Tectonics

All of the information concerning the movement of the earth's crust is combined to form the **theory of plate tectonics.** In this theory, the lithosphere is viewed as a series of rigid plates (Figure 3–6b) that are separated by the earthquake belts of the world—that is, the trenches, ridges, and faults. There are seven major lithospheric plates: the Pacific, Eurasian, African, Australian, North American, South American, and Antarctic plates. Each plate is composed of continental and/or oceanic crust.

At the midocean ridges, where plate boundaries move apart as new lithosphere is formed, divergent plate boundaries occur. Convergent plate boundaries occur at trenches, where plates move toward each other and old lithosphere is destroyed. The plates move past each other at regions known as **faults,** which represent breaks in the earth's crust where one plate can move past the other (see Figure 3–6b).

A transform fault is a special kind of fault that is found in sections of the midocean ridge. Each side of a transform fault is formed by a different plate, and these plates are moving away from each other in opposite directions. The fault zone produced by this movement is quite active and is the site of frequent earthquakes. The motion of the plates along these faults produces a nearly continuous line of cliffs with sharp vertical drops, known as escarpments. In these regions, there are sudden changes in the ocean depth.

Regions where the lithosphere splits, separates, and moves apart as new crust is formed are called **rift zones.** The midocean ridge and rise systems represent the major rift zones at this time. It is generally thought that rifting occurs when rising magma causes enough tension to stretch the overlying crust, creating a sunken rift zone. As this process continues, the rift spreads and the fault deepens and cracks, allowing the magma to seep through and eventually form a ridge. When this happens to the lithosphere under the continents, a sunken rift zone can occur. As the process continues, the fault gets thinner and deeper and can eventually fill with seawater. The Red Sea is an example of this type of formation.

As a plate moves away from the rift zone, it cools and thickens. At the rift, thinning of the crustal plate and increased flow of magma into the rift cause the land mass to separate. A low-lying region of oceanic basalt is formed, and a new ocean basin and ridge system are formed as well.

The Discovery of Rift Communities

In 1977, thriving marine communities were discovered at a depth of 2,500 meters (8,250 feet) in the Galapagos Rift, which lies between the East Pacific Rise and the coast of Ecuador. These communities consist of a variety of animals such as clams, worms, and crabs, some of which are relatively large (Figure 3–7). These organisms depend on the chemosynthetic activity of bacteria for their nutrients. Deep-sea vent communities are unique because they represent food webs that exist in the absence of sunlight. We will examine the biology and ecology of deep sea vent communities in detail in Chapter 18.

Figure 3–7

Vent Community. Thriving communities are found in some rift valleys such as this one in the Galapagos Rift. Since sunlight does not penetrate to this depth, these organisms rely on chemosynthetic bacteria for food. (WHOI/D. Foster/Visuals Unlimited)

ECOLOGY AND THE MARINE ENVIRONMENT

Animal Sculptors of the Seafloor

In the late 1970s while surveying the floor of the Bering Sea for geologic hazards to possible offshore oil platforms, C. Hans Nelson discovered pits and furrows that could not be attributed to any known geologic process. Nelson was aware that many marine mammals either live in or frequent these relatively shallow waters between Alaska and Siberia, and he wondered if any of these animals might be responsible for the disturbances he discovered. In investigating the problem, Nelson and his colleagues discovered that California gray whales (*Eschrichtius robustus*) produce the pits and that Pacific walruses (*Odobenus rosmarus*) produce the furrows. Both animals alter the seafloor in the course of feeding on the continental shelf and, in the process, introduce more sediment into the water of the region than the Yukon River, which annually dumps 60 million metric tons of sediment into the sea.

The gray whales are not permanent residents of this region. They leave their breeding grounds off Baja California in March and migrate up the Pacific coast to feed in the waters of the northeastern Bering Sea. When the winter ice melts, the water is relatively calm and teems with life. The whales spend May to November feeding in these rich waters before returning to their breeding grounds, and, while feeding, they dig up huge

Figure 3–B

An Ampeliscid Amphipod. *Crustaceans such as this are filtered from the sediments by gray whales. (Les Watling, Darling Marine Center, University of Maine)*

amounts of sediment searching for the small, bottom-feeding crustaceans known as amphipods (Figure 3–B), which are their preferred food.

Initially the evidence for gray whales disturbing the sea floor was circumstantial. As far back as the 19th century, biologists had suspected that whales fed on animals in the sea bottom. Whalers frequently reported whales coming to the surface with muddy water streaming from their mouths, and when the stomachs of captured animals were opened, they were full of bottom-dwelling animals that have since been identified as amphipod crustaceans. When the whale feeds, it rolls to one side so that its mouth is parallel to the seafloor. It then retracts its tongue, creating a suction that draws a large amount of food-laden sediment into the mouth, where it is filtered out the other side by a series of fibrous plates, called *baleen*, growing from the jaw.

Two other findings suggested that whales were responsible for the seafloor

THE SHAPE OF THE OCEAN BOTTOM

The ocean bottom has sculptured features just like the land above it. Mountain ranges, canyons, valleys, and great expanses are all part of the underwater landscape (Figure 3–8, pages 42 and 43). These physical features of the ocean bottom are called **bathygraphic features,** and unlike their counterpart topographic features on land, they change relatively slowly. Erosion is slow in the relatively calm recesses of the ocean, and changes mainly involve sedimentation, uplifting, and subsidence.

The Continental Shelves

At the edge of a continent is the **continental shelf** (Figure 3–9). Continental shelves are generally flat areas, averaging 68 kilometers (40 miles) in width and 130 meters (430 feet) in depth, that slope gently toward the bottom of the ocean basin. The width of a continental shelf is frequently related to the slope of the land it borders. Mountainous coasts, like the West Coast of the United States, usually have a narrow continental shelf, whereas low-lying land, like the East Coast of the United States, usually has a wide one. Conti-

text continues on page 44

disturbances. In 1979 aerial observers who were tracking feeding gray whales on the basis of emitted mud plumes found them to be concentrated in the Chirikov Basin, the area where the bottom disturbances were found. Nelson had sampled bottom sediments from several sites on the continental shelf of the Bering Sea and found that a sheet of sand covering the bottom of the Chirikov Basin was inhabited by the amphipods that were the whales' favorite food. In addition, the extent of this sand sheet matched the area where the whales were observed to be feeding. Kirk Johnson and his colleagues confirmed that the pits Nelson found had been caused by the gray whales. Using both direct measurements of the pits and data obtained from sonograms (impressions of the seafloor derived from sound waves), they were able to show that the pits could indeed have been formed by whales and that the area where the pits are located corresponded to the whales' feeding grounds.

It was later realized that walruses were also involved in changing the shape of the seafloor in the region. Pacific walruses disturb the bottom as they forage for clams and other types of bottom-dwelling prey. The walruses are year-round inhabitants of the area and produce grooves or furrows as they feed. In 1972 Samuel Stoker of the University of Alaska saw the furrows from a submarine and noticed walruses feeding nearby. He suggested that the furrows were the feeding tracks of the walruses. Ten years later John Oliver of the Moss Landing Marine Laboratories showed that clam shells that appeared to have recently been excavated and emptied were found along the furrows. These observations confirmed the findings of Eskimos who found sand, gravel, and clam shells in the stomachs of walruses and concluded that they were bottom feeders. Researchers noted that the furrows were generally found in clam beds and that the width of the furrows was approximately the same width as the walrus'

snout. This finding was consistent with Oliver's suggestion that walruses unearth their food with their lips and not with their tusks, as was previously assumed.

The feeding activities of whales and walruses seem to be beneficial to the ecosystem and enhance the area's productivity. The feeding activity of the whales separates mud deposited by the Yukon River from the sand, thus preserving the sandy bottom as an ideal habitat for amphipods. During feeding, only adult amphipods are trapped on the baleen, while the smaller juveniles escape to replenish the supply. The feeding activity also releases nutrients from the bottom sediment and moves them into the water column, where they stimulate the growth of plankton, a primary food source for the amphipods, ensuring a new crop of amphipods the following year.

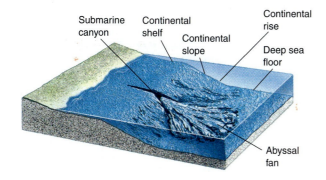

Figure 3–9

Continental Shelf. Continental shelves are generally flat areas that extend from the continental land mass into the sea. Many physical features of continental shelves were formed during glacial periods when these areas were exposed. Submarine canyons are thought to be formed by turbidity currents. As the current slows at the end of the continental slope, sediments fan out to form an abyssal fan.

Figure 3—8

Bathygraphic Chart. Bathygraphic charts such as this one depict the physical features of the ocean bottom. (Marie Tharp)

© Éditions Pierre Charron, 51, rue Pierre-Charron, 75008 Paris, Draeger, Imp.

Figure 3–10

Formation of Continental Shelf.
Continental shelves can be formed by (a) the eroding action of waves, (b) the trapping of sediments by natural dams, (c) the accumulation of sediments deposited by rivers, or (d) the trapping of sediments by coral reefs, islands, or offshore volcanos.

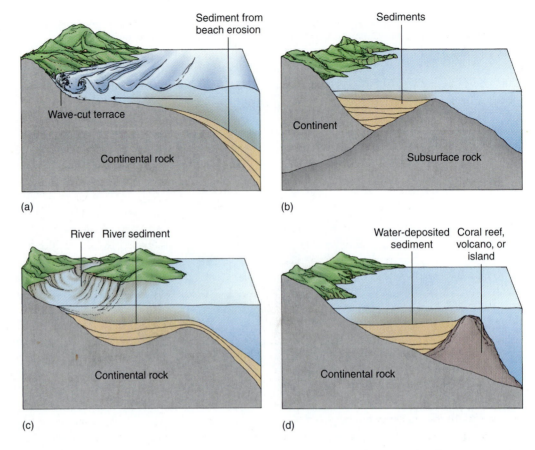

nental shelves are actually part of the continents to which they are attached. They are produced by waves that constantly erode the land mass and by natural dams, such as reefs, volcanic barriers, and rock, that trap sediments between the dam and the coastal land mass (Figure 3–10).

At the boundary of a continental shelf on the ocean side, there is an abrupt change in slope and a rapid change in depth. The extent of the sloping can vary from a gradual drop to a steep decline into an ocean trench, as illustrated by the slope that occurs off the western coast of South America. Because of the steepness of the angle, the continental slope usually has less sediment.

Some continental slopes have submarine canyons (see Figure 3–9) that are similar to canyons found on land. Many of these submarine canyons are aligned with river systems on land and were probably formed by these rivers during periods of low sea level. The Hudson River canyon on the East Coast of the United States is an example of this. Other submarine canyons have ripple marks on the floor, and at the ends of the canyons sediments fan out, suggesting that they were formed by moving sediments and water. Oceanographers believe that these canyons were formed by turbidity currents. **Turbidity currents** are swift avalanches of sediment and water that erode a slope as they sweep

down and pick up speed. At the end of the slope, the current slows and the sediments fan out. Turbidity currents can be caused by earthquakes or the accumulation of large amounts of sediments on steep slopes that overload the slope's capacity to hold them. Oceanographers are currently engaged in producing computer models of this process, and one such model is being tested in the Scripps Submarine Canyon in California.

At the base of a steep continental slope there may be a gentle slope called a **continental rise** (see Figure 3–9). A continental rise is produced by processes such as landslides that carry sediments to the bottom of the continental slope. Most continental rises are located in the Atlantic and Indian Oceans, and around the continent of Antarctica. In the Pacific Ocean, it is more common to find trenches located at the bottom of the continental slopes.

During the glacial periods, much of the ocean water was incorporated into the polar ice sheets, and the continental shelves were periodically uncovered. Erosion carved valleys in the shelves, waves altered their contours, and rivers deposited sediments. When the glaciers melted and the sea level rose, the continental shelves were again covered by water, retaining the surface features that were established during the time they were uncovered.

Some continental shelves still experience changes today. The Mississippi and Amazon Rivers, for instance, continuously carry mud, silt, and sand out to sea to be deposited on the continental shelves. These thick deposits of sediment have an important effect on the productivity of these areas, as we shall see in Chapter 16. However, not all shelves have this thick layer of sediment. For instance, the continental shelf around the eastern tip of Florida is bare of sediment. The swift Florida current continuously sweeps the sediments northward to deeper water, leaving the shelf sediment-free.

The Ocean Floor

The floor of the ocean basin covers slightly more of the earth's surface than the continents. The bottom of many ocean basins is a flat expanse called the **abyssal plain** (Figure 3–11). These plains extend from the seaward side of the continental slope and are formed by turbidity currents as well as sediments falling from above. Dotting the ocean floor are abyssal hills rising as high as 1,000 meters (3,300 feet). Abyssal hills cover as much as 50% of the Atlantic seafloor and as much as 80% of the seafloor of the Pacific and Indian Oceans. They are formed by volcanic action and the movements of the seafloor described earlier.

Another feature of ocean basins is the **seamount,** a steep-sided formation that rises sharply from the bottom. All seamounts are formed from underwater volcanos and are most prevalent in the Pacific Ocean. Some seamounts show evidence of coral reefs and surface erosion, suggesting that at one time they were above the surface. Movements of the ocean floor, the natural process of compaction that volcanic material undergoes, subsidence due to cooling of the ocean floor, erosion, and the increased weight of sediments at the top may be the reasons for the sinking of these structures.

A continuous series of large, underwater, volcanic mountains runs through every ocean of the world and stretches some 68,000 kilometers (40,000 miles) around the earth. The ridges and rises of these mountain ranges separate the ocean basins into a series of smaller, deepwater sub-basins.

Other impressive features of the ocean floor are the trenches. Deepwater trenches are more common and more spectacular in the Pacific Ocean and are usually associated with chains of volcanic islands called *island arcs.* For instance, the deepest of all ocean trenches (and the deepest spot in the ocean) is the Marianas Trench. It is associated with the Marianas Islands chain (Figure 3–12). A portion of the Marianas Trench, the Challenger Deep, measures 11,020 meters (6.85 miles) in depth. The Peru–Chile Trench extends for over 6,120 kilometers (3,600 miles) along the coast of South America and is the longest of the ocean trenches. The Java Trench extends for a distance of almost 4,760 kilometers (2,956 miles) along the coast of the islands of Indonesia. By comparison, the Atlantic has only two, relatively short trenches, the South Sandwich Trench and the Puerto Rico–Cayman Trench.

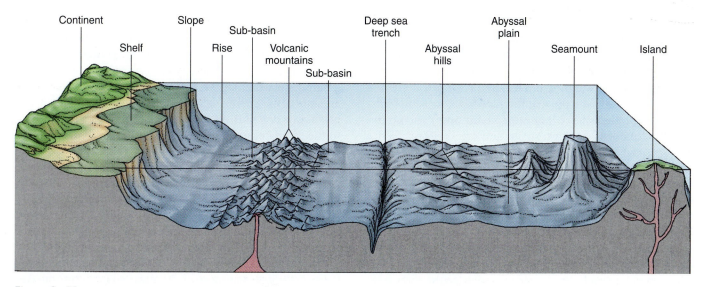

Figure 3–11

Landscape of the Ocean Floor. Like the surface of the continents, the ocean floor displays a variety of physical features, including mountains, hills, trenches, and expansive plains.

Figure 3–12

Ocean Trenches. This map shows the location of the major ocean trenches.

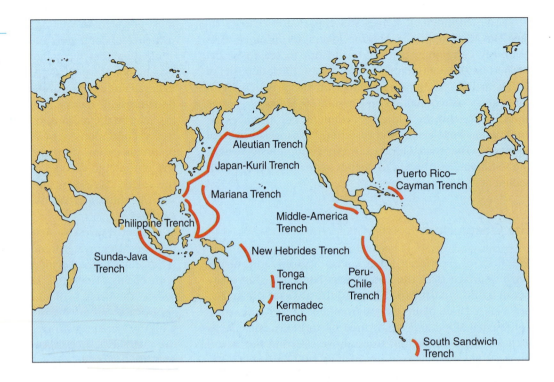

COMPOSITION OF THE SEAFLOOR

Types of Sediments

The amount and type of sediments that are found on continental shelves and the ocean floor are particularly important to the organisms that inhabit these areas. The sediments provide a habitat for many organisms and a source of nutrients for others. The particles that make up these sediments may come from living organisms, the land, the atmosphere, or even the sea itself. Some sediments are produced from terrestrial rocks by the various actions of wind, water, freezing, and thawing. These particles are carried from the land by water, wind, ice, and gravity and are deposited primarily on the continental shelves. Sediments from living organisms (**biogenous sediments**) are mostly particles of corals, mollusc shells, and the shells of microscopic, planktonic organisms.

Some sediments are formed from the seawater as a result of a variety of chemical processes. Carbonates, phosphorites, and manganese nodules are examples of the sediments that may form in salt water and accumulate on the seafloor. In shallow areas, calcium salts, sulfates, and rock salt can form on the bottom of pools when enough water evaporates to cause these minerals to precipitate.

Iron-rich particles from outer space strike the surface of the ocean and slowly drift to the seafloor. Many dissolve before reaching the bottom. Although not as numerous as other sediments, small amounts are found scattered on the bottom of all seas.

Biogenous Sediments

In the deep sea, almost all of the biogenous sediments are composed of the remains of single-celled organisms, such as diatoms, radiolarians, foraminiferans, and coccolithophores, and of molluscs known as pteropods or sea butterflies (Figure 3–13). These organisms produce shells of calcium carbonate (foraminiferans, coccolithophores, and molluscs) or silica (diatoms and radiolarians). When they die their remains settle on the sea bottom. If more than 30% of an area's sediment is made up of these fine biogenous particles, then the sediment is termed an **ooze.** The ooze can be either calcareous (calcium) or siliceous (glass), depending on which of the two shell types is dominant. The amount of biogenous sediment that will be found in any given area depends on the mass of organisms contributing to the sediment and the ability of the water to dissolve the minerals in the shells.

In temperate latitudes, the remains of single-celled organisms called *diatoms* are primarily responsible for siliceous deposits known as *diatomaceous ooze.* Living diatoms are important members of the phytoplankton. They produce their own food by photosynthesis and require a source of inorganic nutrients, such as phosphate and nitrate ions, in addition to sunlight. Sunlight is available at the ocean's surface, but most of the nutrients that are needed are produced in the deeper recesses of the sea. At certain locations, large upward movements of deep water, called *upwellings,* bring these nutrients close to the surface. The combination of plentiful sunlight and nutrients allows diatoms to thrive in these areas. When the diatoms die, their siliceous shells sink to the ocean floor, forming large deposits of diatomaceous ooze.

In the tropics, the water does not contain enough nutrients to support large numbers of diatoms, and siliceous ooze tends to be produced by single-celled, nonphotosynthetic organisms called *radiolarians.* A variety of small organisms such as foraminiferans, coccolithophores, and molluscs, such as pteropods, are the primary sources of calcareous ooze.

Clay and Mud Sediments

Clay, or mud, is found in ocean floors throughout the world wherever marine life is too scarce to form biogenous sediments. Clay is composed of fine powdered rock, and its color depends on the type of minerals that it contains. For instance, red clay contains large amounts of iron compounds. It is believed that this fine powder is blown from land by the wind and/or rinsed by rain from the atmosphere. The particles are so fine that they may remain suspended in the water for many years before settling.

FINDING YOUR WAY AROUND THE SEA

Latitude and Longitude

In order to locate a specific spot on the surface of the earth, or navigate from one place to another, we need a frame of reference. For this we use a grid composed of two sets of lines, called **latitude** and **longitude.** The grid is superimposed on the earth's surface and divides it into sections.

The earth is essentially a sphere, and a circle drawn around the center of the earth perpendicular to its axis of rotation is the **equator** (Figure 3–14a and b). This line is assigned the designation 0 degrees latitude. Each half of the earth is then divided again by other latitude lines at equal intervals ending at the poles. The North Pole is 90 degrees north with reference to the equator and the South Pole is 90 degrees south. Notice that each line of latitude forms a progressively smaller circle, and that it is necessary to state whether the latitude is north or south of the equator. Latitude lines are sometimes referred to as parallels since they run parallel to the equator as well as to each other. In addition to the equator, other important parallels include the Tropic of Cancer (23.5 degrees N) and the Tropic of Capricorn (23.5 degrees S), which define the boundaries of what is known as the **tropical zone,** and the Arctic Circle (66.5 degrees N) and the Antarctic Circle (66.5 degrees S).

Lines of longitude, also known as **meridians,** are at right angles to the latitude lines. The primary line of longitude is an arbitrary one that extends from the North Pole to the South Pole, passing through the Royal Naval Observatory in Greenwich, England. This line is known as the prime

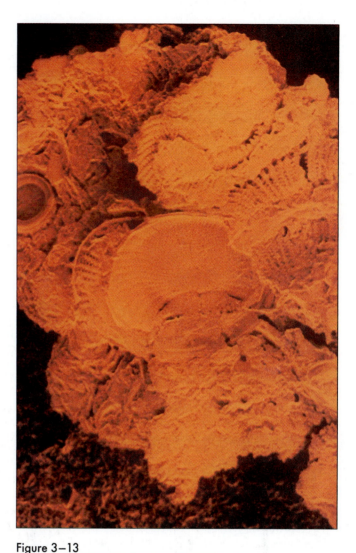

Figure 3–13

Biogenous Sediments. Biogenous sediments are primarily composed of the shells of planktonic organisms such as these diatoms. (Science VU/Visuals Unlimited)

or Greenwich meridian and is designated as 0 degrees. Directly opposite the prime meridian at 180 degrees is a line that approximates the international date line. Longitudinal lines are identified by the angle that they form with respect to the prime meridian and are designated east or west.

Unlike latitude lines, longitude lines all form circles of the same size, since each passes through the earth's poles and equator.

Using latitude and longitude one can locate any point on the earth's surface exactly. For instance, the Canary Islands lie at 15 degrees W and 28 degrees N (Figure 3–14c). Since a degree of latitude or longitude covers a large area, each degree can be subdivided into 60 minutes and each minute into 60 seconds for giving a more precise location. A common unit of distance used in navigation is the **nautical mile,** which is equal to one minute of latitude (1.85 kilometers or 1.15 land miles).

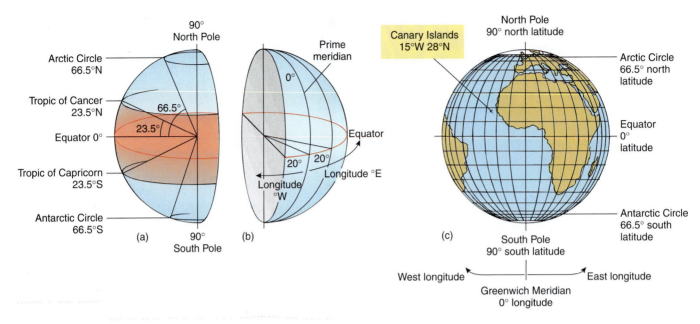

Figure 3–14

Latitude and Longitude. (a) The point of reference for latitude is the equator (0 degrees latitude). Latitude lines are then drawn at equal intervals from the equator to each pole (90 degrees latitude). (b) The prime meridian is the reference point for lines of longitude (0 degrees longitude). Other lines of longitude are designated by the angle they form with respect to the prime meridian and by direction (east or west). (c) Together lines of latitude and longitude form a grid that, when superimposed on the earth's surface, can be used to locate the precise position of any point on the globe.

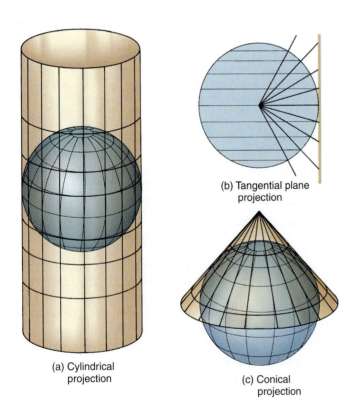

(a) Cylindrical projection

(b) Tangential plane projection

(c) Conical projection

Maps and Charts

A two-dimensional representation of the earth's three-dimensional surface is called a map or chart. Maps usually show the land features of the earth, while the oceans and their features are displayed on charts. Maps and charts are made by projecting the features of the earth's surface together with lines of latitude and longitude, onto a surface to produce a chart projection (Figure 3–15). If you could imagine a clear glass globe with the lines of latitude and longitude drawn on the surface and a light bulb placed inside, you could get an idea of how a projection is made. As the light shines through the globe, it would project an image onto a surface, such as a piece of paper, that is placed in a particular position with reference to the globe. Depending on the position of the paper, one could make three basic types of chart projections: the cylindrical, conical, and tangential planes.

Figure 3–15

Map Projections. If a light were placed inside a transparent globe, the pattern of light and shadow falling on a piece of paper placed in a particular position would form a map. The three most common arrangements for producing a map are the (a) cylindrical projection, (b) tangential plane projection, and (c) the conical projection.

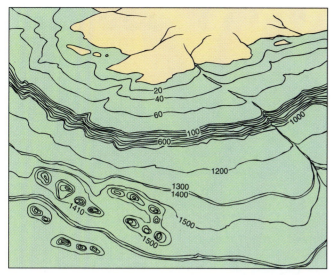

Figure 3–16

Bathymetric and Physiographic Charts. (a) A bathymetric chart indicates variations in ocean depth by lines that connect areas of similar depth. (b) A physiographic chart shows the same information as a bathymetric chart, but it uses coloring or shading rather than lines to indicate areas of similar depth.

The most familiar maps and charts, such as the ones found in texts (including this one) and in classrooms, are modifications of the cylindrical type called a Mercator projection (for examples see Figures 3–1 and 3–6). Even though at the high latitudes there is a great deal of distortion and the poles cannot be shown with this projection, it has the advantage that a straight line drawn on it is a line of true direction (constant compass heading), and as such is useful for navigation. Charts of the ocean that show lines connecting points of similar depth are called **bathymetric charts**

(Figure 3–16a); these are similar to topographic maps showing land elevations. A chart that uses perspective drawing (picturing the sea as it appears to the eye with reference to a relative depth), coloring, or shading, instead of lines, to show the varying depths of the oceans is called a **physiographic chart** (Figure 3-16b).

Principles of Navigation

Early navigators had a difficult time determining their position accurately. Charts and maps were not particularly accurate, and they exhibited a great deal of artistic license. To compensate for this problem, early sailors learned to use the sun and stars as aids to navigation. For instance, they knew that the North Star, Polaris, was positioned approximately over the North Pole. By measuring the angle of the North Star with respect to the horizon, using an instrument called a **sextant** (Figure 3–17), a seaman could get a good idea of his latitude, as long as he was in the Northern Hemisphere. In the early days of sailing, the standard procedure for navigation, once out of sight of land, was to sail north or south to the desired latitude, then sail east or west until reaching land again. When land was sighted, north–south adjustments to the course could be made with reference to landmarks along the coast.

Determining longitude was much more difficult, since the earth rotates and so do the longitudinal lines. Even though as early as the 1500s there were theories for using time to determine longitude, clocks were not reliable enough on sailing vessels to be of any great use. The Flemish astronomer Gemma Frisius proposed that if a clock were

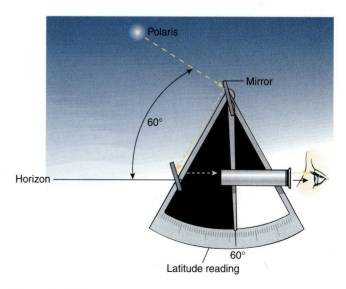

Figure 3–17

A Sextant. Sextants like this one are used to measure the angle of the North Star with respect to the horizon. This information can then be used to determine a ship's latitude.

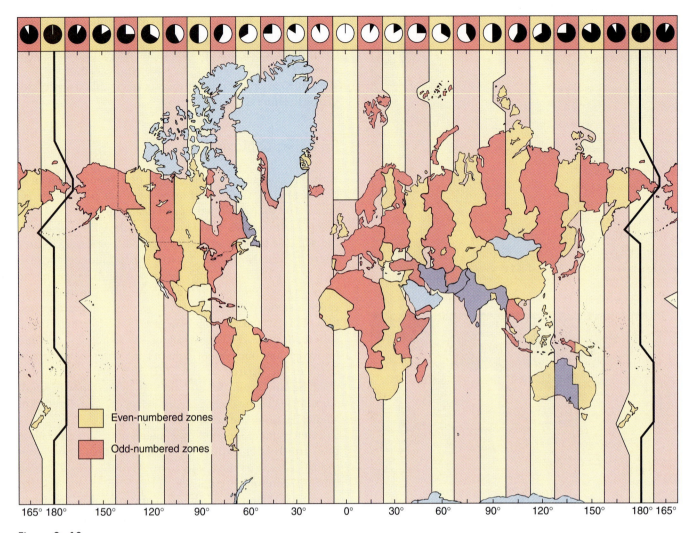

Figure 3–18

The Earth's Time Zones. The reference longitude for time is the prime meridian. Since the earth rotates 15 degrees each hour, each of the world's time zones is 15 degrees wide.

set to exactly noon when the sun was at its zenith (highest point above a reference longitude), and the time taken again when the sun was at its zenith at a different longitude, the difference in time could be used to determine the new longitude. Since the earth rotates 15 degrees each hour, if the zenith of the sun occurred at 1 P.M. instead of noon, the ship would be 15 degrees to the west, and if the zenith at the new longitude occurred at 11 A.M., then the ship would be 15 degrees to the east. In 1735 an Englishman named John Harrison built the first **chronometer,** a seagoing clock that could keep accurate time, and he used it to determine longitude. However, it was not until 1761, with his fourth model, that the instrument proved itself on an 81-day voyage, losing only 51 seconds in that period of time. Today the reference longitude for time is the prime meridian, and the earth is divided into time zones (Figure 3–18), each one 15 degrees of longitude wide.

Even with advances in modern technology, chronometers and sextants are still used as navigational aids. New electronic navigational devices are very important to marine scientists because they help them determine the exact position of their research vessels in any type of weather, permitting them to return again to the same position for more sampling or study.

In the map legend:
- Even-numbered zones
- Odd-numbered zones

Modern Navigational Techniques

Modern research vessels use a variety of techniques for determining their position when at sea. When ships are near land, they can use radar to get a picture of the shoreline and to determine their position relative to the shore. Radar uses high-frequency radio waves sent out by a transmitter. When the waves strike an object on land, they reflect back to an antenna on the ship and display on a screen.

Further out to sea, ships use LORAN (LOng-RAnge Navigation). LORAN employs several stations on land that give off high-frequency radio signals. The ship uses an electronic timing device to measure the difference in arrival time of the signals from two stations. From these data the ship's position can be plotted on a chart that shows the time lines for the stations used. More recently, LORAN receivers have been equipped with computers that can be programmed for the desired latitude and longitude. The LORAN monitor receives the signals from the land-based stations and prints out the course to be sailed and the distance to the desired destination. The computer can also calculate the ship's latitude and longitude from the signals, so the navigator can determine the ship's position at all times.

The most modern, accurate, and sophisticated navigational aid is the satellite navigation system. A special series of satellites orbits the earth and gives off a coded signal with a precise frequency that can be picked up by a receiver on a ship. The receiver monitors the differences in the frequency of the signal as a satellite passes and determines the precise instant in time that the frequency matches what it should be. At this point, the beam of the satellite and the course of the ship are at right angles to each other. With this information and information concerning the satellite's orbit, an onboard computer can determine the ship's position to within 30 meters (100 feet) or less.

All of these modern techniques for navigation require the very accurate measurement of time, and the more sophisticated navigational techniques become, the more precise time measurement must be. With the aid of these modern navigational aids, marine biologists can determine the position of their research vessels in any kind of weather, and the ships can return to the exact spot on a subsequent trip if more sampling or observations are needed. More important, researchers can determine exactly where in the ocean they collected their data and accurately plot this position on a map.

CHAPTER SUMMARY

The world's oceans are believed to have formed more than 3 billion years ago when water vapor escaping from minerals in the earth cooled and condensed on the earth's surface. The position of the earth with respect to the sun and the presence of an atmosphere allows water to exist in a liquid state on the earth's surface and accounts for the earth's relatively constant temperature range. Today there are three major oceans of the world: the Atlantic, Pacific, and Indian Oceans.

The continental masses are not locked into position but are constantly moving at a very slow rate. Geologists have determined that in the past there was only one large continent that fragmented one or more times to form the continents that we have today. Continents move when the plates on which they rest move. The crustal plates move horizontally when molten magma from the earth's core moves to the crust and breaks through, pushing the plates laterally and forming new crust. Old crust is removed in the deep trenches and other subduction zones of the ocean, preventing the earth from getting larger.

The seafloor has many surface features. Sloping away from the edge of the continental masses are the continental shelves. Continental shelves have been alternately submerged and exposed over the history of the earth. During the times they have been exposed, forces such as rivers and turbidity currents have carved submarine canyons. At the bottom of the continental shelves is the abyssal plain, which stretches along the floor of the ocean and is marked by mountain ranges, ridges, valleys, trenches, and seamounts.

The seafloor is covered with a variety of sediments. Sediment particles are formed by the action of physical processes on rocks; some are formed from the seawater itself, and some rain down from space. An important source of ocean sediments are small organisms that have shells of silica or calcium carbonate. When these organisms die, their shells are deposited on the seafloor and make up biogenous sediments.

Navigators use lines of latitude and longitude to locate their position while at sea. Lines of latitude run perpendicular to the long axis of the earth, which extends from pole to pole. Lines of longitude run perpendicular to the lines of latitude. Maps and charts are two-dimensional representations of the earth's three-dimensional surface. Charts show the features of the ocean and can be used for navigation or to display the shape and characteristics of the ocean bottom.

SELECTED KEY TERMS

seafloor spreading, *p. 35*

continental drift, *p. 35*

theory of plate

tectonics, *p. 39*

bathygraphic

  features, *p. 40*

continental shelf, *p. 40*

abyssal plain, *p. 45*

seamount, *p. 45*

biogenous

  sediments, *p. 46*

ooze, *p. 46*

latitude, *p. 47*

longitude, *p. 47*

equator, *p. 47*

QUESTIONS FOR REVIEW

MULTIPLE CHOICE

1. What portion of the earth's surface is covered by oceans today?
   a. 25%
   b. 50%
   c. 66%
   d. 71%
   e. 88%

2. How many principal oceans exist today?
   a. 1
   b. 2
   c. 3
   d. 4
   e. 5

3. A line drawn around the center of the earth perpendicular to the earth's north–south axis is the
   a. prime meridian
   b. Tropic of Cancer
   c. Tropic of Capricorn
   d. equator
   e. international date line

4. One minute of latitude is equal to
   a. one minute of time
   b. one degree of angle
   c. one nautical mile
   d. one minute of longitude
   e. one meridian

5. Charts that use lines connecting points of similar depth to show the physical characteristics of the ocean bottom are
   a. topographic charts
   b. bathymetric charts
   c. physiographic projections
   d. Mercator projections
   e. depth charts

6. The edge of a continent where land meets water is called the
   a. continental shelf
   b. continental ridge
   c. continental rise
   d. continental trench
   e. subduction zone

7. The bottom of many ocean basins is a flat
   a. ridge
   b. seamount
   c. trench
   d. valley
   e. plain

8. Biogenous sediments are formed from
   a. sediments washed into the sea by rivers
   b. chemical reactions between elements in seawater
   c. meteorite particles showering down on oceans
   d. the shells of plankton and small animals
   e. all of the above

9. The outermost layer of the earth is called the
   a. mantle
   b. lithosphere
   c. asthenosphere
   d. atmosphere
   e. crust

10. Regions where the lithosphere splits, separates, and moves apart as new crust is formed is called a
    a. rift zone
    b. subduction zone
    c. deep-sea vent
    d. trench
    e. continental rise

SHORT ANSWER

1. What evidence supports the theory of continental drift?
2. What processes are responsible for the formation and sculpturing of the continental shelves?
3. What is a seamount and how is it formed?
4. Explain how the oceans were originally formed.
5. Explain how time can be used to help determine a ship's longitude.
6. Explain how biogenous sediments are formed.
7. Describe how the continents are thought to move apart.

THINKING CRITICALLY

1. How would you expect continental drift to affect the distribution of bottom-dwelling marine organisms?
2. Would you expect to find more pelagic (actively swimming) fish species above the continental shelves or in the large expanses of open sea? Explain.
3. While doing research along a coastal area, you discover the bottom sediments are predominantly calcareous ooze. What does this imply about the local conditions?

SUGGESTIONS FOR FURTHER READING

Allegre, C. J. and S. H. Schneider. 1994. The Evolution of the Earth, *Scientific American* 271(4):66–75.

Ballard, R. 1975. Dive into the Great Rift, *National Geographic* 147(5):604–615.

Bloxham, J. and D. Coubbins. 1989. The Evolution of the Earth's Magnetic Field, *Scientific American* 261(6):68–75.

Bonati, E. 1994. The Earth's Mantle below the Oceans, *Scientific American* 270(3):44–51.

Continents Adrift and Continents Aground: Readings from Scientific American, 1976. Freeman, San Francisco.

Curtsinger, B. 1996. Realm of the Seamount, *National Geographic* 190(5):72–80.

Marine Geology and Geophysics, *Oceanus* 35(4), Winter 1992/1993.

Chapter 4

Water, Waves, and Tides

Living cells contain about two-thirds water by mass and have a concentration of salts similar to that of seawater. The similarity between the salt concentration in cells and seawater reflects the fact that life evolved in the oceans. From the beginning, water has played a central role in the evolution of living organisms on this planet and continues to play a vital role in the maintenance of life. Marine organisms are constantly influenced by water and the forces associated with it, such as waves and tides. In this chapter, you will learn about the characteristics of water, especially seawater, that make it unique and vital to life. You will also learn about currents, waves, and tides so that you can better understand the role these forces play in the biological communities that we will study in later chapters.

THE NATURE OF WATER

Physical Properties of Water

Water is not only necessary for life, but it is also the most abundant component of living things. For instance, most marine organisms are between 70% and 80% water by mass. By comparison, terrestrial organisms are about 66% water by mass. The importance of water to living organisms is related to its unique chemical and physical properties (Table 4–1). Water is an excellent solvent (medium for dissolving other substances). It has a high boiling point and freezing point compared to similar chemical compounds. It is denser in the liquid form than it is as a solid. Water helps to support marine organisms by buoyancy and provides a medium for the different chemical reactions that are necessary for life.

Water molecules are composed of two atoms of hydrogen bonded to one atom of oxygen. Since the electrons that form the chemical bond between the hydrogen and oxygen atoms are attracted more toward the oxygen atom, the oxygen carries a slight negative electrical charge. By the same token, the hydrogen atoms carry a slight positive electrical

KEY CONCEPTS

1. The polar nature of water accounts for many of its important physical properties.

2. Seawater contains a number of salts, the most abundant being sodium chloride.

3. Salts are constantly being added to and removed from the oceans.

4. The exchange of energy between oceans and the atmosphere produces winds that drive ocean currents and weather patterns.

5. The density of seawater is mainly determined by temperature and salinity.

6. Vertical mixing of seawater carries oxygen to the deep and nutrients to the surface.

7. Waves are the result of forces acting on the surface of the water.

8. The gravitational pull of the moon and the sun on the oceans produces tides.

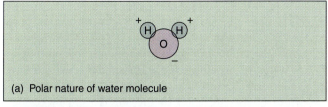

(a) Polar nature of water molecule

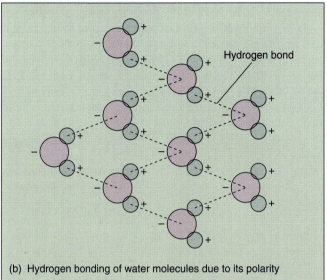

Hydrogen bond

(b) Hydrogen bonding of water molecules due to its polarity

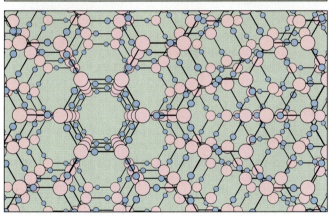

(c) Structure of water molecules in a solid state (ice)

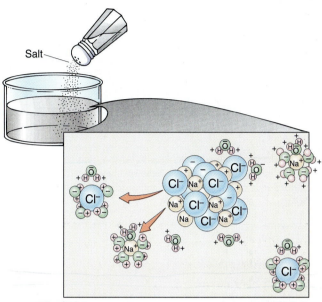

Salt

(d) Salt crystals dissolving in water

Figure 4–1

Water Molecules. (a) A water molecule contains two atoms of hydrogen and one atom of oxygen. Water molecules are polar because the hydrogen atoms have a slight positive charge and the oxygen has a slight negative charge. (b) The polar nature of water molecules causes them to be attracted to each other and form hydrogen bonds. The hydrogen bonding in water accounts for its relatively high freezing point and boiling point. (c) Because of the polar nature of water molecules, when water freezes the molecules spread out. As a result solid water (ice) is less dense than liquid water and floats. (d) The ions in salts are attracted to polar water molecules and readily move into solution. In solution, the ions are surrounded by water molecules so there is less chance of them being attracted to each other and reforming crystals.

charge. This separation of electrical charges makes the water molecule **polar,** meaning that it has a positive end and a negative end (Figure 4–1a).

Because water molecules are polar, they have a tendency to come together and form hydrogen bonds with each other. Hydrogen bonds are weak attractive forces that occur between the positively charged hydrogen atoms of one molecule and the slightly negative oxygen ends of a neighboring molecule (Figure 4–1b). The ability of water molecules to form hydrogen bonds accounts for many of water's special properties. For instance, the relatively high boiling point of water (100°C) reflects the substantial amount of energy that is required to overcome the attractive forces of the hydrogen bonds so that individual water

Table 4–1
Physical Properties of Water

Boiling point	100°C
Freezing point	0°C
Heat capacity	1.00 cal/g/°C
Density (at 4°C)	1.00 g/cm³
Latent heat of fusion	80 cal
Latent heat of vaporization	540 cal

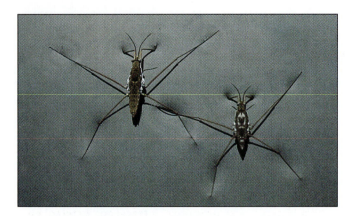

Figure 4–2

Surface Tension. Since water molecules have a greater attraction for each other than for molecules in the air, they form a tight layer at the surface that is able to support small organisms like this water strider. (Dennis Drenner)

molecules can separate from each other and go into the gaseous state. The comparatively high freezing point of water (0°C) is also due to the presence of hydrogen bonds. Because the molecules have a natural attraction to each other, less energy is required to fix them into position so that they will form a solid. When water does freeze, the molecules move away from each other because of repulsive electrical forces between electrons of neighboring atoms. As a result, solid water (ice) is less dense than liquid water, and it floats (Figure 4–1c). This characteristic is important because a body of water freezes at its surface, forming an insulating layer, and the water below stays liquid. This characteristic of water keeps the oceans from freezing solid and allows living organisms to survive, even when the ocean surface is frozen.

The polar nature of water molecules also accounts for water's remarkable solvent properties. Other polar substances, such as salts, readily dissolve in water. Salt crystals are made up of individually charged particles called **ions,** which are attracted to the polar water molecules. As ions become separated from the crystal, they dissolve in the water. The attraction of polar water molecules for charged ions then helps to keep the ions in solution once they have dissolved (Figure 4–1d). Water is not able to dissolve molecules that are nonpolar, such as oil and petroleum products.

The hydrogen bonds in water cause water molecules to be very cohesive, that is, they stick together. This cohesiveness gives water a high **surface tension**. The water molecules at the surface of the liquid have a greater attraction for other water molecules than they do for molecules in the air. As a result, the molecules at the surface form a tight, closely packed layer that is held tightly by the hydrogen bonds of the water molecules beneath the surface. The high surface tension contributes to the high boiling point of water, since

water must take in enough energy to overcome the surface tension before the molecules move into the vapor state. Many small organisms such as water striders (*Halobates*) have become adapted to supporting themselves on the surface of water, taking advantage of the surface tension to give them support (Figure 4–2).

Water is also attracted to the surface of objects that carry electrical charges. This property is called **adhesion** and accounts for the ability of water to make things wet. It also accounts for the ability of water to rise in narrow spaces, a property known as **capillary action**. Microscopic organisms that live in the spaces between bottom sediments, such as nematode worms, and some burrowing organisms rely on adhesion and capillary action to supply the water necessary to support life (Figure 4–3).

Because the presence of hydrogen bonds greatly restricts the movement of water molecules, it takes significantly more heat energy to raise the temperature of 1 gram of water 1°C than other common liquids. The amount of heat energy required to change the temperature of 1 gram of a substance 1°C is called the **specific heat** or **thermal capacity**. Large bodies of water, such as oceans, maintain a more or less constant temperature over time because of the relatively large change in heat energy required to raise or lower the temperature of water. Since the temperature of the ocean is relatively stable, with the exception of intertidal plants and animals, marine organisms have not evolved mechanisms for adapting to rapid fluctuations in temperature, as have terrestrial plants and animals.

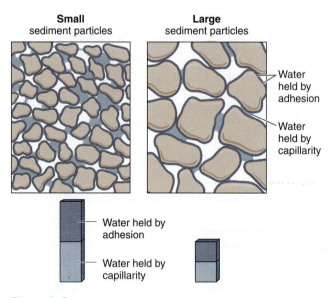

Figure 4–3

Adhesion and Capillary Action. Water moves through the tiny spaces between sediment particles and adheres to the surface of the particles, supplying moisture for numerous organisms that live in bottom sediments. The smaller the particles, the smaller the spaces between them and the more capillary water they can hold.

The pH scale is an indicator of the number (concentration) of hydrogen ions in a volume of a solution. It ranges from 0 to 14, with numbers below 7 indicating an acidic solution. The pH of pure water is 7, which is considered the neutral point. At a pH of 7, the number of hydrogen ions in solution is equal to the number of hydroxide ions; thus pure water is neither acidic nor basic. The lower the pH, the more hydrogen ions there are in solution and the stronger the acid. A decrease of one pH unit indicates a tenfold increase in acidity. On the other hand, the higher the pH, the more hydroxide ions there are in solution and the stronger the base. An increase of one pH unit indicates a tenfold decrease in acidity. The pH of an organism's internal and external environment is a vital factor in determining the distribution of marine life. For example, corals cannot grow in water that is acidic since the low pH inhibits their ability to form external skeletons.

SALT WATER

The Composition of Seawater

When salts dissolve in water they form ions, and most of the salts that are present in seawater are present in their ionic form (Table 4–2). A total of six ions are responsible for 99% of the dissolved salts in the ocean. These ions are the major constituents of seawater and include sodium (Na^+), magnesium (Mg^{2+}), calcium (Ca^{2+}), potassium (K^+), chloride (Cl^-), and sulfate (SO_4^{2-}). Other elements that are dissolved in seawater but are present in concentrations less than one part per million are called *trace elements*.

Seawater is approximately 3.5% salt and 96.5% water by mass. Sodium chloride, or table salt, is the most common salt present. Since **salinity** (the concentration of salt in a given volume of water) is usually expressed in either grams of salt/kilogram of water or parts per thousand (‰), seawater would have an average salinity of 35 g/kg or 35‰. The salinity of surface seawater varies with latitude and the

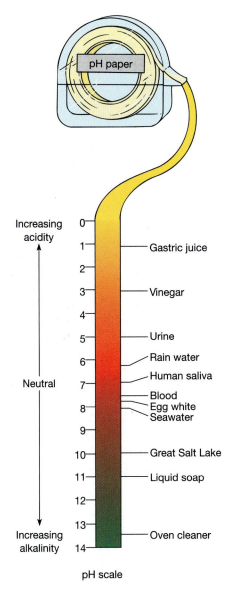

Figure 4–4

pH Scale. The pH scale is a convenient way of expressing the acidity of a solution. A pH of 7 is neutral. The more hydrogen ions a solution contains, the more acidic it is and the lower the pH. The more hydroxide ions a solution contains, the more alkaline it is and the higher the pH.

Table 4–2
Major Ions Found in Seawater

Ion	g/kg of Seawater	Percentage by Weight
Chloride (Cl^-)	19.35	55.07
Sodium (Na^+)	10.76	30.62
Sulfate (SO_4^{2-})	2.71	7.72
Magnesium (Mg^{2+})	1.29	3.68
Calcium (Ca^{2+})	0.41	1.17
Potassium (K^+)	0.39	1.10
Bicarbonate (HCO_3^-)	0.14	0.40
Total		99.76

Chemical Properties of Water

Compounds are classified as acids or bases on the basis of how they ionize when placed into water. Acids are compounds that form hydrogen ions (H^+) when they are added to water, whereas bases or alkaline substances form hydroxide ions (OH^-) when added to water. To indicate whether a solution is acidic or alkaline (basic), biologists use a shorthand notation called the **pH scale** (Figure 4–4).

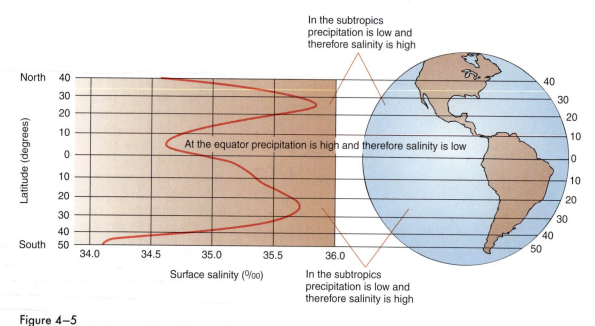

Figure 4–5

Ocean Salinity. The salinity of the surface water of the ocean varies with latitude as a result of regional differences in evaporation and precipitation.

topographical features of an area. These variations are the result of evaporation, precipitation, freezing, thawing, and freshwater runoff from nearby land masses. As you might expect, areas of ocean that are close to the mouths of rivers (fresh water) have comparatively low salinity, whereas surface waters in subtropical areas of the world where evaporation is high and precipitation is low have comparatively high surface salinities (Figure 4–5).

The processes of evaporation and precipitation contribute heavily to the salinity of surface water in the mid-ocean. In the region between 10 degrees N and 10 degrees S of the equator, rainfall is heavy, and the surface water has a relatively low salinity. The regions of ocean at approximately 30 degrees N and S, the latitudes that correspond with many of the world's deserts, exhibit a pattern that is characterized by evaporation exceeding precipitation. As a result, these areas of ocean have a relatively high salinity. Farther north and south, from 50 degrees latitude, precipitation is again heavy, and the surface water is relatively less salty. Finally, in the regions of the poles, the process of freezing removes water from the sea, leaving the salt behind. Thus, the water beneath the ice has a relatively high salinity.

The ions and minerals in seawater are vitally important to the lives of marine organisms. Snails, clams, corals, and various crustaceans, for instance, require calcium to form their shells. Even minerals that are present in trace quantities are important to many marine organisms. Several marine organisms can concentrate trace minerals in their tissues at much higher levels than exist in seawater. For example, kelps (*Macrocystis*) and shellfish (shrimp, oysters, etc.) are able to concentrate high levels of iodine in their

tissues. Kelps have even been commercially harvested as a source of this element, an important nutrient in the human diet.

The Cycling of Sea Salts

The original sources of sea salts were rocks and other constituents of the earth's crust and interior. It is currently thought that these salts entered the ocean dissolved in water that seeped up through the ocean floor. Currently several processes continue to contribute salts to the ocean. Rocks previously formed on the seafloor release their ions as they are broken down by physical and chemical processes. Volcanic eruptions produce gases such as hydrogen sulfide and chlorine, which then dissolve in rainwater and enter the ocean with precipitation. River water carries large amounts of ions, formed by the weathering of rocks, into the sea.

Geologists believe that the composition of the oceans has been the same for about the last 1.5 billion years, despite the fact that each year runoff from the land adds 2.5×10^{12} kilograms (about 250 million dumptruck loads) of salt to the sea. In order for the salinity of the ocean to remain the same over time, the amount of salt added to the oceans by runoff must be balanced by salt removal. There are several ways in which salt is removed from seawater. Some salt ions can react with each other to form insoluble complexes that precipitate to the ocean floor. Alternatively, when waves strike a beach, the sea spray is carried off by air cur-

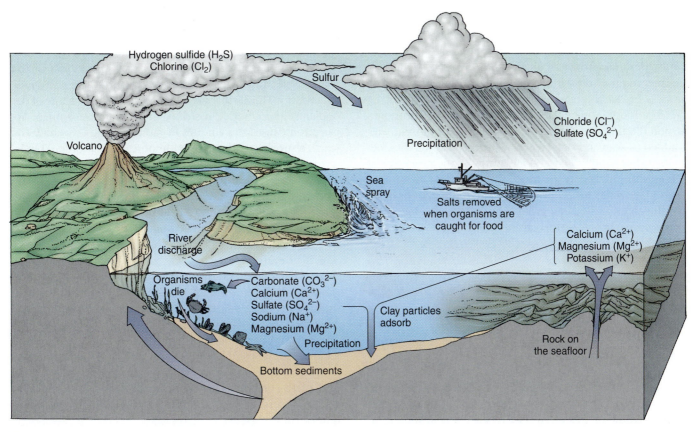

Figure 4–6

Sea Salt Cycling. Salts are constantly being added to and removed from seawater by a variety of processes.

rents and forms a layer of salt on surrounding objects such as rocks, land, buildings, and vehicles.

During the development of the earth, shallow extensions of the ocean became cut off from the sea, and the water evaporated. This process produced salt deposits called *evaporites* (the salt flats of southern California are an example). Similar situations exist today when small, shallow bodies of water become separated from the ocean and the water evaporates, leaving the salt behind. Biological processes also play a role in salt cycling. Certain ions absorbed by marine organisms are later returned to the environment when the organisms excrete them or die and decompose. Salts concentrated in the body tissues of marine organisms are permanently removed if the organisms are harvested for food.

The most important process for the removal of salt, however, involves the sticking of ions to the surface of fine particles, a process known as **adsorption.** These particles then settle to the bottom and become trapped in the sediment. Clay particles formed from weathered rock and carried to the oceans by rivers attract ions and attach them to their surfaces by the process of adsorption. Ions and minerals that become deposited in this manner in the ocean's sediments can later be moved by geological processes to areas well above sea level where weathering can remove the minerals and return them to the oceans (Figure 4–6).

Gases in Seawater

The gases of the atmosphere, primarily oxygen (O_2), carbon dioxide (CO_2), and nitrogen (N_2), are also found in seawater. The proportion of oxygen and nitrogen in seawater is about the same as it is in the atmosphere. The concentration of carbon dioxide is slightly higher than it is in the atmosphere since it dissolves in seawater more readily than the other two (Table 4–3). These gases dissolve at the surface of the sea and are distributed throughout the water by mixing processes and currents.

Table 4–3
Gases Found in Seawater

Gas	Percentage by Volume in Atmosphere	Percentage by Volume in Surface Seawater[a]	Percentage by Volume in Ocean Total
Nitrogen (N_2)	78.08	48	11
Oxygen (O_2)	20.99	36	6
Carbon dioxide (CO_2)	0.03	15	83
Other gases	0.95	1	

[a]salinity 35‰ and temperature = 20°C

Oxygen and carbon dioxide play important roles in the lives of marine organisms. Bacteria, phytoplankton, algae, and plants use carbon dioxide in the process of photosynthesis, producing oxygen as a by-product. Most living organisms require oxygen and produce carbon dioxide. The process of decomposition, which requires respiration by bacteria or other organisms, also uses oxygen and produces carbon dioxide. Oxygen as a by-product of photosynthesis is added to seawater only near the surface. Carbon dioxide, on the other hand, is added at all depths by the processes of decomposition and respiration. The levels of these two gases in seawater have a profound effect on the kinds of organisms that can live in a given area and the number of organisms a particular region can support.

The amount of gas that water can hold depends on the temperature, salinity, and pressure of the water. For instance, cold water holds more gas than warm water, and more gas will dissolve if the salinity of the water is low and the gas pressure is high.

OCEAN HEATING AND COOLING

The sun's radiant energy is responsible for warming the earth's surface. The latitudes between the Tropic of Cancer and the Tropic of Capricorn receive the greatest amount of radiant energy, while the middle latitudes receive moderate amounts, and the poles receive the least. This difference in the amount of radiant energy striking the earth's surface is the result of the earth's spherical shape and the presence of the atmosphere (Figure 4–7). The greatest amount of radiant energy is received where the sun's rays strike the earth at right angles, the area around the equator. As you move to the north or the south, the angles of the sun's rays relative to the earth's surface increase because of the earth's curved surface. As a result, at the middle latitudes and the poles, the same amount of sunlight falls on a larger area of the earth. Thus, these regions receive proportionately less radiant energy. The earth's atmosphere also absorbs some of the radiant energy before it strikes the earth. Atmospheric absorption causes the amount of radiant energy reaching the earth's surface to decrease with increasing latitude. The decrease is due to the longer path that the sun's rays must travel to reach the earth's surface at higher latitudes. The combined effect of the earth's curved surface and atmosphere is a larger amount of heat being received around the equator, less in the temperate regions, and least at the poles.

Since the earth's temperature is not constantly increasing, heat must be lost as well as gained. Some of the incoming solar radiation is reflected back to space by the atmosphere and the surface of the earth and plays no role in heating the earth's surface. Of the remaining energy, most is absorbed by the earth's surface, and the rest is absorbed by the atmosphere. Ultimately, an equivalent amount of heat energy is reradiated back into space. The atmosphere loses more energy back to space than it gains from the sun, while the earth, on the other hand, loses less energy back to space. As a result, the atmosphere cools, and the surface of the earth warms.

In order to maintain the earth's temperature, the excess energy must be transferred from the earth to the atmosphere, and this is accomplished by evaporation and radiation (Figure 4–8). Energy from the earth's surface is used to produce water vapor from liquid water by evaporation. This energy is then transferred to the atmosphere when the vapor condenses to form rain or freezes to form hail or snow. The heat that is radiated from the earth's surface is absorbed by the water content of the atmosphere. Approximately two thirds of the heat that is transferred to the atmosphere is by evaporation, and the other one third is by radiation and other processes.

When considering the amount of heat energy in an area of ocean, we must take into account several factors, such as the amount of energy absorbed at the sea's surface, the loss of heat energy by evaporation, the transfer of energy into or out of the area by ocean currents, the warming or cooling of the overlying atmosphere by heat from the sea's surface, and heat lost from the sea back to space by radiation. Since all of these processes change with time, regions of the ocean show both daily and seasonal variations in temperature (Figure 4–9). More heat is gained than lost in the region around the equator, whereas more heat is lost than is gained at the higher latitudes. Winds and ocean currents are responsible for removing the excess heat from the tropics and moving it to the higher latitudes so that the overall pattern of surface temperatures across the earth is maintained.

The amount of solar radiation reaching the earth also undergoes an annual cycle of seasonal variations. These

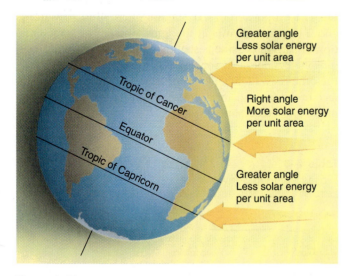

Figure 4–7

Distribution of Solar Energy. As a result of the earth's curved surface, more radiant energy reaches the tropics than the temperate regions or the regions surrounding the poles.

text continues on page 62

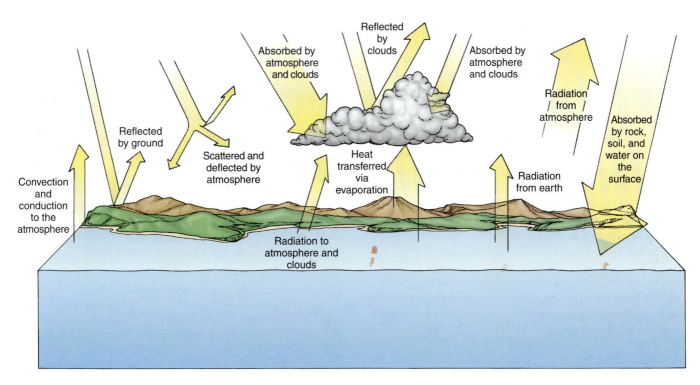

Figure 4–8

The Earth's Energy Budget. Only about 47% of incoming solar radiation is absorbed by the earth's surface. The remainder is absorbed by the atmosphere or reflected back to space. Heat is lost from the surface of the earth by the processes of evaporation, radiation, conduction, and convection. The width of the arrows in the diagram indicates the relative amount of energy being transferred.

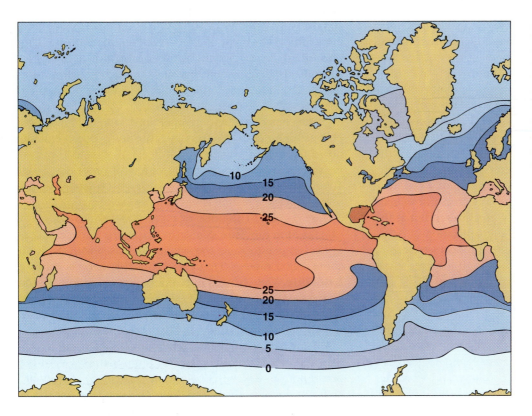

Figure 4–9

Average Surface Temperatures of the World's Oceans. The surface temperature of the sea varies from one region to the next as the result of differences in the amount of solar radiation received, loss of heat via evaporation, and heat energy transferred by ocean currents and the atmosphere.

variations are most pronounced between the latitudes 40 degrees and 60 degrees N and 40 degrees and 60 degrees S because the angle of the sun's rays changes so dramatically at these latitudes with changes in season. This seasonal change in radiant energy produces seasonal changes in the surface temperature of the sea at these latitudes. Between the Tropic of Cancer and the Tropic of Capricorn, the amount of solar radiation remains relatively constant, and the surface temperature of the ocean also remains fairly uniform. As the sun makes its annual migration between the Tropic of Cancer and the Tropic of Capricorn, its rays cross this area twice, producing only a small, semiannual variation in the intensity of solar radiation. As mentioned previously in this chapter, oceans have a very high heat capacity, absorbing and releasing large amounts of heat at their surface with very little change in ocean temperature. During the summer, when the ocean receives more radiant energy, heat that is absorbed at the surface is moved deeper by the action of winds, waves, and currents, helping to stabilize the surface temperature.

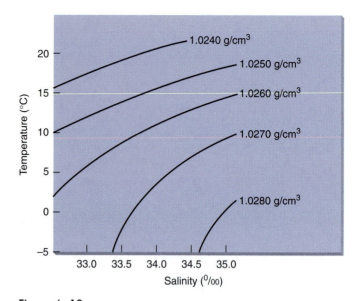

Figure 4–10

Effects of Temperature and Salinity on the Density of Seawater. Increases in salinity or decreases in water temperature result in an increase in the density of seawater. This graph shows the combined effects of salinity and temperature on seawater densities.

OCEAN LAYERS AND VERTICAL MIXING

The Role of Salinity and Temperature in Forming Ocean Layers

Density is a physical property of a substance: the mass of the substance in a given volume. It is generally measured in grams per cubic centimeter (g/cm^3). For instance pure water has a density of 1 gram per cubic centimeter. Since seawater contains dissolved salts, it has a higher density than pure water ($1.0270 \, g/cm^3$). Variations in the surface temperature and salinity of the oceans control the water's density. Density increases with an increase in salinity and a decrease in temperature (water reaches its maximum density at a temperature of 4°C). Density decreases when there is a decrease in salinity or an increase in temperature. Salinity increases when evaporation occurs or when some of the seawater comes out of solution in freezing to form ice. Precipitation, influx of river water, melting ice, or a combination of factors can all contribute to a decrease in salinity. Pressure can also affect the density of seawater, but at the surface these effects are minor. The combination of different surface temperatures and salinities produces regions of ocean with different densities (Figure 4–10). Denser water sinks until it joins water of similar density, whereas less dense water rises. This situation produces the effect of layered water in the ocean.

Seawater that is not very dense, like the warm, low-salinity water around the equator ($1.0230 \, g/cm^3$) remains at the surface. Surface water at 30 degrees N and 30 degrees S is also warm, but it has a higher salinity and is denser ($1.0267 \, g/cm^3$) than the equatorial waters. This difference in density causes the water at 30 degrees N to extend from

the surface at this latitude to below the less dense surface water at the equator and then back to the surface again at 30 degrees S. The combination of colder temperatures and higher salinities produces even denser surface water at 60 degrees N and 60 degrees S ($1.0276 \, g/cm^3$). This water extends from the surface in one hemisphere below the other surface waters to the surface of the other hemisphere. During winter at the poles, the lower water temperature and increased salinity that results from the formation of sea ice results in very dense water that sinks toward the floor of the ocean. The variations in the density of surface water, then, produce a layered effect in the ocean.

The characteristics of temperature, salinity, and density change with depth. The surface layer of the ocean extends down to about 100 meters (330 feet), is warmed by solar heating, and is well-mixed by a variety of processes. From 100 meters (330 feet) to 1,000 meters (3,300 feet) temperatures decrease, creating layers of water of increasing density (Figure 4–11a). This zone of rapid temperature change is called a **thermocline**. Similarly, below the surface waters in the temperate zone, the salinity increases with depth to about 1,000 meters (Figure 4–11b). This zone is called a **halocline**. The changes in temperature and salinity in the region from 100 meters to 1,000 meters produce a **pycnocline** (PIK-nuh-klyn), a zone where density increases rapidly with depth (Figure 4–11c).

Below the thermocline, temperatures are relatively stable, with only small decreases in temperature toward the ocean bottom. Likewise, below the halocline, salinities are constant down to the ocean floor. As a result, water that is deeper than 1,000 meters has relatively constant density.

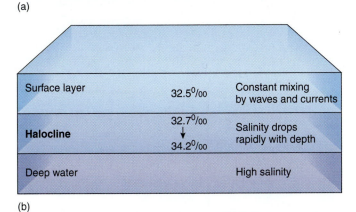

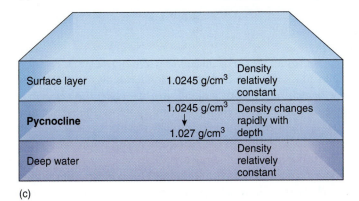

Figure 4–11

Changes in Temperature, Salinity, and Density of Seawater with Depth. (a) Changes in water temperature with depth. (b) Changes in salinity with depth. (c) Changes in density with depth.

Vertical Mixing

When the density of water increases with depth, the water column from the surface down is said to be stable. If the top water in a water column is more dense than the water below it, then the water column is unstable. Unstable water columns do not persist, because the denser water at the top sinks and the less dense water below rises to the surface. This change over in the water column produces a vertical overturn (Figure 4–12). If the water column has the same density from top to bottom it is **isopycnal** (eye-soh-PIK-

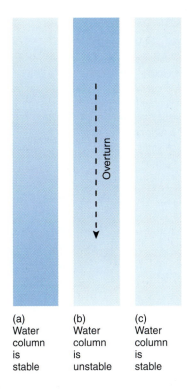

Figure 4–12

Vertical Overturn. (a) In a stable water column, water density increases with depth. (b) In an unstable water column, the water at the top of the column is denser than the water beneath. As a result, the denser surface water sinks and the less dense water beneath is displaced to the surface. This process is called *vertical overturn*. (c) An isopycnal water column has the same density from top to bottom and is stable.

nuhl). Such neutral stability means there is no tendency for water in the column to either sink or rise. A water column that is neutrally stable can be easily mixed in the vertical direction by such forces as wind, wave action, or currents.

Any process that increases the density of surface water will cause vertical movement of the water or vertical mixing. Vertical mixing is an important process for the exchange of water from top to bottom throughout the world's oceans. Since bottom water usually contains abundant nutrients from the settling of organic matter from above and the process of decomposition, vertical mixing provides a means of exchanging this nutrient-rich bottom water with oxygen-rich surface water. Since density is generally controlled by temperature and salinity, this type of circulation is also known as *thermohaline circulation*.

The open ocean in temperate latitudes exhibits a seasonal effect. During the summer months, the surface water is warm, and the water column is relatively stable. In the fall as the surface water cools, its density increases, the water column becomes unstable, and the surface water sinks. The mixing process continues with the aid of winter storms and the continued cooling of surface waters. The thermocline developed during the previous summer is ultimately elimi-

Figure 4—13

Seasonal Changes and Vertical Mixing. Seasonal changes in the temperature and salinity of surface waters of temperate seas produce a mixing effect that brings nutrient-rich bottom water closer to the surface and oxygen-rich surface water closer to the bottom.

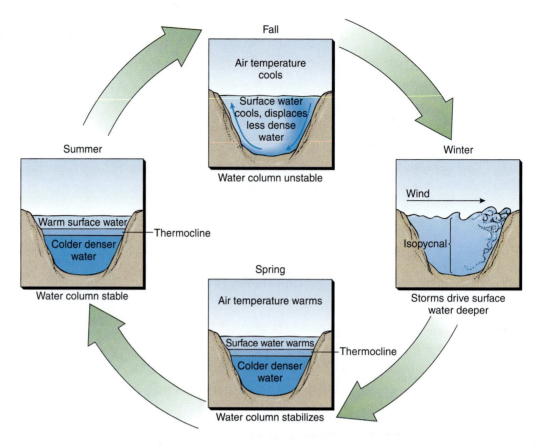

nated, and the upper regions become isopycnal to the deeper water (Figure 4–13). With spring come warmer temperatures, and the thermocline begins to reestablish itself, stabilizing the water column. This process of stabilization then continues through the summer.

In the open ocean, surface temperature is more important than salinity in determining the water's density. For instance, surface water in the tropics has the highest salinity in the open ocean, but the water is so warm that it remains less dense than the water below it, and it does not sink. On the other hand, water in the North Atlantic has a lower salinity but is much colder, and, as a result, the surface water is denser and sinks.

Salinity becomes a more important factor in determining the density of water close to shore. This is especially evident in semi-enclosed bays that receive a large amount of freshwater runoff. Wave action also contributes to the density and mixing of ocean water close to shore, where the water is relatively shallow.

Upwellings and Downwellings

Since the quantity of water in the oceans is essentially fixed, any movement of water from one place to another would cause a similar but opposite movement to replace the water that leaves an area. For instance, when dense surface water sinks to a depth at which it is no longer denser than the water beneath it, it stops sinking. At this point, the water begins to move horizontally, making room for denser water that is sinking behind it. Eventually this water rises back to the surface, replacing the surface water that has been sinking. At the same time, surface water is also moving horizontally into areas where the surface water is sinking.

Downwelling zones represent areas where surface water is sinking, whereas **upwelling zones** are areas where bottom water rises to the surface (Figure 4–14). These processes are extremely important for the organisms that live at the various depths. Because downwelling carries oxygen-rich surface water to deeper areas, many organisms can live in the deep water where downwellings occur. This process is especially important for organisms that live below the photic zone (deeper than sunlight can penetrate). Upwelling, on the other hand, brings water rich in nutrients from sedimentation and decay processes from great depths to the surface, where it supplies the needs of photosynthetic organisms and the zooplankton that are vital parts of oceanic food chains.

Upwelling and downwelling can also be produced by wind-driven surface currents. When currents drive two masses of surface water together, water is forced downward. In other regions of the ocean, surface currents may

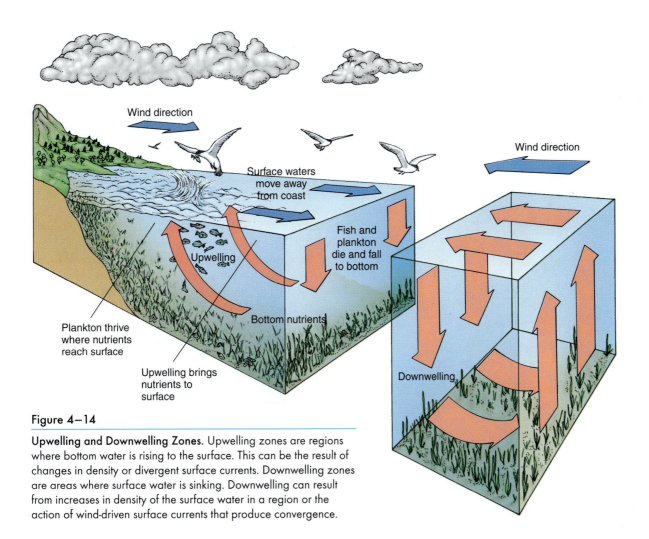

Figure 4—14

Upwelling and Downwelling Zones. Upwelling zones are regions where bottom water is rising to the surface. This can be the result of changes in density or divergent surface currents. Downwelling zones are areas where surface water is sinking. Downwelling can result from increases in density of the surface water in a region or the action of wind-driven surface currents that produce convergence.

push two water masses in opposite directions, drawing deeper water to the surface. During the slow movement produced by these currents, adjacent layers of ocean water are mixed and exchange nutrients and chemicals.

WINDS AND CURRENTS

How Winds Are Produced

The surface currents of the world's oceans are driven by wind. Winds are the result of horizontal air movements that in turn are caused by some of the same factors that are responsible for vertical movement in ocean water, namely temperature and density. As air heats, its density decreases and it rises; and as it cools, its density increases and it falls toward the earth. The density of air is determined by its temperature, water vapor content, and atmospheric pressure. Air density decreases when it is warmed, when its va-

por content increases, or when the atmospheric pressure decreases. Air becomes more dense when it is cooled, its vapor content decreases, or the atmospheric pressure increases. These forces produce two sets of winds that circulate in opposite directions: surface winds at the earth's surface and upper winds such as the jet stream.

Wind Patterns

The region of the earth around the equator receives large amounts of solar energy that heats the air and water. This combination of warm air with a high water vapor content produces less dense air that rises. At the poles, on the other hand, the air is cooling, moisture has been removed by precipitation, and there is increased atmospheric pressure. This combination of factors causes the air to become denser and sink. The denser air from the poles is then sucked toward the equator to replace the less dense air that has moved away. These processes establish a basic pattern of

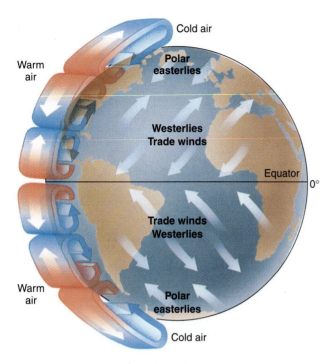

Figure 4–15

North–South Air Flow. As warm air from the equator moves north and south, it displaces colder, denser, polar air that then flows toward the equator. This produces the basic pattern of upper air currents.

upper air flow (winds) from the equator to the north, in the Northern Hemisphere, and from the equator to the south, in the Southern Hemisphere (Figure 4–15).

If the atmosphere were attached tightly to the earth's surface and the earth and its atmosphere moved in unison, then these north and south movements would be the primary air movements. The atmosphere, however, is not rigidly attached to the earth's surface, and the amount of friction between the moving atmosphere and the moving earth is so small that the movement of the atmosphere is somewhat independent of the movement of the earth's surface. For instance, a mass of air that lies over the equator and appears to be stationary is actually moving eastward with the rotating earth.

As the earth rotates on its axis, a point at the equator moves faster than a point at a higher latitude. Both points make one full revolution per day, but the point at the equator has to travel a longer distance, so therefore it must be moving faster. For this reason, air moving north or south away from the equator toward the poles moves progressively faster than the earth beneath it. This causes the air mass to appear to curve relative to the earth's surface. In the Northern Hemisphere this effect causes the air to deflect to the right of the direction of the air movement, and in the

Southern Hemisphere the deflection of the air is to the left of the direction of air flow.

This apparent deflection of the path of the air is called the **Coriolis effect** (Figure 4–16a and b). The Coriolis effect is named after Gaspard Coriolis, who developed a mathematical model for deflection in frictionless motion when the motion is referenced to a rotating body. The Coriolis effect is applicable to any rotating body, and it plays an important role not just in air movements but also in the movement of surface water and currents.

Because of the Coriolis effect, air that rises at the equator does not move straight to the north or the south, but is deflected as it moves (Figure 4–16c). At 30 degrees N and S, the equatorial air sinks and either moves back along the surface of the water to the equator or to 60 degrees N or S. The air that reaches the poles cools and sinks and begins to move back toward the equator. As the air flows back toward the equator, it warms and picks up water vapor, and by the time it reaches 60 degrees N or S, it rises again. The net result of these processes is the division of the earth's surface into three convection cells in each hemisphere.

Between the equator and 30 degrees N, the surface winds are deflected to the right, producing the "northeast tradewinds." In the Southern Hemisphere, the winds are deflected to the left, producing the "southeast tradewinds." (Winds are designated with reference to the direction from which they are coming, not to which they are blowing.) Between 30 degrees N and 60 degrees N, the deflected surface air produces winds that blow from the south and the west, producing the "westerlies." In the Southern Hemisphere, the westerlies blow from the north and the west. Between 60 degrees N and the North Pole, the winds blow from the north and east, forming the "polar easterlies." Between 60 degrees S and the South Pole, the easterlies blow from the south and east.

Low-density air rises at 0 degrees, and 60 degrees N and S, whereas high-density air descends at 30 and 90 degrees N and S. Areas where the air rises are zones of low atmospheric pressure and usually exhibit clouds and rain. Areas where the air descends are associated with high atmospheric pressure, clear skies, and low precipitation. Surface winds move from areas of high pressure to areas of low pressure (Figure 4–17). In the areas of vertical movement between the wind belts, wind movement is unsteady and unreliable. In the early days of sailing ships, these areas were a cause of great concern to sailors, since they could be stranded for days without a breeze to move them. The area of rising air at the equator is known as the "doldrums," and the areas of descending air at 30 degrees N and S are known as the "horse latitudes." It is said that the horse latitudes received their name from early sailing ships carrying horses. If the ship became becalmed for a long period of time, there would not be enough drinking water to support both the crew and the horses, so the horses were thrown overboard.

text continues on page 68

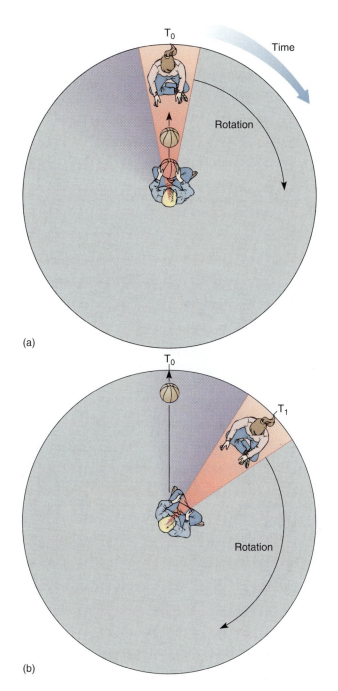

(a)

(b)

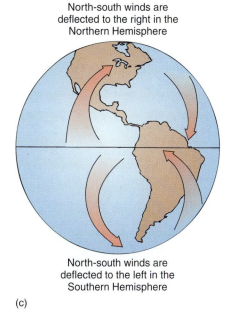

North-south winds are deflected to the right in the Northern Hemisphere

North-south winds are deflected to the left in the Southern Hemisphere

(c)

Figure 4–16

The Coriolis Effect. (a) The Coriolis effect can be demonstrated on a playground merry-go-round. The center of the merry-go-round is analogous to the South Pole and the outer edge to the equator. (a) If you throw a ball to a friend at time zero (T_0) when the merry-go-round is rotating clockwise, the ball at the time T_1 (b) appears to curve to your left instead of going straight. (c) In a similar fashion, wind patterns and ocean currents do not move directly north or south but appear to curve due to the earth's rotation.

Figure 4–17

Surface Wind Patterns. Differences in air temperature and density are responsible for producing global wind patterns.

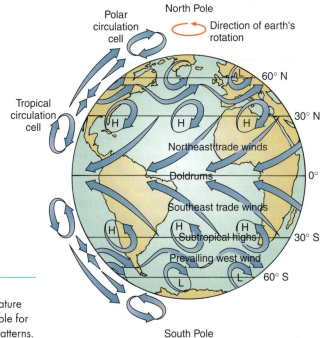

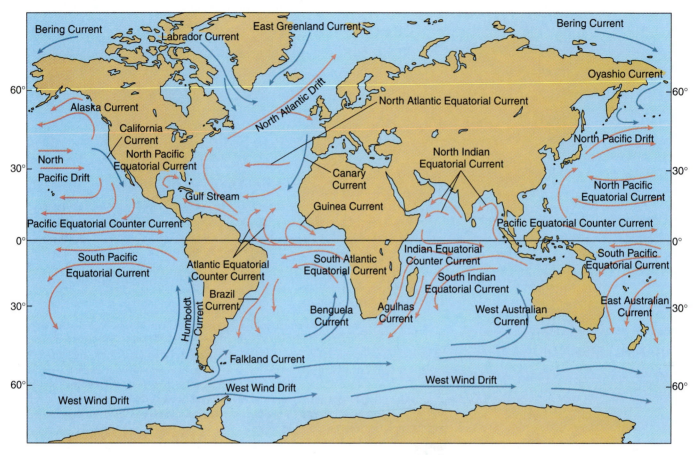

Figure 4–18

Major Ocean Currents. Ocean currents are produced by the driving force of wind. Notice the similarities in the patterns of ocean currents and the surface winds shown in Figure 4–17.

Ocean Currents

Ocean currents are produced when wind blowing across the surface of the ocean pushes and pulls the water, causing it to move. With few exceptions these currents follow a nearly constant pattern (Figure 4–18). Just as with air, the friction between the moving water and the earth is very small, and the moving water is deflected by the Coriolis effect. Since water moves more slowly than air, it takes the water longer to move the same distance than air would. During this longer time, the earth rotates farther out from under the water than it would from under air. Therefore, the slower moving water is deflected to a greater degree than the air. In the Northern Hemisphere, the current is deflected to the right of the prevailing wind direction, and in the Southern Hemisphere, the deflection is to the left. In the open sea, this deflection can be as much as a 45° angle from the wind direction. It is the combined action of wind and the Coriolis effect that produce the major surface currents of the oceans. Like terrestrial rivers, these major cur-

rents maintain their course, making only slight, though important, changes in strength and location in response to seasonal wind and climate patterns.

WAVES

Waves are the result of forces acting on the surface of the water. A force that disturbs the water's surface, such as a stone dropped into the water or wind blowing across the water's surface, is called a **generating force.** The disturbance produced by the generating force moves outward, away from the point of the disturbance. Consider, for instance, a rock that is dropped into water. As the rock hits, the surface water is pushed aside. Then, as the rock sinks to the bottom, the water moves back into the space left behind. The momentum of the returning water forces it upward so that it is raised above the surface. The raised surface then falls back down and creates a depression in the surface. The depression is filled with more water, and the

El Niño Southern Oscillation

Not only do oceans supply the water that allows life to exist on land, they also influence other aspects of terrestrial climate as well. In 1983, it appeared that a dramatic change was occurring in the earth's climate. These changes were manifested in a variety of ways. Some areas, such as California and the coasts of South America, were drenched with rains that caused severe flooding. Other areas, such as Australia and Indonesia, suffered terrible droughts, while parts of Polynesia were hammered by severe typhoons. The effects of these changes were also seen in marine organisms. Off the coast of South America, large numbers of fish and sea birds died for no apparent reason, and many marine mammals, such as whales, dolphins, and seals, disappeared from their usual feeding grounds.

Scientists studying these events found that they were the result of changes in wind patterns and ocean currents that occur periodically in the winter months. Farmers and fishermen in Peru had been aware of the phenomenon for years and had named it El Niño ("the Child"), in honor of the Christ child, since the phenomenon usually occurred around Christmas time. An El Niño is usually a short-term, local phenomenon, but in 1983, it was particularly severe.

The coastal waters of Peru and Ecuador are highly productive fishery areas because of the nearly constant upwelling of deep, nutrient-rich water. The upwelling is driven by the Pacific tradewinds that move cold, dense water to the coastal area, where it sinks and displaces less dense water to the surface. Sometimes the tradewinds lessen and warm tropical water moves eastward across the Pacific to accumulate along the coasts of North and South America.

The warm water that accumulates causes the death of cold-water organisms that are the basis of food chains for many marine fishes, mammals, and birds. The loss of food causes the death of these animals. In severe cases the number of decaying organisms is so great that the surface water contains large quantities of hydrogen sulfide. Although the increased temperature of the coastal waters usually ends by April, sometimes it may persist for as long as a year. In the past, severe El Niños occurred in 1953, 1957–1958, 1965, 1972–1973, 1976–1977, 1982–1983, and 1990–1991.

No one is certain of the exact causes of El Niños, but certain processes have been correlated with the phenomenon's appearance. One process is the

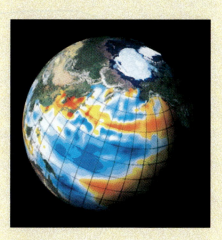

Figure 4–A

El Niño Southern Ocean Oscillation. The orange and yellow band in the lower right of this picture is warm water of the 1991 El Niño. The orange and yellow band further north is a sea temperature change caused by the 1982–1983 El Niño. El Niños are thought to be the result of changes in atmospheric pressure over the eastern and western Pacific Ocean that produce changes in the tradewinds, causing a mass of warm water to move east across the Pacific from Asia. This changes the position of the atmospheric jet streams and results in climate shifts across the globe. (G. Jacobs, Stennis Space Center/Geosphere Project/Science Photo Library/Photo Researchers, Inc.)

Southern Ocean Oscillation, in which the atmospheric pressure on one side of the Pacific increases while the pressure on the opposite side decreases. The pressure changes then reverse themselves. Normally there is a high pressure system over Easter Island in the eastern Pacific and a low pressure system over Indonesia in the western Pacific. Under these conditions the west-to-east tradewinds are strong and constant, and upwellings occur along the coasts of Peru and Ecuador. When the normal pressure system reverses, the tradewinds break down and warm water from the western Pacific moves eastward, ultimately warming the surface waters along the coast of South America and depressing the upwelling. The elevated surface temperature is the result of both warm water moving in and colder, denser water not being able to rise. The warm surface water carries the low pressure zone of rising air and precipitation along with it, producing large amounts of precipitation in normally dry areas. The effects of El Niño usually lessen by midsummer. Another slight warming is often observed in November and December before the Southern Ocean Oscillation reverses again and the tradewinds return to normal. A severe El Niño can last for 15 months.

Attempts have been made to forecast El Niños based on changes in the Southern Ocean Oscillation and the strength of the tradewinds, both of which appear to precede El Niños. The models used to study previous El Niños did not predict some events that occurred, while predicting others that did not occur. As researchers continue to search for more clues that might increase their forecasting accuracy, the phenomenon of El Niños continues to remind us of the close interrelationship between marine and terrestrial ecosystems.

process is repeated, producing a series of waves that ripple outward from the point of disturbance.

The force that causes the water to return to the undisturbed level is called the **restoring force.** If the amount of water that is displaced is small, the restoring force is the surface tension of the water (surface tension was discussed earlier in this chapter), and the small waves are referred to as **capillary waves.** When the amount of water displaced is quite sizable, the restoring force is gravity and the waves are referred to as **gravity waves.** A wave, then, results from the interactions between generating forces and restoring forces. Generating forces can be any event that adds energy to the surface of the sea. Geological events, such as earthquakes and volcanic eruptions, objects dropped into the water, the movement of ships, and disturbances from beneath the surface, such as breeching whales, are all possible generating forces, but the most common factor in generating waves is the wind.

As wind blows across the surface of still water, it creates drag (friction) that lifts some of the water away from the surface (Figure 4–19). If the amount of water displaced is very small, the surface tension of the water pulls it back to restore a smooth surface, and a series of ripples are formed. If the force of the air is greater than a small breeze, more friction is created. As the water is stretched by the wind, more of it is pulled away from the surface, and a combination of surface tension and gravity acts to pull the water back to the surface. As the surface becomes rougher, it becomes easier for the wind to add more energy. The frictional drag between the air and the water is increased, and the waves become progressively larger. A wave in the ocean does not represent a flow of water but a flow of energy or motion. The energy that is added to the water to cause the wave is either dissipated at sea or is transferred to a beach or structures on the beach when the wave strikes.

Most ocean waves are generated by wind and restored by gravity, and they progress in a particular direction. This type of wave is called a **progressive wave.** Progressive waves can be formed by local storm centers or by the prevailing winds of the wind belts such as the tradewinds or westerlies. Waves produced by storms at sea move outward from the storm center in all directions. As wind waves are formed by the storm, they are forced to increase in size and speed by the input of energy from the storm; for this reason they are also known as *forced waves.* When the energy from a storm or other generating force no longer has an effect on the waves, they become free waves, moving at speeds that are determined by the wave's length and **period,** the time required for one wavelength to pass a fixed point (wave speed = wavelength/period).

The various parts and characteristics of waves are shown in Figure 4–19. Waves with long periods and long wavelengths move faster than those with short periods and short wavelengths. Eventually, they escape a storm and appear as a regular pattern of wave crests on the ocean's surface. These long-period, uniform waves are called **swells.** Swells carry a considerable amount of energy and can travel for thousands of kilometers.

Waves that occur in water that is deeper than one half of a wave's wavelength are called **deepwater waves.** The height of a deepwater wave increases with wind speed, the duration that the wind blows, and the fetch, or distance over the water that the wind blows. If the wind speed is low, the waves will be small, regardless of the duration or the fetch. A strong wind blowing for a short period of time does not produce large waves even if the fetch is large. If a strong wind blows for a long time over a short fetch, the waves are again small. Large waves are only produced when all three factors (wind speed, wind duration, and fetch) are of high magnitude.

When a deepwater wave enters shallow water it becomes a shallow-water wave. As the waves approach the shore, the decreasing depth begins to affect their shape and speed, and their crests become flatter. The friction that results from the orbit of the wave dragging the bottom slows the forward movement of the wave. Whereas the length and speed of a deepwater wave is determined by the wave period, the speed and length of a shallow-water wave is determined by the depth of the water.

The **surf zone** is the area along a coast where waves slow down, become steeper, break, and disappear. **Breakers** form in the surf zone when the lower part of a wave is slowed by friction with the bottom but its crest continues moving toward the shore at a speed faster than that of the wave. As a result, the crest overtakes the base of the wave in front of it and eventually falls into the preceeding trough and breaks up (Figure 4–20). The two most common types of breakers are called *plungers* and *spillers.*

Plunging breakers form when the beach slope is steep. The crest curls and curves over the air below it and outruns the rest of the wave. It then breaks with a sudden loss of energy and a splash. Spilling breakers are more common and

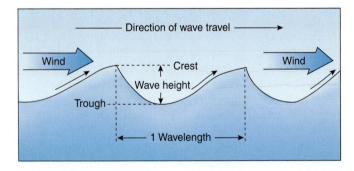

Figure 4–19

Characteristics of Waves. The drag produced by wind blowing across the surface of still water lifts some of the water away from the surface, producing a wave. Important wave characteristics include wavelength (the distance between the crests of two consecutive waves) and wave height (amplitude).

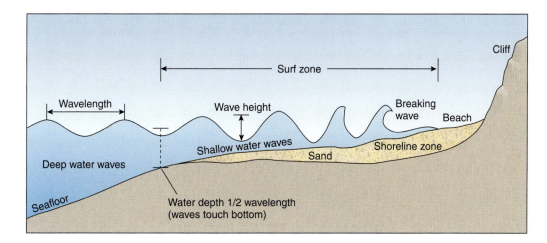

Figure 4–20

Breakers. As a wave approaches shallow water, the lower part of the wave may become slowed by friction with the bottom, although the top of the wave is not affected much. The result is a wave known as a *breaker.*

are found on flatter beaches, where the energy of the waves is dissipated more gradually as the wave moves over the shallow bottom. This action produces waves that are less spectacular and consist of turbulent water and bubbles that flow down the face of the wave. Spilling breakers last longer than plungers because they lose energy more gradually. If you were a surfing enthusiast, you would get a longer ride on spillers and a more exhilarating ride on plungers.

Sudden movements of the earth's crust produce earthquakes, which may produce large seismic sea waves or **tsunamis.** These waves are sometimes referred to as *tidal waves,* a term that is used by oceanographers to describe a totally unrelated event. If a large area of the seafloor rises or falls, it will cause a proportional movement in the surface of the sea. The disruption of the surface forms waves with long wavelengths, long periods, and low height. Because tsunamis are such long waves, they behave like shallow-water waves when they move out from the point of the seismic disturbance. As the wave approaches a coast or an island, the wave energy is compressed into a smaller volume of water as the depth decreases. The sudden increase in energy causes the height of the wave to increase dramatically, and the energy rapidly dissipates as the water races over the land mass. The surge can cause mass destruction, wrecking buildings and docks and depositing vessels high on dry land (Figure 4–21).

Figure 4–21

Tsunami. This damage to the town of Aonae on Okushiri Island, west of Japan's main northern island of Hokkaido, was caused by a tsunami that struck following an earthquake on July 14, 1993. Tsunamis form when activity along fault lines causes large areas of seafloor to rise or fall, producing proportional movement at the surface of the ocean. (Reuters/Corbis-Bettmann)

Figure 4–22

Tides. Tides occur when the gravitational force of the moon pulls ocean water toward it, while the centrifugal force of the earth–moon system forces a mass of ocean water to move in the opposite direction.

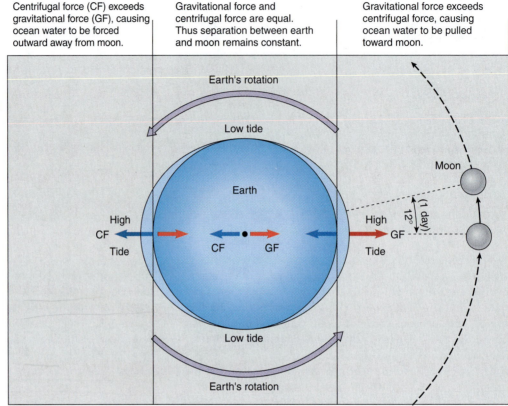

Centrifugal force (CF) exceeds gravitational force (GF), causing ocean water to be forced outward away from moon.

Gravitational force and centrifugal force are equal. Thus separation between earth and moon remains constant.

Gravitational force exceeds centrifugal force, causing ocean water to be pulled toward moon.

Earth's rotation

Low tide

Earth's rotation

Earth

High Tide — CF

CF GF

High Tide — GF

Low tide

Moon

(1 day) 12°

Scale has been exaggerated

TIDES

The shifting patterns of water level that we refer to as *tides* are the result of the gravitational pull exerted on the water of the oceans by the moon and the sun. The easiest way to explain the effects of the sun and the moon on tides is to imagine the earth as being completely covered with water and the tides behaving as ideal waves, ones that are not influenced by friction. In this model, the moon is held in orbit around the earth by the earth's gravitational force. There is also a force called *centrifugal force* acting in the opposite direction and pulling the moon away from the earth (Figure 4–22). Together the earth and moon are held in their orbit by the gravitational attraction between the sun and the center of mass of the earth–moon system. An opposing centrifugal force pulls the center of mass of the earth–moon system away from the sun. For the earth–moon system to remain in orbit around the sun, the gravitational force must equal the centrifugal force. Because the moon also exerts a gravitational pull on the earth, there must be an opposing centrifugal force to prevent the earth from moving toward the moon. This centrifugal force is created by the earth's center of mass rotating around the center of mass of the earth–moon system.

Calculations using Newton's law of gravity indicate that the opposing forces are in balance at the earth's center, but at the earth's surface, the forces are not in balance. Although the mass of the moon is much less than the mass of the sun, it is much closer to the earth. When the forces are calculated for the effects of the sun and the moon on the water at the earth's surface, the moon's gravitational pull is found to have the greater effect.

The water on the side of the earth facing the moon is acted on by a gravitational force that is larger from the moon than from the center of the earth. Because water is fluid, it moves to a point directly under the moon and produces a bulge. On the opposite side of the earth, the centrifugal force acting on the water is larger than that present at the earth's center, producing another bulge on the side of the earth opposite the moon. At the same time that the bulges are formed, areas of low water are formed between the bulges. Since the earth rotates on its axis, a location on the earth's surface would experience a high tide when it is under one of the bulges and a low tide when it is not under one. Therefore, most locations experience two high tides and two low tides each day. Although one rotation of the earth takes 24 hours, the moon also moves slightly in its orbit each day so that a location on the earth's surface

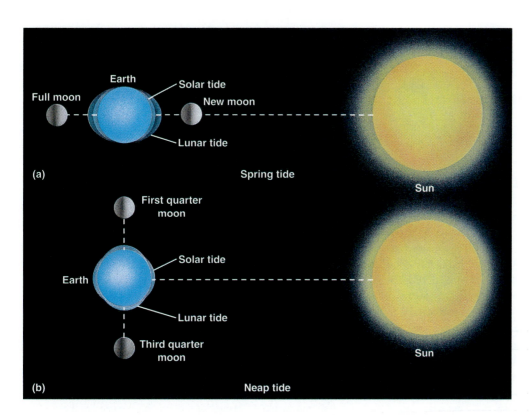

Figure 4–23

Spring and Neap Tides. (a) Spring tides occur when the moon and the sun are in line with each other and the gravitational pull of the two are working together. (b) Neap tides occur when the sun is at a right angle to the position of the moon. The small gravitational pull of the sun cancels some of the moon's gravitational pull, producing smaller tides.

needs an extra 50 minutes to come in line with the moon again. For this reason a full tidal cycle takes 24 hours and 50 minutes.

Although the moon plays the greater role in producing tides, the sun also participates. Even though the sun's mass is much larger than that of the moon, it is so far away from the earth that its tide-producing gravitational force is only about 46% that of the moon. The bulges that are produced by the sun are smaller and usually masked by those produced by the moon.

Twice a month at the full moon and new moon, the earth, moon, and sun are all in a straight line. As a result, the gravitational pull of the sun is added to that of the moon. These are the times of the highest and lowest tides. These tides are referred to as **spring tides,** not for the season of the year but for the water "springing up" at the shore. During the first and last quarter of the moon, the sun and moon are at right angles to each other. The pull of the sun cancels out some of the moon's pull and produces **neap** (from an Anglo-Saxon word meaning "napping") **tides,** which have the smallest change between the high and low tide marks (Figure 4–23).

Tidal range is greatly influenced by geographic factors, such as the contours of shorelines and the depth of coastal waters. Some coastal areas have only one high tide and one low tide each day. This condition is referred to as a **diurnal tide.** Most areas experience two high tides and two low tides each day, a condition called a **semidiurnal tide.** In the typical semidiurnal tide, the two high tides are the same height and the two low tides are at about the same level. If the high and low tides are at different levels, the tide is referred to as a **mixed semidiurnal tide.**

In a tidal system, the greatest height to which the high tide rises is called *high water* and the lowest point is called *low water.* Observations of tides over a long time are used to determine the average tide. A rising tide is called a **flood tide,** because the water floods the coast, and a falling tide is referred to as an **ebb tide.** Currents called **tidal currents** are associated with the rising and falling of the tides. Tidal currents can be very swift and sometimes dangerous as they move water onshore and offshore. During the change of tide from high to low or low to high, there is a period called **slack water,** during which the tidal currents slow down and then reverse.

Tides play an important role in the life of many marine organisms, especially those that live in the area that is exposed at low tide and covered by high tide (the intertidal zone). These organisms exhibit adaptations that allow them to survive whether they are exposed to air at low tide or are submerged by water during high tide. We will examine this habitat and the organisms that inhabit it in detail in Chapter 14.

MARINE BIOLOGY AND THE HUMAN CONNECTION

Harnessing Energy from Waves and Tides

It is estimated that the energy available in ocean waves is about 2.7×10^{12} watts or 3,000 times the generating capacity of the Hoover Dam. Unfortunately, this energy source is widely distributed and not constant in location or time. It is quite a challenge to tap this supply of energy, even to harness small quantities of it.

There are three basic ways that wave energy can be captured. One is to use the energy of waves to lift an object. The wave energy is then stored in the object (potential energy) and can be recovered when the object drops back down (kinetic energy). The energy that is released when the object changes position can then be converted into electricity or some other form of useful energy. A second method is to harness the orbiting or rocking motion of waves to push an object back and forth. Finally, the rising water could be used to compress air in a chamber. The compressed air could then be used to turn a turbine or power some other device. In any of these cases, electrical energy could be generated if the wave energy were directly or indirectly used to mechanically turn a generator (Figure 4–B).

There is also the possibility of obtaining energy from tidal fluctuations. This process would be most practical in coastal areas where tidal differences are quite large or in narrow channels that have swift tidal currents. The first step would be to build a dam across a bay or an estuary so that the water can be held in the bay at high tide. When the tide drops, the water level behind the dam will be higher than in front of it, and the water can be released through turbines to generate electrical power. The reservoir behind the dam would refill during the next high tide when flood gates opened to allow water to enter. This type of system could produce power only during a portion of each ebb tide and would require a tidal range of at least 7 meters (23 feet) to function efficiently.

Another system has been devised to produce power during both the ebb and the flood tides (see Figure 4–B). In this system a dam is built across an estuary as before. This time, once the water has been released back to sea, the flood gates of the dam are closed and more water is pumped seaward, lowering the level of the water behind the dam even more. Although this requires an expenditure of energy, it is worthwhile if the difference in water levels will generate more energy on the next cycle. The gates remain closed until the tide rises. Once this occurs, the water is allowed to enter the reservoir through turbines, thus generating electrical energy. At the end of the cycle, the flood gates are closed and more water is pumped into the reservoir to raise the water level to the maximum. When the tide drops, the water is released, generating more energy, and the cycle is then repeated. Special turbines that can be driven by water in either direction to generate electrical energy are needed for this type of system to work.

Although these systems appear rather simple and cost effective, there are few places in the world with sufficient tidal range or that have natural bays or estuaries that can be dammed at their entrance at a reasonable cost. Conversely, the areas that do satisfy the requirements for tidal range and bays or estuaries are not always located near population centers that require energy. The costs of installing and distributing the power as well as periods of low production due to normal, monthly tidal fluctuations make this type of power expensive relative to other sources. Another consideration is the negative impact that damming an estuary would have on the estuarine ecosystem. There are currently two tidal power stations functioning in Europe. One is on La Rance River Estuary in France, producing 5.4×10^{10} watt-hours per year, and the other is located in Kislaya Bay in Russia.

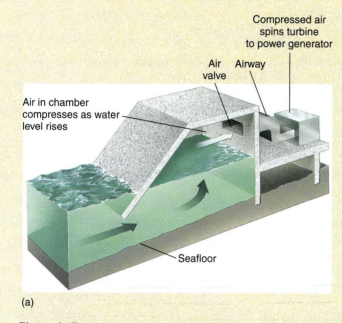

(a)

Air in chamber compresses as water level rises

Air valve

Airway

Compressed air spins turbine to power generator

Seafloor

Figure 4–B

Devices for Capturing Energy from Waves and Tides. *(a) Wave energy is used to compress air that in turn runs a turbine generator. (b) The changing water levels that occur with tides can be used to turn a turbine and produce electricity.*

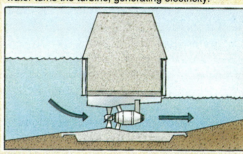

As the tide rises, gates are opened and moving water turns the turbine, generating electricity.

High tide– no generation of electricity.

As the tide recedes, gates are opened and moving water turns turbine, generating electricity.

(b)

CHAPTER SUMMARY

The unique physical and chemical properties of water make it an important component of all living cells. Many of water's properties are due to the polar nature of the water molecules and their ability to form hydrogen bonds.

Seawater contains a number of salts; sodium chloride is the most abundant. The salinity of seawater varies with latitude because of processes such as evaporation, precipitation, freezing, thawing, and freshwater runoff. Evaporation and surf spray return salts to the land, and erosion processes return the salt to the sea.

Seawater also contains gases such as oxygen and carbon dioxide. Oxygen levels are highest and carbon dioxide levels lowest in the upper regions of the ocean, where physical processes and photosynthesis play an important role in exchanging the two gases. Deeper in the ocean, there is more carbon dioxide as the result of respiration and decay processes.

In some areas of the ocean, seasonal turnover mixes the deeper nutrient-rich, oxygen-poor water with the oxygen-rich, nutrient-poor surface waters. Wave action and currents play important roles in mixing deep and surface waters. Evaporation, precipitation, and temperature affect the density of seawater. These changes in density contribute to the process of vertical mixing.

Winds are produced by differences in the density of air. Warmer air at the equator rises and moves toward the poles, while colder air at the poles sinks and returns to the equator. Air masses in the Northern Hemisphere move to the right, while air masses in the Southern Hemisphere move to the left. This apparent deflection of air masses is called the *Coriolis effect.* Wind patterns are responsible for driving ocean currents. Like air masses, ocean currents also exhibit the Coriolis effect. Ocean currents play an important role in nutrient mixing and water turnover.

Waves are produced by forces acting at the surface of the water, most commonly wind. The force raises the water up, and either capillary action or gravity acts to restore the water to its original position. This combination of events creates a wave. The energy that is transferred to the water remains with the wave until it can be dissipated, as in crashing against the shore.

The periodic rise and fall in the level of coastal waters are known as *tides.* Tides are the result of the gravitational pull of the moon and the sun acting on the waters of the ocean.

SELECTED KEY TERMS

ions, *p. 56*

surface tension, *p. 56*

capillary action, *p. 56*

specific heat, *p. 56*

pH scale, *p. 57*

salinity, *p. 57*

density, *p. 62*

downwelling zone, *p. 64*

upwelling zone, *p. 64*

Coriolis effect, *p. 66*

surf zone, *p. 70*

tsunami, *p. 71*

QUESTIONS FOR REVIEW

MULTIPLE CHOICE

1. The relatively high boiling and freezing points of water are due to
   a. the size of the water molecules
   b. the ability of water molecules to form hydrogen bonds
   c. the low thermal capacity of water
   d. the neutral pH of water
   e. the shape of water molecules

2. Compounds that release hydrogen ions when they dissolve in water are called
   a. acids
   b. bases
   c. salts
   d. alkaline compounds
   e. organic compounds

3. Carbon dioxide gas is produced by living organisms during the process of
   a. photosynthesis
   b. digestion
   c. respiration
   d. chemosynthesis
   e. anabolism

4. A zone that exhibits a rapid change in temperature with depth is called a
   a. pycnocline
   b. halocline
   c. barocline
   d. thermocline
   e. mesocline

5. The *Coriolis effect* refers to
   a. the process by which ocean currents are formed
   b. the movement of nutrient-rich bottom water to the surface in exchange for oxygen-rich surface water

c. the apparent deflection of the path of air currents and ocean currents that results from the earth and atmosphere moving at different speeds

d. the vertical exchange of water due to differences in density

e. the horizontal exchange of water between the poles and the equator that occurs as the result of changes in the temperature and salinity of the water

6. Ocean currents are produced by
   a. earthquakes
   b. winds
   c. differences in water density
   d. upwellings and downwellings
   e. tides

7. The restoring force for large waves is
   a. wind   d. friction
   b. capillary action   e. gravity
   c. adhesion

8. Long-period, uniform waves that can travel for thousands of kilometers are called
   a. plungers   d. tsunamis
   b. spillers   e. spoilers
   c. breakers

9. _____ are periodic changes in sea level due to the gravitational pull of the moon and the sun.
   a. waves   d. tides
   b. ocean currents   e. tsunamis
   c. winds

10. When the sun, moon, and earth are all in a line, _____ tides occur.
    a. ebb   d. seasonal
    b. neap   e. no
    c. spring

SHORT ANSWER

1. What role do photosynthesis and respiration play in the distribution of gases in seawater?
2. What factors are responsible for the prevailing wind patterns of the earth?
3. What combination of factors produce neap tides?
4. Explain how the polar nature of water molecules influences water's physical characteristics.
5. Explain how salt from the sea is returned to the land.
6. Explain how vertical mixing of seawater occurs.
7. Describe how winds are produced.
8. Explain how waves are formed.
9. Explain how breakers are formed.
10. Why are upwelling and downwelling zones important?

THINKING CRITICALLY

1. Would it be easier for a planktonic organism to float in water with a high salinity or a low salinity? Explain.
2. Why do upwelling zones and downwelling zones support more biomass than areas of the open sea where these zones don't exist?
3. Why do coastal cities usually experience cooler summers and warmer winters than cities of the same latitude that are inland?
4. Sometimes the discharge of organic wastes from land into coastal waters results in a population explosion of phytoplankton. How would this affect organisms that live in the water below the surface and on the bottom?

SUGGESTIONS FOR FURTHER READING

Bascom, W. 1980. *Waves and Beaches: The Dynamics of the Ocean Survey,* Rev. ed. Doubleday, Garden City, NJ.

Canby, T. Y. 1984. El Niño's Ill Wind, *National Geographic* 165(2):144–183.

Curtsinger, B. 1996. Realm of the Seamount, *National Geographic* 190(5):72–80.

MacLeish, W. H. 1989. The Blue God, Tracking the Mighty Gulf Stream, *Smithsonian* 19(11):44–59.

McCredie, S. 1994. When Nightmare Waves Appear Out of Nowhere to Smash the Land, *Smithsonian* 24(12):28–39.

Open University. 1989. *Ocean Circulation.* Pergamon Press, Oxford, England; Open University, Milton Keynes, England.

Open University. 1989. *Seawater: Its Composition, Properties, and Behavior.* Pergamon Press, Oxford, England; Open University, Milton Keynes, England.

Pickard, G. L. and W. J. Emery. 1982. *Descriptive Physical Oceanography, An Introduction,* 4th ed. Pergamon Press, NY.

Weller, R. A., and D. M. Farmer. 1992. Dynamics of the Ocean Mixed Layer, *Oceanus* 35(2):46–55.

The leafy sea dragon of South Australia represents one of many examples of adaptations that allow marine organisms to survive in their habitats. This animal's camouflage helps it to both avoid predators and ambush prey. (Copyright Norbert Wu)

Marine Organisms

Evolution and Biological Classification

1. Evolution is the process by which the genetic composition of populations of organisms changes over time. Evolution is generally the result of natural selection.

2. The process of natural selection favors the survival and reproduction of those organisms that possess variations that are best suited to their environment.

3. A species is a group of physically similar, potentially interbreeding organisms that share a common gene pool, are reproductively isolated from other such groups, and are able to produce viable offspring.

4. The binomial system of nomenclature uses two words, the genus and the species epithet, to identify an organism.

5. Most biologists classify organisms into one of five kingdoms, categories that reflect theories about evolutionary relationships.

I f you spend time observing marine organisms, you will surely notice that all of their activity seems to be directed toward two objectives: survival and reproduction. In animals, for instance, behaviors and body structures are specialized to gather food, produce reproductive cells, and use those reproductive cells to produce new individuals. Other behaviors, such as defense against predators, enhance an organism's ability to survive so that it can continue to feed and reproduce. All organisms that are alive today descended from ancestors that were well-adapted to their environments and survived long enough to reproduce successfully.

The sea is home to large numbers of living organisms, and each one seems to be well-suited for surviving and reproducing in its particular habitat. How and when did these organisms evolve, and what role does the environment play in determining the characteristics of organisms that can live in any given area? The answers to these and other questions lie in **evolution,** the process by which populations of organisms change over time.

By studying and comparing the fossilized remains of extinct organisms and modern organisms, biologists try to piece together the evidence for evolution and discover the natural processes that have produced and continue to produce the diversity of species that we find today. While studying living organisms, biologists find it helpful, and indeed necessary, to use a universal system of naming the more than one million known extinct and existing species. We use several schemes to classify the organisms we study in marine biology. The terms *plankton, nekton,* and *benthos* classify organisms based on where and how they live. The terms *herbivore, carnivore,* and *decomposer* classify them according to ecological roles. Placing organisms into kingdoms and their subdivisions is a classification scheme that indicates the evolutionary relationships of the organisms being studied. In this chapter you will be introduced to the evolutionary process. You will also be introduced to scientific names and the ways in which organisms are classified to reflect their evolutionary relationships. The concepts presented in this chapter will help us to organize the information about living organisms that will come in later chapters.

EVOLUTION AND NATURAL SELECTION

Darwin and the Theory of Evolution

In 1831 Charles Darwin set sail aboard HMS *Beagle* for a voyage of discovery around the world. During the next five years, he observed and collected specimens from many parts of the globe and began to formulate his ideas on the mechanism by which evolution operates.

In the years before Darwin's great voyage, geologists were finding that the earth was much older than most people believed and that layers of rocks of different ages contained fossils of life forms that had existed in the past that were different from, but obviously related to, living species. Observations such as these laid the groundwork for the theory of evolution by natural selection that Darwin would eventually propose. Darwin was aware of these fossil findings as he set sail on the *Beagle*. As he traveled, he read

books by the geologist Charles Lyell in which Lyell outlined his hypotheses for geological change. Lyell proposed that the physical features of the earth, such as mountains and valleys, were formed over long periods of time by the same slow geological processes, such as uplifting and erosion, that occur today. Darwin was greatly influenced by two conclusions drawn from the observations of geologists such as Lyell. First, if geological change is slow and continuous, then the earth must be very old. Second, slow and subtle changes that occur over a long period of time can produce substantial changes. Darwin realized that the age of the earth would have allowed ample time for gradual changes to occur in different populations of organisms and for new forms of organisms to arise.

As he observed and collected specimens from around the world, Darwin was amazed at the diversity of living organisms. He marveled at how each organism he observed seemed to have what he referred to as the "perfection of structure" for doing whatever was necessary to survive and produce offspring. He wondered what natural processes might be responsible for the evolution of the diverse life forms and their beneficial characteristics (Figure 5–1).

Following his return to England, Darwin began to document his ideas on the process of evolution. In 1838, he read a 40-year-old essay by the English mathematician Thomas Malthus. Malthus had observed that the human population was continuing to grow larger and, if allowed to grow unchecked, would eventually run out of space and food. Malthus believed that the consequences of uncon-

Figure 5–1

Beneficial Adaptations. This triggerfish is adapted for the aquatic life on a coral reef. Fins help the fish maneuver in water, and the laterally compressed body allows the fish to easily navigate among the corals. Powerful jaws allow the triggerfish to feed on small animals with protective body coverings that are found on the reef. (Fred McConnaughey 1988/Photo Researchers, Inc.)

trolled population growth would be famine, disease, and war and that these external controls would keep the human population from growing too large. If this were true of humans, who have relatively few offspring, Darwin reasoned, it must be even more true of plants and animals that produce numerous offspring. After years of work, Darwin came up with a scientific explanation for why populations generally do not exhibit unchecked growth and how they can change over time. He called his hypothesis "evolution by natural selection."

For many years Darwin discussed his ideas with only a few scientific colleagues. It was not until 1858, when he was asked to read a paper by the naturalist Alfred Russel Wallace, that he was encouraged to present his views. Wallace had been working in the country now known as Malaysia and was presenting a theory of evolution similar to the one Darwin had been working on for years. Wallace's paper, as well as a paper by Darwin, were presented at a meeting of the Linnaean Society in 1858. A year later, Darwin published his classic work, *On the Origin of Species by Means of Natural Selection.*

Evolution by Natural Selection

Darwin was quite familiar with the process known today as *artificial selection,* which is practiced by farmers and animal breeders. In **artificial selection,** only animals and plants with certain desirable traits are selected for breeding in an effort to produce more animals or plants with the same desirable traits. Darwin believed that a similar process was occurring in nature as the result of what he called **natural selection,** a process that favors the survival and reproduction of those organisms that possess variations that are best suited to their environment.

Unlike artificial selection, natural selection acts without the purpose of a farmer or a breeder. In fact, the process of evolution operates without any ultimate goal, selecting those forms that are best able to survive and reproduce under the environmental conditions in which they live. It is important to note that natural selection does not have the ability to cause variations that are better suited than others. The variations either occur or do not occur due to chance mutations. Only after the variations appear can they be affected by natural selective forces. **Selective forces** are the physical and biological characteristics of the environment (such as temperature, salinity, predation, availability of food, etc.) that favor the survival of one species over another. According to Darwin's theory, selection occurs over many generations. During this time, organisms become better adapted by a process that can involve all aspects of an organism, such as anatomy, physiology, and behavior.

Darwin's theory of evolution by natural selection contains four basic premises.

1. All organisms produce more offspring than can possibly survive to reproduce.
2. There is a great deal of variation in traits among individuals in natural populations. Many of these variations can be inherited.
3. The amount of resources necessary for survival (food, light, living space, etc.) is limited. Therefore organisms must compete with each other for these resources.
4. Those organisms that inherit traits that make them better adapted to their environment are more successful in the competition for resources. They are more likely to survive and produce more offspring. The offspring inherit their parents' traits, and they continue to reproduce, increasing the number of individuals in a population with the adaptations necessary for survival.

Not all of an organism's traits are necessarily advantageous. Some modern evolutionary biologists believe that many traits become fixed in a population without necessarily being the best traits. Some traits may remain not because they are beneficial but because they confer no disadvantage. Others may be linked to traits that are beneficial, and the linked trait is "extra baggage" that is not harmful. To put it another way, the combination of traits exhibited by a successful organism represents a balance among several selective forces. Table 5–1 summarizes some key points concerning the evolutionary process.

Genes and Natural Selection

At the time Darwin proposed his theory of evolution, cell division, genes, and chromosomes had not yet been discovered. Since that time, the discovery of DNA (deoxyribonucleic acid) and its role in heredity has completely revolutionized the field of biology. **DNA** is a double-stranded molecule with the shape of a helix. It contains genetic information (genes) and is capable of copying itself so that the information can be passed from one generation to another. Modern discoveries in the fields of genetics (the study of heredity) and molecular biology (the study of the structure and function of nucleic acids like DNA) have not contradicted Darwin's theory but have expanded and deepened our understanding of the evolutionary process.

Our modern view of how evolution occurs, the Modern Synthetic Theory of Evolution, is essentially Darwin's 1858 idea as it has been refined by modern genetics. We now know that the variation in natural populations is the result of the different genes present in the individual organisms. A

Table 5-1
Evolution: What, Who, How

	Random Genetic Changes	Natural Selection	Evolution
What Is It?	The raw material acted upon by natural selection	Difference in reproductive success between genetically different individuals of a population	Genetic change in a population from generation to generation
Whom Does It Affect?	Only individuals	Individuals and populations	Only populations
How Does It Work?	Produces an organism with an individual genetic makeup that: 1. is unique 2. may be inherited by the organism's offspring	Produces an organism with genetic differences that: 1. are advantageous under the existing environmental conditions 2. make it more likely to survive and reproduce, passing these advantages on to offspring	Natural selection affects generations and produces a population that is better adapted to existing environmental conditions

Wrong: "Individuals can evolve."
Right: "Individuals do not evolve."

The genes that determine the anatomical features of the mouth cannot be modified in an individual. Therefore, this seahorse has no way to change its mouth to bite or chew prey if the small organisms that it feeds on were to become scarce or disappear completely.

Wrong: "Evolution produces perfection."
Right: "Organisms are not perfectly adapted."

Almost all adaptations represent a compromise in which the adaptation has advantages and disadvantages. For example, a turtle's shell provides good protection against the attack of sharks and other predators, but this same shell slows locomotion, decreases maneuverability, and makes reproduction challenging.

Wrong: "Evolution has a purpose."
Right: "Evolution involves chance."

Selective factors are unpredictable, as is the course of evolution. For example, when shoreline rocks are continuously submerged, this rock anemone survives and reproduces. If geological changes or human intervention causes these same rocks to be raised above the water line, the same genes that allowed the anemone to survive previously will not help it survive in the dry environment, and the anemone will die.

Bacterium magnified 1,000,000

Wrong: "More evolved equals better."
Right: "New species are not better than older species."

Any species alive today is successful (so far), whether it originated 200 years ago or 200 million years ago. Is a bacterium that evolved a billion years ago better than a blue whale? It is less beautiful to the human eye, perhaps, but the bacterium is just as success-ful. Indeed, bacteria may very well outsurvive the blue whale and all of its relatives.

---

gene is a unit of hereditary information. It contains the chemical instructions for making proteins, and these proteins are largely responsible for the structure and function of organisms (i.e., their traits). Genes produce traits such as longer fins or better vision when the genetic information is translated into proteins by the process of protein synthesis. The protein molecules can then form parts of the organism (structural proteins), direct the organism's biological activities (chemical messengers), or function in the production of other molecules necessary for the organism's structure and function (enzymes). Not all of an organism's genes are activated. The genes that are activated and provide the information for protein synthesis are determined randomly by a variety of processes. Thus several offspring from the same parents can receive the same genes but express different combinations as the result of different combinations' being activated. This accounts for the variations among offspring of the same parents.

Genes are located in an organism's chromosomes (Figure 5–2). **Chromosomes** are structures consisting of DNA and protein. They are found in a cellular structure called the **nucleus**, which acts as the control center of a cell. Genes are transmitted from one generation to another when DNA in the form of chromosomes is passed from parents to offspring in the process of reproduction.

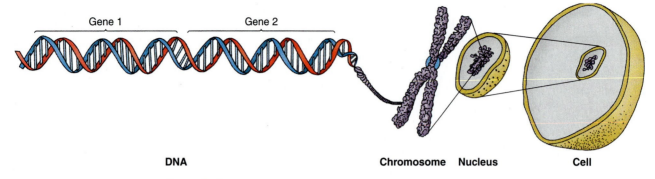

DNA   Chromosome Nucleus   Cell

Figure 5–2

DNA and the Cell. DNA contains the chemical directions (genes) for producing proteins. This double-stranded molecule is combined with protein to form structures called *chromosomes* that are found in the nucleus of a cell.

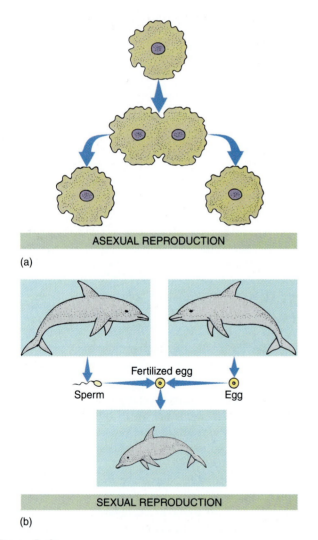

(a)

(b)

Figure 5–3

Asexual and Sexual Reproduction. (a) Asexual reproduction requires only one parent and produces offspring that are genetically identical to the parent (clones). (b) Sexual reproduction requires two parents to contribute genes to the next generation. This results in offspring that have new combinations of genes (half from each parent) and are not identical to their parents.

In **asexual reproduction** (Figure 5–3a), offspring receive their genes from only one parent, and as a result, they are identical to the parent in all of their characteristics. This type of reproduction introduces no variety into the population (the only source of variation is mutation). It is common in organisms that live in stable environments where there is a benefit to reproducing many individuals in a short period of time. Asexual reproduction is very efficient since every member of the population can reproduce and mating is not required.

In **sexual reproduction** (Figure 5–3b), genes are randomly assorted into new combinations during the formation of sex cells. Sex cells (eggs, sperm, pollen, etc.) are called **gametes.** During fertilization, a male gamete fuses with a female gamete to produce a new individual that has inherited half its genes from one parent and half from the other. The sexual reproduction that occurs in each generation produces a constant reshuffling of genes in the species' **gene pool** (all of the genes in a given population that exist at a given time). This random mixing of genes by sexual reproduction along with the random sorting of genes during gamete production and other genetic processes, such as mutation, provide the variations among different individuals of a species that are acted upon by the process of natural selection.

Environmental factors are not static but are constantly changing. For a population of organisms to survive and reproduce under these changing conditions, it must be able to adapt to them. A population's ability to adapt is limited by the genes that are carried by the individuals in the population. Only those populations that have combinations of genes that provide adaptations to their particular surroundings are likely to survive and leave offspring. An organism's biological success or **fitness** is measured in terms of the number of its own genes that are present in the next generation of a population. The more offspring an individual contributes to the population, the more biologically successful that individual is.

WHAT IS A SPECIES?

The Definition of Species

Each different kind of plant, animal, fungus, protist, and bacterium is a different **species.** What exactly, though, is a species? In the past, a species was defined as an organism with a definable set of characteristics that is visibly different from other similar organisms. Two fishes could be the same size and shape but have a different color pattern, and they would be considered different species in the historical sense. This definition of species is called *typological* and is based on **morphology,** or the structure and appearance of the organism. For the purpose of identification, a museum specimen considered to be representative of the species is designated the type specimen, and other specimens are then compared to the type. If they appear similar upon examination of several important morphological characteristics, they are considered to be the same species, and, if they are different, they are considered to be different species.

Unfortunately, this is not the best way to define a species. As we noted previously, there is a large amount of variation in natural populations, and not all members of the same species are exactly alike. For instance, in some marine animals, such as fiddler crabs (*Uca pugnax*) and rosy razorfish (*Hemipteronotus martinicensis*) (Figure 5–4a), the

(a)

(b)

(c)

(d)

(e)

Figure 5–4

Morphological Diversity Within a Species. Some species, such as this rosy razorfish (*Hemipteronotus martinicensis*), exhibit sexual dimorphism, that is, the (a) male and (b) female look distinctively different. (c) The juvenile bluehead wrasse (*Thalassoma bifasciatum*) does not look the same as (d) the adult of the species. (e) Snails such as *Nerita communis* exhibit a large range of variation as adults. (a, b, and c, Fred McConnaughey 1984/Photo Researchers, Inc.; d, Nancy Sefton/Photo Researchers, Inc.; e, Jon L. Hawker)

males and females look quite different—a condition known as **sexual dimorphism.** In other species, such as the blue-head wrasse (*Thalassoma bifasciatum*) and the spotted hagfish (*Bodianus pulchellus*), the juveniles look distinctly different from the adults (Figure 5–4b). In many marine snails, such as the communal nerite (*Nerita communis*), which is found in the Philippine Islands, individuals in a population show a variety of colors and patterns and are quite dissimilar (Figure 5–4c). According to the strict typological concept of species, these animals would be considered as different species, when indeed they are really the male and female of the same species, the juvenile and adult forms, or ecological variants. The typological definition of a species has caused a great deal of confusion over the years, and many organisms have been erroneously identified as separate species. Today, whenever possible, several specimens are defined as types to demonstrate the range of natural variation.

By modern definition, a species is one or more populations of potentially interbreeding organisms that are reproductively isolated from other such groups. This definition is not as easy to apply as the typological definition and is often not practical, but it is more useful in establishing relationships among organisms. Modern research techniques allow biologists to analyze and compare the genes (DNA) and the proteins these genes produce. Such analysis provides the information necessary for applying the modern genetic definition of a species to organisms being studied.

Reproductive isolation means the members of different species are not in the same place at the same time and/or are physically incapable of breeding. This inability to reproduce with individuals outside the gene pool prevents the genes of one species from mixing with the genes of another. There are several ways that organisms of different species are prevented from reproducing in nature. These preventive mechanisms, called **isolating mechanisms,** can be divided into two categories: those that prevent fertilization from ever occurring and those that prevent successful reproduction following fertilization.

Isolating mechanisms that prevent fertilization include differences in habitat, breeding time, anatomy, and behavior. For instance, if one species of snail lives on sandy beaches in Florida, while a similar species lives in mangrove swamps, they will not be able to interbreed because they never encounter each other. In invertebrate animals, different species frequently have incompatible copulatory organs that prevent similar species from reproducing with each other. This is an example of anatomical isolation.

Some animals exhibit special behaviors during the breeding season, and only members of the same species recognize the behavior as courtship. The male common cuttlefish (*Sepia officinalis*), for instance, presents a particular striped color pattern that identifies him to a female *Sepia officinalis* as a prospective mate. Only female *Sepia officinalis* recognize the color pattern as a mating display. Members of different species do not recognize or respond to this courtship display and thus do not attempt to mate with the displaying individual.

Most organisms reproduce only at certain times of the year or under certain conditions (temperature, salinity, hours of light). If the time that members of one species are ready to reproduce does not coincide with the time members of a related species are reproducing, then the two species will be separated by temporal isolation.

Members of closely related species that succeed in copulating are frequently infertile because biochemical or genetic differences between their eggs and sperm prevent successful fertilization. Incompatible genes or biochemical differences can prevent a fertilized egg from developing further. In some instances, fertilization and development may be successful, but the resulting offspring may be infertile or ecologically weak (unable to successfully compete with the well-adapted parental species) and thus quickly die. These mechanisms prevent successful reproduction even if mating or fertilization does take place.

The Process of Speciation

Speciation refers to the various mechanisms by which new species arise. Speciation frequently begins when two or more populations of the same species become geographically isolated from each other or when segments of a single large population become geographically isolated from other segments. This type of speciation is called **allopatric** ("different homeland") **speciation.** Although allopatric speciation is not the only type of speciation, it is thought to be responsible for the formation of the greatest number of species. Geological changes such as earthquakes, mountain building, and the formation of islands can act to separate populations. This geographic isolation due to geological change is often the first step in the process of evolution. Geographic isolation is followed by the appearance of biological isolating mechanisms. This occurs because once two populations have become physically isolated, there is no longer any gene flow between them. From that point on, the processes of natural selection and reassortment of genes during reproduction will operate on each population independently. This is especially true when the separated populations are subject to different environmental conditions.

For example, over 4 million years ago the Caribbean Sea and Pacific Ocean were connected between North and South America. The geological changes that formed Central America between 3 and 4 million years ago separated many species of fish and invertebrates into separate Pacific and Caribbean populations. These isolated populations have evolved independently as they adapted to the differences in temperature, salinity, sediments, food availability, and other conditions in the two seas. Although some species from the two sides of Central America are morphologically similar,

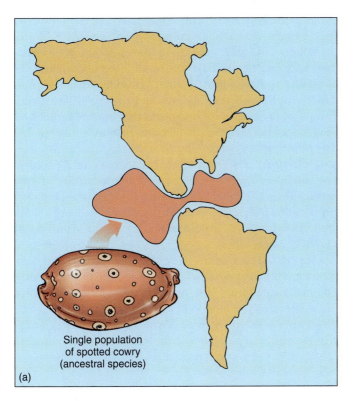
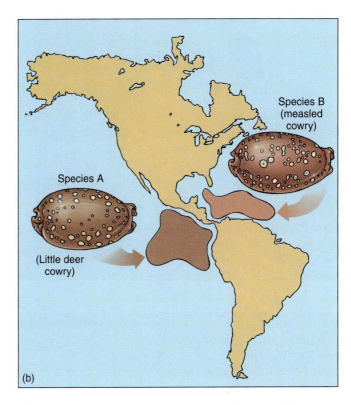

Figure 5—5

Allopatric Speciation. (a) Over 4 million years ago the Pacific Ocean and the Caribbean Sea were connected and home to many populations, such as the ancestral population of spotted cowries. (b) When the land mass known as Central America was formed between 3 and 4 million years ago, many animal populations, including the spotted cowry population, became geographically isolated from each other, resulting in the evolution of similar yet distinct species. The measled cowry (Cypraea zebra) is found in the Gulf of Mexico, while the little deer cowry (Cypraea cervinetta) is found along the Pacific coast of Central America. These two species are quite similar and very likely share a common ancestor. When brought together today in an aquarium, they can no longer mate successfully.

others have diverged in such traits as color, habitat preference, breeding behavior, and breeding season. These differences act as reproductive isolating mechanisms. Once this happens, the populations can no longer interbreed, even if the physical barrier between them is removed (as in the building of the Panama Canal) or if they jointly colonize a common area. These populations, formerly the same species, have become separate species (Figure 5—5).

LINNAEUS AND BIOLOGICAL CLASSIFICATION

Living organisms need to be named so that scientists all over the world can exchange information about them. Unfortunately, most organisms do not have common names, and those that do have names that vary from one country to another and even among geographical regions within a

country. Biologists need a common system of naming things, so that every researcher in the field is calling the same organism by the same name. To deal with the problems of naming and classifying the large number of organisms that inhabit the earth, biologists have developed the science of **taxonomy,** which deals with naming, describing, and classifying organisms.

The system that biologists now use for naming organisms was introduced by Karl von Linné (or, in the Latin form, *Carolus Linnaeus*), a Swedish botanist, in 1758. During Linnaeus's time, organisms were classified on the basis of a lengthy description. Linnaeus found this method to be too cumbersome and developed a standardized method that he refined over the years. In 1758, he introduced the idea of **binomial nomenclature,** a system of naming that uses two words to identify an organism. The first word of a proper scientific name is the **genus** (plural **genera**), and Linnaeus placed all organisms that shared a certain number of common traits in the same genus. For instance, almost all of the different butterflyfishes belong to the genus

(a)

(b)

Figure 5–6

Butterflyfishes. Two different species of butterflyfish, (a) *Chaetodon longirostris* (long-nose butterflyfish), and (b) *Chaetodon ocellata* (spotfin butterflyfish). (a, Jon L. Hawker; b, Fred McConnaughey 1985/Photo Researchers)

Chaetodon (Figure 5–6). The second part of the two-part name is the **species epithet.** The species epithet specifies a particular kind (species) of organism in the genus. For instance, *Chaetodon longirostris* is the long-nose butterfly-fish, whereas *Chaetodon ocellata* is the spotfin butterfly-fish. The first letter of the genus name is always capitalized and the species name is always in lowercase. When the sci-

entific name appears in print, it is in italics, or the genus and species name are each underlined.

When Linnaeus developed the system of binomial nomenclature, he placed organisms into genera on the basis of similar morphology. He then grouped organisms into larger categories called *classes* and *orders.* These larger groups were made up of organisms that bore superficial resemblances but were not in all cases related. Today we still use the binomial system of nomenclature, but organisms are now classified on the basis of evolutionary relationships, not solely on the basis of structural similarities. If two species share the same genus name, it means they not only appear similar, but it is thought that they also share a common ancestor. This is determined by the distribution of shared characteristics.

Several categories are currently used to show the complex evolutionary relationships between related organisms. The major categories from most inclusive to least inclusive (most specific) are: kingdom, phylum (in classifying plants, *division* is used in place of *phylum*), class, order, family, genus, and species. Table 5–2 shows examples of complete classification for some representative marine organisms.

THE FIVE KINGDOMS

During Linnaeus's time, and until recently, biologists grouped all living organisms into one of two kingdoms: Plantae and Animalia. The kingdom Animalia contained the organisms that fed on other organisms (heterotrophs) and that were capable of independent movement. All other organisms (autotrophs and saprotrophs, organisms such as fungi that absorb food from their surroundings) were classified in the kingdom Plantae.

With the invention of the microscope and the accumulation of more information from all fields of biology, it became apparent that many single-celled organisms are more closely related to each other than they are to multicellular organisms. There were also major differences among organisms within the groups of multicellular organisms. Today most biologists recognize a five-kingdom scheme of classification (Figure 5–7).

The kingdom **Prokaryotae** (previously called *Monera*) contains all of the unicellular organisms that lack a nucleus. Members of this kingdom are called **prokaryotes** and consist chiefly of the bacteria.

Members of the remaining four kingdoms are all **eukaryotes,** organisms whose cells contain a nucleus. The kingdom **Protista** contains a wide variety of organisms. Protists are defined by exclusion. If a eukaryotic organism is not an animal, not a plant, and not a fungus, then it is a protist. The kingdom Protista includes most of the eukaryotic microorganisms, all algae including seaweeds, water molds,

Table 5-2
Classification of Some Representative Marine Organisms

	Atlantic Cowry	**Pacific Seahorse**	**Killer Whale**		**Giant Kelp**
Kingdom	Animalia	Animalia	Animalia	**Kingdom**	Protista
Phylum	Mollusca	Chordata	Chordata	**Division**[a]	Phaeophyta
Class	Gastropoda	Actinopterygii	Mammalia	**Class**	Phaeophyceae
Order	Caenogastropoda	Gastrosteiformes	Cetacea	**Order**	Laminariales
Family	Cypraeaidae	Syngnathidae	Delphinidae	**Family**	Lessonaciae
Genus	*Cypraea*	*Hippocampus*	*Orca*	**Genus**	*Macrocystis*
Species	*spurca*	*ingens*	*orcinus*	**Species**	*pyrifera*

[a]Bacteria, fungi, plants and plantlike protists use *division* rather than *phylum*.

slime molds, and protozoans. Of all five kingdoms, the kingdom Protista appears to have the most diverse ancestry, and many biologists believe that eukaryotic cells probably evolved more than once, giving rise to the different lines of protistan cells.

The kingdom **Fungi** is predominantly composed of multicellular organisms. Fungi are eukaryotic organisms that are not capable of photosynthesis and that have cells with cell walls containing a polysaccharide (*poly,* meaning "many"; *saccharide,* meaning "sugar") known as *chitin* (KY-tin). This is the same molecule that strengthens the external covering of shrimp, lobsters, and crabs. Most fungi are decomposers, although there are some parasitic species. Yeasts, molds, and mushrooms are a few examples of fungi.

The kingdoms Plantae and Animalia contain only multicellular organisms. The kingdom **Plantae** is made up of eukaryotic organisms that have cells with cell walls containing another polysaccharide, cellulose, and that are capable of photosynthesis. The kingdom **Animalia** is composed of the eukaryotic organisms that have cells without a cell wall and that do not photosynthesize but are heterotrophs instead.

Students should always keep in mind that classification is a human endeavor, and that the important thing is not the memorization of kingdoms or other categories but a basic understanding of why each organism is classified as it is. In the next few chapters, we will examine in some detail the marine organisms that represent these five kingdoms, paying particular attention to their ecological relationships.

Animalia—Animals are multicellular, eukaryotic, heterotrophic organisms that obtain food mainly by ingestion. Most animals can move, and this permits them to acquire food from their environment by *going* for it—in contrast to plants and fungi, which must either *wait* for it or grow toward it.

Plantae—All members of this kingdom are eukaryotes with cell walls containing cellulose. They carry on photosynthesis, a process in which sunlight is used to convert nutrients from water, air, and soil into food molecules.

Fungi—Although fungi have external cell walls, they are not classified as plants because they cannot make their own food. Instead, fungi are saprotrophs, absorbing food from a living or nonliving source.

Protista—One-celled and colonial eukaryotic organisms and multicellular algae. Protists have a defined nucleus and distinct organelles, and they exihibit a variety of feeding types.

Prokaryotae—All prokaryotic organisms, the bacteria and cyanobacteria, have no true nuclei and contain no organelles. Most prokaryotes are heterotrophs, but some are autotrophs.

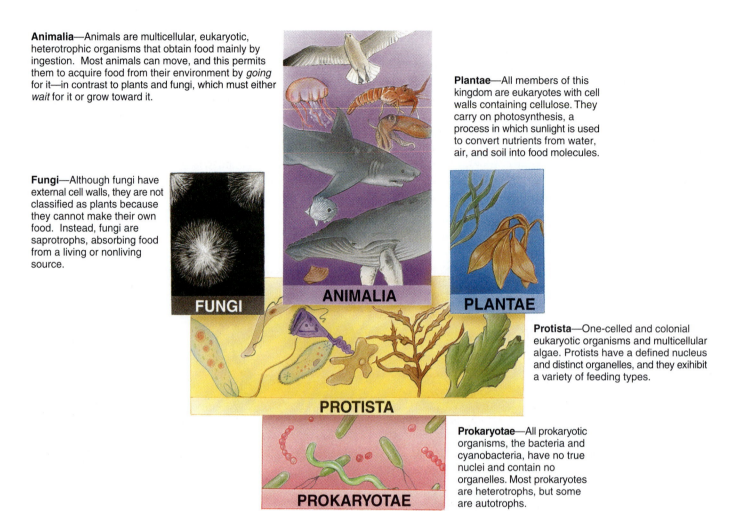

Figure 5–7

The Five-Kingdom Scheme of Classification. The relative size of the boxes relates to the number of known marine species in each kingdom, not the number of individuals. Thus in the marine environment, while prokaryotes are the most numerous, the kingdoms Protista and Animalia contain the greatest variety of species.

CHAPTER SUMMARY

Charles Darwin introduced the theory of evolution by natural selection in 1859 in his book *On the Origin of Species by Means of Natural Selection*. Darwin recognized that the amount of resources necessary for survival are limited. His theory suggests that only those organisms possessing adaptations that make them more successful in competing for these resources will survive, while those that cannot cope will die out. Natural selection functions by favoring the survival of individuals with the combination of traits best suited to their particular place and time. These individuals are also the most likely to reproduce successfully, passing their advantageous traits to their offspring. As environments change, new sets of characteristics are favored. Over long periods of time, new characteristics accumulate in populations, and new, better adapted species gradually replace older ones that are not as well adapted.

Although the genetic mechanisms that govern inheritance play an important role in evolution, they were unknown during Darwin's time. The Modern Synthetic Theory of Evolution is basically Darwin's idea refined by modern genetics. The variations among sexually reproducing individuals in natural populations occur because, in general, each individual possesses a unique combination of genes. Individuals that possess genes for traits that best adapt them to their environment are the most likely to survive and reproduce, thus contributing more of their genes to the population's gene pool. Biological success is measured in terms of the number of genes an organism contributes to the population.

In the past, a species was defined as a group of organisms with an observable set of characteristics that was different from other, similar organisms. Today a species is defined as one or more populations of potentially interbreeding organisms that are reproductively isolated from other such groups. New species may form when environmental or genetic changes cause portions of a population to become reproductively isolated from the rest of the group.

The science of taxonomy deals with naming, describing, and classifying organisms. The categories in the taxonomic scheme are organized to reflect evolutionary relationships. The broadest, most inclusive category is the kingdom. Organisms are subdivided into one of five kingdoms based on several characteristics such as cell type, number of cells, and feeding type. Kingdoms are divided into progressively less inclusive taxonomic categories. From most inclusive to least inclusive, they are: phylum (or division in plants), class, order, family, genus, and species.

SELECTED KEY TERMS

evolution, *p. 80*

natural selection, *p. 82*

gene, *p. 82*

asexual reproduction, *p. 84*

sexual reproduction, *p. 84*

gamete, *p. 84*

morphology, *p. 85*

species, *p. 85*

reproductive isolation, *p. 86*

taxonomy, *p. 87*

binomial nomenclature, *p. 87*

prokaryote, *p. 88*

eukaryote, *p. 88*

protist, *p. 88*

QUESTIONS FOR REVIEW

MULTIPLE CHOICE

1. Darwin proposed that evolution occurs as the result of
   a. cosmic forces
   b. human intervention
   c. artificial selection
   d. natural selection
   e. inherent need

2. Biological success is measured by
   a. how long an organism lives
   b. how large an organism grows
   c. how many times an organism mates
   d. how many of an organism's genes are in a population
   e. how successful an organism is at avoiding predation

3. A population of potentially interbreeding organisms that is reproductively isolated from other populations defines
   a. a kingdom
   b. a community
   c. a family
   d. a genus
   e. a species

4. Living organisms are currently divided into _____ kingdoms.
   a. 2
   b. 3
   c. 4
   d. 5
   e. 6

5. Most unicellular, eukaryotic organisms belong to the kingdom
   a. Fungi
   b. Prokaryotae
   c. Protista
   d. Plantae
   e. Animalia

SHORT ANSWER

1. What are the four points of Darwin's theory of evolution by natural selection?

2. What are the higher categories in modern classification?

3. Describe how new species are formed.

4. Describe how populations can become reproductively isolated.

5. Describe the characteristics of organisms in each of the five kingdoms. Give some examples of marine organisms in each kingdom.

6. Compare and contrast asexual and sexual reproduction.

THINKING CRITICALLY

1. While studying plankton samples, a friend asks you how you can distinguish protozoans from small animals like crustaceans. How would you answer?

2. While on a field trip, you collect two snails that look very similar. How would you determine if they were the same or different species using the methods available in 1858? How would you determine it today?

3. While working in the field you collect several similar specimens of marine fishes from populations in two different localities. What experiments would you perform to determine if these fishes were the same or different species? What results might you expect from these experiments and what might the results prove? (You may want to refer back to the box on The Scientific Method in Chapter 1.)

SUGGESTIONS FOR FURTHER READING

Margulis, L., K. Schwartz, and M. Dolan. 1994. *The Illustrated Five Kingdoms: A Guide to the Diversity of Life on Earth.* Harper Collins, New York.

Mayr, E. 1978. Evolution, *Scientific American* 239(3):46–55.

Price, P. W. 1996. *Biological Evolution.* Saunders College Publishing, Philadelphia.

Scheltema, R. S. 1996. Describing Diversity, *Oceanus* 39(1):16–18.

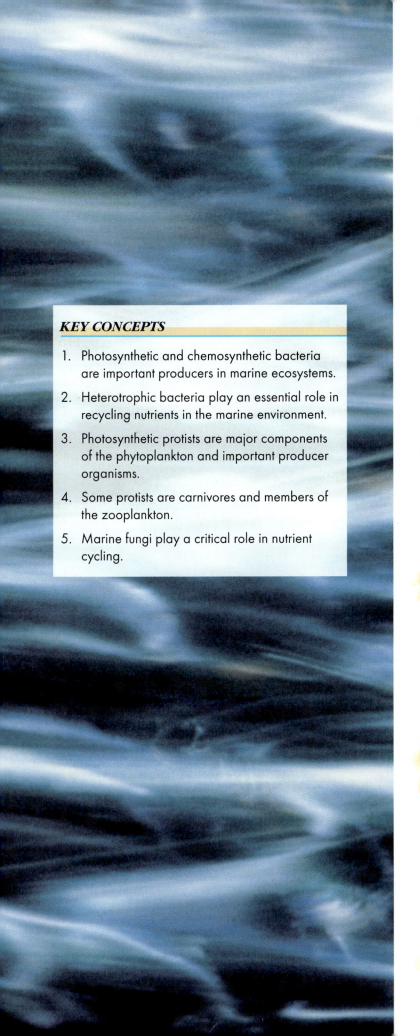

Chapter 6

Marine Microorganisms

Microscopic organisms such as bacteria and protists are the most numerous organisms in the sea. Many are responsible for producing the food on which larger, more complex organisms rely. Many others are decomposers that play an important role in recycling nutrients. In this chapter you will become acquainted with these organisms and the important and diverse roles they play in the marine environment.

BACTERIA

General Characteristics

Bacteria belong to the kingdom Prokaryotae and are found throughout the marine environment. They have a cell wall composed of polysaccharides that surrounds their cell membrane, and, compared to the more advanced eukaryotic cells, they are relatively simple. Bacteria exhibit more metabolic diversity than eukaryotes. For instance, while all of the photosynthetic reactions in plants are similar, bacteria exhibit several very different photosynthetic reactions. This diversity is in part responsible for the great success of these organisms. Bacteria reproduce rapidly by **binary fission,** a form of asexual reproduction (see Chapter 5). In this process, a cell's genetic material (DNA) is duplicated, a membrane forms between the duplicated DNA molecules, and ultimately the cell splits into two daughter cells. Each daughter cell then grows rapidly and repeats the process. Because of the important role bacteria play in producing food and recycling nutrients, life as we know it would be impossible without them.

The Role of Bacteria As Producer Organisms

Chemosynthetic bacteria (also known as *chemoautotrophs*) can produce their own food from inorganic compounds without sunlight, a process called **chemosynthe-**

KEY CONCEPTS

1. Photosynthetic and chemosynthetic bacteria are important producers in marine ecosystems.

2. Heterotrophic bacteria play an essential role in recycling nutrients in the marine environment.

3. Photosynthetic protists are major components of the phytoplankton and important producer organisms.

4. Some protists are carnivores and members of the zooplankton.

5. Marine fungi play a critical role in nutrient cycling.

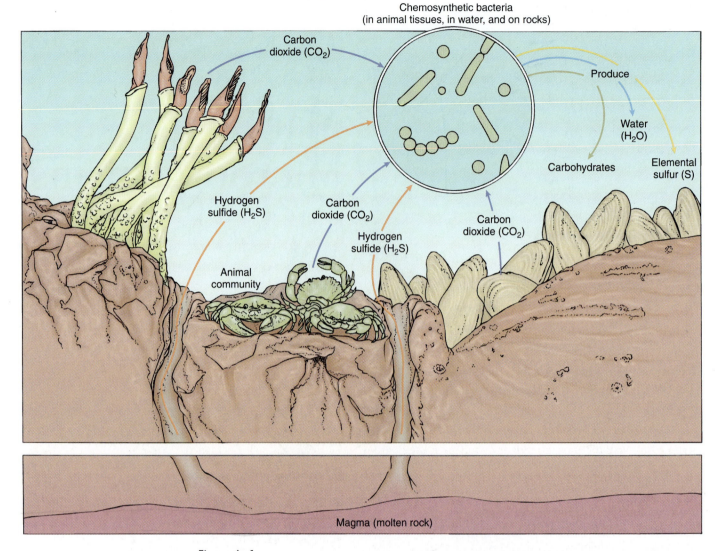

Figure 6–1

Chemosynthesis. Bacteria that live in deep-sea vent communities can produce food molecules from carbon dioxide and hydrogen sulfide, using the energy derived from chemical reactions.

sis. They use energy derived from chemical reactions that involve substances such as ammonia (NH_3), sulfides (S^{2-}), nitrates (NO_3^-), and sulfates (SO_4^{2-}). This energy is then used to manufacture organic food molecules. Chemosynthetic bacteria are found in areas where there is little or no sunlight and where the inorganic substances that they require are in abundance. The bacteria that live around deep-sea hydrothermal vents are examples of chemosynthetic bacteria (Figure 6–1). These bacteria use sulfide ions that spew forth from the vents. They oxidize the sulfide to sulfur and sulfates and use the energy released by this process to produce food. Like other autotrophs, they form the base of a productive food chain that consists of a diverse assembly of organisms, including worms, clams, and crabs, that constitutes the hydrothermal vent community.

Several groups of bacteria, including the purple bacteria, the green bacteria, and cyanobacteria, are able to photosynthesize. Purple bacteria and green bacteria are usually strict anaerobes (organisms that live in environments that lack oxygen) and are found in areas of the marine environment such as mud flats that provide light, anaerobic conditions, and sulfur-containing compounds. These organisms cannot use water in their photosynthetic process, and therefore they do not produce any oxygen. Instead, they often use hydrogen sulfide (H_2S) and produce elemental sulfur or sulfate (Figure 6–2a). Purple bacteria and green bacteria are an important source of food for some zooplankton and filter feeders.

Cyanobacteria, or blue-green bacteria, are another group of important photosynthetic prokaryotes. Near the

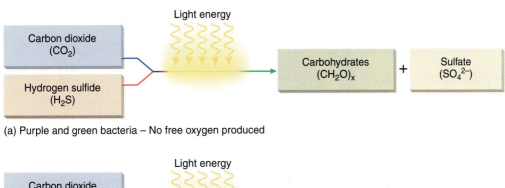

(a) Purple and green bacteria – No free oxygen produced

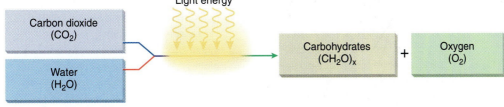

(b) Cyanobacteria – Free oxygen produced

Figure 6–2

Bacterial Photosynthesis. (a) Purple bacteria and green bacteria do not use water when they perform photosynthesis. Instead, hydrogen is supplied by compounds such as hydrogen sulfide (H_2S). (b) Cyanobacteria carry out photosynthesis in the same way that green plants do, using water as a source of hydrogen atoms and giving off oxygen.

surface, they are the most abundant members of the phytoplankton. They are also important members of the **benthos** (the organisms that live on the sea bottom) in regions of the ocean that receive enough sunlight to power photosynthesis. Cyanobacteria contain chlorophyll a, a green pigment that absorbs light in the violet and red regions of the visible spectrum. Chlorophyll a is also the primary pigment of all photosynthetic eukaryotes. Like eukaryotic autotrophs and unlike other photosynthetic bacteria, cyanobacteria are able to use water in their photosynthetic process and release oxygen as a by-product (Figure 6–2b). Cyanobacteria may exist as single cells or form dense mats or long filaments (Figure 6–3). The common name for these organisms comes from the blue-green color of some species, which is due to the pigments they contain. Not all cyanobacteria, however, are blue-green in color. They can be brown, black, purple, yellow, or red. The Red Sea was named for the red cyanobacteria that sometimes become so numerous that they turn the water reddish in color.

Closely related to cyanobacteria are the Prochlorophyta. Prochlorophytes are the only group of photosynthetic bacteria to have both chlorophyll a and chlorophyll b. Chlorophyll b is a yellow-green pigment that absorbs light energy that is not absorbed by chlorophyll a, mainly in the blue region of the spectrum. Some members of Prochlorophyta can be found as symbionts in the tissues of marine invertebrates called *ascidians* (sea squirts). Other species are found in clear water, where they may make up as much as 25% of the primary producers.

Figure 6–3

Stromatolites. Stromatolites are formed by dense mats of cyanobacteria. Some fossil stromatolites date back 3.4 billion years. (Fred Bavendam/Peter Arnold, Inc.)

The Role of Bacteria in Nutrient Cycling

All of the elements that are necessary for life, such as carbon, hydrogen, nitrogen, oxygen, sulfur, and phosphorus, to name a few, are available in the biosphere in fixed amounts and must be recycled from one organism to an-

MARINE ADAPTATION

Purple Bacteria

Purple bacteria belong to an ancient group of bacteria known as *halophiles* or "salt lovers." They received their name because they grow best in extremely salty environments like salt lakes and salt evaporation ponds. In fact if the salt concentration of their environment drops below 165‰ (five times the concentration of normal seawater), purple bacteria will die.

Like many other species, purple bacteria obtain their energy by the process of aerobic respiration. Since warm, salty water holds very little oxygen, purple bacteria frequently experience a shortage of this important gas. Aerobic respiration requires oxygen; therefore, a shortage of oxygen greatly decreases the amount of ATP that these bacteria can synthesize. In order to produce enough ATP to survive, purple bacteria employ a backup system that uses light energy to synthesize ATP under low-oxygen conditions.

Light energy is captured by patches of purple pigment, called *bacterial rhodopsin,* a pigment very similar to the visual pigment in vertebrate eyes. The patches of rhodopsin may occupy as much as 50% of the surface of the cell membrane, and the pigment molecules extend through the membrane such that there is a portion of the molecule exposed at both surfaces. When light strikes the rhodopsin, the molecule undergoes a series of reactions in which a hydrogen ion is pumped from inside of the cell to the outside. The flow of hydrogen ions back into the cell provides the energy for ATP synthesis. Unlike other photosynthetic organisms, the purple bacteria do not synthesize organic molecules but rely on other organisms to produce the nutrients they absorb. Only when oxygen levels in their surroundings are low and there is plenty of sunlight do these bacteria utilize sunlight to synthesize ATP.

Photosynthesis in the purple bacteria is interesting because it represents a very simple hydrogen ion pump. In higher organisms such as algae and plants, a variety of pigments work together to trap light energy and then pass it to a complex system that synthesizes the ATP used in photosynthesis. Purple bacteria, on the other hand, use only the single pigment, rhodopsin. The existence of such a mechanism suggests that primitive prokaryotic cells may have evolved similar systems to trap sunlight and synthesize ATP long before the evolution of modern systems. If this is the case, early cells had access to a great deal more energy than biologists had previously thought. The ability to use the energy of sunlight certainly would have had a tremendous effect on the evolution of early life.

---

other and from one generation to the next. Photosynthetic and chemosynthetic organisms withdraw inorganic substances such as carbon dioxide (CO_2), nitrates (NO_3^-), and phosphates (PO_4^{3-}) from the environment and incorporate them into living tissue. In the marine environment, bacteria and protists, such as diatoms, dinoflagellates, and coccolithophores, are the major producers of organic material in an oceanic food web. The organic molecules produced by these photosynthetic organisms supply energy to almost all of the other organisms in the oceans.

When an organism dies, the elements that are in the organic molecules of its cells are returned to inorganic forms, which producer organisms can use again to make more organic molecules. The same is true of animal excretory products. Organic molecules in animal wastes contain elements that must be returned to the inorganic state so they too can

be recycled. In marine ecosystems, heterotrophic bacteria play the major role in recycling these vital nutrients as they decompose the bodies, tissues, cells, and excretory molecules of producers and consumers.

When an organism dies and is not eaten by another organism, microbes begin to digest and break down the organic molecules present in the dead tissue. The process of breaking down the organic molecules initially requires oxygen, and sometimes all of the available oxygen is consumed from the region around the decomposing material (the region becomes anaerobic).

Certain bacteria are adapted to live in such anaerobic environments. These bacteria carry out the process of fermentation to release energy from their food. In **fermentation,** organic nutrients, such as glucose, are systematically broken down into simpler molecules, releasing energy in

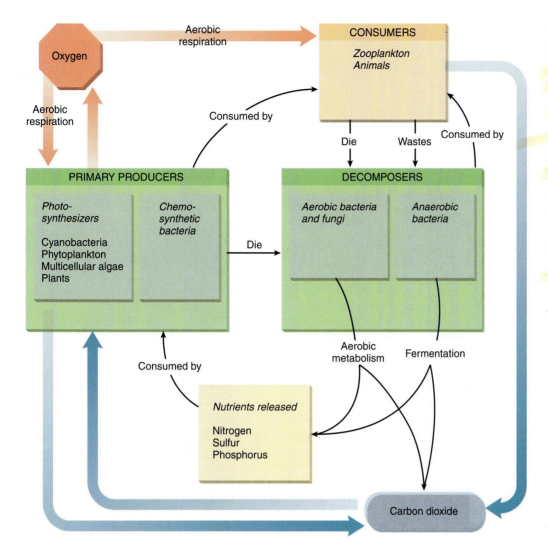

Figure 6–4

Bacteria and Nutrient Cycling. Bacteria decompose the bodies of dead organisms as well as animal waste products, releasing nutrients such as carbon, nitrogen, and sulfur back into the environment. Photosynthetic and chemosynthetic bacteria can recycle carbon dioxide and other nutrients back into food molecules for themselves and consumers. Most marine organisms metabolize their food molecules by the process of aerobic respiration, which consumes oxygen and returns carbon dioxide to the environment.

the process. Fermentation takes place in the absence of oxygen and does not capture very much energy in the form of ATP molecules. Cells use molecules of ATP (adenosine triphosphate) to store energy for use in metabolism. The products of fermentation may diffuse to other areas where oxygen is present, or they may be oxidized by bacteria that can use the oxygen found in inorganic substances such as sulfate (SO_4^{2-}), nitrate (NO_3^-), or carbonate (CO_3^{2-}). Eventually, most of the carbon in the organic compounds will be converted into carbon dioxide, which can be used by photosynthetic and chemosynthetic organisms to produce new organic molecules.

In the presence of oxygen, the metabolic process called **aerobic respiration** breaks down organic molecules to release energy. In this process, carbon dioxide and water are produced as organic molecules (such as sugar) and oxygen are consumed. The majority of marine organisms respire aerobically, that is, they consume oxygen as they break down their organic food molecules. In photosynthesis, oxy-

gen is released from water, and the remaining hydrogens are combined with carbon dioxide to form organic molecules (see Figure 2–9). Carbon dioxide, water, and oxygen are constantly being recycled through respiration, decomposition, fermentation, and photosynthesis (Figure 6–4).

Bacteria also play an important role in the recycling of nitrogen, an element that is particularly important in the growth of algae and plants. Although the earth's atmosphere contains mostly nitrogen gas (N_2), algae and plants cannot use it in this form. Some groups of bacteria, such as cyanobacteria, are capable of converting the nitrogen of nitrogen gas into usable forms. This process is called nitrogen fixing. In the process of **nitrogen fixation,** atmospheric nitrogen is combined with other elements to produce inorganic substances, such as ammonia (NH_3), that can then be used by plants and algae to produce proteins. Some species of cyanobacteria have thick-walled chambers called *heterocysts* that contain enzymes for fixing nitrogen (Figure 6–5a). The thick walls of the heterocysts protect the

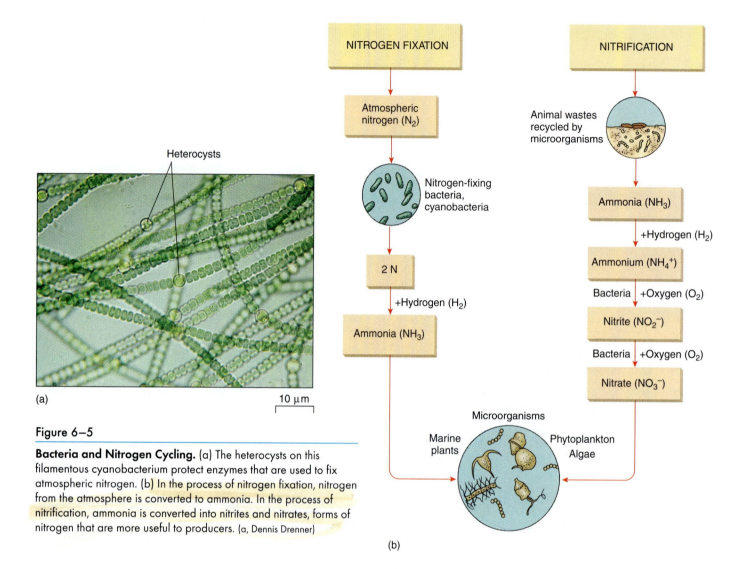

Heterocysts

(a)

10 μm

Figure 6–5

Bacteria and Nitrogen Cycling. (a) The heterocysts on this filamentous cyanobacterium protect enzymes that are used to fix atmospheric nitrogen. (b) In the process of nitrogen fixation, nitrogen from the atmosphere is converted to ammonia. In the process of nitrification, ammonia is converted into nitrites and nitrates, forms of nitrogen that are more useful to producers. (a, Dennis Drenner)

enzymes needed for nitrogen fixation from breakdown by oxygen and ultraviolet radiation and provide a correct environment for the nitrogen-fixing reactions.

Many aquatic animals in the marine environment excrete ammonia as their primary, nitrogen-containing waste product. If the ammonia, a very toxic molecule, were allowed to accumulate in the environment, it would eventually kill the organisms that produced it. Most algae, however, can use one form of ammonia (ammonium ions NH_4^+) as a source of nitrogen for synthesizing protein. In addition, **nitrifying bacteria** are capable of taking the ammonia from animal wastes and dead tissue and converting it into nitrate ions (NO_3^-), which then can be used by algae and plants as a nitrogen source. This process is called **nitrification** (Figure 6–5b).

Another element important to life is sulfur. Microorganisms and eukaryotic autotrophs use sulfur in the form of sulfate ions (SO_4^{2-}) to produce important components of proteins. Animals obtain the sulfur that they need from eat-

ing other organisms. When an organism dies, certain bacteria convert organic sulfur (the sulfur contained in an organic molecule such as protein) into hydrogen sulfide (H_2S), a gas that has the odor of rotten eggs. Hydrogen sulfide is responsible for the characteristic odor of the sediments of many swamps and marshes. Sulfur bacteria then convert the hydrogen sulfide into elemental sulfur, and other bacteria (Thiobacillus) oxidize the sulfur to synthesize sulfate ions, making it available to algae, plants, and microorganisms once again (Figure 6–6). Since the sulfate that is produced forms sulfuric acid when it is combined with water, these oxidizing bacteria often cause pH changes in their environment.

Seawater contains a high concentration of sulfate ions (see Table 4–2), and the conversion of sulfate to hydrogen sulfide by bacteria is very important in marine ecosystems. The dark black mud and ooze on the ocean floor and in mud flats and the odor of hydrogen sulfide are signs that this process is taking place.

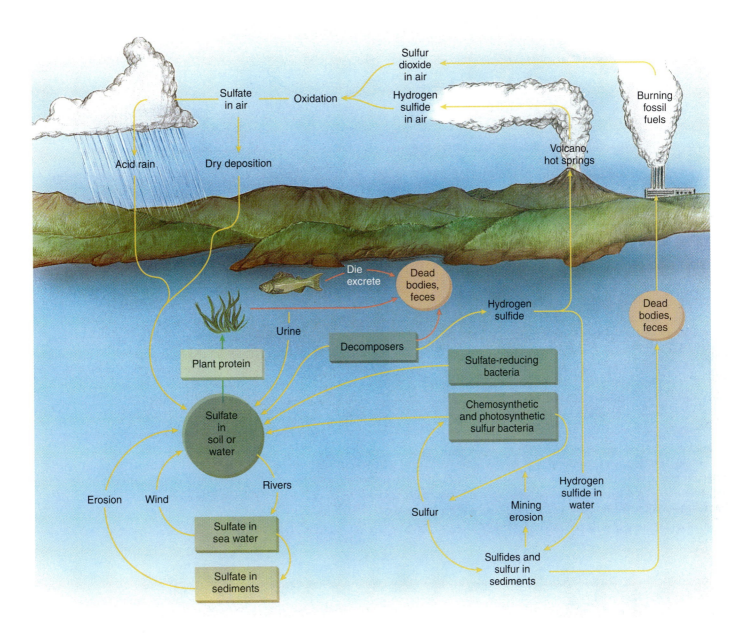

Figure 6—6

The Sulfur Cycle. Sulfur is necessary for the production of proteins. It enters the marine environment from the air, from soil and runoff, and from the action of decomposers. Sulfur bacteria can recycle sulfur from dead organisms and animal wastes. Some species of chemosynthetic bacteria can convert elemental sulfur into sulfate, making it available for producer organisms.

Symbiotic Bacteria

Symbiosis refers to the biological phenomenon in which two different organisms form a very close association with each other (see also Chapter 2). Many bacteria have evolved symbiotic relationships with a variety of marine organisms. In deep-sea vent communities, chemosynthetic bacteria live within the tissues of marine tube worms and clams. These bacteria supply their hosts with organic food molecules as the result of chemosynthetic processes, while the hosts supply the bacteria with carbon dioxide and other essential nutrients.

ECOLOGY AND THE MARINE ENVIRONMENT

The Unusual Blue Dragon

Many marine organisms rely on dinoflagellates known as *zooxanthellae* for their survival. Coral polyps and giant clams, for instance, host large numbers of these symbiotic protists to help them supply the nutrients needed for everyday life. William Rudman, a researcher from Australia, has discovered that certain species of shell-less marine molluscs, called *nudibranchs* or "sea slugs," have also developed symbiotic relationships with zooxanthellae. Rudman has studied a nudibranch called the *blue dragon* (*Pteraeolidia ianthina*), a delicate animal with many finger-like structures called cerata projecting from its body surface (see Figure 6–A).

Nudibranchs feed on cnidarians such as coral polyps, sea anemones, hydrozoans, and small jellyfish. Early in its life, the blue dragon feeds on hydrozoan colonies that have symbiotic zooxanthel-

Figure 6–A

Blue Dragon. *The nudibranch known as the blue dragon harbors symbiotic zooxanthellae. The zooxanthellae produce enough food for their host so that the nudibranch no longer needs to feed on its usual prey, hydrozoans.* (Fred Bavendam/Peter Arnold, Inc.)

lae. As the blue dragon grows, it traps some of the zooxanthellae in pouches that are connected to its intestine. The zooxanthellae thrive within these

pouches, producing food that is released directly into the nudibranch's intestine. After some time, thin, zooxanthellae-filled tubes run from the intestinal pouches through the cerata just beneath the animal's skin. The cerata are arranged so that they do not shade each other, and so the zooxanthellae gain maximum exposure to sunlight for photosynthesis. Once the colony of zooxanthellae becomes fully established in the blue dragon, it appears that the animal no longer feeds on hydrozoans but is fully maintained by the nutrients produced by its symbionts. The result of this intimate relationship between a producer and a consumer is a mutual partnership that frees the organisms from the usual competition for food and nutrients, increasing their chances of survival in a competitive environment.

Another example of symbiosis involves various species of bioluminescent bacteria. These bacteria are capable of producing light using the energy from ATP. A number of squid species have bioluminescent bacteria in glands that are embedded in the ink sac. These glands are partially surrounded by a reflective tissue, and on the surface of the glands are lenses composed of cells that are able to transmit light. In some species, the animal can control the emission of light by contracting muscles associated with the ink sac and forcing the light glands closer to the lens. It is thought that squid use the light as a means of species recognition and that it may play a role in mating.

Many species of deepwater fish have also developed symbiotic relationships with bioluminescent bacteria. These fishes live in the ocean's depths where sunlight is dim or cannot penetrate. The bacteria are usually found in pits located in the animal's skin, and the pattern and arrangement of these pits produce a visual cue that is used for

species recognition. Fishes in the genera *Photoblepharon* and *Anomalops* contain light glands beneath the eye (Figure 6–7). In *Photoblepharon,* the emission of light can be controlled by covering it with a black tissue, much like an eyelid, and in *Anomalops* by rotating the gland so that the light is directed into a pocket of black tissue.

PROTISTS

General Features

The kingdom Protista is a very diverse group of eukaryotic organisms (see Chapter 5). Some protists are unicellular and others multicellular. Some are plantlike photosynthesizers; others are animal-like consumers. Photosynthetic protists are important members of the phytoplankton, whereas the animal-like protists, known as **protozoa,** are important

Figure 6–7

Bioluminescent Bacteria. Bioluminescent bacteria frequently live in the tissues of deepwater fishes, where they provide light in an otherwise dark environment. This flashlight fish houses symbiotic bacteria in an organ below the eye. The fish can block the light by raising a black partition. By "blinking," the fish can lure prey, confuse predators, and communicate with other flashlight fish. (John B. Corliss)

members of the zooplankton (the animal component of plankton) and the benthos (organisms living in and on the bottom). Most protists are free-living, but many species are symbiotic, living within the cells or organs of larger organisms. All protists can reproduce asexually, and many employ sexual reproductive processes too.

Dinoflagellates

Dinoflagellates (phylum Pyrrophyta) (Figure 6–8) are an interesting group of protists. Most are single-celled, although some colonial forms exist. These organisms fre-

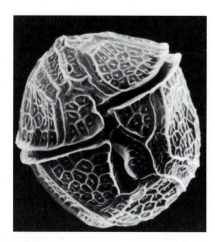

Figure 6–8

Dinoflagellate. This scanning electron micrograph of the dinoflagellate *Gonyaulax* shows the intricate plates that protect the single cell. (John D. Dodge)

quently have cell walls composed mainly of cellulose, as well as two flagella that are used for locomotion. Some species are bioluminescent and, when disturbed, give off a blue-green glow. Most dinoflagellates are photosynthetic and are important members of the marine phytoplankton. They contain chlorophylls a and c as well as carotenoid pigments (pigments that absorb light from the yellow-orange portion of the spectrum). These pigments give some species their characteristic color. Most species of dinoflagellate reproduce only asexually.

Dinoflagellates play a very important role in marine ecosystems. Together with the protists known as *diatoms* and *coccolithophores*, they are a major component of the phytoplankton that provides food directly or indirectly to many marine animals. Some dinoflagellates are parasitic and live in the intestines of marine crustaceans called *copepods*. Other species, collectively called **zooxanthellae** (ZOH-oh-zan-thel-ee), are important symbionts of jellyfish, corals, and molluscs. They lack the cell walls and flagella found in most other dinoflagellate species. The zooxanthellae are photosynthetic and provide food for their host organisms, while the hosts provide carbon dioxide, other important nutrients, and shelter. We will examine the role of the zooxanthellae in more detail in Chapters 8 and 15.

Some dinoflagellates are responsible for the phenomenon known as **red tide.** Red tides occur when certain species of dinoflagellate undergo a population explosion or "bloom." During a bloom, the number of organisms can be so great that they impart a red or orange color to the water (Figure 6–9). The specific cause of the development of these blooms is not known, but blooms are usually associated with shallow, stratified water that is rich in nutrients. Some of the species that cause red tides produce potent toxins, such as saxitoxin, that affect the nervous systems of fishes, causing massive fish kills. As bacteria decompose the dead fish, the oxygen content of the water may be depleted, leading to more fish kills. Dead fish wash up on beaches and decay, creating problems of sanitation and making the beaches unfit for recreational use.

Some dinoflagellate species produce toxins that are taken up by molluscs—in particular, clams and mussels. The molluscs themselves are not harmed by the toxin, but the toxin is concentrated in their tissues. Other animals may feed on these contaminated shellfish and be injured or killed as a result. Humans who consume contaminated shellfish exhibit neurological problems such as loss of balance and coordination, a tingling sensation or numbness in the lips and extremities, slurred speech, and nausea. Death can occur if an amount of toxin sufficient to paralyze the respiratory muscles is ingested. This condition is known as **paralytic shellfish poisoning** or **PSP.** Since dinoflagellate toxins are not proteins, they cannot be destroyed during the cooking process. Contamination of commercial shellfish by red tides results in a substantial loss of revenue to the shellfishing industry.

Figure 6–9

Red Tide. Red tides such as this one are caused by population explosions of certain species of dinoflagellates. (Visuals Unlimited/Sanford Berry)

Diatoms, Coccolithophores, and Silicoflagellates

Diatoms, coccolithophores, and silicoflagellates (all of which are members of the phylum Chrysophyta) are another major component of the marine phytoplankton. These photosynthetic organisms contain chlorophyll a and c and large amounts of a xanthophyll (ZAN-thuh-fil) pigment called *fucoxanthin* that is primarily responsible for their brown to golden-brown coloration. Some of these organisms store their food reserves as lipids (oils and fatty acids), which helps to make them more buoyant.

Diatoms are unicellular organisms that have a glassy covering composed of two parts called frustules, one larger and one smaller, that fit together like a box and lid (Figure 6–10). The frustules are composed of silica and have very intricate geometric patterns that are useful in identifying different species. Diatoms mainly reproduce asexually. During cell division, the frustules separate, the cell divides, and each of the old frustules becomes the larger half of a new cell. As a result of this pattern, with each cell division one of the two new cells will be successively smaller.

When the diatoms reach a certain minimum size (about one-half of their original size), they reproduce sexually. The gametes shed their frustules and fertilize each other, forming a zygote. The zygote grows considerably before forming a new set of frustules. This restores part of a diatom population to the maximum species size. Then the process of asexual reproduction begins again for these cells.

Because the frustules contain silica, a material that bacteria cannot decompose and recycle, when diatoms die, their frustules sink and accumulate on the bottom of the sea. Eventually, some of these accumulations may form sed-imentary rock. Over millions of years, geological events have moved some of these deposits to the surface. The deposits are called *diatomaceous earth,* and they are mined commercially. Because of the small size of the diatoms and the small pores in their frustules, diatomaceous earth makes an excellent filtering material that is used in swimming pool filters and in filtering beer and champagne. It is a mild abrasive used in products such as silver polish and toothpaste. It is also used in soundproofing and insulation products.

The nutrient reserves that are stored in diatoms as lipids also collect in sediments at the ocean bottom when diatoms die. Most of the world's petroleum reserves formed in this way from millions of years of diatom productivity and death.

Coccolithophores (kahk-oh-LITH-oh-forz) are photosynthetic protists that are covered by calcareous (calcium carbonate) plates called *coccoliths* (Figure 6–11a). These organisms are mostly found in the tropics, where they are important members of the phytoplankton. Like the dinoflagellates, they have two flagella that are used for locomotion and to help them maintain their position in the water column. When the cells die, their calcareous plates are deposited on the ocean bottom as sediment.

Since the shells of both diatoms and coccolithophores are relatively inert and do not decompose (although limestone dissolves at depths greater than 4,000 meters), the amount of these sediments in the seafloor can be used to estimate what surface conditions were like in the past. Large drills are used to take core samples from the seafloor. These samples allow researchers to analyze the diatom and coccolith content and estimate the rate at which they accumulated in the bottom sediments. When surface waters were warm and carbon dioxide was plentiful for photosynthesis, there were large numbers of these organisms, and their remains rapidly accumulated on the bottom as they died. On the other hand, when climates were cool or carbon dioxide was low, there were fewer organisms and less remains accumulated in the ocean sediments.

Silicoflagellates (sil-i-koh-FLAJ-uh-layts; Figure 6–11b) are tiny members of the phytoplankton that are extremely abundant in cold water. They have internal shells composed of silica, and one or two silica rods usually extend from the shell. Silicoflagellates also possess one or two flagella. The silica rods and flagella create drag in the water and help to prevent them from sinking.

Protozoans

The term *protozoa* ("first animals") is usually applied to a diverse group of animal-like protists. Protozoans are heterotrophs and rely on producer organisms such as phyto-

100 μm

(a)

Asexual Reproduction

Mitosis

Mitosis

Mitosis

Mitosis

Mitosis

Sexual Reproduction

New cell

Frustule formation

Growth of the cell

Zygote

Gamete from another

Gametes formed

Gametes released

Cells' division continues until cells become too small to divide

(b)

Figure 6–10

Diatoms. (a) A sample of representative diatoms shows the amazing diversity of their glassy shells called *frustules*. (b) Asexual and sexual reproduction in diatoms. During asexual reproduction, diatoms divide by mitosis, each time producing a smaller cell. When a cell reaches a certain critically small size, it stops dividing, forms gametes, and sexual reproduction occurs. (a, The Stock Market/Phillip Harrington)

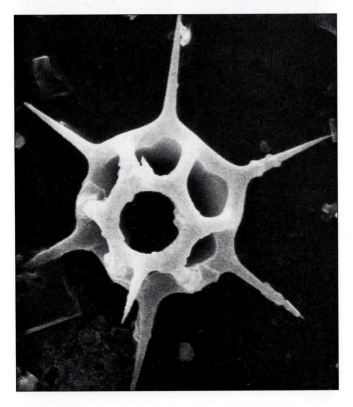

Figure 6–11

Coccolithophores and Silicoflagellates. (a) Coccolithophores are photosynthetic protists that have cells covered by calcareous plates called *coccoliths*. (b) Silicoflagellates are tiny phytoplankton with internal shells composed of silica. (Dr. Elizabeth Venrick/Scripps Institute of Oceanography)

plankton for their food. Some protozoans are members of the zooplankton, the animal component of plankton; others are important members of the benthos, where they feed on other microorganisms. Protozoans form important links between microscopic producers and multicellular consumers in the ocean's food webs.

Amoebas are single-celled, heterotrophic organisms that can be found in the mud and sandy bottoms of marine habitats (Figure 6–12a). They move by pushing forward finger-like projections of membrane and cytoplasm called *pseudopods* (*pseudo,* meaning "false," *pod,* meaning "foot"). As the pseudopod pushes forward, the rest of the cytoplasm flows toward it, and the cell effectively moves. The pseudopods are also used to trap prey. An amoeba surrounds its prey with pseudopods and draws the prey into its cell, forming a membrane-bound sac called a *food vacuole.* Cellular enzymes then digest the engulfed food. Amoebas reproduce asexually by binary fission.

Foraminiferans (for-uh-muh-NIF-uh-ranz; Figure 6–12b) are shelled amoebas that are found in large numbers in the marine environment. They often produce elaborate, multi-chambered shells made of calcium carbonate. As foraminiferans grow, they add more chambers, and the resulting shell frequently resembles a microscopic snail shell. The shells contain many small openings through which pseudopods project. Foraminiferans use their long, thin pseudopods to trap prey. Organisms that come into contact with the sticky surface of the pseudopod become attached and are then moved into the cell where they are digested. Foraminiferans reproduce both sexually and asexually, and their life cycles are somewhat complex. Although most foraminiferan species are bottom dwellers, a few species with enormous numbers of individuals are members of the zooplankton. The shells of planktonic species are usually thin and light and bear many spiny processes that help to reduce the sinking rate. The shells of dead foraminiferans are a major constituent of special sediments on the ocean floor. These sediments are called *globigerina* (gloh-bij-eh-REE-nuh) *ooze* because of the large number of shells of the genus *Globigerina,* a planktonic foraminiferan. Over millions of years, geological change has brought some of this sediment to the surface, where it forms large chalk deposits. The White Cliffs of Dover on the English Channel, for example, are formed from foraminiferan sediments (Figure 6–12c). Foraminiferans also contribute to the sand found on the beaches of many islands, including Tonga in the Pacific Ocean and the pink sands of Bermuda and the Bahamas.

Radiolarians are primarily members of the zooplankton (Figure 6–12d). They have intricate silica shells with openings that allow their pseudopods to project outward. They use their pseudopods to gather food, and these long, thin pseudopods also reduce their sinking rates. Radiolarians reproduce sexually, forming gametes of similar size that are released into the surrounding water, where they fuse and develop into adults.

text continues on page 106

Pseudopod

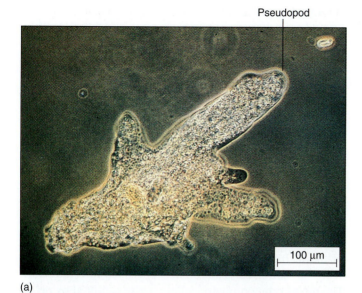

(a)

100 µm

Pseudopods

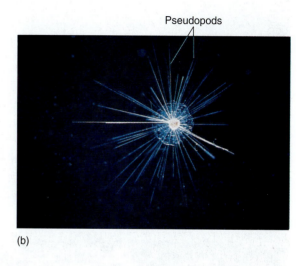

(b)

(c)

(d)

Figure 6–12

Protozoans. (a) Amoebas use pseudopods for locomotion and to capture food. (b) Foraminiferans produce shells of calcium carbonate that contain chambers and resemble microscopic snail shells. They use their pseudopods to capture prey. (c) Over a period of millions of years, foraminiferan shells accumulated in the bottom sediments of the seas around Great Britain. Geological events ultimately moved a large amount to the surface, forming the White Cliffs of Dover. (d) Radiolarians are covered by a glassy shell composed of silica. (a, Biophoto Associates; b, Peter Parks/Peter Arnold, Inc.; c, Lynn McLaren/Photo Researchers, Inc.; d, M. I. Walker/Photo Researchers Science Source)

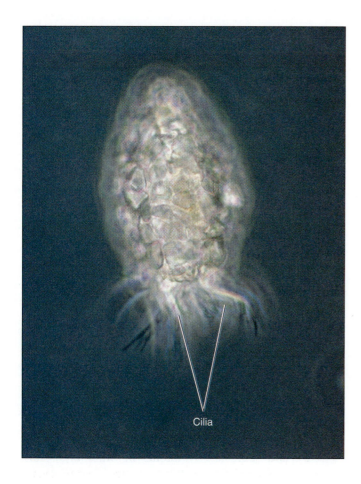

Cilia

Radiolarians exhibit stratification, a layered type of distribution, with some species being found only in the upper zones of the ocean and others existing as deep as 5,000 meters (16,500 feet). Still other species migrate from one zone to another, depending on the season. Like foraminiferans, the shells of dead radiolarians contribute greatly to the bottom sediment of the ocean. Radiolarian ooze is found at greater depths than the foraminiferan ooze. This is because under the physical conditions found at greater depths, the calcium carbonate of foraminiferan shells dissolves, but the silica of radiolarian shells does not.

Ciliated protists are protozoans that possess hairlike structures called *cilia* on their outer surface. The cilia function in locomotion and sometimes aid in feeding. Ciliated protists are widely distributed in the marine environment. Some are members of the zooplankton, some are benthic, and others are symbionts living on or in other organisms. Some of the most common planktonic forms are the **tintinnids** (Figure 6–13), which are important intermediates in

marine food chains. These protozoans form a body covering called a *lorica* that can either be secreted by the cell or formed from foreign particles that are cemented together. Around the "oral" end of tintinnids is a ring of cilia that function in feeding and locomotion.

PLANKTON SAMPLING

Many of the microorganisms introduced in this chapter are important members of phytoplankton. Because of its importance at the base of marine food webs, the composition of phytoplankton in various regions of the ocean is of great interest to marine biologists. The standard method for obtaining plankton samples is to tow a plankton net behind a boat. These nets are made of fine-mesh nylon and filter the plankton from the water. Plankton that can be captured with standard-sized plankton nets are called **net plankton.** The most common phytoplankton to be captured by plankton nets are colonial cyanobacteria, dinoflagellates, and diatoms. These organisms are also referred to as **microplankton** because of their small size (20–200 microns). Smaller phytoplankton such as coccolithophores and silicoflagellates (5–20 microns) are referred to as **nanoplankton.** Nanoplankton cannot be captured by plankton nets and are removed from samples of seawater by very fine mesh filters. In recent years, new filtering materials have allowed marine biologists to obtain samples of phytoplankton that are even smaller (ultraplankton, 2–5 microns, and picoplankton, less than 2 microns). These are mainly unicellular bacteria. The importance of these very tiny forms to ocean productivity is still being investigated.

FUNGI

General Features

Members of the kingdom Fungi have cells with a cell wall in addition to their cell membrane. Fungi are not photosynthetic. Like most bacteria, fungi are heterotrophic decomposers and play an important role in recycling organic material. Most are capable of both sexual and asexual reproduction. When they reproduce asexually, they form special cells called *spores* that are released in large numbers into the environment. If the spore lands in a suitable place, it germinates and forms a new fungus.

Marine Fungi

Fungi found in marine environments are most abundant in the intertidal zone, shallow water, salt marshes, and mangrove swamps, where they are primarily bottom dwellers.

Some species with reproductive structures called *fruiting bodies* as large as 3 millimeters have been reported from deep ocean habitats. In salt marshes and mangrove swamps, fungi play an important role in the decomposition of dead plants and animals. In addition to the role of decomposers, some marine fungi act as parasites, infecting diatoms, algae, and a variety of higher plants and invertebrate animals. Although poorly studied, marine fungi are turning out to be very abundant and important in most benthic habitats.

CHAPTER SUMMARY

Bacteria are prokaryotes and important primary producers and decomposers in marine ecosystems. Chemosynthetic and photosynthetic bacteria extract inorganic nutrients, such as nitrogen, phosphorus, and carbon dioxide, from the environment and incorporate them into organic molecules (produce their own food). Chemosynthetic bacteria use energy derived from chemical reactions to produce their food molecules, while photosynthetic bacteria use the radiant energy from the sun. Some producers, such as cyanobacteria and eukaryotic autotrophs, release oxygen during the photosynthetic process.

Animals and other consumers gain their nutrients by eating producer organisms or each other. Most marine organisms require oxygen to metabolize their food, a process known as *aerobic respiration*. Aerobic respiration consumes oxygen and returns carbon dioxide to the environment. When an organism dies, its remains are broken down by decomposers. The components of their cells and tissues are utilized by decomposers and later returned to the environment and recycled. Decomposers also recycle the components of animal wastes.

Bacteria play an essential role in recycling nitrogen and sulfur in the marine environment. Cyanobacteria can fix atmospheric nitrogen, making it available for producers. Nitrifying bacteria can convert the ammonia in animal wastes to nitrate, a form of nitrogen that can be used by producers. Some sulfur bacteria can recycle the gas hydrogen sulfide, while others can convert elemental sulfur to sulfate for use by producers.

Many marine bacteria have formed symbiotic relationships with other marine organisms. The chemosynthetic bacteria of hydrothermal vent communities and the bioluminescent bacteria found in association with deep-sea organisms are some examples of this phenomenon.

In marine environments, protists make up a large portion of the plankton and benthos. Dinoflagellates, diatoms, coccolithophores, and silicoflagellates are photosynthetic producers and important members of the phytoplankton. One group of dinoflagellates, the zooxanthellae, are important symbionts of marine invertebrates. Other species of dinoflagellates are harmful, causing red tides or producing toxins that can injure or kill vertebrate animals, including humans.

Amoebas, foraminiferans, radiolarians, and ciliated protists are called *protozoans* because they are more animal-like. Protozoans are members of the zooplankton and the benthos, and some are symbionts of other organisms. Amoebas, foraminiferans, and radiolarians move by means of pseudopods. They also use these organelles to capture food. Ciliated protists use their cilia for locomotion.

Plankton samples are usually taken with a nylon plankton net. Plankton that are large enough to be trapped in plankton nets are called *net plankton*. Many species of phytoplankton are too small to be captured by plankton nets and must be removed from samples of seawater using special filters.

Fungi have cells with cell walls. They are not photosynthetic but are important decomposers helping to recycle nutrients back into the environment.

KEY TERMS

bacteria, *p. 93*

binary fission, *p. 93*

chemosynthesis, *p. 93*

cyanobacteria, *p. 94*

benthos, *p. 95*

nitrogen fixation, *p. 97*

nitrification, *p. 98*

protozoa, *p. 100*

dinoflagellate, *p. 101*

zooxanthellae, *p. 101*

red tide, *p. 101*

paralytic shellfish poisoning (PSP), *p. 101*

diatom, *p. 102*

coccolithophore, *p. 102*

silicoflagellate, *p. 103*

amoeba, *p. 104*

foraminiferan, *p. 104*

radiolarian, *p. 104*

tintinnid, *p. 106*

net plankton, *p. 106*

QUESTIONS FOR REVIEW

MULTIPLE CHOICE

1. The bacteria that are primary producers in hydrothermal vent communities are
   a. purple bacteria
   b. cyanobacteria
   c. chemosynthetic bacteria
   d. nitrogen-fixing bacteria
   e. prochlorophytes

2. As a result of the process of decay, organic molecules are converted into
   a. ammonia and nitrates
   b. carbon dioxide and water
   c. glucose
   d. water and oxygen
   e. oxygen and carbon dioxide

3. In the process of nitrification
   a. ammonia is converted into nitrogen and water
   b. nitrogen is converted into ammonia
   c. ammonia is converted into ammonium ion
   d. ammonia is converted into nitrates
   e. nitrate is converted into ammonia

4. Which of the following is not a member of the marine phytoplankton?
   a. dinoflagellates
   b. foraminiferans
   c. diatoms
   d. coccolithophores
   e. cyanobacteria

5. _____ are symbiotic dinoflagellates found in the tissues of some jellyfish, clams, and coral.
   a. zooxanthellae
   b. diatoms
   c. tintinnids
   d. cyanobacteria
   e. coccolithophores

6. Globigerina ooze is composed of the shells of dead
   a. diatoms
   b. coccolithophores
   c. foraminiferans
   d. radiolarians
   e. tintinnids

7. Most of the world's petroleum was produced by
   a. cyanobacteria
   b. dinoflagellates
   c. radiolarians
   d. diatoms
   e. prochlorophytes

8. Plankton that are too small to be trapped by plankton nets are called
   a. net plankton
   b. phytoplankton
   c. zooplankton
   d. nanoplankton
   e. megaplankton

9. The phenomena known as *red tides* are associated with population explosions of
   a. amoebas
   b. radiolarians
   c. dinoflagellates
   d. diatoms
   e. silicoflagellates

10. Which of the following organisms would you not expect to find in benthic habitats?
    a. cyanobacteria
    b. fungi
    c. amoebas
    d. diatoms
    e. foraminiferans

SHORT ANSWER

1. What vital role do the nitrifying bacteria play in the marine environment?

2. What are three important roles of dinoflagellates in the marine environment?

3. What important ecological role do protozoans play in marine ecosystems?

4. Compare photosynthesis in the green bacteria with photosynthesis in cyanobacteria and eukaryotic organisms.

5. Describe how bacteria recycle organic material from dead tissue.

6. Explain how some species of deep-sea fish benefit from the presence of symbiotic bacteria.

7. Why are dinoflagellates and diatoms usually found in different regions of the ocean?

8. Distinguish between net plankton and nanoplankton.

THINKING CRITICALLY

1. How could a comparison of photosynthetic pigments be used to determine relationships among various species of phytoplankton?

2. Many planktonic microorganisms have projections of some sort from their cells. What is the probable advantage of these structures?

SUGGESTIONS FOR FURTHER READING

Anderson, D. A. 1994. Red Tides, *Scientific American* 271(2):62–68.

Dale, B., and C. M. Yentsch. 1978. Red Tide and Paralytic Shellfish Poisoning, Oceanus 21:41–49.

Davis, C. S., C. J. Ashijian, and P. Alatalo. 1996. Zooplankton Diversity: A Bizarre—and Changing—Array of Life Forms, *Oceanus* 39(1):7–11.

Dunlap, P. V. 1995. New Insights on Marine Bacterial Diversity, *Oceanus* 38(2):16–19.

Hoover, R. B. 1979. Those Marvelous, Myriad Diatoms, *National Geographic* 155(6):870–878.

Huyghe, P. 1993. Killer Algae, *Discover* 14(4):70–75.

Krogmann, D. W. 1981. Cyanobacteria (Blue-Green Algae)—Their Evolution and Relation to Other Photosynthetic Organisms, *Bioscience* 31:121–124.

Rudman, W. B. 1987. Solar-Powered Marine Animals, *Natural History* 96(10):50–53

Waite, A. M. 1996. Phytoplankton Biodiversity: Boxes, Spheres, Spirals, and Sunbursts, *Oceanus* 39(1):2–5.

Chapter 7

Multicellular Producers

Although most of the primary production in marine ecosystems is carried out by phytoplankton and autotrophic bacteria, seaweeds and flowering plants also contribute to this important process, especially in coastal habitats. Not only do they provide food directly to herbivores, but more important, decaying plant parts are a significant source of detritus for detrital food chains. In addition to their role as producers, these organisms also provide habitats for other marine organisms. Marine plants play an important role in keeping the water clear by trapping sediments, and their root systems stabilize the bottom sediments. In this chapter we will explore the important ecological roles performed by these organisms.

MULTICELLULAR ALGAE (SEAWEEDS)

The Distribution of Multicellular Marine Algae

Multicellular marine algae are a diverse group of organisms that are commonly referred to as *seaweeds*. They are divided into three major groups, or divisions: the red algae (division Rhodophyta); the brown algae (division Phaeophyta); and the green algae (division Chlorophyta). Each group has different accessory photosynthetic pigments that give them their characteristic color.

Most species of multicellular algae are benthic, growing on materials such as rock, sand, and coral on the sea bottom. The distribution of benthic algae is limited by the availability of sunlight at various depths. The evolution of a variety of accessory pigments that absorb those wavelengths of light that are able to reach the deeper zones of the ocean allows the algae to survive in these bottom habitats. Seawater selectively absorbs light with longer wavelengths (Figure 7–1), such as the reds and yellows, so the light that penetrates to the greatest depths is the short-wavelength blues and greens. Pigments such as fucoxanthin

---

KEY CONCEPTS

1. Multicellular marine algae are mostly benthic organisms that are divided into three major groups based on their accessory photosynthetic pigments.

2. Marine algae supply food and shelter for many marine organisms.

3. Plants that live in the sea exhibit adaptations for survival in saltwater habitats.

4. Seagrasses are important primary producers and sources of detritus, and they provide habitat for many animal species.

5. Salt marsh plants and mangroves stabilize bottom sediments, filter runoff from the land, provide detritus, and provide habitat for animals.

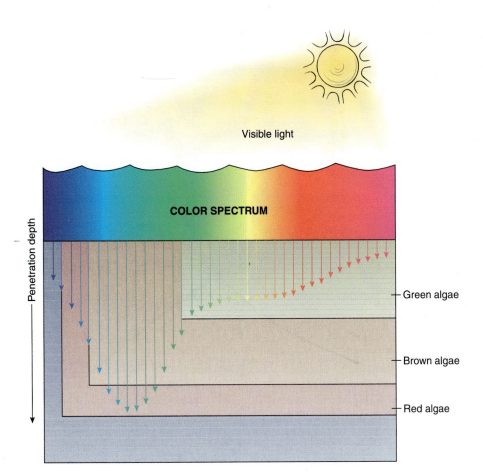

Visible light

COLOR SPECTRUM

Penetration depth

Green algae

Brown algae

Red algae

Figure 7–1

Pigments and Light Absorption. Light of different colors penetrates to different depths in the ocean. Low-energy red and yellow light penetrate the least, while high-energy blue and green light penetrate the deepest. The evolution of a variety of pigments that absorb different wavelengths of light has allowed photosynthetic organisms to occupy all areas of the photic zone. The colored boxes in this diagram represent the range of wavelengths absorbed by each group of algae to power photosynthesis. Notice that algae with pigments allowing them to absorb heavily in the blue and green range can live in deeper water than those that absorb heavily in the red and yellow range of the spectrum. The actual colors of the different algae represent the wavelengths of light that are reflected and not absorbed.

and phycoerythrin (fy-koh-e-RITH-rin), which absorb blue and green light, allow algae to grow at greater depths than those algae that do not possess these or similar accessory pigments.

The distribution of algae is also affected by temperature. The greatest diversity of algal species is in tropical waters. Farther north or south of the equator, the number of species decreases, and the species themselves are different. Many marine algae found at the colder latitudes are perennials, which means they live longer than two years. During the colder seasons only part of the algae remains alive, sometimes a few cells but most often a mass of stemlike structures. When the temperature warms up in the spring, this body part initiates new growth. Temperature is not usually a limiting factor for algae that live in tropical and subtropical seas, although for some species, the temperature in intertidal areas may be too warm.

General Characteristics of Marine Algae

Algal cells have cell walls in addition to their cell membranes, and their photosynthetic pigments are found in complex, membrane-bound organelles called *chloroplasts*

(Figure 7–2). Many algae secrete a slimy, gelatinous mucilage (MYOO-suh-lij) that covers their cells. The mucilage can hold a great deal of water and may act as a protective covering that retards desiccation (drying out) in intertidal algae that are exposed at low tide. It can also be sloughed off to remove any organisms that may have become attached to the algae. Some algae have a thick, multilayered covering of carbohydrate called a *cuticle*. The cuticle is a protective layer. In some species the cuticle is so thick that it gives the algae an iridescent sheen.

Algae can reproduce both asexually and sexually. In some, asexual reproduction can occur when the body of the alga, called the **thallus,** breaks into pieces, and each new piece develops into a new alga (Figure 7–3a). This type of asexual reproduction is called **fragmentation.** Another type of asexual reproduction involves spore formation. The spores are dispersed in the environment and, if they settle in a suitable habitat, germinate and form new algae (Figure 7–3b).

Sexual reproduction involves the formation of gametes (sex cells) that must fuse to form a new alga. Compared to a typical animal life cycle, most algal life cycles are much more complex, sometimes involving as many as three generations.

(text continues on page 113)

Figure 7–2

A Generalized Algal Cell. An algal cell is surrounded by a rigid cell wall. In addition, many marine algae also possess a cuticle that protects the cells and a gelatinous layer called *mucilage* that covers the cell wall and helps prevent desiccation. Algal cells contain numerous chloroplasts, organelles that function in photosynthesis, and a large vacuole that stores water.

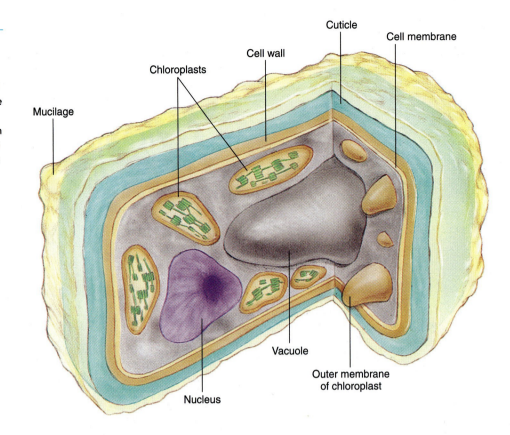

Figure 7–3

Asexual Reproduction in Algae. Algae can reproduce asexually by (a) fragmentation and (b) spore formation.

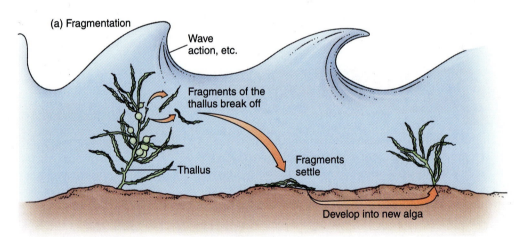

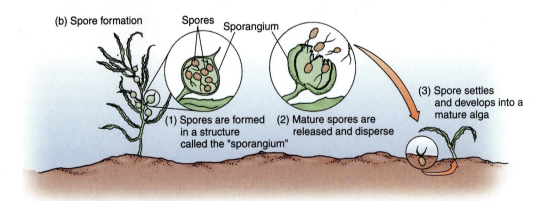

RED ALGAE

Red algae (division Rhodophyta) are primarily marine organisms. Although they are most widespread in tropical oceans, they can also be found in cooler waters as well (Figure 7–4a). In addition to chlorophyll a, red algae contain the accessory pigments phycoerythrin and phycocyanin (fy-koh-SY-uh-nin). These pigments are most effective at absorbing blue and green light. The normally reddish color of these algae comes from the large amount of phycoerythrin, a red pigment that masks the green color of chlorophyll. Red algae are mostly benthic in distribution. Since their pigment systems are adapted to absorbing virtually all wavelengths of light, some species can survive at depths as great as 200 meters (660 feet) in clear water.

Most red algae are less than 1 meter (3.3 feet) long. Some species develop a large, blade-like thallus, while others are small, delicate organisms. A few of the smaller species are epiphytes (*epi* meaning "upon"; *phyte* meaning "plant"; Figure 7–4b), that is, they grow on other algae or plants. Some of these will grow on a variety of hosts, while others will grow only on specific host organisms. Some species of smaller red algae are parasitic on larger species of algae (usually other reds), relying on nutrients from their hosts for their survival.

Species of red algae known as *coralline algae* possess a coating of calcium carbonate and resemble coral (Figure 7–4c). Together with the animals known as corals, coralline algae play an important role in precipitating calcium from seawater. Coralline algae help to cement bits and pieces of loose coral together. This process is necessary to hold the many parts of a reef together and eventually forms large areas of nearly solid rock. More than half of the mass of some reefs is contributed by coralline algae. We will examine the role of coralline algae in reef formation in more detail in Chapter 15.

Red algae exhibit a variety of life cycles, some of which can be quite complex (Figure 7–4d). The stage in the life cycle that produces gametes is called the *gametophyte* (ga-MEE-toh-fyt; "gamete plant"), and it is a haploid organism (has only one copy of each chromosome). The male gametophyte produces small, nonmotile "male" gametes, while the female gametophyte produces egg cells. When the male gametes are released from the male, they are carried by water currents to the female, where they become attached. The nucleus of the male cell fertilizes the egg cell, forming a diploid (two copies of each chromosome) zygote (ZY-goht). The zygote usually divides while still attached to the parent, forming a stage that is unique to red algae. This stage produces a number of diploid spores, and each of these spores is capable of forming a new adult alga when it germinates. The diploid adults that develop from these spores form the sporophyte (SPOR-oh-fyt; "spore plant") stage of the life cycle. This stage releases haploid spores that

disperse, germinate, and form a new (haploid) gametophyte generation, and the life cycle continues.

Red algae produce large amounts of polysaccharides around their cells. Several of these polysaccharides are commercially important. Agar, for example, forms thick gels at very low concentrations and is resistant to being degraded by most microorganisms. Its primary human use is in making solid culture media for growing bacteria in microbiology laboratories. It is also used in foods and pharmaceuticals as a thickening agent. Another polysaccharide from red algae, carrageenan (ka-ruh-JEE-nuhn), is also used as a thickening and binding agent in making ice cream, pudding, and salad dressings, and by the cosmetic industry for making various creamy preparations.

Some red algae are used by humans as a source of food. Irish moss (*Chondrus crispus*) is widespread on the shores of Ireland and many other parts of the North Atlantic. The Irish make a traditional pudding by boiling small amounts of Irish moss in milk. The dessert was adopted in New England, Canada, Great Britain, and France, where it is called *blanc mange*. A red seaweed called *Dulce* has been used as a foodstuff in England for hundreds of years. In Japan and other parts of the orient, the red alga *Porphyra* is used in making sushi, soups, and seasonings. In addition to being a source of human food, red algae are also cultivated to produce animal feed and fertilizer in many parts of Asia.

BROWN ALGAE

Brown algae (division Phaeophyta) are represented by such familiar forms as rockweeds, giant kelps, and sargassum weed. Brown algae range in size from microscopic, filamentous forms to the largest of all algae, the giant kelp, which can attain lengths of 100 meters (330 feet). The characteristic olive brown color of these algae is due to the pigment fucoxanthin, which is also found in diatoms. Much of the familiar seaweed that is found along the shore and in shallow water is brown algae (Figure 7–5a). Brown algae are almost exclusively marine organisms and for the most part prefer colder water. They are found in abundance along the coastlines of cold and temperate regions. A notable exception is the sargassum weed (*Sargassum*). Some free-floating species are quite abundant in the subtropical open ocean of the North Atlantic.

Most species of brown algae develop large, flat, leaflike structures called **blades.** The larger blades frequently have gas-filled structures called **gas bladders** that help to buoy the large blade, allowing it to gain maximal exposure to sunlight. The gas bladders and blades are supported by a large, stemlike structure called the **stipe** that is attached to the bottom by a branching system of fibers called the **holdfast** (Figure 7–5b). Although there a is superficial resemblance,

(text continues on page 116)

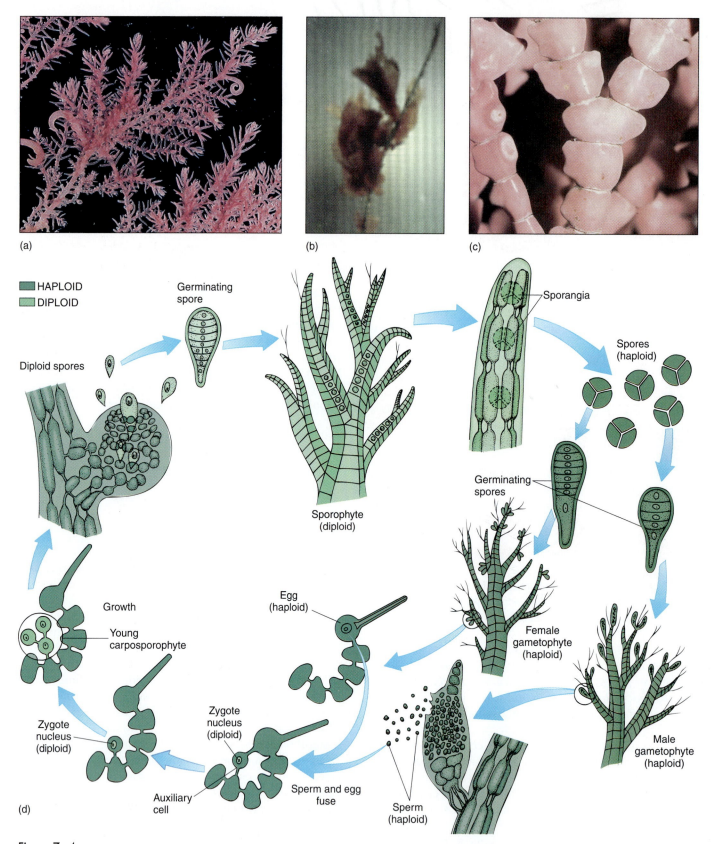

(a)

(b)

(c)

HAPLOID
DIPLOID

Diploid spores

Germinating spore

Sporophyte (diploid)

Sporangia

Spores (haploid)

Germinating spores

Female gametophyte (haploid)

Male gametophyte (haploid)

Growth

Young carposporophyte

Egg (haploid)

Zygote nucleus (diploid)

Zygote nucleus (diploid)

Auxiliary cell

Sperm and egg fuse

Sperm (haploid)

(d)

Figure 7–4

Red Algae. (a) Most red algae are multicellular with complex bodies. (b) *Smithora* is an epiphytic red alga. (c) Coralline red algae have cell walls containing calcium carbonate, and their bodies are hard and brittle. (d) The generalized life cycle of a red alga. Nonmotile, haploid spores produced by the sporophyte germinate and develop into either male or female gametophytes. The sperm produced by the male gametophytes are nonmotile and are carried by water currents to the female gametophytes to fertilize their eggs. The zygote nucleus that results from fertilization is then transferred to an auxiliary cell that develops into a new generation, the carposporophyte. The carposporophyte produces nonmotile, diploid spores that germinate and form the next sporophyte generation. (a, D. P. Wilson and David Hosking/Photo Researchers, Inc.; b, J. Robert Waaland/Biological Photo Service; c, Sea Studios, Inc./Peter Arnold, Inc.)

(a)

(b) Reproductive structures

Air bladder

Blade

Stipe — Holdfast

■ HAPLOID
■ DIPLOID

Zygote (diploid)

Sperm and egg fuse

Young sporophyte (diploid)

Sperm (haploid)

Eggs (haploid)

Sperm packet

Egg

Female gametophyte containing eggs (haploid)

Male gametophyte containing sperm (haploid)

Sperm

Magnified view

Sporophyte (diploid)

Cross-section of reproductive structure

(c)

Figure 7–5

Brown Algae. (a) *Fucus*, or rockweed, is a familiar seaweed on rocky shores. (b) The body or thallus of *rockweed* consists of a holdfast, stipe, and blades. (c) The life cycle of *rockweed*, a common brown alga. The sporophyte is the large, conspicuous stage in the life cycle. Gametophytes develop within receptacles at the tips of the sporophyte blades. The male and female gametophytes release their gametes into the surrounding water, where fertilization occurs. The zygote then settles to the bottom and develops into a new sporophyte. (a, M. Graybill/Biological Photo Service)

seaweeds are not true plants, and the blades are not leaves, the stipe is not a stem, and the holdfasts are not roots. Seaweeds do not have flowers and do not produce seeds.

The life cycle of most brown algae consists of an alternation of generations between sporophyte and gametophyte stages (Figure 7–5c). In the giant kelps, the large kelp is the sporophyte stage. It reproduces by releasing haploid, motile spores into the surrounding water. These spores are produced in extremely large quantities and are an important food source for a variety of filter-feeding animals and zooplankton. It is interesting that such a large alga as the kelp has a gametophyte stage that is microscopic.

Fucus, or rockweed, produces haploid sperm or egg cells at the tips of the thallus in specialized chambers. The eggs and sperm are then shed into the water, where fertilization takes place. The resulting zygote develops into either a male or female adult alga that continues the life cycle.

In both rockweed and kelp, after the egg is fertilized it attaches to the bottom with rootlike structures called *rhizoids*. The holdfast eventually develops from the rhizoids and puts forth a shoot. The shoot grows larger and develops one or more growth zones from which further growth proceeds. In rockweed, the growth zone is located at the tips of the thallus. In kelp species, growth zones may be located farther back from the blade. As wave action breaks off the tips of kelp or herbivorous animals devour them, the central growth area produces new tissue that is pushed upward and outward to take the place of lost tissue.

Although the rockweed *Fucus* is abundant on intertidal rocks, the majority of brown algae are found from the low tide line to a depth of about 10 meters (33 feet). The larger forms, the kelps, grow so profusely that they form offshore kelp forests (Figure 7–6). These forests are very efficient at capturing sunlight and are extremely productive. They are home to a large and diverse group of marine animals including sea urchins, fishes, crustaceans, molluscs, sea lions, and many more. (Kelp ecosystems are discussed in Chapter 16.)

The kelps exhibit a degree of thallus organization that is usually not found in algae. They have a primitive type of vascular system, a system composed of tubular structures that help to physically support these large algae and to transport materials through the thallus. The lower stipes and holdfasts of some species cannot produce enough food for themselves by photosynthesis because the blades above absorb and block so much of the available sunlight. In these algae, chains of cells called *trumpet cells,* which resemble the vascular tissue in plants, carry sugars from the blade to the stipe and the holdfast to support their metabolic needs.

Since brown algae live primarily in shallow water or on shoreline rocks, they exhibit adaptations that protect them from the relentless pounding of the waves, or wave shock. The body of a brown alga is tough but also flexible so that it can bend with the wave action. Unlike red and green algae, brown algae usually do not reproduce by fragmentation because the tissues are too highly specialized to regenerate

Figure 7–6

Kelp. Large kelp, like these off the coast of California, form massive undersea forests. (Richard Herrmann)

new parts. A major exception to this rule is sargassum weed (*Sargassum*). Large masses of this seaweed are found floating in the Atlantic Ocean, where they tend to accumulate and grow in an area known as the Sargasso Sea. These free-floating clumps of sargassum weed form a complex, three-dimensional habitat that is home to a variety of unique organisms (Figure 7–7).

Like red algae, brown algae secrete a variety of polysaccharide substances that form jelly-like coatings on their cells. Some of these substances, such as algin, are harvested commercially and used by the cosmetic and food industries as thickening agents. Brown algae, such as kelps, concentrate iodine from seawater in their tissues. The amount of iodine in some species of kelp may be ten times that of the surrounding seawater. Before cheaper methods of obtaining iodine were developed, seaweeds were the main source of this trace nutrient that is added to iodized table salt to prevent goiter, a thyroid gland disorder. In some areas of the world, especially the Orient, brown algae are used as food. Some coastal countries also use brown algae as cattle feed.

Seaweeds and Medicine

Seaweeds have been used for centuries to treat a variety of illnesses. Asiatic cultures used seaweeds as long ago as 300 B.C. to treat glandular disorders, such as goiter. A goiter is an enlargement of the thyroid, a gland located in the neck. The thyroid gland requires iodine to produce its secretions, and when an individual's diet is deficient in iodine, the gland frequently enlarges. Since many seaweeds concentrate iodine from seawater in their tissues, consumption of seaweed treats the condition.

The ancient Romans used seaweeds for treating burns and rashes and healing wounds. The slimy mucilage that coats the blades of many seaweeds effectively blocks air and microorganisms from reaching an affected area. This relieves some of the discomfort, helps prevent infection, and promotes healing. The red alga *Porphyra* contains vitamin C and was used by English sailors to prevent scurvy, a disease caused by vitamin C deficiency. *Porphyra* was more readily available than citrus fruits during ocean voyages and did not spoil as quickly. Several species of red algae were used to eliminate parasitic worms from the intestines of affected individuals. Kaenic acid, which is extracted from the red alga *Digenia* is still used for this purpose.

In the past, polysaccharides called *phycocolloids* (fy-koh-KAHL-loydz) that were isolated from red algae were used to treat a variety of intestinal ailments. Phycocolloids dissolve slowly and are not digested. These characteristics allow them to coat the lining of the stomach and intestines so that material in the digestive tract, such as acids, will not irritate it, thus alleviating some of the discomfort related to ulcers and stomachaches. Phycocolloids were used to treat constipation because they promote retention of fluid in the large intestine, which helps make fecal material easier to move. Today products from red algae, such as agar and carrageenan, are used in the treatment of ulcers. The pharmaceutical industry uses a variety of polysaccharides from red algae in the coating of pills and the production of time-release capsules.

Probably the most important algal product used in medicine and research today is agar, which is used for culturing microorganisms. Agar can withstand high temperatures so it can be sterilized. It is porous so that it allows the movement of nutrients. It is solid at room temperature and resists decomposition by most microorganisms. These characteristics make agar an ideal medium for growing bacteria and fungi for study and research.

Figure 7–7

Sargassum Weed. *Sargassum* forms large, floating masses that are home to many unique species of marine life such as this sargassum fish. (Larry Lipsky/Tom Stack and Associates)

GREEN ALGAE

Green algae (division Chlorophyta) are a diverse group of organisms that contain the same kinds of pigments found in vascular plants: chlorophyll a and b and carotenoids. The majority of green algae are small, unicellular, or filamentous. Very few species of green algae are marine; most are freshwater species. A common marine form is sea lettuce, *Ulva* (Figure 7–8a). The thallus of this alga resembles a large leaf of lettuce that may exceed 30 centimeters (12 inches) in length. It is found in intertidal and shallow coastal waters, where it attaches to the bottom by means of a tiny holdfast. The life cycle of *Ulva* is interesting because the sporophyte and gametophyte stages are nearly identical (Figure 7–8b). The gametes produced by the haploid gametophytes are nearly identical, being small ovoid cells with a pair of flagella. Mating types are designated + and −, and gametes of opposite mating types must fuse in order for fertilization to occur.

(text continues on page 119)

Figure 7—8

Sea Lettuce. (a) Sea lettuce is a common marine green alga. (b) The life cycle of sea lettuce. The sporophyte produces two different kinds of motile spores designated + and −. When the spores germinate, they develop into two different types of gametophyte, one that produces large flagellated gametes (+) and one that produces small flagellated gametes (−). The gametes are released into the water, where a large gamete fuses with a small gamete to produce a zygote. The zygote settles on solid bottom, germinates, and forms a new sporophyte. (a, Dennis Drenner)

(a)

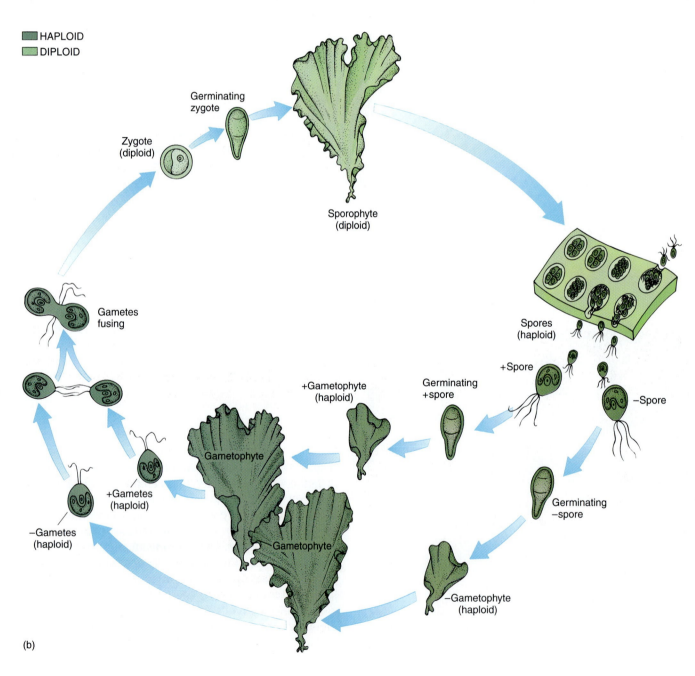

(b)

Many marine green algae exhibit a **coenocytic** (see-nuh-SIT-ik) body plan. A coenocytic thallus consists of a single giant cell, or a few large cells, containing more than one nucleus. Such algae go through a developmental process in which the cell grows and the nucleus divides, but the cell does not divide. This process is repeated many times, resulting in a large, multinucleate cell. Some examples of this body plan are shown in Figure 7–9. The simplest form is found in *Valonia,* which has a saclike thallus. Although the body form at first appears quite simple, closer examination shows that it maintains its shape with an elaborate cell wall composed of cellulose. The cellulose is deposited in multiple layers for strength, similar to the arrangement of plies in a multiple-ply automobile tire.

(a)

(b)

Figure 7–9

Coenocytic Green Algae. Examples of coenocytic green algae include (a) *Caulerpa* with its feathery thallus and (b) *Acetabularia,* which is unusual not only because it is single-celled but also because it has only one nucleus. (a, William C. Jorgensen/Visuals Unlimited; b, Dr. Dennis Kunkel/Phototake)

The alga *Caulerpa* (Figure 7–9a) forms elegant, feather-shaped thalli, while others form structures that look like stems and leaves. One of the most beautiful marine algae is *Acetabularia* (Figure 7–9b). This alga has a slender stalk with a delicate, round, umbrella-like cap at the top. The cap functions in photosynthesis. This single-celled alga was and is used by cell biologists to study the role of the nucleus in controlling cellular processes. Another interesting coenocytic alga is the seaweed *Codium.* This alga has a thallus composed of many long coenocytic filaments wrapped together like a rope. The green alga *Halimeda* has a thallus that contains many segments filled with deposits of calcium carbonate. Calcareous green algae such as *Halimeda* play an important role, along with coralline red algae and coral animals, in the formation of coral reefs.

MARINE PLANTS

General Characteristics

The plant kingdom is divided into several major groups. The presence of specialized vessels that carry fluids and give structural support (vascular tissue) separates the more advanced plants, such as trees and flowers (vascular plants), from more primitive types, such as mosses. The most advanced of the vascular plants reproduce by means of seeds and are called *seed plants.* The seed is a specialized structure containing the embryo plant and a supply of nutrients surrounded by a protective outer layer. There are two groups of seed-bearing plants: the conifers, such as pine trees, which bear seeds in structures known as *cones,* and the flowering plants, which bear seeds in structures known as *fruits.* The fruit is composed of several layers of tissues; it protects the seed and aids in its dispersal. Conifers are exclusively terrestrial plants, but a small number of salt-tolerant flowering plants, known as **halophytes** (HAL-uh-fyts), are adapted to the marine environment.

Seagrasses

Seagrasses are marine plants that generally inhabit the protected shallow waters of temperate and tropical coastal areas. These plants are not true grasses, and they represent several genera that appear to be more closely related to members of the lily family. The most extensive areas of seagrasses are found in the tropics. They are concentrated in two major regions of the Indo-Pacific, as well as both coasts of the Americas. Seagrasses do not thrive in areas of low light intensity. If the water in a seagrass bed becomes too turbid (clouded with sediments), it can destroy the seagrasses and their dependent organisms. A massive die-off of seagrasses took place in Maryland's Chesapeake Bay estuary during the

1960s, largely due to diminished light associated with excessive sediment runoff from the land surrounding the bay.

Seagrasses are **hydrophytes** (HYD-roh-fyts), which means they generally live submerged beneath the water. To survive in their subtidal habitat, these plants must be adapted to a saline environment with wave action and tidal currents. They must also be able to carry out pollination and seed dispersal under water.

The major stems of seagrasses, called **rhizomes** or *long shoots,* grow horizontally, usually just below the surface of the bottom sediments (Figure 7–10). The rhizomes and roots of seagrasses help to stabilize the bottom and, along with the leaves, help trap large amounts of sediment. The leaves are either flat and oval, ribbon-shaped, or cylindrical and are flexible to better withstand water movement while remaining erect. Leaves either develop directly from the rhizomes or from small vertical side-stems called *short shoots.* Roots develop from both the rhizomes and the base of short shoots. Unlike the fibrous roots of terrestrial grasses, the roots of seagrasses are usually thicker and more fleshy. The flowers, normally inconspicuous, small, and pale white, are usually located at the bases of the leaves. The stamens (male flower parts) and pistils (female flower parts) generally ex-

Figure 7–11

Turtle Grass. A white sea anemone thrives in a Florida turtle grass bed. Turtle grass is the most common species of seagrass in the Caribbean. (Dr. David Campbell)

tend above the petals of the flower. As a rule, the pollen is released from the stamens in long, gelatinous strands that are carried by water currents to the pistils.

The internal structure of seagrasses is typical of a hydrophyte. One of the most characteristic features is the regular arrangement of air spaces in the soft tissues of the plant. This tissue aids in flotation of the leaves and in gas exchange throughout the plant. The outer cell layer of the plant is called the *epidermis.* Unlike most land plants, the epidermal cells of seagrasses contain chloroplasts, and it has been suggested that specialized cells in the epidermis may play a role in maintaining the proper amount of salts in the cells' environment. Since seawater helps to buoy the leaves of seagrasses, supporting tissues in these plants are much reduced and not very rigid.

Eelgrass (*Zostera marina*) is widely distributed in the North Atlantic and Pacific Oceans and even extends into the Arctic Circle. It is generally a subtidal species but may be found in the lower intertidal regions of shallow lagoons. It grows well on bottoms of sand, mud, or a mixture of sand, gravel, and mud. **Turtle grass** (*Thalassia testudinum*; Figure 7–11) is the most common and abundant seagrass in the Caribbean Sea; it is found in calm water to depths of about 10 meters (33 feet). The second most abundant and important Caribbean seagrass is **manatee grass** (*Syringodium filiforme*). Manatee grass grows in pure stands as well as mixed with turtle grass. It is usually found at depths of 10 meters (33 feet) or less. Manatee grass tolerates lower salinities than other species, and its reproduction is controlled by temperature.

Species of the genus *Phyllospadix,* commonly known as **surf grasses,** are unusual because they grow in intertidal areas where there is high surf. The various species often grow together in tight clusters with their rhizomes tightly packed and strongly attached to rocks or other solid sur-

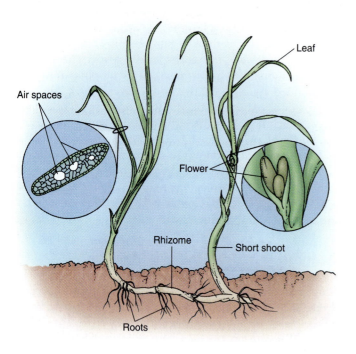

Figure 7–10

Seagrass Anatomy. The stems of seagrasses are called *rhizomes;* they grow horizontally beneath the bottom sediments. Roots, short shoots, and leaves grow directly from the rhizomes. The leaves contain many air spaces that help make them buoyant and that function in gas exchange. The flowers of seagrasses are small, white, and inconspicuous.

faces. The roots have branching root hairs that form a dense mat that anchors the plant on the rocks and helps prevent *Phyllospadix* from being dislodged by the surf.

Salt Marsh Plants

Plants such as cordgrass (*Spartina*), rushes (*Juncus*), and pickle weed (*Salicornia*; Figure 7–12) are adapted to life in salt marshes. These plants generally grow in the middle to upper intertidal zones, where they are protected from the action of waves and tidal currents, and where they are not completely submerged. Like seagrasses and mangroves, they are considered *facultative halophytes,* meaning that they can tolerate salty conditions as well as freshwater. Since they live in sediments with a high salt content, salt marsh plants tend to lose water to their environment by osmosis and display adaptations similar to those of desert plants to help them retain water. Both cordgrass and rushes have leaves covered by a thick cuticle to help retard water loss, and the stems and leaves contain well-developed vascular tissues for the efficient transport of water within the plant's tissues. Many of the plants that inhabit salt marshes, such as pickle weed, have thick, water-retaining, succulent leaves similar to those of plants that live in hot, arid terrestrial habitats.

Plants of the salt marsh have two types of root: one that acts to anchor the plant and another, short-lived type that functions in nutrient absorption. Stems develop from rhizomes and produce the leaves. Like seagrasses, salt marsh plants usually have shallow roots and rhizomes. This arrangement helps to stabilize coastal sediments and pre-

Figure 7–12

Salt Marsh Plants. Cordgrass, *Spartina*, dominates this Maryland salt marsh. (Connie Toops)

vent shoreline erosion. The rhizomes of cordgrass also play an important role in recycling the nutrient phosphorus by transferring phosphates produced in the bottom sediments into the leaves and stems of the plant.

Mangroves

Along protected tropical shores with limited wave action and usually high rates of sedimentation, plants known as **mangroves** thrive. The word *mangrove* comes from the Portuguese *mangue* which means "tree" and the English *grove* and is used to refer to trees and shrubs that are found in shallow, sandy, or muddy areas. Mangroves occur in the Caribbean Sea, Atlantic Ocean, Indian Ocean, and western Pacific Ocean. They are frequently associated with saline lagoons and are commonly found in tropical estuaries, as well as on the protected sides of islands and atolls. Mangroves form the dominant vegetation in communities called *mangrove swamps,* or *mangals,* which we will study in more detail in Chapter 13. The most common mangroves found in mangals are the red mangrove (*Rhizophora*), black mangrove (*Avicennia*), white mangrove (*Laguncularia*), and buttonwood (*Conocarpus*).

Mangroves grow in sediments that have a high salt concentration, and they are usually surrounded by seawater. As a result of these conditions, they must conserve water and they exhibit adaptations similar to those found in salt marsh plants. The leaves of mangroves are thick and leathery. The epidermis of the leaf is usually covered with a thick cuticle, and the stomata (stoh-MAH-tuh), openings in the leaves for gas exchange and water loss, are sunken and usually confined to the undersurface. These adaptations help to retard the loss of water by evaporation from the leaves. The black mangrove has salt glands on the upper surface of the leaf that exude a concentrated salt solution. This allows the plant to eliminate excess salt that is taken in through the roots.

Since the muddy sediments in which mangroves grow are soft and loose, anaerobic, and high in salt concentration, these plants do not have large tap roots (single roots that penetrate deep into the soil to anchor a plant). Rather, they produce a variety of support roots such as prop roots, drop roots, and cable roots (Figure 7–13a) that serve both as structural supports and as a means of aerating the buried root tissues with a system of gas-filled spaces similar to those found in seagrasses.

Another interesting adaptation in many mangroves species involves the seeds. Unlike in most other plants, the embryo in the seed begins to germinate while the seed is still attached to the parent plant (Figure 7–13b). This lack of seed dormancy is important since the developing root and stem can establish more quickly after the seed, seedling, or fruit is dropped from the tree. The embryonic root can at-

(a)

(b)

Figure 7–13

Mangroves. (a) This mangrove swamp contains mostly red and black mangroves. Since the sediments in which these trees grow are loose, mangroves have many supports, such as prop roots, above the mud. These above-ground roots not only help to support the large plant but also aid in gas exchange for the parts of the root buried in the anaerobic sediments. (b) Seeds of this red mangrove are beginning to germinate while still attached to the parent. When they drop off, they either take root in the surrounding soft sediments or are carried by water to other areas. (a and b, Jon Hawker)

tach quickly to loose sediments, preventing it from being washed back out to sea, and the stem can quickly produce leaves to support the metabolic demands of the plant. The seeds of mangroves often float for several weeks before becoming waterlogged and attaching to the bottom. This aids in dispersing the seeds away from the parent plants.

Ecological Roles of Marine Plants

Marine plant communities serve several important ecological functions in the marine environment. They act as sediment traps, helping to collect and stabilize bottom sediments and improving the clarity of the water. Improved clarity permits more sunlight to penetrate the water to support photosynthesis. Mangroves function in soil formation because their complex root systems are particularly good at trapping and accumulating sediments and debris such as twigs and leaves.

Marine plants filter runoff from coastal areas, removing some of the potentially toxic organic pollutants of terrestrial origins and preventing them from entering the sea. Marine plants may also play an important role in maintaining water quality in shallow water areas by removing excess nutrients entering from terrestrial sources that might contribute to algal blooms. When parts of the plants break off or when a plant dies, the decaying plant material contributes to detritus food webs. Marine plants are major producers of detritus, which contributes to offshore productivity as well as to productivity within the communities themselves. They also provide habitats and shelter for a multitude of diverse organisms from invertebrates to mammals.

Seagrass communities provide food for animals in a number of ways. Seagrasses are primary producers with high rates of production. They are a direct food source for some animals, including marine turtles, some sea urchins, and snails. More important, seagrass beds fulfill a critical

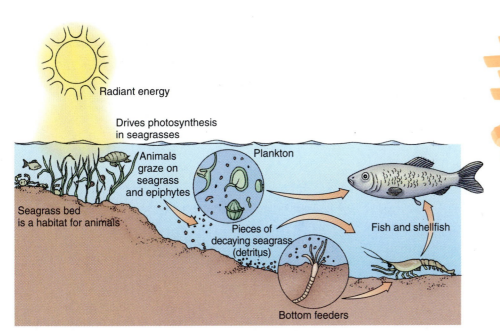

Figure 7—14

Ecological Role of Seagrasses. Seagrasses are important primary producers and, like other marine plants, supply large amounts of nutrients to other marine organisms in the form of detritus. Marine plants such as seagrasses also provide habitat for a variety of marine animals.

Radiant energy

Drives photosynthesis in seagrasses

Animals graze on seagrass and epiphytes

Plankton

Seagrass bed is a habitat for animals

Pieces of decaying seagrass (detritus)

Fish and shellfish

Bottom feeders

role in detritus production and export, especially in estuarine communities (Figure 7–14). In addition, they provide attachment sites for epiphytes, which are an important source of nutrients for certain invertebrate and vertebrate animals.

The detritus from salt marsh plants supplies the nutrient needs of a variety of fishes and shellfish. One study suggests that as many as 95% of sport and commercial fish species, many of which spend part or all of their lives in coastal and estuarine waters, are supported by the productivity of detrital food chains in salt marshes.

CHAPTER SUMMARY

Multicellular marine algae are mostly benthic organisms. They are divided into three major groups (red algae, brown algae, and green algae) based on the accessory pigments they contain. Algal cells have cell walls in addition to their membranes, and their pigments are located within organelles called *chloroplasts*. Many algae secrete a slimy covering that both protects and retards desiccation. Algal life cycles tend to be complex, involving both sexual and asexual generations.

Red algae are most widespread in the tropics. Their red color is due to the presence of the pigment phycoerythrin, which allows them to survive at great depths. Brown algae are almost exclusively marine organisms. With the notable exception of sargassum weed, most brown algae can be found along continental shorelines. Many brown algae have bodies that are quite flexible, an advantage for avoiding wave shock. Green algae are found in shallow coastal waters. Many marine species of green algae exhibit a coenocytic body plan. Algae are an important food source for many organisms, including humans, and they provide a habitat for many species. They are also a source of many commercial products.

The plants living in marine environments are vascular plants that produce seeds. Seagrasses are mainly found in shallow waters of temperate and tropical oceans. Salt marsh plants generally grow in the middle to upper intertidal zones, where they are protected from wave action. Mangroves are tropical trees and shrubs that grow in sandy and muddy coastal areas that are protected from wave action and have high sedimentation rates. All marine plants exhibit adaptations that allow them to survive in salt water.

Marine plant communities play many important ecological roles in the environment. They act as sediment traps, help to stabilize sediments, and filter runoff from the land. Marine plants are important primary producers and sources of detritus that contributes to local and offshore productivity. They also provide habitats for many organisms.

KEY TERMS

thallus, *p. 111*

red algae, *p. 113*

brown algae, *p. 113*

gas bladder, *p. 113*

holdfast, *p. 113*

green algae, *p. 117*

coenocytic, *p. 119*

hydrophyte, *p. 120*

rhizome, *p. 120*

eelgrass, *p. 120*

turtle grass, *p. 120*

manatee grass, *p. 120*

mangrove, *p. 121*

QUESTIONS FOR REVIEW

MULTIPLE CHOICE

1. Some algae are able to live in deep water because
   a. they have holdfasts to fasten them to the bottom
   b. they have accessory pigments that absorb blue and green light
   c. their chlorophyll is more efficient at trapping light
   d. there are fewer herbivores to graze on them
   e. they tolerate the cold

2. The body of a multicellular alga is called a
   a. rhizome
   b. holdfast
   c. thallus
   d. stipe
   e. blade

3. _____ is a product derived from red algae that is used to produce media for culturing microorganisms.
   a. agar
   b. algin
   c. carrageenan
   d. mucilage
   e. polyan

4. The stage in the algal life cycle that produces gametes is called the
   a. sporophyte stage
   b. adult stage
   c. hydrophyte stage
   d. gametophyte stage
   e. vascular stage

5. A type of brown algae that grows quite large and forms undersea forests is
   a. sargassum weed
   b. Irish moss
   c. kelp
   d. rockweed
   e. sea lettuce

6. A coenocytic thallus consists of a single cell or several large cells that contain more than one
   a. cell wall
   b. chloroplast
   c. thallus
   d. holdfast
   e. nucleus

7. Plants that live submerged beneath the water are called
   a. halophytes
   b. hydrophytes
   c. anthophytes
   d. vascular plants
   e. algae

8. Salt marsh plants and mangroves exhibit adaptations similar to those of terrestrial plants that grow in the
   a. tropical rainforests
   b. prairies
   c. swamps
   d. deserts
   e. tundra

9. Each of the following is an important ecological role of marine plants except
   a. improving water clarity
   b. trapping nutrients
   c. stabilizing bottom sediments
   d. fixing nitrogen
   e. providing a habitat

10. Plant adaptations for life in the marine environment include
    a. thick epidermis in leaves
    b. stems that project above the water level
    c. roots with low salt tolerance
    d. very little chlorophyll in leaves
    e. all of the above

SHORT ANSWER

1. What factors affect the distribution of algae in the marine environment?

2. List some of the important ecological roles of marine plants.

3. Describe the alternation of generations that occurs in the life cycle of an alga or plant.

4. Describe the adaptations that have evolved in algae to protect against wave shock.

5. Describe the adaptations that have evolved in salt marsh plants to help them survive in areas where the salt content is high.

THINKING CRITICALLY

1. Salt marshes play an important role as nurseries for many commercially important fishes and shellfish. What characteristics of salt marshes make them such ideal nurseries?

2. Predict the effects of coastal zone development on sea-grass communities.

3. In Chapter 2 you learned that overgrazing of kelp by sea urchins led to a decline in several other animal species associated with kelp forests. Based on what you have learned in this chapter, why would a decrease in the amount of kelp affect animals that do not feed on kelp?

SUGGESTIONS FOR FURTHER READING

Cox, P. A. 1993. Water Pollinated Plants, *Scientific American* 269(4):68–74.

Dawes, C. J. 1981. *Marine Botany.* John Wiley & Sons, New York.

Dring, M. J. 1983. *The Biology of Marine Plants.* University Park Press, Baltimore.

Duffy, J. E., and M. E. Hay. 1990. Seaweed Adaptations to Herbivory, *Bioscience* 40(5):368–375.

Jacobs, W. P. 1994. Caulerpa, *Scientific American* 271(6):100–105.

Rutzler, K., and I. C. Feller. 1996. Caribbean Mangrove Swamps, *Scientific American* 274(3):94–99.

Schafer, K. 1988. Mangroves, *Oceans* 21(6):44–49.

Teal, J. M. 1996. Salt Marshes: They Offer Diversity of Habitat, *Oceanus* 39(1):13–15.

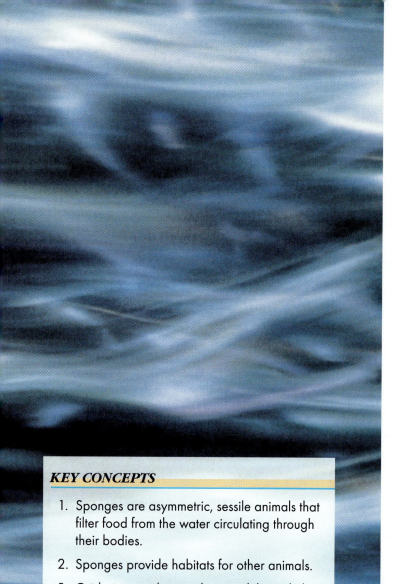

Chapter 8

Sponges, Cnidarians, Comb Jellies, and Marine Worms

KEY CONCEPTS

1. Sponges are asymmetric, sessile animals that filter food from the water circulating through their bodies.

2. Sponges provide habitats for other animals.

3. Cnidarians and ctenophores exhibit radial symmetry.

4. Cnidarians possess a highly specialized stinging cell used to capture prey and for protection.

5. Marine worms exhibit bilateral symmetry.

6. Turbellarians are free-living flatworms; flukes and tapeworms are parasitic flatworms.

7. Nematodes are abundant and important members of the meiofauna.

8. Polychaete diversity stems from the evolution of a segmented body that allows for increased motility.

9. In addition to being important consumer organisms, polychaetes are the primary prey of many marine animals and play an important role in recycling nutrients.

10. Several other groups of wormlike animals, including ribbon worms, spiny-headed worms, peanut worms, acorn worms, and beardworms, play important ecological roles in the marine environment.

It may seem easy to distinguish animals from other forms of life, but this is not always the case. The animals known as sponges, for instance, were generally thought to be plants until 1765, when a British naturalist named Ellis observed that sponges could expel water from their bodies. This observation prompted Ellis to suggest that sponges were animals. Their place as animals, however, was not fully accepted until 1825, when R. E. Grant of Edinburgh University, using improved microscope technology, demonstrated that sponges could take tiny colored particles from the water, circulate them through their bodies, and then expel them.

The earth is populated by a wide variety of animals, and it is difficult to arrive at a definition of an animal that applies to them all. Even though there is great diversity among animals, there are three characteristics that they all share.

1. Animals are multicellular. This distinguishes them from bacteria and most protists, which are unicellular.
2. Animal cells are eukaryotic, and they lack cell walls. These characteristics distinguish animals from bacteria, whose cells are prokaryotic and have cell walls; and fungi, multicellular algae, and plants; all of which have rigid cell walls.
3. Animals cannot produce their own food, so they depend on other organisms for nutrients. In other words, they are heterotrophs.

The sea is inhabited by a large number of animal species, and most of them are invertebrates. **Invertebrates**

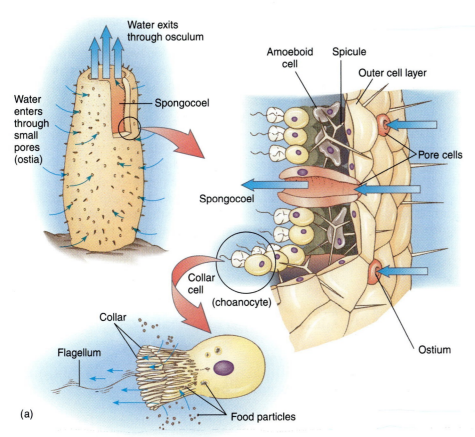

Figure 8–1

Sponges. (a) Anatomy of a sponge. (b) Sponges exhibit one of three possible body plans: asconoid, syconoid, or leuconoid. Blue arrows represent the direction of water flow. Areas occupied by collar cells are shown in black.

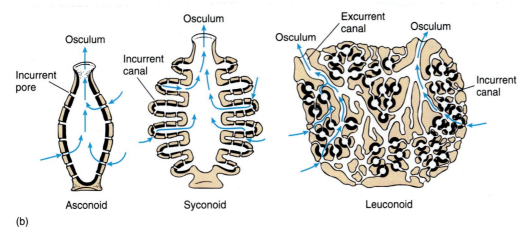

are animals that lack a vertebral column (backbone). These animals range in size from microscopic creatures to giant squids. Regardless of their size, each one plays an important role in the fabric of life in the sea. In this chapter and the next, we will examine more closely the ecological roles played by invertebrate animals.

SPONGES

Sponges (phylum Porifera) are simple, asymmetric, sessile (permanently attached) animals that exhibit a wide variety

of sizes and shapes. Their shape is frequently determined by the shape of the bottom sediments or material on which they are growing and by the water currents flowing over them. Although many living sponges are drab, some species are brightly colored. Red, yellow, green, orange, and purple specimens are common.

Sponge Structure

A sponge's body is full of tiny holes or pores called *ostia* (AHS-tee-uh; singular: ostium; Figure 8–1a) through which large amounts of water circulate. The water is a source of nutrients and oxygen, and it carries away the animal's wastes. Water enters a sponge's body through the ostia and eventually flows into a spacious cavity called the *spongocoel* (SPUN-joh-seel). Water then exits the spongocoel through a large opening called the *osculum* (AHS-kyuh-luhm). Many species have several spongocoels and oscula. The spongocoel and internal canals of a sponge can be home to a variety of organisms such as shrimp and fish that take advantage of the protection and continuous flow of water available to them. For instance, one specimen of loggerhead sponge (*Spheciospongia*) from Florida was found to contain 16,000 snapping shrimp (Alphaeidae)!

Sponges have several cell types that perform specific functions within the animal. These are shown in Figure 8–1. A particularly important cell type is the **collar cell** (**choanocyte**), which is found lining the spongocoel and other chambers in a sponge. This oval cell has a single flagellum and a collar of finger-like cellular projections that acts to strain food from the water. Collar cells create the current of water that moves through the sponge. The beating of their flagella creates a current that draws water in through the ostia and expels it through the osculum. As the water circulates through the sponge, bacteria, plankton, and detritus are engulfed and digested by various cells.

Since sponges feed on material that is suspended in seawater, they are called **suspension-feeders.** They are also referred to as **filter-feeders** since they filter their food from the water. Approximately 80% of a sponge's food is trapped by the collar cells and consists of microplankton (organisms 20 microns or less). Large particles (1 to 50 microns) are engulfed and digested by specialized cells along the sponge's system of canals that carry water from the ostia to spongo-

coels. The smaller particles (0.1 to 1.0 microns) are trapped by the collars of the collar cells and digested.

The size of a sponge is limited by its ability to circulate water through its body (Figure 8–1b). Sponges with simple body plans, like **asconoid sponges,** can filter only small volumes of water, and thus they do not grow very large. **Syconoid sponges** have bodies with many invaginations along the spongocoel. This modification greatly increases the surface area available for collar cells. Thus, the sponge can circulate a greater volume of water. **Leuconoid sponges** exhibit the most complex body plan, with several spongocoels and many chambers leading to them. This plan offers the greatest amount of internal surface area and allows the greatest filtering capacity because there is more room for collar cells. For example, even *Leuconia,* one of the smaller leuconoid sponges, 10 centimeters (4 inches) tall and 1 centimeter (.375 inch) wide, contains approximately 2,250,000 chambers and pumps about 22.5 liters (5 gallons) of water per day through its body. It is no wonder that all of the largest sponges have a leuconoid body type.

Reproduction in Sponges

Sponges can reproduce both sexually and asexually (Figure 8–2). Asexual reproduction can involve either budding or fragmentation (essentially the same process as we saw in algae in Chapter 7). In budding, a group of cells develops on the outer surface of the sponge. After attaining a certain size, this bud drops off and can form a new sponge near the original or float with the currents to settle and mature else-

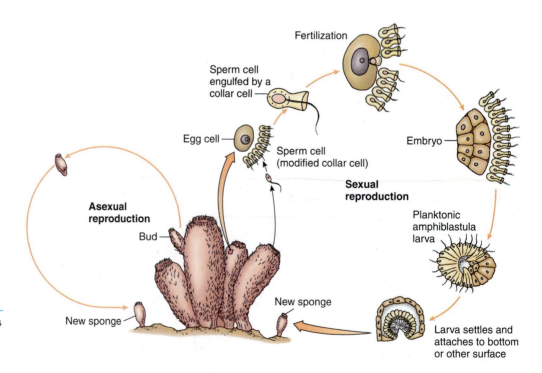

Figure 8–2

Sponge Reproduction. Sponges can reproduce asexually (by budding) or sexually.

where. Fragmentation involves the production of a new sponge from pieces that are broken off.

Since the sponge is rather loosely organized anatomically, pieces of sponge can form new sponges if they contain enough of the various cell types. For years, sponge fishermen have taken advantage of this characteristic to replenish sponge beds. Large commercial sponges are cut into pieces, tied to concrete blocks or other material that will anchor them, and then thrown back into the water. The pieces of sponge will then eventually grow into new adult sponges. Sponges exhibit marvelous powers of regeneration, and when a piece is broken off or eaten, the sponge can readily replace the missing part.

Classes of Sponges

Sponges are divided into four classes based on the characteristics of supporting structures known as **spicules** (SPIK-yoolz; see Figure 8–1). Sponges in the class Calcarea are usually small and have hard spicules made of calcium carbonate. Sponges in the class Hexactinellida have spicules that are usually fused to form elaborate networks. They are commonly called *glass sponges* because their spicules are made of silica and resemble glass fibers. The class Demospongia, which contains most of the larger sponge species, often has flexible fibers that are made up of a protein called *spongin* (SPUN-jin) as well as (or instead of) siliceous spicules. The class Sclerospongia is a small group of sponges that live in shaded or dark habitats. They secrete massive skeletons of calcium carbonate, and it is thought that they were the original reef-forming organisms, having evolved before the corals.

Commercial Use of Sponges

Large species of the class Demospongia are harvested for use as commercial sponges (Figure 8–3). Their dried spongin skeletons can hold as much as 35 times their weight in liquid. In the early days of sponge fishing, divers would weigh themselves down with heavy rocks while harvesting the sponges. Although today most sponge fishermen use diving equipment, sponge divers who fish off the coast of Tunisia still use a lead weight and a rope. For thousands of years, Kalynos, Greece, was a major port for sponge fishermen, until a disease killed off the rich sponge beds. Many Greek sponge fishermen emigrated to Tarpon Springs, Florida, where their descendants still fish the sponge beds off the Florida coast. The Tarpon Springs sponge industry is just recently recovering from a double blow of killer fungi and red tides (see Chapter 6).

It takes approximately five years for a sponge in the wild to reach a marketable size (12.5 centimeters or 5 inches), and it will retail for about $4.00. Although this is expensive when compared to the price of synthetic sponges, many people are willing to pay the extra money for a sponge that is superior to any synthetic product. For some applications, such as polishing metal, there is no good substitute for a natural sponge.

Ecological Relationships of Sponges

Very few animals feed on sponges, possibly because it would be like eating a mouthful of needles. There are a few species of bony fishes and molluscs that will eat sponges, and the hawksbill sea turtle (*Eretmochelys*) feeds almost

Figure 8–3

Commercial Sponge Fishing. A sponge fisherman at Tarpon Springs, Florida, is drying some commercial sponges. (Copyright George Billiris Sponge Merchant International)

(a)

(b)

(c)

Figure 8–4

Types of Sponges. (a) Venus's flower basket, a deepwater glass sponge from the Philippines. (b) Boring sponges in a mollusc shell. (c) Sponge crabs camouflage themselves by attaching sponges to their bodies. (a, courtesy American Museum of Natural History; b, Copyright 1995 Joshua Singer; c, E. R. Degginger/Animals Animals)

exclusively on them. The tough lining of the turtle's mouth and digestive tract prevents the spicules from injuring the animal. Sponges also produce antibacterial chemicals and are thus generally free from bacterial infections.

Many species of sponge that live on coral reefs contain symbiotic cyanobacteria in their bodies. A sponge provides protection and a sunlit habitat, and the cyanobacteria provide the sponge with nutrients and oxygen.

The glass sponge known as Venus's flower basket (*Euplectella*) that is found in the deep waters of the tropical Pacific Ocean exhibits an interesting symbiotic relationship with certain species of shrimp (*Spongicola*; Figure 8–4a). A male and a female shrimp enter the sponge's spongocoel when they are young. They feed on plankton in the water that circulates through the sponge's body. Eventually the shrimp grow so large that they cannot escape through the network of spicules that covers the osculum. Their entire life is spent in the sponge. A specimen of this sponge along with the shrimp used to be given in Japanese wedding ceremonies to symbolize "until death do us part."

Members of the family Clionidae (class Demospongia), the boring sponges (Figure 8–4b), are important in recycling calcium in the marine environment. These small sponges burrow into coral and mollusc shells. In the process they convert the calcium into a soluble form that is returned to the seawater for use by other organisms.

The biggest problem that sponges face is finding enough suitable solid material for attachment. Consequently, their primary competitors for space are corals and bryozoans. Some sponge species produce chemicals that either kill corals or inhibit their growth. Other species, such as the boring sponges, create their own habitat by boring into the corals. Some species of crabs known as *sponge crabs* (family Dromiidae) attach pieces of sponge to themselves for camouflage and protection and in the process provide a solid surface for the transplanted sponge species (Figure 8–4c).

RADIAL SYMMETRY

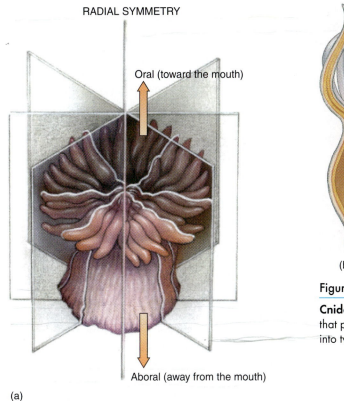

(a)

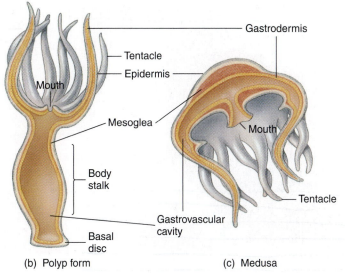

(b) Polyp form

(c) Medusa

Figure 8–5

Cnidarian Body Plans. (a) Cnidarians exhibit radial symmetry. Any plane that passes longitudinally through the central axis of the animal will divide it into two equivalent halves. (b) A typical polyp. (c) A typical medusa.

HYDROZOANS, JELLYFISH, CORALS, AND ANEMONES

Jellyfish, corals, anemones, and their relatives belong to the phylum Cnidaria (ny-DEHR-ee-uh). The phylum is named for the **stinging cell,** or **cnidocyte** (NYD-uh-syt), that is characteristic of this group of animals. Cnidarians have bodies that exhibit **radial symmetry** (Figure 8–5a). This means that many planes can be drawn through the central axis that will divide the animal into two equivalent halves. The body parts of these animals are arranged in a circular pattern around a central axis, much like the spokes of a wheel. Radial symmetry is particularly beneficial to sessile organisms and organisms that are not active swimmers since it allows them to meet and respond to their environment equally well from all sides.

Cnidarian Structure

Cnidarians often exhibit two different body plans in their life cycles (Figure 8–5b and c). The **polyp** (PAHL-uhp) is generally a benthic form characterized by a cylindrical body that has an opening at one end, the mouth, which is usually surrounded by a ring of tentacles. The **medusa** is a free-floating stage that is commonly known as a *jellyfish.* Many

cnidarians exhibit both body plans during their life cycles, although some such as corals and anemones exist only in the polyp stage.

One of the most important features of cnidarians is their stinging cells (Figure 8–6a). The stinging cell contains a stinging organelle called the **nematocyst** (neh-MAT-oh-sist) that is used for capture of prey, for defense, and in some cases, for locomotion. A short, bristle-like structure called the *cnidocil* extends from one end of the cell and acts as a trigger. When the cnidocil comes into contact with prey or some other object, it causes the nematocyst to discharge. The stinging cell can also be triggered through the action of certain chemical substances released by prey organisms. Once the stinging cell has fired, it is reabsorbed by the animal's body, and a new one is formed to take its place. Stinging cells are most common on the tentacles of cnidarians, but they can also be found on the outer body wall and lining the digestive cavity in some species.

Many species of cnidarian, such as the Portuguese man-of-war (*Physalia*), can cause painful stings. The symptoms include an immediate, intense, burning pain with possible redness and swelling in the regions of contact (Figure 8–6b). Some individuals may also experience weakness, nausea, headaches, pain and spasms of abdominal and back muscles, dizziness, and increased secretions from the nose and eyes. The sting is generally not fatal to humans unless they are allergic to the toxin.

(a)

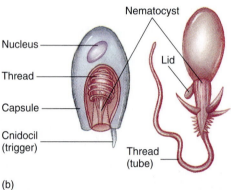

Nematocyst

Nucleus

Lid

Thread

Capsule

Cnidocil
(trigger)

Thread
(tube)

(b)

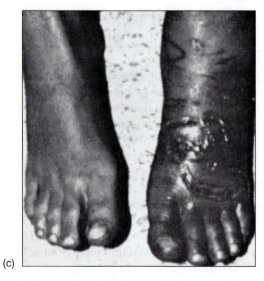

(c)

Figure 8–6

Cnidarian Stinging Cells. (a) An electron micrograph of a stinging cell (cnidocyte) showing a nematocyst that has not been discharged. (b) The structure of a stinging cell and nematocyst. (c) This victim's left foot shows five-day-old injuries produced by a small sting from a jellyfish called the sea wasp. (a, G. B. Chapman, Cornell University Medical College; c, courtesy J. H. Barnes. From Rees, W. J., ed. 1966. *The Cnidaria and Their Evolution*, Zoological Society of London)

Cnidarians are mostly carnivorous and feed on a variety of animals, depending on their size. The prey is paralyzed by a toxin on the nematocyst, and, in some cases, the nematocyst has a long fiber that entangles the prey. The prey is then drawn into the mouth and forced into the gastrovascular cavity, a central cavity that functions in both digestion and the movement of materials within the animal. Waste products and undigestible material are forced back out through the mouth after digestion is completed.

Classes of Cnidarians

There are three classes of cnidarians: hydrozoans (hydroids), scyphozoans, and anthozoans. Much of the marine growth that is seen on pilings and rocks along the seashore is composed of hydrozoans (Figure 8–7a). **Hydrozoans** (class Hydrozoa) are mostly colonial organisms that generally exhibit both an asexual polyp stage and a sexual medusa stage in their life cycles. Colonial forms of hydrozoans usually contain two types of polyp. The **feeding polyp** (gastrozooid) functions in capturing food and feeding the colony. The other type of polyp is the **reproductive polyp** (gonangium), which is specialized for the process of reproduction (Figure 8–7b).

Not all hydrozoans form sessile colonies. Members of the hydrozoan orders Siphonophora and Chondrophora produce floating colonies. Perhaps the best known of these is the Portuguese man-of-war (*Physalia;* Figure 8–8). The colony is suspended from a large gas sac (up to 30 centimeters or 12 inches) that acts as a float. It is believed that this structure develops from the original larval polyp, and specialized polyps are responsible for keeping the gas sac inflated with carbon dioxide produced by respiration. Like sessile colonies, these floating colonies contain several types of specialized polyps. Feeding polyps have a single long tentacle and function in digestion. Fishing polyps have tentacles strewn with stinging cells and are responsible for capturing prey, usually surface fish such as mackerel or flying fish. After the prey is paralyzed, the fishing polyps bring it to the feeding polyps for digestion. Other parts of the colony are modified medusae that contain ovaries or testes and function in reproduction.

Scyphozoans (class Scyphozoa) comprise most of the animals commonly known as true jellyfish. Although jellyfish are capable of swimming by pulsating their bell, many of them are not strong swimmers, and most float with the currents and are thus part of the plankton. The predominant stage in the life cycle of these animals is the medusa (Figure 8–9). Although most medusae are small (less than 10 centimeters or 3 inches), some, such as the medusae of the jellyfish *Cyanea*, have a bell diameter of 2 to 3 meters (7 to 10 feet) and tentacles that are 60 to 70 meters (200 to 230 feet) long!

(a)

Figure 8–7

Hydrozoans. (a) *Obelia* is an example of a colonial hydrozoan. (b) A general hydrozoan anatomy and life cycle. (a, Zig Leszczynski/Animals Animals)

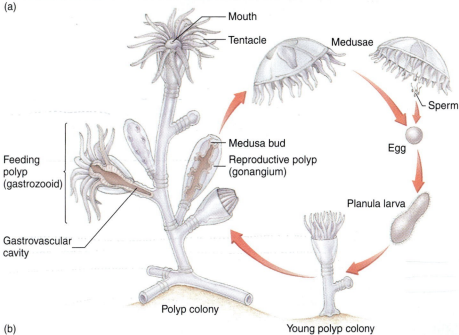

(b)

Mouth
Tentacle
Medusae
Sperm
Medusa bud
Reproductive polyp (gonangium)
Egg
Feeding polyp (gastrozooid)
Gastrovascular cavity
Planula larva
Polyp colony
Young polyp colony

Scyphozoans are carnivorous, feeding mostly on fish and larger invertebrates. Some, however, such as the upside-down jellyfish (*Cassiopeia*; Figure 8–10a), feed on plankton that are trapped in mucus produced by modified tentacles. These animals lack a true mouth. The food is passed along fused tentacles by the action of cilia to multiple openings of the gastrovascular cavity, where it is digested. Unlike typical jellyfish, the upside-down jellyfish spends little time swimming. Instead, it usually lies upside-down in shallow lagoons, an ideal place for sunlight to reach the symbiotic algal cells that provide nutrition and oxygen to this species of jellyfish. The margin of its bell pulsates regularly, drawing a current of water containing plankton across its tentacles. The contractions of the bell also help to hold the animal in place so it is not swept away by strong tidal currents.

Figure 8–8

Portuguese Man-of-War. Although this animal appears to be a single jellyfish, it is really a colony of many specialized individuals. (Oxford Scientific Films/Animals Animals)

Figure 8–9

Moon Jellyfish. The life cycle and anatomy of the moon jellyfish, *Aurelia,* a typical jellyfish.

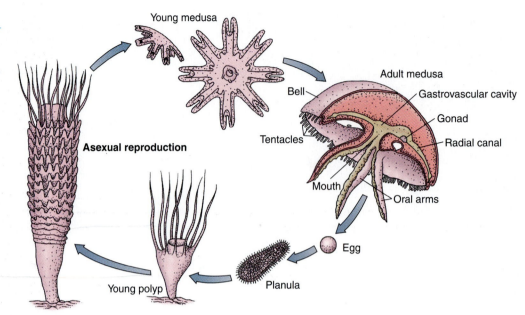

Young medusa

Asexual reproduction

Adult medusa

Bell
Gastrovascular cavity
Gonad
Tentacles
Radial canal
Mouth
Oral arms
Egg

Planula
Young polyp

Another interesting group of jellyfish are the box jellyfish (Cubozoa) (Figure 8–10b). Box jellyfish are tropical animals that have box-shaped bells and are all strong swimmers. They are voracious predators, feeding primarily on fish. The stings of some, such as the sea wasp (*Chironex fleckeri*), can be fatal to humans. The sea wasp is seasonally abundant around parts of Australia, causing the closing of some swimming beaches.

Representative **anthozoans** (class Anthozoa) include sea anemones, corals, sea fans, sea pens, and sea pansies. These animals exhibit only the polyp stage in their life cycle, and many of them bear a resemblance to brightly colored flowers. **Sea anemones** are polyps that are larger, heavier, and more complex than hydrozoan polyps (Figure 8–11a). Sea anemones contain a groove within their body that is lined by cells with hairlike projections on their free

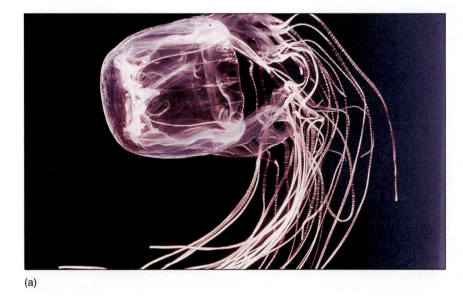

(a)

(b)

Figure 8–10

Jellyfishes. (a) The sea wasp from Australia is an example of a box jellyfish. (b) An upside-down jellyfish from the Caribbean. (a, David J. Wrobel/Biological Photo Service; b, Neville Coleman/Visuals Unlimited)

(a)   (b)

(c)   (d)

Figure 8—11

Anthozoans. (a) A sea anemone from the coast of British Columbia. (b) Sea anemones attached to the shell of a hermit crab from the Red Sea, an example of symbiosis. (c) Live coral polyps from the Caribbean with their tentacles extended for feeding. (d) An orange sea pen from Vancouver Island, British Columbia. (a, Neil G. McDaniel/Photo Researchers, Inc.; b, Jeffrey Rotman; c, Mike Bacon/Tom Stack and Associates; d, Fred Bavendam/Peter Arnold, Inc.)

surface called *cilia* (SIL-ee-uh). The beating of the cilia creates a water current through the animal's gastrovascular cavity that brings in oxygen and removes waste. A few species of sea anemone feed on fish, but most eat invertebrates or plankton, and nutrition from symbiotic algae is important to shallow-water species.

Sea anemones have well-developed muscles and are capable of expanding, contracting, and reaching out their tentacles to trap and subdue their prey with toxic stings. When disturbed, sea anemones will withdraw their tentacles into their oral openings and contract their bodies.

Certain clownfishes (family Pomacentridae) form symbiotic relationships with some of the large species of sea anemone that occur in the tropical Pacific Ocean. A special mucus that covers the fish's skin prevents the discharge of

the anemone's nematocysts. The anemone provides shelter and protection for the fish. The fish in turn protects the anemone from predatory butterflyfishes and sea stars. The clownfishes may also help ventilate the sea anemone by their movements, keep the anemone free of sediment, and attract other fish for the anemone to feed on. It has also been suggested that the fecal pellets from clownfishes may provide some nourishment to these large polyps.

One of the more unusual symbiotic relationships involving anemones is between the hermit crab, *Eupagurus prideauxi,* and the anemone, *Adamsia palliata* (Figure 8–11b). A young sea anemone will attach itself to the mollusc shell that is occupied by the young hermit crab. The sea anemone is then carried around by the crab. The anemone's mouth is located just above and behind the crab's, and it feeds on the scraps the crab leaves. Eventually the anemone overgrows the mollusc shell and absorbs it, replacing it with its own body. At this point, the sea anemone and crab both grow at the same rate so that the crab remains covered. Not only does the crab benefit by being camouflaged and protected, it also does not have to find new shells as it grows, as do other species of hermit crab. The sea anemone benefits by being carried from place to place and sharing the crab's food. If the sea anemone is removed from the crab, it will soon die, and the undisguised crab is easy prey.

Coral animals (Figure 8–11c) are polyps that, unlike sea anemones, secrete some kind of skeleton, which may be hard or soft, external to their bodies. The hard or stony corals produce a skeleton of calcium carbonate, and some of these species form large colonies of polyps. Coral reefs are the products of these reef-building corals, coralline red algae, and calcified green algae. We will discuss corals and coral reef formation in more detail in Chapter 15.

Soft corals form colonies that look more like plants than animals. Sea fans and sea pens (Figure 8–11d) are just two examples of this type of coral.

CTENOPHORES

Ctenophores (TEEN-uh-forz) or **comb jellies** (phylum Ctenophora) are named for the eight rows of comb plates (ctenes) that are used by the animal for locomotion (Figure 8–12). The comb plates are made of very large cilia, and

Figure 8–12

Ctenophore. Ctenophores, also known as *comb jellies,* propel themselves through the water with specialized structures called *comb plates.* (Robert Brons/Biological Photo Service)

when the cilia of all eight rows beat, the animal travels in a forward direction, mouth first. Ctenophores are weak swimmers and are mostly found in surface waters. The delicate bodies of ctenophores are iridescent during the day. At night, almost all ctenophores give off flashes of luminescence, possibly to attract mates or prey. Along with other bioluminescent plankton, they are responsible for the luminescence viewed in many seas.

Like cnidarians, ctenophores exhibit radial symmetry, but they lack the stinging cells that are the hallmark of cnidarians. One species, however, *Haeckelia rubra*, does have stinging cells that it gets from the jellyfish upon which it feeds. *Haeckelia* then uses the stolen stinging cells for its own defense.

Most ctenophores feed on small planktonic organisms that are trapped by a sticky substance secreted by glue cells on the tentacles. The tentacles carry the prey to the mouth, where they are digested in the gastrovascular cavity. Ctenophores that lack tentacles either rely on cilia to bring food to the mouth or feed on other ctenophores by sucking them up whole.

MARINE WORMS

Flatworms

Flatworms have flattened bodies with a definite head and tail. Some flatworms, such as the turbellarians (class Turbellaria), are free-living, while others, such as flukes (class Trematoda) and tapeworms (class Cestoda), are parasites. Flatworms exhibit **bilateral symmetry** (Figure 8–13a). In bilateral symmetry, the body parts are arranged in such a way that only one plane through the midline of the central axis (midsagittal plane) will divide the animal into similar right and left halves. Bilateral symmetry allows for a more streamlined body shape and is more advantageous than radial symmetry for animals with active lifestyles.

The active lifestyle of turbellarians favored the concentration of sense organs at one end of the animal, the end that would first meet the environment. Turbellarians have sensory receptors that can detect light, chemicals, and movement and that function in maintaining balance. This evolutionary trend toward concentration of sense organs in the head region of an animal is known as **cephalization** (sef-uh-luh-ZAY-shun).

Turbellarian flatworms range in size from a few millimeters to 50 centimeters (20 inches) (Figure 8–13b). They are mainly found under rocks, in bottom sediments, or as symbionts of larger animals. The flatworm's body is covered with a layer of cells called the *epidermis.* The ventral surface of the epidermis is frequently ciliated and contains gland cells that produce mucus. Very small flatworms can swim or crawl by using their cilia. Larger flatworms secrete slime trails of mucus over which they glide, using their cilia

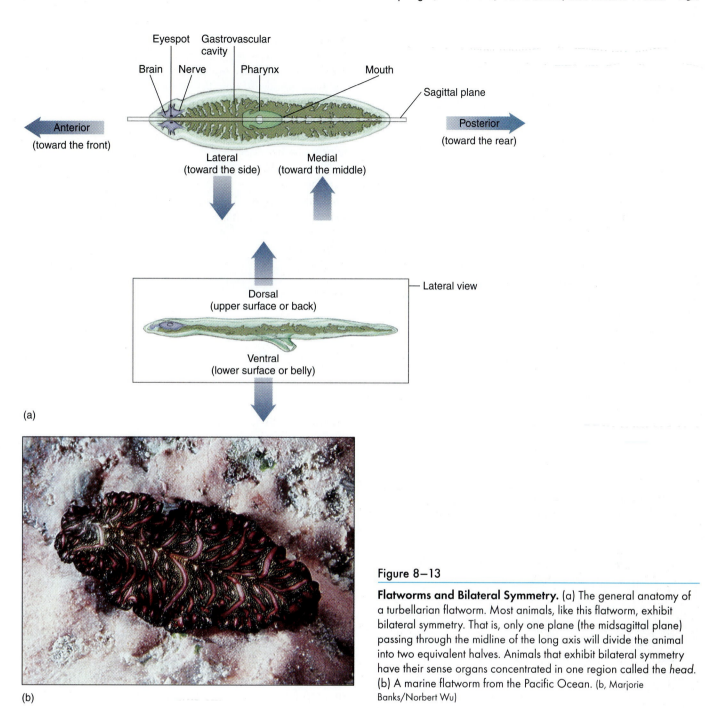

Eyespot   Gastrovascular
               cavity

Brain   Nerve   Pharynx   Mouth

Sagittal plane

◄ Anterior
(toward the front)

Posterior ►
(toward the rear)

Lateral
(toward the side)

Medial
(toward the middle)

Dorsal
(upper surface or back)

Lateral view

Ventral
(lower surface or belly)

(a)

(b)

Figure 8–13

Flatworms and Bilateral Symmetry. (a) The general anatomy of a turbellarian flatworm. Most animals, like this flatworm, exhibit bilateral symmetry. That is, only one plane (the midsagittal plane) passing through the midline of the long axis will divide the animal into two equivalent halves. Animals that exhibit bilateral symmetry have their sense organs concentrated in one region called the *head*. (b) A marine flatworm from the Pacific Ocean. (b, Marjorie Banks/Norbert Wu)

to propel them. This form of locomotion is not efficient for larger forms, which also use muscle contractions to propel themselves.

Most turbellarians feed on detritus or are carnivorous, feeding on small invertebrates that they locate with chemical-detecting organs called *chemoreceptors*. Some species subdue their prey by entangling it in mucus and pinning it against a solid surface until it suffocates. The animal then extends a muscular tube called the *pharynx* (FA-ringks), pumps out enzymes onto the prey, and sucks out body fluids or sucks off pieces of food that are further digested in the gastrovascular cavity.

Excretion of metabolic wastes and gas exchange are accomplished by diffusion through the animal's skin. This is made possible because the thin and flat body allows wastes and gases to diffuse quickly enough to meet the animal's needs. Although this method works well for small animals such as flatworms, it limits their size. Without gills and some means of internal fluid circulation, diffusion cannot occur rapidly enough to support the metabolic needs of a large animal. It is for this reason that free-living flatworms are so small.

Flukes and **tapeworms** are parasitic flatworms. Flukes usually have complex life cycles, frequently involving one or more species as an intermediate host. Tapeworms are parasites that live in the digestive tract of their host. The largest tapeworms are those that infest whales. Individuals of the species *Polygonoporous giganteus* found in the intestines of sperm whales have measured as long as 30 meters (99 feet)!

Nematodes: The Roundworms

Nematodes (phylum Nematoda), also known as *roundworms,* are the most numerous animals on earth. Although the number of species described is only about 12,000, the number of individuals representing those species is staggering, and nematodes can be found in virtually every conceivable habitat. They are crucial scavengers, and many are important parasites of plants, animals, and humans. The great numbers of nematodes that are found in estuaries are sometimes the result of excess detritus accumulation. This is especially true in small harbors where human waste is abundant. Free-living nematodes feed on bacteria, detritus, algae, fungi, and a variety of small invertebrates, including each other. Most nematodes are less than 5 centimeters (2 inches) long, although some parasitic species have been reported that are over 1 meter (3.3 feet) in length.

Marine nematodes are the most abundant members of the **meiofauna** (MY-oh-fawn-uh), the tiny invertebrates adapted to living in the spaces between sediment particles. Their bodies are round, slender, elongated, and tapered at both ends (Figure 8–14). They feed on microorganisms and detritus that are too small for larger animals. The nematodes in turn provide food for larger heterotrophs such as fishes, birds, and large invertebrates. In food chains, free-living nematodes are important in channeling nutrients from microscopic producers and consumers to larger consumers.

Annelids: The Segmented Worms

Annelids (phylum Annelida) are worms whose bodies are divided internally and externally into segments that allow them to be more mobile by enhancing leverage (Figure

Figure 8–14

Marine Nematode. Nematodes are prominent members of the meiofauna. This specimen was taken from bottom sediments in the North Sea. (Robert Brons/Biological Photo Service)

8–15a). Although the most familiar annelids are earthworms, nearly two thirds of the known species are marine. The body wall of annelids contains both longitudinal and circular muscles, allowing them to crawl, swim, and burrow efficiently. The skin of many annelids contains small bristles called **setae** (SEET-ee) that are used for locomotion, anchorage, and protection. The body cavity, or coelom, is fluid filled. The coelom is a closed space and the fluid it contains is under pressure, contributing to the support and shape of the animal's body and locomotion. This arrangement is known as a **hydrostatic skeleton.**

The most common annelids in marine environments are **polychaetes** (PAHL-eh-keets; class Polychaeta). Polychaetes are known by common names such as *sandworms, clamworms, featherduster worms, tube worms,* and *lugworms.* They range in size from 1 millimeter to over 3 meters (10 feet), and some are brightly colored or iridescent. Polychaetes live burrowed in sand and mud, under rocks and corals, in crevices, in shells of other marine organisms, and in tubes they produce themselves. These animals are an important source of food for other marine organisms, both invertebrate and vertebrate.

Many polychaetes are burrowing animals and, like other burrowing organisms, play an important role in the cycling of nutrients. Most oceanic production occurs above the bottom sediments in the photic zone, yet most decomposition occurs on or in the benthos. Over time, many of the nutrients released by the process of decomposition become buried in the bottom sediments. As burrowing organisms tunnel through sediments, many of these nutrients are

(text continues on page 140)

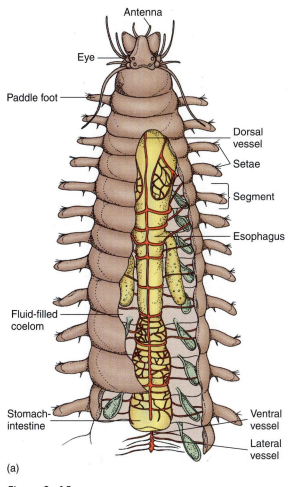

Antenna

Eye

Paddle foot

Dorsal vessel

Setae

Segment

Esophagus

Fluid-filled coelom

Stomach-intestine

Ventral vessel

Lateral vessel

(a)

(b)

(c)

(d)

Figure 8–15

Polychaete Worms. (a) The general anatomy of a polychaete worm. (b) This bristle worm from the Red Sea is a carnivorous polychaete. (c) *Amphitrite* is an example of a deposit-feeder. (d) These featherduster worms from the Caribbean are filter-feeders. (b, Jeffrey Rotman; c, G. I. Bernard/Animals Animals; d, Biological Photo Service)

brought back to the surface, where currents and other natural processes can move them to areas where they can be used by producers to make new organic molecules.

Traditionally, polychaetes are divided into two groups: the **errant polychaetes** that move actively and the **sedentary polychaetes** that are sessile forms. Members of the errant polychaetes have mouths that are equipped with jaws or teeth, and they are active predators (Figure 8–15b). Most of the sedentary polychaetes are filter-feeders or suspension-feeders that use cilia or mucus to trap detritus and plankton. Some errant and many sedentary polychaetes are **deposit-feeders.** These animals feed on the bottom sediments, which include deposit material (organic material that settles to the bottom). The organic material is digested, and the mineral portion is eliminated as castings. These organic nutrients are then channeled to larger organisms when polychaetes are consumed by predators.

Polychaetes can be either direct or indirect deposit-feeders. Direct deposit-feeders (such as earthworms) consume sand and mud directly along with the organic material it contains. Polychaetes such as *Capitella* do this as they ingest the sediments through which they burrow. Indirect deposit-feeders, such as *Amphitrite* and *Terebella,* use special head structures that extend out over the bottom sediments (Figure 8–15c). Deposit material adheres to mucous secretions on the surface of these feeding structures and is then conveyed to the mouth by way of ciliated grooves.

Tube worms are sedentary polychaetes that construct their tubes from a variety of materials. Some species sort sand grains from the food particles that they trap and use the sand to enlarge their tubes. Some tubes are made of protein and have a consistency similar to stiff paper. Others are solid tubes composed of calcium carbonate, while still others are built from stones, coral, or sand granules that are held together by mucus. Featherduster worms (Sabellidae and Serpulidae) live in tubes that they form in or on the bottom sediments or other solid surface. From the opening of the tube emerges a ring of tentacles, each of which is very delicate and frequently brightly colored (Figure 8–15d). They are ciliated and are used for straining food from seawater as well as providing a surface for gas exchange. The tentacles are equipped with light-sensitive organs, and when the worm is disturbed or when a shadow falls on it, it quickly withdraws its tentacles into its tube and out of sight.

Errant polychaetes usually reside under rocks or coral or in mucus-lined burrows in the sand. They tend to be most active at night, when they come out of hiding to search for small invertebrates that they capture with their jaws. Polychaetes in the genus *Hermodice* are commonly called *fireworms.* Their setae are hollow and contain a poisonous secretion. On contact, the setae break off in the victim's skin, releasing a toxin that causes a severe burning sensation in the area of the injury. These animals generally feed on a variety of cnidarians.

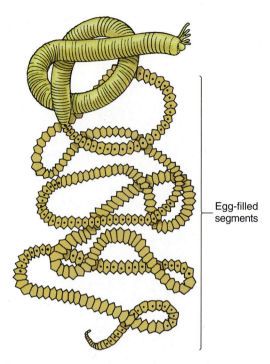

Egg-filled segments

Figure 8–16

Epitoky. The epitoke (pelagic reproductive form) of the Samoan palolo worm. In this species, the anterior part of the worm remains unchanged, while the posterior portion grows to form a chain of egg-filled segments.

Some errant polychaetes exhibit an interesting reproductive phenomenon known as *epitoky.* **Epitoky** (EP-i-toh-kee) is the formation of a pelagic reproductive individual, or epitoke, that is different from the nonreproducing form of the worm. The epitoke is usually most modified in the posterior segments that contain reproductive structures (Figure 8–16).

In the palolo worm, *Palola viridis,* an errant polychaete found in the waters of the South Pacific around the islands of Samoa, the posterior segments increase in number in the epitoke and are filled with either eggs or sperm. The males and females come to the surface in large numbers at night to reproduce, a phenomenon known as *swarming.* Swarming occurs only at specific times of the year and appears to be related to the lunar cycle and tides.

During swarming, the female epitoke releases a substance called *fertilizin* that stimulates the male to shed his sperm. The sperm, in turn, stimulate the females to release their eggs, and fertilization occurs externally. The large numbers of worms reproducing during swarming help to ensure the species' chances for survival and dispersal of the young. The epitokes also provide food for many fish, birds, and some indigenous peoples of tropical islands.

(a)

(c)

(b)

Figure 8–17

Other Marine Worms. (a) A ribbon worm from the Pacific coast of Panama. (b) A peanut worm (sipunculid) from the Great Barrier Reef of Australia. (c) An acorn worm from the Atlantic coast of North America. (a, Kjell B. Sandved; b, Animals Animals Copyright 1991 Kathie Atkinson/Oxford Scientific Films; c, C. R. Wyttenbach, University of Kansas/BPS)

Other Marine Worms

Many other wormlike animals inhabit the marine environment, such as ribbon worms, horsehair worms, spiny-headed worms, peanut worms, beardworms, and acorn worms. They differ from those previously discussed in certain anatomical and developmental respects, and as such they are classified into different phyla. The representatives of these phyla are not as numerous as the others discussed in this chapter, but each fills a particular niche in the overall marine ecosystem.

Ribbon worms (phylum Nemertea) are named because of their ribbon-like bodies (Figure 8–17a). Some ribbon worms are found coiled under stones at low tide or buried in bottom sediments. Others live in empty mollusc

shells, in seaweed, or swim near the surface. Ribbon worms are carnivorous and feed on a variety of organisms. Prey are captured by a tubelike structure called a *proboscis* that can extend from the mouth. The proboscis is used to ensnare prey in mucus or to wrap around the prey, preventing its escape. Some species have a sharp, pointed structure called a *stylet* at the tip of the proboscis. The stylet carries a toxin that may aid in preventing prey from escaping.

Horsehair worms (phylum Nematomorpha) received their name because of their resemblance to the hairs of a horse's tail. As adults, these animals are free-living, but the juveniles are parasites of arthropods such as deep-sea shrimp. **Spiny-headed worms** (phylum Acanthocephala) are parasites, mostly of fish, birds, and mammals. They get their name from their cylindrical proboscis that has several rows of spines. The proboscis is used to penetrate the intestine of the host, and the damage can be quite traumatic. Spiny-headed worms are frequently troublesome parasites in seals.

Sipunculids (phylum Sipuncula; Figure 8–17b) are solitary marine worms that live in burrows in mud or sand, in empty mollusc shells, or in coral crevices. These animals have an extensible proboscis with a ring of tentacles. Organic material is trapped in mucus that coats the tentacles, and the food is then drawn into the mouth by ciliary action. When disturbed, these animals will contract their bodies so that they resemble a peanut kernel, which gives rise to a common name for some of these species, *peanut worms.*

Beardworms (phylum Pogonophora) live in tubes that are buried in the ocean bottom at depths of 200 meters (656 feet) or more. The body of these animals is cylindrical with a ring of tentacles around the anterior end. Beardworms are unusual in that they lack a mouth and a digestive tract. Evidence suggests that they derive nutrition by actively transporting dissolved nutrients from the seawater into their tentacles. Large beardworms (*Riftia*) of up to 3 meters (10 feet) in length and 5 to 8 centimeters (2 to 3 inches) in diameter are found exclusively in deepwater vent communities. These animals obtain all of their nourishment from large numbers of chemosynthetic bacteria that live in their tissues (see Chapter 18).

Acorn worms (phylum Hemichordata) are sessile bottom dwellers (Figure 8–17c). They live burrowed in the sediments of intertidal mud or sand flats or under stones. An acorn worm uses its large proboscis to collect food from its surroundings. The food is trapped in mucus and then transported by cilia, located in grooves, to the collar and then into the mouth, where it is swallowed. Some species use their proboscis to dig burrows, which usually have two openings. The head extends from one opening for feeding, and the anus deposits a characteristic mound of fecal material around the second opening. As with other burrowing organisms, the activity of acorn worms helps to make organic nutrients more accessible to other organisms.

CHAPTER SUMMARY

Sponges depend on their ability to filter large amounts of water through their bodies in order to survive. Their bodies are asymmetrical and contain several cell types that perform specific functions. Sponges provide habitats for many organisms and play a role in recycling calcium.

Cnidarians and ctenophores exhibit radial symmetry. Cnidarians have evolved a highly specialized stinging cell that they use for capturing prey and for defense. Ctenophores lack the stinging cell of cnidarians and move by rows of cilia called *comb plates.*

Marine worms display bilateral symmetry and cephalization. Flatworms are represented by both free-living and parasitic forms. Turbellarians are free-living flatworms. Flukes and tapeworms are parasitic flatworms. Nematodes are the most numerous of marine worms. They can be free-living or parasites and are important members of the meiofauna. Annelids are worms whose body is divided internally and externally into segments, allowing them to be more mobile. The most abundant annelid worms in the marine environment are polychaetes. Polychaetes may be either active (errant polychaetes) or sessile (sedentary polychaetes). Some polychaetes are carnivores, while others are filter-feeders, suspension-feeders, or deposit-feeders. Polychaetes are important sources of food for other organisms, and burrowing polychaetes play an important role in recycling nutrients.

Other wormlike invertebrates include ribbon worms, sipunculids, beardworms, and acorn worms. Each of these free-living worms contributes to the flow of energy in marine ecosystems. Horsehair worms and spiny-headed worms have members that are important marine parasites.

SELECTED KEY TERMS

invertebrate, *p. 126*	polyp, *p. 131*	ctenophore, *p. 136*	nematode, *p. 138*
sponge, *p. 127*	medusa, *p. 131*	bilateral symmetry, *p. 136*	meiofauna, *p. 138*
suspension-feeder, *p. 128*	nematocyst, *p. 131*	cephalization, *p. 136*	annelid, *p. 138*
filter-feeder, *p. 128*	hydrozoan, *p. 132*	turbellarian	polychaete, *p. 138*
spicule, *p. 129*	scyphozoan, *p. 132*	flatworm, *p. 136*	deposit-feeder, *p. 140*
stinging cell, *p. 131*	anthozoan, *p. 134*	fluke, *p. 138*	
radial symmetry, *p. 131*	sea anemone, *p. 134*	tapeworm, *p. 138*	

QUESTIONS FOR REVIEW

MULTIPLE CHOICE

1. Each of the following statements concerning sponges is true except one. Identify the exception.
   a. Sponges are living animals.
   b. Sponges have asymmetric bodies.
   c. Sponges reproduce only sexually.
   d. Spicules help to give a sponge support.
   e. The size of a sponge is related to the amount of water it can circulate through its body.

2. If a sponge lacked collar cells it would not be able to
   a. form spicules
   b. produce colored pigments
   c. feed
   d. exchange respiratory gases
   e. protect itself

3. Jellyfish use their nematocysts to
   a. capture prey
   b. digest food
   c. coordinate swimming movements
   d. circulate water through their bodies
   e. all of the above

4. Organs known as *comb plates* are found on animals known as
   a. flatworms
   b. nematodes
   c. cnidarians
   d. ctenophores
   e. annelids

5. Which of the following animals exhibits radial symmetry as an adult?
   a. sponge
   b. flatworm
   c. annelid
   d. nematode
   e. jellyfish

6. Worms with segmented bodies are called
   a. flatworms
   b. nematodes
   c. annelids
   d. acorn worms
   e. ribbon worms

7. An example of a parasitic flatworm would be a
   a. polychaete
   b. nematode
   c. planarian
   d. fluke
   e. ribbon worm

8. Most sedentary polychaetes are
   a. herbivores
   b. carnivores
   c. filter-feeders
   d. parasites
   e. symbionts

9. Nematodes play an important role in
   a. recycling calcium in marine environments
   b. removing carbon dioxide from seawater
   c. producing oxygen
   d. recycling organic matter
   e. controlling marine parasites

10. Tiny invertebrates that live in the spaces between sediment particles are collectively called
    a. polyps
    b. plankton
    c. medusae
    d. polychaetes
    e. meiofauna

SHORT ANSWER

1. What role do boring sponges play in marine environments?

2. What are the advantages of bilateral symmetry?

3. What important ecological contribution do burrowing organisms make to the environment?

4. What is a hydrostatic skeleton?

5. Describe the ecological role of meiofauna.

6. Describe how sponges feed and reproduce.

7. Describe how the cnidarian stinging cell (cnidocyte) functions.

8. Why do the largest sponges always exhibit a leuconoid body plan instead of an asconoid body plan?

9. Why is radial symmetry advantageous to a sessile organism?

10. Distinguish between hydrozoans and scyphozoans.

THINKING CRITICALLY

1. Development of beachfront property frequently increases the amount of sediment in coastal water. How would you expect this to affect sponges and hydrozoans that are living in the shallow water close to shore?

2. Based on the information in this chapter, construct a food web that includes meiofauna, marine worms, and larger predators such as fishes.

3. Explain with examples how symbiotic relationships can allow marine animals to live in habitats where they normally could not survive.

SUGGESTIONS FOR FURTHER READING

Brownlee, S. 1987. Jellyfish Aren't Out to Get Us, *Discover,* August, 42–54.

Dorit, R. L., W. F. Walker, Jr., and R. D. Barnes. 1991. *Zoology.* Saunders College Publishing, Philadelphia.

Hammer, W. M. 1994. Australia's Box Jellyfish: A Killer Down Under, *National Geographic* 186(2):116–130.

Rudloe, J. and A. Rudloe. 1991. Jellyfishes Do More with Less Than Almost Anything Else, *Smithsonian* 21(11): 100–111.

Ruppert, E. E. and R. D. Barnes. 1994. *Invertebrate Zoology,* 6th ed. Saunders College Publishing, Philadelphia.

Chapter 9

Molluscs, Arthropods, Lophophorates, Echinoderms, and Tunicates

KEY CONCEPTS

1. Molluscs have soft bodies that are usually covered by a shell.

2. Molluscs are important herbivores and carnivores in the marine environment.

3. Arthropods have external skeletons, jointed appendages, and sophisticated sense organs.

4. Crustaceans make up a majority of the zooplankton that are a major link between phytoplankton and higher order consumers in oceanic food webs.

5. Echinoderms exhibit radial symmetry as adults.

6. Echinoderms have internal skeletons and a unique water vascular system that functions in locomotion, food gathering, and circulation.

7. Lophophorates and tunicates are important filter-feeders.

T he invertebrate animals that will be introduced in this chapter are among the most familiar creatures of sea and shore. If you have ever picked up a seashell or sea star or watched a crab scurry across the beach, you have encountered some of these animals. Not only are these animals very familiar, many are commercially important. Scallops, oysters, mussels, crabs, lobsters, and shrimp are only some of the invertebrate animals that provide food for large numbers of people. Many of these animals are also threatened as a result of human activities. In this chapter, you will learn more about these animals and the roles they play in marine environments.

MOLLUSCS

Molluscs (phylum Mollusca) are one of the largest and most successful groups of animals. The term *Mollusca* (Latin for "soft") refers to the animals' soft bodies, which in most cases are covered by a shell made of calcium carbonate. Molluscs are represented by familiar animals such as chitons, snails, clams, octopods, and squids. They range in size from microscopic snails to the giant squid, *Architeuthis*. Molluscs exhibit a variety of feeding types, including herbivory, carnivory, filter-feeding, detritus-feeding (detritivores), deposit-feeding, and parasitism. They can be found in most marine habitats and exhibit a variety of lifestyles. Some molluscs, such as oysters, scallops, and

Figure 9–1

Generalized Molluscan Body Plan. The generalized molluscan body consists of a head-foot composed of the animal's head and muscular foot and a visceral mass containing the animal's other organs.

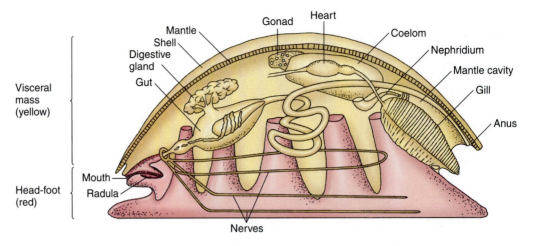

conchs, are important food animals for humans, while others, such as shipworms, cause commercial damage.

The generalized molluscan body plan consists of two major parts: the head-foot and the visceral mass (Figure 9–1). As the name implies, the head-foot region contains the head with its mouth and sensory organs and the foot, which is the animal's organ of locomotion. The dorsal visceral mass contains the other organ systems, including the circulatory, digestive, respiratory, excretory, and reproductive systems.

The soft parts of a mollusc are covered by a protective tissue called the **mantle.** The mantle extends from the visceral mass and hangs down on each side of the body, forming a space between the mantle and the body known as the *mantle cavity.* The mantle is responsible for forming the animal's shell, in those species that have one. Molluscs such as squids and octopods use the mantle in locomotion, and those molluscan species without gills also use the mantle for gas exchange.

A structure unique to molluscs is the **radula** (Figure 9–2). The radula is a ribbon of tissue that contains teeth and

is found in all molluscs except bivalves. Depending on the species, the radula is adapted for scraping, piercing, tearing, or cutting pieces of food, such as algae or flesh, which are then moved into the digestive tract. As the anterior portions of the radula are worn out, new teeth are produced at the posterior end. The characteristics of the radula are important in the classification of molluscan species.

The molluscan shell is secreted by the mantle and is normally composed of three layers (Figure 9–3). The outermost layer, the **periostracum** (per-ee-AHS-treh-kuhm), is composed of a protein called *conchiolin* that protects the shell from dissolution and boring organisms. The middle

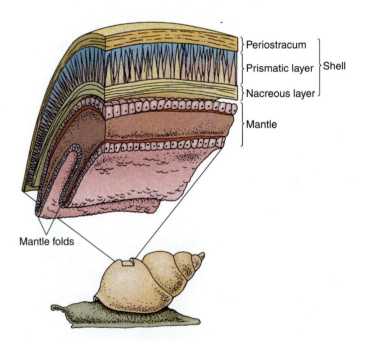

Figure 9–3

The Mollusc Shell. The molluscan shell is produced by the animal's mantle and is composed of three layers: the periostracum, prismatic layer, and nacreous layer.

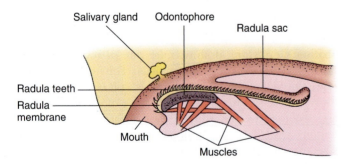

Figure 9–2

The Radula. The radula is a ribbon of teeth that is unique to molluscs. It is formed in the radula sac and supported by a mass of muscle and cartilage called the *odontophore.* Muscles attached to the odontophore move the radula in and out of the animal's mouth.

layer, the **prismatic layer,** is composed of calcium carbonate and protein and makes up the bulk of the shell. The innermost **nacreous layer** is also composed of calcium carbonate, but its crystal structure is different from that found in the prismatic layer, and it is formed in thin sheets.

As the animal grows, new periostracum and prismatic layers form at the margin of the mantle. The nacreous layer is secreted continuously and accounts for the increased thickness of the shell in older animals. The iridescent nature of the nacreous layer is due to the arrangement of the calcium carbonate crystals. The thin sheets act as a prism, refracting light and producing an iridescent effect. The nacreous layer of oysters is also known as the *mother-of-pearl* layer. Pearls are formed when the nacreous material is layered over irritating particles, such as sand grains, beneath the animal's mantle.

Chitons

Chitons (KY-tuhnz; class Polyplacophora) have flattened bodies that are most often covered by eight shell plates (Figure 9–4). These plates are held together by a tough girdle that is formed from the mantle, and in some animals the plates are hidden below in the mantle tissue. Chitons are usually found attached to intertidal rocks, where they feed on algae that they rasp off the surface with their radulae. The animal's large flat foot allows it to attach tightly to rocks, and this, combined with its low body profile, helps it survive wave shock and prevents it from becoming dislodged by wave action. When chitons are removed from a rock, they roll up into a ball for protection. Chitons that live

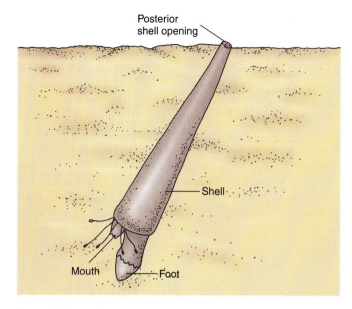

Figure 9–5

Scaphopods. Scaphopods, or tusk shells, use their foot to bury into bottom sediments. They feed primarily on foraminiferans.

on intertidal rocks are periodically exposed to air. When this occurs, they pull their bodies tightly against the surface of the rock with the muscular foot, sealing out the air and preventing their gills from losing moisture and drying out. Some species can breathe air for short periods of time until the tide returns to submerge them.

Scaphopods

Scaphopods (SKA-foh-pahdz; class Scaphopoda; *scapho* meaning "sheath" and *poda* meaning "foot") are commonly called *tusk shells.* They derive their name from the shape of their shell, which resembles an elephant's tusk (Figure 9–5). The tubelike shell is open at both ends, and the animal's foot, which is used for burrowing, protrudes from the larger end. Water enters and exits at the smaller end, bringing in oxygen and removing wastes. They are found buried in bottom sediments from the intertidal zone to several thousand meters deep, where they feed primarily on foraminiferans. Scaphopods use their foot or special tentacles that emerge from the head to capture their prey.

Gastropods

Gastropods (class Gastropoda; *gastro* meaning "stomach" and *poda* meaning "foot") exhibit a tremendous amount of diversity. Most have shells, but some, such as nudibranchs, have lost them. Some are so small that they are microscopic,

Figure 9–4

Chitons. Chitons, like these from the Caribbean, are usually found on intertidal rocks, where they feed on algae. Their shells are composed of eight separate plates held together by a leathery girdle. (Ray Coleman/Photo Researchers, Inc.)

while others, like the marine snail, *Syrinx aruanus,* reach lengths approaching 1 meter (3.3 feet).

Gastropods exhibit a wide variety of feeding styles. Some are herbivores, such as limpets and cowries that scrape algae from rocks, or conchs that graze on seagrass (Figure 9–6a). Some are scavengers, such as members of the genus *Ilyanassa,* that feed on dead and decaying material. Others are carnivores. Tulip snails (*Fasciolaria*), for instance, feed on a variety of clams by pulling the clams' shells apart and then thrusting in their mouths to devour the prey. Oyster drills of the genus *Urosalpinx* can devastate entire oyster beds, boring holes in the shells of their prey and sucking out the contents. The flamingo tongue (*Cyphoma*) and its relatives feed on colonial cnidarians, and large snails such as Triton's trumpet (*Charonia*) and helmet snails (*Cassis*) feed on echinoderms.

Snails in the genus *Conus* have a radula shaped like a harpoon (Figure 9–6b). The radula is coated with a toxin that is produced in a poison gland located near the mouth. These snails feed on worms, fish, and other molluscs, paralyzing their prey with a venomous sting. Six of the approximately 400 species of *Conus* are known to have caused human fatalities.

Some gastropods feed on microorganisms, while others feed on organic material in sand or mud. Slipper limpets (*Crepidula*) feed by ciliary action, using their gills to trap

(a)

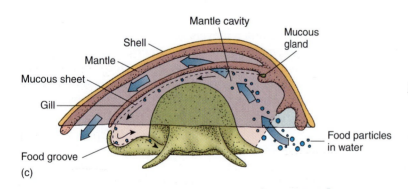

Mantle cavity

Shell

Mantle

Mucous sheet

Gill

Food groove

Mucous gland

Food particles in water

(c)

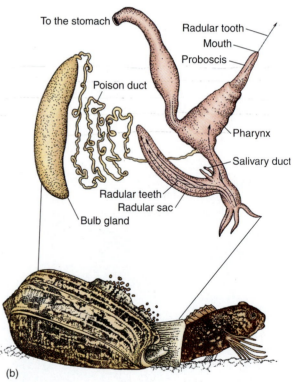

To the stomach

Radular tooth

Mouth

Proboscis

Poison duct

Pharynx

Salivary duct

Radular teeth

Radular sac

Bulb gland

(b)

Figure 9–6

Gastropod Feeding Strategies. (a) Cowries are herbivores that graze on algae. (b) Cone snails use their harpoon-like radula to inject a paralyzing toxin into their prey. (c) Slipper limpets use their gills to filter food from the water. (a, Biological Photo Service)

MARINE BIOLOGY AND THE HUMAN CONNECTION

Deadly Snails and Medicine

The marine snail *Conus geographus*, or geography cone, is one of nature's most sophisticated predators (Figure 9–A). It has a long, brightly colored proboscis that it uses to attract the attention of fish. A fish apparently mistakes the proboscis for an edible worm. When the fish moves in to bite, the snail fires its poisonous radula into the fish's mouth. The radula penetrates the soft tissues in the mouth and paralyzes the fish. After its prey has succumbed to the venom, the snail begins the leisurely process of digesting its meal.

To date, at least 20 unwary swimmers in the South Pacific region have been killed by this animal. Humans are usually stung when they come into contact with the snail while swimming or when they pick it up to admire its colorful shell. Since humans are much larger than the fishes the snail usually preys upon, it takes longer for the venom to move through their body. Death, usually due to respiratory paralysis, occurs within one hour of the sting.

In studying the mode of action by which the venom paralyzes its prey, re-searchers have discovered chemicals that can be used to probe the mysteries of the human nervous system, especially the relationship between nerves and

Figure 9–A
The Geography Cone Snail. *The geography cone* (Conus geographus) *from the Indian and Pacific Oceans is one of the six species of cone snails known to cause human fatalities. The toxin produced by this animal causes death by paralyzing the victim's respiratory muscles.* (W. Gregory Brown/Animals Animals)

muscles. Researchers have isolated three different toxins from the snail's venom, one of which is quite similar to the neuro-toxin that is produced by cobras. Investigators have found that some components of the venom interfere with the action of nerve cells, and others interfere with the movement of stimuli from nerve cells to muscles. The toxins are highly specific and bind only to certain receptors on cell surfaces. Scientists can use these tox-ins to locate the receptors in different cells and to study the role receptor mole-cules play in normal and abnormal func-tion of nerves and muscles. By studying the effects of the toxin on nerves and muscles, researchers hope to gain new insight into how some neuromuscular dis-eases, such as myasthenia gravis (a pro-gressive disease that causes muscle weakness in the face, neck, and chest) and muscular dystrophy (a group of dis-eases that causes progressive wasting of the muscles), produce their effects.

---

food particles that are then formed into a mucous ball and carried to the mouth for digestion (Figure 9–6c). Plank-tonic gastropods, such as sea butterflies (*Gleba* and *Corolla*), form a mucous net that they use to catch the small organisms on which they feed.

Gastropods can be found in a wide variety of habitats from the littoral zone (shore) to the ocean's bottom, and even pelagic (open ocean) species are known. When a shell is present, it is always one-piece (univalve), and it may be coiled (as in snails) or uncoiled (as in limpets; Figure 9–7). As the animal grows, the whorls or turns of the shell in-crease in size around a central axis. The most common coil-ing pattern is clockwise (*dextral,* meaning "to the right"), although some species normally coil counterclockwise (*sinistral,* meaning "to the left").

The weight of the shell makes these animals slow movers; the shell is a major means of defense, along with protective coloration and secretive behaviors. Many snails can withdraw into the safety of their shells and close the opening, or aperture, with a cover called the **operculum.** The operculum can be horny or calcareous in composition, and it is not found in all species of snail.

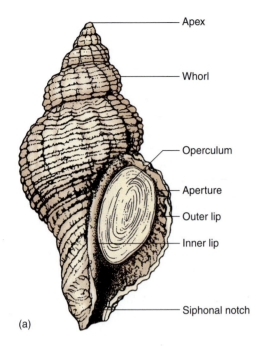

Apex

Whorl

Operculum

Aperture

Outer lip

Inner lip

Siphonal notch

(a)

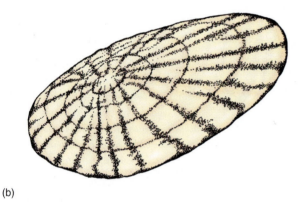

(b)

Figure 9–7

Gastropod Shells. Gastropod shells may be coiled (a) as in this oyster drill *Urosalpinx cinerea* or uncoiled (b) as in the limpet *Acmaea.* Each turn of a coiled shell is called a *whorl,* and the whorls wrap around a central axis. Many species have an operculum that is used to close the opening or aperture when the animal withdraws within its shell.

Nudibranchs (NOO-di-brangks) are marine gastropods that lack a shell (Figure 9–8). They are usually brightly colored and sometimes have bizarre shapes due to their gills and appendages. The name *nudibranch* means "naked gill" and refers to the projections from the animal's body called **cerata** (sir-AH-tuh) that increase the surface area available for gas exchange.

Some nudibranchs feed on cnidarians. Instead of digesting their prey's stinging cells, the nudibranch moves them intact by means of ciliated tracts in its digestive system. These tracts extend into the cerata. The stinging cells are

deposited in the tips of the cerata, where they are used by the nudibranch for its own defense. The bright colors of many species act as warning coloration, advertising that the animals are poisonous or distasteful. Potential predators recognize the bright colors or patterns and avoid them.

Most gastropods have separate sexes, although some are hermaphroditic. In some species, sperm and eggs are swept by water currents from the mantle cavity, and fertilization occurs externally. Most species, however, exhibit internal fertilization. Males of these species use a tentacle-like penis to deposit sperm in or near the female genital opening. Females that exhibit internal fertilization usually lay their eggs surrounded by a jelly-like substance or in protective egg cases.

In primitive molluscs that shed their eggs directly into the sea, a free-swimming larval stage, called the **trochophore** (TROHK-eh-fohr) **larva,** hatches from the egg. More characteristic, however, of marine gastropods is a free-swimming larval stage called the **veliger** (Figure 9–9). Veligers can sometimes travel great distances on currents, allowing the otherwise slow-moving gastropods to disperse to other suitable habitats. In some species of marine snails, there is no free-swimming larva, and a juvenile that resembles the adult hatches from the egg.

One of the more interesting reproductive strategies occurs in slipper limpets of the genus *Crepidula.* These more or less sessile animals are hermaphroditic and tend to congregate in groups in which several individuals are stacked on top of each other (Figure 9–10). The right margin of the shells are next to each other, allowing the penis of the individual on top to reach the female opening of the individual

Figure 9–8

Nudibranchs. Nudibranchs, like this one from California, are gastropods that lack a shell. The projections from the animal's body are called *cerata* and function in gas exchange. (David Wrobel 1993/Biological Photo Service)

Figure 9—9

Molluscan Larvae. The veliger larvae of a slipper limpet. (Robert Brons/Biological Photo Service)

Figure 9—10

Crepidula fornicata. Slipper limpets are normally found in groups such as this. The male is on the top, and the next two individuals are in transition from male to female. When the transition is complete, all of the females will be fertilized by the male at the top.

below. The young of these species are always male. This initial male phase is then followed by a transition period during which the male reproductive tract degenerates, and the animal either develops into a female or another male depending upon the sex ratio of the group. If males are in short supply, the animals will develop into males. On the other hand, if females are limited, the animals will develop into females. The sex of the individual is probably controlled by hormones called *pheromones* that are released into the environment by members of the group to control the development and behavior of other members. Older males remain males longer if they remain attached to a female. If an older male is removed from the female or isolated, it will develop into a female. Once an individual becomes a female, it cannot return to being a male.

Bivalves

Bivalves (class Bivalvia, also known as class Pelecypoda; *pelecy,* meaning "hatchet" and *poda,* meaning "foot") are molluscs that have shells divided into two jointed halves called **valves.** This group includes clams, oysters, mussels, scallops, and shipworms. They range in size from tiny, 1-millimeter clams to the giant clam *Tridacna gigas* that can grow as large as 1 meter (3.3 feet) and weigh 230 kilograms (507 pounds) or more.

Bivalves have no head and no radula. Most are sedentary filter feeders, depending on the ciliary action of the gills to bring in food. Their bodies are laterally compressed, and the two halves of the shell that cover it are usually attached dorsally at a hinge by ligaments (Figure 9–11). The

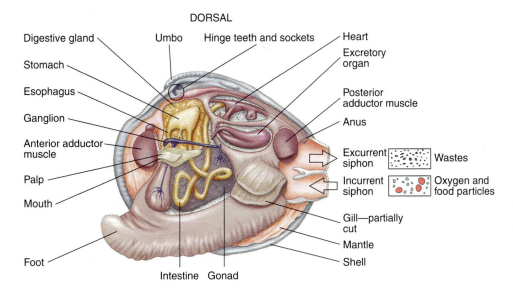

DORSAL

Digestive gland — Umbo — Hinge teeth and sockets — Heart

Stomach — Excretory organ

Esophagus — Posterior adductor muscle

Ganglion — Anus

Anterior adductor muscle —

Palp — Excurrent siphon :::: Wastes

Mouth — Incurrent siphon :::: Oxygen and food particles

Gill—partially cut

Mantle

Foot — Shell

Intestine   Gonad

Figure 9—11

Bivalve Anatomy. The internal anatomy of a typical marine clam, *Mercenaria mercenaria.* Notice that food and oxygen enter through the incurrent siphon while wastes are eliminated by way of the excurrent siphon.

oldest part of the shell is the area around the hinge called the *umbo,* and growth generally occurs in a concentric fashion around it at the edge of the shell. Large muscles called **adductor muscles** open and close the two valves. When the muscles relax, the elasticity of the ligaments and the weight of the valves cause them to open.

The animal's foot, which is somewhat hatchet-shaped, is located ventrally and, in clams, functions in burrowing and locomotion. In many species, the mantle frequently forms tubular structures called **siphons.** Cilia on the surface of the animal's gills provide the force that moves water through the animal's body, and the siphons direct the water flow into and out of the mantle cavity. The siphons in some marine species can be quite long, allowing the animal to remain burrowed deep in the bottom-sediments. For example, the geoduck (GOO-ee-duhk; *Panopea generosa*) from the Pacific coast can burrow as deep as 1 meter (3.3 feet) into the mud and can have a siphon over 1 meter in length. The enormous siphon can account for as much as one half of this animal's 2.75-kilogram (6-pound) body.

Scallops and their relatives are able to move by a type of jet propulsion. The animal contracts its adductor muscle, causing the two valves to close forcefully and emit two streams of water backwards and to each side. This causes the animal to move forward in a series of jerky movements and helps it avoid one of its major predators, the sea star.

Some bivalves are capable of burrowing into wood or stone and can cause a great deal of commercial damage. Members of the genera *Teredo* and *Bankia* are known as *shipworms* and can cause damage to wooden ships and wharves. These bivalves resemble tiny worms with a shell, and the valves bear microscopic teeth that help the animal

burrow (Figure 9–12). Pieces of wood that are removed by the burrowing action are swallowed and digested with the aid of enzymes produced by symbiotic bacteria. These bacteria live in a special organ located in the animal's digestive system. The bacteria also fix nitrogen, which helps their hosts compensate for their low-protein diet. Since few marine organisms are able to break down wood, wood-boring bivalves play an important ecological role. Their feeding strategy helps to prevent the accumulation of dead wood, such as driftwood, in the marine environment.

Another group of boring clams belongs to the genus *Pholas.* These bivalves can burrow into soft rocks such as limestone and play a role in recycling calcium.

Cephalopods

Cephalopods (SEF-uh-loh-pahdz; class Cephalopoda; *cephalo,* meaning "head" and, *poda,* meaning "foot") are represented by familiar animals such as the octopus, squid, and cuttlefish. The term *cephalopod* means "head-footed" and refers to the animal's foot, which is modified into a headlike structure. A ring of tentacles emerges from the anterior edge of the head, and they are used in capture of prey, for defense, in reproduction, and in some they aid in locomotion.

Cephalopods are the most structurally complex molluscs and are among the most advanced of all invertebrates as well. They range in size from 2 centimeters (.75 inch) to over 18 meters (59 feet). Early cephalopods, known as *ammonites,* had bodies that were covered by straight (uncoiled) shells. Later some species evolved coiled shells sim-

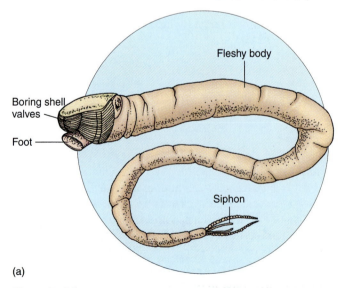

(a)

(b)

Figure 9–12

Destructive Bivalves. (a) Shipworms are bivalves with long wormlike bodies and small shells that are used to bore through wood. (b) Damage caused by shipworms. (b, William Jorgensen/Visuals Unlimited)

Labels in figure (a): Fleshy body, Boring shell valves, Foot, Siphon

(a)

Figure 9–13

Shelled Cephalopods. (a) This chambered nautilus from the tropical Pacific Ocean has changed very little since it first evolved. (b) The nautilus occupies only the last chamber of its shell. The other chambers are filled with gases from the blood, such as carbon dioxide. The animal can regulate the gas content of the chambers with its siphuncle, allowing it to rise toward the surface, sink to the bottom, or maintain neutral buoyancy as necessary. (a, Animals Animals Copyright 1991 W. Gregory Brown)

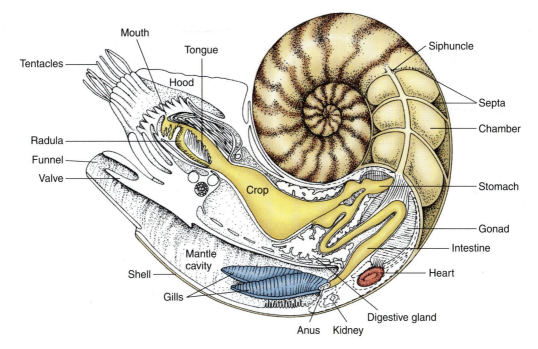

(b)

ilar to those of the chambered nautilus (*Nautilus pompilius*). With the exception of the nautiloids, modern cephalopods either have small internal shells, such as some squids, or lack shells entirely, such as the octopods.

There are two major groups of living cephalopods. One group, the nautiloids, is represented by the genus *Nautilus* (Figure 9–13a). These animals produce large, coiled shells composed of chambers. The chambers are filled with gas and aid in maintaining buoyancy for swimming. The animal inhabits only the last chamber but remains connected to the others by means of a cord of tissue called the **siphuncle**

(SY-fuhn-kuhl; Figure 9–13b). The siphuncle removes the seawater from each new chamber as it forms. The living tissue of the siphuncle removes salts from the seawater in the new chamber by active transport. The water that remains is then more dilute than the animal's body fluids and readily moves by osmosis into the animal's body. The excess water is ultimately eliminated by the excretory system. The gases that replace the water in the chambers are produced by the respiration of the siphuncle's living tissue. The amount of gas in the chambers can be regulated, allowing the animal to rise or sink in the water column. The animal's head has

(a)

(b)

Figure 9–14

Cephalopods. (a) The cuttlefish resembles a squid and has a small internal shell. (b) Squids have streamlined bodies and are active swimmers. The part of the animal's body that looks like a head is actually its foot. (c) The internal anatomy of a male squid. (a, Biological Photo Service; b, Jeffrey Rotman)

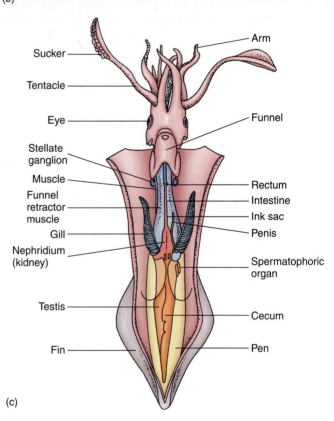

(c)

60 to 90 tentacles. Although they do not have suckers, they are coated with a sticky substance that helps them to adhere to what they touch.

Like other cephalopods, nautiloids move by means of jet propulsion. Water is drawn in through an incurrent siphon and forced out through an excurrent siphon, pushing the animal shell-first through the water.

Nautiloids tend to come to the surface at night and dwell on the bottom during the day. Originally it was thought that they came to the surface at night to feed, but recent evidence indicates that they prey on hermit crabs and scavenge other food on the bottom.

The other group of living cephalopods includes squids, octopods, and cuttlefish. Cuttlefish (Figure 9–14a) have a bulky body, fins, and ten appendages. The eight short, heavy appendages are called *arms,* and the two larger appendages are called *tentacles.* They have small, internal shells, some of which have chambers like those of the nautilus shell. The shells of cuttlefish, however, are embedded in the mantle.

Squids (Figure 9–14b) have large cylindrical bodies with a pair of fins derived from mantle tissue. They have eight arms, two tentacles, and an internal strip of hard protein that lends support to the mantle, called the **pen,** which represents a degenerate shell (Figure 9–14c). Octopods have eight arms and no tentacles, and their bodies are saclike and lack fins. The arms and tentacles of this group of cephalopods bear suckers. The suckers of squids have a rim of toothlike structures that are not present in the suckers of octopods. All of these animals are extremely adept predators, feeding on a variety of fish and invertebrates.

Like nautiloids, squids, octopods, and their relatives can swim by jet propulsion produced by forcing water through a ventrally located siphon. Most slow swimming in squid species, however, is achieved by fin undulation. Squids and cuttlefish are very good swimmers, and their bodies are streamlined for efficient movement. Although octopods can swim, they are better adapted to crawling over the bottom.

Cephalopods have the most advanced and complex nervous system of any invertebrate. Octopods can be trained to perform simple tasks and are used as model systems in the study of neural development associated with learning and memory. Except for the nautiloids, cephalopods have highly developed eyes and can recognize shapes and colors. Octopods seem to have a highly developed tactile sense and can discriminate objects on the basis of touch.

Cephalopods can communicate with each other through movements of their arms, bodies, and color changes. Color changes involve special cells in the skin called **chromatophores** that contain pigment granules. When the granules are dispersed, the color of the animal's skin becomes darker, and when the granules are concentrated, the color is lighter due to the permanent light background color (Figure 9–15). Not only can the general body color change, but the animal can produce stripes and patterns that communicate information to other members of

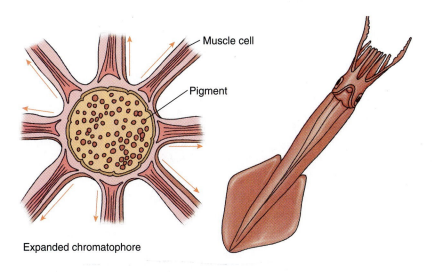

Muscle cell

Pigment

Expanded chromatophore

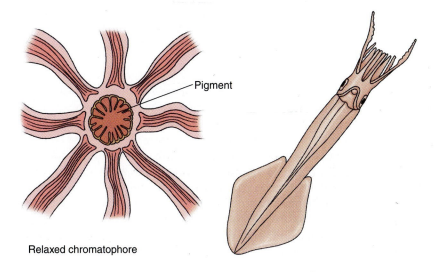

Pigment

Relaxed chromatophore

Figure 9–15

How Cephalopods Change Color. Tiny muscle cells pull at the edges of the chromatophore, causing the cell to expand and the pigment to disperse, producing a darker color. When the muscle cells relax, the chromatophore returns to its original size, concentrating the pigment; the color of the animal is then lighter.

Figure 9–16

How Cephalopods Avoid Predation. (a) The blue-ring octopus, *Hapalochlaena maculosa*, from Australia advertises its toxicity with bright blue rings. (b) Many cephalopods, such as this octopus, escape predators by releasing an inky substance into the water. (a, Fred Bavendam; b, Jeffrey Rotman)

its species, as well as to other species. Color can be used as a warning. For instance, when agitated, the blue-ring octopus (*Hapalochlaena maculosa*) of the Pacific Ocean produces blue rings on its skin to warn potential antagonists (Figure 9–16a). The bite of this octopus is extremely toxic and can be fatal to humans.

Another means of protection found in many cephalopods is the ink sac. An ink gland produces a dark fluid called *sepia* (SEE-pee-uh) that contains a high concentration of a brown-black pigment called *melanin* (in deep-water squid the fluid is white or luminescent). When the animal is disturbed or frightened, it can release the ink into the water (Figure 9–16b). The ink cloud resembles the general shape of the animal and distracts predators while the animal escapes.

Sexes are separate in cephalopods, and mating frequently involves some form of courtship display. Male squid have a modified arm that is used to take a package of sperm, the spermatophore, from its own mantle cavity and place it in the mantle cavity of the female near the opening of the oviduct. The eggs then are fertilized as they are released from the oviduct. Some species, such as members of the genus *Argonauta,* lay their eggs in delicate shells secreted by modified tentacles of the females. Other species usually attach their eggs to stones or other objects. The females of some octopus species incubate their eggs until they hatch, constantly pumping a stream of water over the eggs to keep

them oxygenated and to prevent fungus and other microorganisms from infecting them. Cephalopods usually reproduce only once in their life cycle and then die.

THE ARTHROPODS: ANIMALS WITH JOINTED APPENDAGES

Arthropods (phylum Arthropoda) represent the most successful group of animals in the animal kingdom, and almost 75% of all identified animal species (most of them insects) belong to this phylum. Several factors have played a role in the enormous success of this group, including the evolution of a hard exterior or exoskeleton, jointed appendages, and sophisticated sense organs.

Arthropods have evolved a hard, protective exterior skeleton (**exoskeleton**) composed of protein and a tough polysaccharide called **chitin** (KY-tin). In many marine species, the exoskeleton is impregnated with calcium salts to give it extra strength. The exoskeleton is flexible enough in the region of the joints to allow movement, and it provides points of attachment for muscles, allowing more efficient movement.

The exoskeleton does have its drawbacks, however. Since the exoskeleton is not living, it does not grow with the animal. As the animal grows, it must shed its old exoskeleton and produce a new, larger one. This process is called **molting.** While the animal is molting, it is very vulnerable and usually seeks a hiding place until a new exoskeleton has hardened. As the exoskeleton increases in size, so does its weight, making it more difficult for the animal to carry around. It is not surprising that the largest arthropods are found in aquatic habitats where the buoyancy of the water helps to counteract the weight of their heavy exoskeletons.

Another factor that has contributed greatly to the success of arthropods was the evolution of jointed appendages. Generally, each body segment of the animal has a pair of jointed appendages, and many of these function in locomotion, allowing the animals to move quickly and efficiently. Other appendages have been modified into mouthparts for more efficient feeding, sensory structures, and body ornamentation that helps to attract a mate or to make the animal more difficult to see.

Arthropods have highly developed nervous systems. Their sophisticated sense organs allow them to respond quickly to changes in their environment. The high degree of development of the arthropod nervous system has also given rise to a number of behavior patterns that play an important role in their daily activities. Experiments have shown that some species are capable of learning.

Chelicerates

Chelicerates (keh-LI-suh-rehts; subphylum Chelicerata) are a primitive group of arthropods that includes spiders, ticks, scorpions, horseshoe crabs, and sea spiders. These animals have six pairs of appendages. One pair, the **chelicerae** (keh-LI-suh-ree), is modified for the purpose of feeding and takes the place of mouthparts.

Horseshoe crabs (*Xiphosura*; Figure 9–17) are not true crabs but chelicerates that live in shallow coastal waters. They have not changed much since they first evolved over 230 million years ago. The body of the animal is in the shape of a horseshoe and is composed of three regions: the cephalothorax (sef-uh-loh-THOR-aks), the abdomen, and the telson. The cephalothorax is the largest region of the body and contains the more obvious appendages. The abdomen is smaller than the cephalothorax and contains the gills. The telson is a long spike that is used for steering and defense. The entire body is covered by a hard outer covering called the *carapace.*

Horseshoe crabs can move by walking or by swimming. They swim by flexing the abdomen. They are scavengers that feed primarily at night on worms, molluscs, and other organisms, including algae. They pick up food with their chelicerae and pass it to the walking legs. The walking legs have structures at their bases that crush the food before passing it to the mouth.

Males are smaller than females, and during the mating season, a male or several males will attach to the carapace of

Figure 9–17

Horseshoe Crabs. Horseshoe crabs are not really crabs but are more closely related to spiders. They come to shallow water to reproduce. The smaller male attaches himself to the carapace of the larger female. As the female releases her eggs, the male sheds sperm to fertilize them. (Peter J. Bryant/Biological Photo Service)

a female. The animals come to shore during high tide to mate, and the female will dig up the sand with the front of her carapace, depositing eggs in the depression. The males riding on her back shed their sperm onto the eggs before they are covered by the female. The eggs are then incubated by the sun and hatch into larvae that return to the sea during another high tide to grow into adults.

Sea spiders (Figure 9–18; class Pycnogonida) are chelicerates that can be found from intertidal waters to depths

Figure 9–18

Sea Spider. Sea spiders are marine chelicerates that use their proboscis to suck fluids from their prey, much like terrestrial spiders. (Animals Animals Copyright Doug Allan/Oxford Scientific Films)

of over 6,000 meters (19,700 feet) in all oceans, but especially the polar seas. They have small, thin bodies and usually four pairs of walking legs, although some species may have more. Males in several species have an extra pair of appendages that are used to carry the developing eggs. These are the only marine invertebrates known where the male carries the eggs. Other appendages include chelicerae for capturing prey and sensory structures called *palps.* Sea spiders feed on the juices of cnidarians and other soft-bodied invertebrates that they extract by means of a long, sucking proboscis.

Crustaceans

Crustaceans (subphylum Crustacea) are represented by a variety of animals that range in size from microscopic zooplankton to large lobsters. Many are important food sources for other marine animals, and some, such as crabs, lobsters, and shrimp, are important food sources for humans. Crustaceans have three main body regions: the head, thorax, and abdomen (Figure 9–19). In some species the head and thorax are fused to form a cephalothorax. Each body segment usually bears a pair of appendages, although in some species, abdominal appendages are lacking.

Crustacean appendages include sensory antennae (they are the only arthropods that have two pairs) and mandibles and maxillae that are used for feeding. Depending on the species, other appendages may include walking legs and legs modified for swimming, reproduction, feeding, and defense.

Small crustaceans exchange gases through their body surface, especially areas at the base of their legs, where the exoskeleton is thin. Larger crustaceans have gills for gas exchange. These are feathery structures that are usually located beneath the carapace or at the base of specialized appendages.

Crustaceans exhibit a variety of reproductive styles. Most have separate sexes, and the males usually have special appendages modified for clasping the female and delivering the sperm. Fertilization is generally internal. Some crustaceans, such as barnacles, are hermaphroditic. Others, such as some species of ostracods and branchiopods, reproduce by **parthenogenesis,** a process by which eggs develop without sperm.

Most crustaceans brood their eggs and have brood chambers or modified appendages for this purpose. In some, development is direct, and a miniature adult, or juvenile, hatches from the egg. In most crustaceans, however, a larva called the **nauplius** (NAW-plee-uhs) **larva** hatches from the egg and gradually matures into an adult.

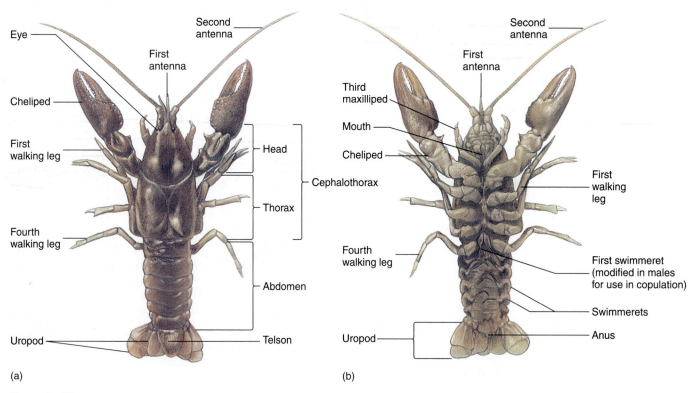

Figure 9–19

Crustacean Anatomy. The external anatomy of a lobster as seen from (a) a dorsal view and (b) a ventral view.

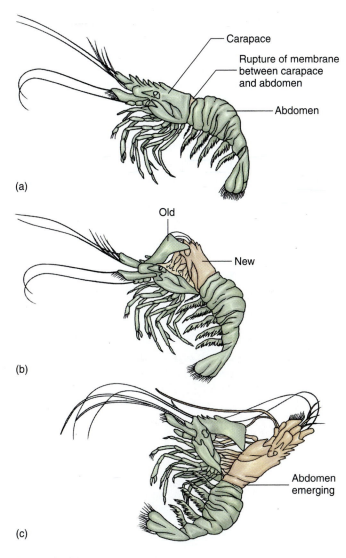

Carapace

Rupture of membrane between carapace and abdomen

Abdomen

(a)

Old

New

(b)

Abdomen emerging

(c)

Figure 9–20

Molting. The nonliving exoskeleton of crustaceans must be shed periodically. (a) Old membranes rupture along a joint line. (b) The animal pulls itself free of the anterior portion of the exoskeleton and then (c) crawls out of the posterior part. Once the old exoskeleton has been loosened, it usually takes about 15 minutes for the animal to shed it. After molting, the crustacean's body is soft, and the animal is particularly vulnerable until, in several days, the new exoskeleton is formed.

Molting is a very important part of the crustacean life cycle. Stages in this process are diagrammed in Figure 9–20. During the molting process, the animal is defenseless and remains hidden. Young crustaceans molt frequently, and the time between molts is relatively short. As the animal ages, molting is less frequent, and some species eventually reach an age where they no longer molt. Molting is directly controlled by specific hormones produced by glands in the crustacean's head. The process appears to be initiated by changes in environmental conditions, such as temperature

and day length, that alter the level of specific hormones and set the molting process into action.

Crustaceans exhibit almost every type of feeding style. Some, such as barnacles, are filter-feeders, using their feathery modified antennae to strain plankton, microorganisms, and detritus from the water. Others are scavengers, such as some crab species, feeding on dead animals. Another group of crustaceans are carnivores, feeding on worms, molluscs, fish, and other crustaceans. Predators use walking legs, especially pincers (chelipeds), to capture their prey, while most crustaceans use their mandibles to crush their food. Some species have plates in their stomachs that will further grind and process the food for digestion.

Some of the more important groups of crustaceans found in the marine environment are decapods, krill, copepods, barnacles, isopods, and amphipods.

Crabs, lobsters, and true shrimp are called *decapods* (order Decapoda; *deca*, meaning "ten," and *poda*, meaning "feet") because they have five pairs of walking legs. The first pair of walking legs is modified to form chelipeds that are used for capturing prey and for defense.

Although most decapods are relatively small, some can get quite large. The largest of all crustaceans is the giant spider crab (*Macrocheira kaempferi*) from Japan (Figure 9–21). Specimens are known to have exceeded 4 meters

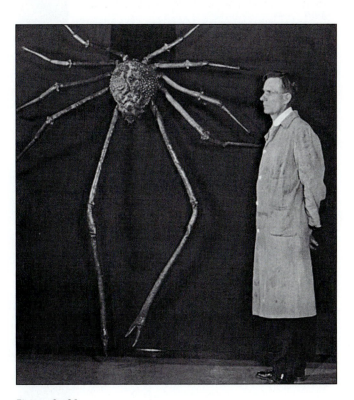

Figure 9–21

The Japanese Spider Crab. *Macrocheira kaempferi* from the northern Pacific is the largest living arthropod. (Neg. Trans. No. 312007. Courtesy of the Department of Library Services, American Museum of Natural History)

(13 feet) in width and weighed over 18 kilograms (40 pounds). The North Atlantic lobster (*Homarus americanus*) can also get quite large. Two record specimens were captured off the coast of Virginia in 1934. One measured over 1 meter (3.3 feet) in length, and one weighed 19 kilograms (42 pounds) and the other 17 kilograms (37 pounds).

Many decapods display interesting adaptations and behaviors that help them to survive in their habitat. Hermit crabs (Paguroidea) have a soft abdomen. They inhabit empty gastropod shells or other suitable enclosures, and, as they grow, they not only molt but must also find a larger shell to accommodate their ever-increasing bodies. When disturbed, these animals will withdraw into their shells and close the opening with a large cheliped. Decorator crabs (Majidae) attach bits of sponge, anemones, and hydrozoans to their carapace for camouflage (Figure 9–22a).

Several species of decapod are important sources of food for humans as well as other animals. Blue crabs (*Callinectes sapidus;* Figure 9–22b), shrimp (*Penneus*), and lobster are fished commercially along many parts of the eastern United States and the Gulf of Mexico. Several species of decapod, including the Alaskan king crab (*Paralithodes camtschatica*), are fished commercially along the Pacific coast of the United States.

Figure 9–22

Representative Decapods. (a) This decorator crab from the Pacific Ocean attaches a variety of sessile organisms to its shell for camouflage. (b) The blue crab *Callinectes sapidus* (the name means "beautiful swimmer that tastes good") is fished commercially for food along the East Coast of the United States. (a, Copyright 1996 Beverly Factor; b, H. Wes Pratt/Biological Photo Service)

(a)

(b)

Figure 9–23

Krill. Krill are important members of the zooplankton and the favorite food of many marine mammals and penguins. Most species of krill, like this one, are bioluminescent. (Flip Nicklin/Minden Pictures)

Krill (order Euphausiacea) are pelagic, shrimplike creatures that are about 3 to 6 centimeters (1 to 2 inches) long (Figure 9–23). They are filter-feeders that feed primarily on zooplankton. Most species of krill are bioluminescent, producing light in a specialized organ called a **photophore** (FOH-toh-for). It is thought that the luminescence acts as a signal to attract individuals into large masses called *swarms.* It may also function in reproduction to attract mates. Many Antarctic species such as *Euphausia superba* live in large swarms. Swarms of krill may cover an area of several hundred meters and be as thick as 5 meters (16.5 feet). The density of krill in these swarms can be as great as 60,000 individuals per cubic meter of water. Krill such as *Euphausia superba* constitute the main diet of some whales, seals, penguins, and many fishes. A single blue whale (*Bal-*

aenoptera musculus) can consume a ton of krill in one feeding and may feed four times a day. *Euphasia superba* can molt so quickly that, when alarmed, individuals will literally jump out of their skins. The shed molt may then function as a decoy.

Copepods (class Copepoda) are the largest group of small crustaceans. Copepods in the genus *Calanus* (Figure 9–24) comprise a major portion of the zooplankton and are a major food source for several species of commercially important fishes as well as some whales, sharks, and birds. Copepods represent the major link between phytoplankton and higher levels of the oceanic food web.

Barnacles (class Cirripedia) are crustaceans that live a sessile existence. The bodies of most barnacles are covered by a carapace that secretes a shell of calcium carbonate. The

Figure 9–24

Copepod. Copepods in the genus *Calanus* comprise a major portion of the zooplankton that channels energy from phytoplankton to larger consumers. (David Wrobel 1994/Biological Photo Service)

(a)

Cirripeds

Stalk

(b)

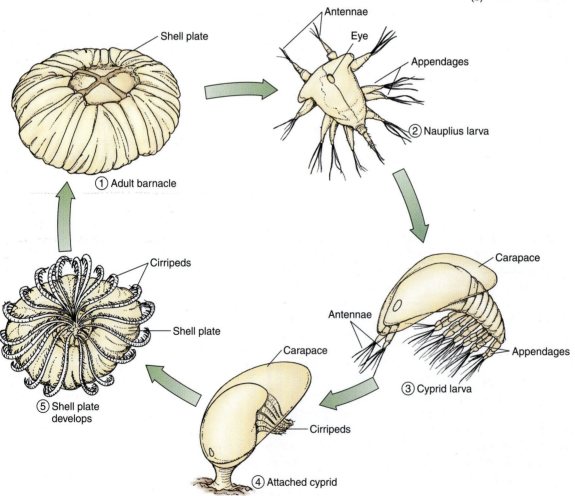

Shell plate

① Adult barnacle

Antennae

Eye

Appendages

② Nauplius larva

Carapace

Antennae

Appendages

③ Cyprid larva

Carapace

Cirripeds

④ Attached cyprid

Cirripeds

Shell plate

⑤ Shell plate develops

◀ Figure 9–25

Barnacles. (a) Acorn barnacles attached to rocks. (b) Goose barnacles attach to solid surfaces by means of a long, fleshy stalk. (c) A generalized life cycle of an acorn barnacle. (1) After fertilization, eggs are brooded in the adult's shell until they develop into nauplius larvae. A single adult may release as many as 13,000 larvae. (2) Like other crustaceans, the nauplius has an eye, antennae, jointed appendages, and an exoskeleton. (3) After several molts, the nauplius develops into a cyprid larva. The cyprid has larger antennae and more body segments and appendages than the nauplius and a thin, but tough, carapace that protects its body. (4) Shortly after becoming a cyprid, the larva settles to the bottom and attaches to a solid surface, using specialized cement glands located on the antennae. Once attached, the cyprid molts and rotates its body so that its appendages, now called *cirripeds*, face upward. (5) The cyprid's carapace acts as a form around which the animal's shell develops. The feathery cirripeds filter food from the surrounding water. (a, William S. Ormerod, Jr./Visuals Unlimited; b, Fred Bavendam/Peter Arnold, Inc.)

shell can be attached directly to a hard surface, as in acorn barnacles (*Balanus;* Figure 9–25a), or be attached by a stalk, as in the goose barnacles (*Lepas;* Figure 9–25b). When the shell is open, feathery appendages known as **cirripeds** (SER-uh-pedz) extend into the water. The cirripeds filter food from the surroundings.

Barnacles are hermaphroditic and usually cross-fertilize utilizing a long, extensible penis to copulate. A nauplius larva hatches from the brooded egg and develops into a **cyprid** (SIP-rid) **larva** that has compound eyes and a carapace composed of two shell plates. When the cyprid larva finds a suitable surface, it attaches by its antennae, which contain adhesive glands, and metamorphoses into an adult (Figure 9–25c).

Some species of barnacle pose a problem for ships by attaching to the hulls in large numbers and causing increased friction as the ships move through the water. Other barnacles attach to plants or animals such as whales and large fishes.

Crustaceans known as *isopods* (order Isopoda) and *amphipods* (order Amphipoda) are frequently seen scurrying around on sand beaches and in the beach drift that washes ashore. They are mainly scavengers and detritus-feeders, although some members of both groups are carnivorous. Some amphipods are suspension-feeders. Many larger animals, including birds, feed on these crustaceans.

LOPHOPHORATES

Lophophorates (lohf-uh-FOHR-ayts) are sessile animals that lack a distinct head and possess a feeding device called a **lophophore** (LOHF-uh-fohr). The lophophore is an arrangement of ciliated tentacles that surrounds the mouth and functions in feeding and respiration. It can be withdrawn when the animal is disturbed. There are three phyla of lophophorate animals: the phoronids, the bryozoans, and the brachiopods.

Phoronids (phylum Phoronida) are small, wormlike animals that range in size from a few millimeters to 30 centimeters (12 inches) in length. They secrete a tube of leathery protein or chitin that can be attached to rocks, shells, or pilings or buried in bottom sediments. Phoronids feed on plankton and detritus that are caught in the mucus on their tentacles, then carried in ciliated grooves to the mouth.

Bryozoans (phylum Ectoprocta) are small animals (average, 0.5 millimeters or 0.02 inches) that are extremely abundant (Figure 9–26a). They live in shallow water, forming colonies on a wide variety of solid surfaces such as rock, shell, algae, mangroves, and ship bottoms. The colonies may appear as white encrustations or as fuzzy growths, while still others form colonies that resemble hydrozoans. Colonies are composed of thousands of tiny individuals, each of which inhabits a boxlike chamber that it secretes. When feeding, the lophophore is extended; at other times it is withdrawn into the chamber.

Some species of **brachiopod** (phylum Brachiopoda), or lamp shells, have changed very little since they first evolved over 400 million years ago (Figure 9–26b). Most species are benthic and are found in shallow water. They range in size from 5 millimeters to 8 centimeters (0.02 to 3 inches) and have bivalved shells that resemble the shells of molluscs. The two valves of the brachiopods are different in size and shape and are dorsal and ventral, whereas molluscan bivalves have right and left valves that are generally similar in size and shape. Another difference between molluscs and brachiopods is the fleshy stalk called the *pedicle* found in many brachiopods. The pedicle attaches the brachiopod shell to a hard surface or is buried into soft sediments. The animal's body only occupies a portion of the shell, and an extension of the body wall, the mantle, lines the shell and is responsible for secreting it. Lying in the mantle cavity is a large, horseshoe-shaped lophophore that is used for feeding and respiration. Brachiopods feed on detritus and algae that are swept into a groove leading to the mouth by the ciliated lophophore.

(a)

Figure 9–26

Representative Lophophorates. (a) A colony of bryozoans from Narragansett Bay, Rhode Island, with lophophores extended for feeding. (b) A brachiopod or lamp shell. Notice that the two valves are not symmetrical as they are in bivalve molluscs and that the animal is fastened to the bottom by a fleshy stalk (pedicle). (a, H. W. Pratt/Biological Photo Service; b, James R. McCullagh/Visuals Unlimited)

(b)

ECHINODERMS: THE ANIMALS WITH SPINY SKINS

The **echinoderms** (phylum Echinodermata) are represented by well-known animals such as sea stars, sea urchins, and sea cucumbers. The name *echinoderm* means "spiny skinned" and refers to the spiny projections found on many species. This condition is the result of the endoskeleton (internal skeleton) that lies just beneath the epidermis. The endoskeleton is composed of plates of calcium carbonate (ossicles) that are held together by connective tissue, and the spines and tubercles that produce the spiny surface of the echinoderm project outward from these plates. Around the bases of the spines are tiny, pincerlike structures called **pedicellariae** (ped-uh-suh-LEHR-ee-eh). These structures are found only in some echinoderms and function to keep the surface of the body clean and free of parasites and the settling larvae of various fouling species. In some species, they may also aid in obtaining food.

It is thought that echinoderms evolved from a bilateral ancestor, since all of the larval forms still exhibit bilateral symmetry. Most of the adults, however, exhibit a modified form of radial symmetry. The move toward radial symmetry may have been an adaptation to a sessile lifestyle similar to that exhibited by some of the living echinoderms such as sea lilies. As mentioned previously, radial symmetry benefits a sessile animal since it allows the animal to meet its environment equally well on all sides.

The **water vascular system** is a feature unique to echinoderms (Figure 9–27). It is a hydraulic system that functions in locomotion, feeding, respiration, and excretion. Water enters the water vascular system by way of the **madreporite** (mad-ruh-POHR-ryt) and passes through a system of canals that runs throughout the animal's body. Attached to some of these canals are tube feet. Each tube foot is hollow and has a saclike structure, the **ampulla,** that lies within the body, and a sucker at the end that protrudes from the ambulacral groove. In some species, however, the terminal suckers are lacking. Later in this section we will examine how different species use their tube feet for locomotion and feeding.

Echinoderms are mostly benthic organisms and can be found at virtually all depths. In fact, sea cucumbers and brittle stars are frequently the most common form of animal life in deep-sea dredgings. Echinoderms are frequently brightly colored, although the colors fade rapidly when the animals are removed from the water.

Because of their spiny skins, echinoderms are not preyed upon by many animals. Some molluscs relish sea urchins and sea stars, and sea otters feed on sea stars and sea urchins. Many spider crabs feed on echinoderms, which they tear apart or crush with their chelipeds. Humans eat sea urchin gonads and sea cucumbers, and some fishes have mouthparts modified for feeding on echinoderms. On the other hand, many echinoderms are predators of molluscs, other echinoderms, cnidarians, and crustaceans.

Sea Stars

A typical sea star (class Asteroidea) is composed of a central disc and five arms or rays. The mouth is located on the underside, and radiating from the mouth along each ray is a groove (ambulacral groove) that contains tiny tube feet (see Figure 9–27). The aboral surface (the side opposite the mouth) is frequently rough or spiny.

Sea stars move when water is pumped into the tube feet from the ampullae, causing them to project out of the ambulacral groove. The suckers on the end hold firmly to solid surfaces, and muscles in the tube feet contract, forcing water back into the ampullae and causing the tube feet to shorten. As a result, the animal is pulled over the surface in the desired direction. This type of locomotion is best suited for hard surfaces like rocks. When the animal is moving across sand, the tube feet engage in a walking type of motion. This type of locomotion is generally slow, moving the animal along at the rate of a few centimeters per minute.

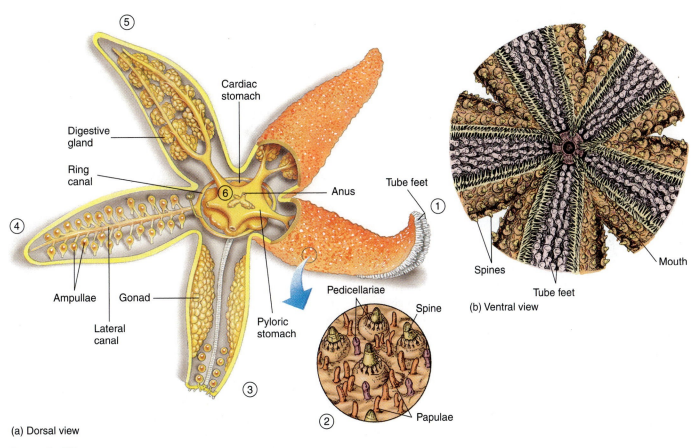

(a) Dorsal view

(b) Ventral view

Figure 9–27

Sea Star. (a) A dorsal view showing the general internal and external anatomy of a sea star, a representative echinoderm. (1) The dorsal surface of a sea star ray with the end turned up to show the tube feet on the ventral surface. (2) A close-up view of the dorsal surface of a sea star ray. The spines are outward projections of the bony plates (ossicles) beneath the skin. The pincer-like pedicellariae keep the surface free of parasites and larvae. The papulae function in gas exchange and excretion. (3) Paired gonads are located in each ray. (4) Radiating from the ring canal, like spokes on a wheel, are the lateral canals. The lateral canals are attached by short side branches to the ampullae of the tube feet. (5) Each ray contains digestive glands. (6) The central disk contains the stomachs, anus, madreporite, and ring canal. (b) An enlarged ventral view of a sea star. The mouth is centrally located. Notice the rows of tube feet within the grooves of each ray.

Figure 9–28

Sea Star Feeding. Some sea stars use their tube feet to pry open the shells of bivalves. When the valves are opened, the sea star everts a portion of its stomach and secretes enzymes to break down the flesh of its prey. (Richard Chesher/Planet Earth Pictures, Ltd.)

Sea stars also use their tube feet in feeding (Figure 9–28). When feeding on large bivalves, such as mussels, the sea star wraps around its prey and pries the valves apart. The sea star can evert a portion of its stomach out of its mouth, and the stomach is inserted into the bivalve, where it digests the animal. Digestive enzymes are supplied by pairs of digestive glands that are located in each of the sea star's rays. When the sea star has finished feeding, it withdraws its stomach back into its mouth and moves away from its prey.

Sea stars will feed on a variety of animals including other echinoderms, molluscs, worms, coral, and small fish. When feeding on other echinoderms and small animals, the prey is frequently swallowed whole, and, after it has been digested, the ossicles and other undigestible materials are regurgitated through the mouth.

The crown-of-thorns sea star (*Acanthaster planci*), which is found in the Pacific Ocean, feeds on coral polyps (Figure 9–29). Over the past twenty years, there has been great concern that population explosions of these animals were causing tremendous damage to Pacific reefs, including the Great Barrier Reef of Australia. Although these animals have caused extensive damage to reefs, the damage is not as widespread or as permanent as originally feared. Many theories have been proposed for the population explosion, but the actual cause is still unknown.

Sea stars have great powers of regeneration. They can produce new rays if one or more are removed. Some species are able to cast off injured rays and then grow new ones. Several species are capable of regenerating entire individuals from a single ray as long as a portion of the central disc is present. Regeneration is a slow process, and it usu-ally requires one year before the missing parts are completely reformed.

In the past, oyster fishers would try to kill the sea stars they would find in their oyster beds by chopping them in two. Subsequently, they would discover that instead of fewer sea stars there would be more, and most of them had deformed rays. The two pieces of sea star, when returned to the water, simply grew into two new individuals.

Brittle Stars

Brittle stars (Figure 9–30) (class Ophiuroidea) contain the greatest number of echinoderm species. They are benthic organisms that can be found from shallow water to the ocean's depths. Like sea stars, they have five arms, but the arms are very slender and distinct from the central disc, and they are covered with many spines. Because their arms tend to writhe in a serpentine fashion when they move, brittle stars are often called *serpent stars*. Brittle stars lack pedicellariae, and their ambulacral grooves are closed. The tube feet play a role in feeding and locomotion, but do not have suckers.

Brittle stars get their name because they will drop off one or more arms when disturbed, and some species may even shed a portion of the central disc. Once an arm is shed, it undulates wildly, no longer controlled by the central nervous system. The undulating arm distracts potential predators, while the brittle star is able to move away to safety and regenerate the missing part.

Brittle stars are mostly suspension-feeders, filtering food from the water, or deposit-feeders, feeding on organic material that they find on the bottom. A few are carnivorous. When filter-feeding, they lift their arms from the bottom and wave them through the water. Strands of mucus that are strung between the spines on adjacent arms trap plankton and organic material. The food is then either

Figure 9–29

The Crown-of-Thorns Sea Star. *Acanthaster planci* feeds on coral polyps and has caused extensive damage to some Pacific reefs, including the Great Barrier Reef of Australia. (Jon L. Hawker)

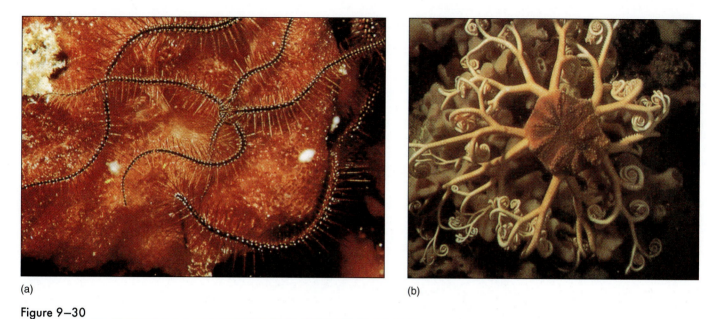

(a)
(b)

Figure 9–30

Brittle Stars. (a) Several serpent stars on the surface of a sponge. (b) The basket star *Gorgonocephalus arcticus* from the coast of Maine. (a, Charles Seaborn; b, Wes Pratt/Biological Photo Service)

swept to the mouth by ciliary action or collected from the spines by tube feet and passed to the mouth. Brittle stars tend to avoid light, coming out at night to feed and hiding under rocks and in crevices during the day. Many species are burrowers. They bury themselves in bottom sediments with only their arms trailing over the sea floor to trap food.

Sea Urchins and Their Relatives

Sea urchins and their relatives (class Echinoidea) have a body that is enclosed by a hard endoskeleton or **test.** Tube feet project from five pairs of ambulacral areas that are homologous to the arms of sea stars, and spines project from the test (Figure 9–31). Sea urchins have relatively long spines, and they move mostly by means of their tube feet, with some help from their spines. The spines function in protection and, in some species, contain a venom that creates a severe burning sensation. When a person comes into contact with one of these stinging urchins, the tip of the spine breaks off in the skin, injecting the venom. Some spines have barbed spinelets along their length that make them difficult to remove. The spines also allow some species to live along the shore in regions pounded by surf. The spines help to dissipate the energy of the waves and protect the animal from wave shock. Sand dollars and heart urchins have very short spines that give them a fuzzy appearance. They use these spines instead of tube feet for locomotion.

There are several types of jawlike appendages called *pedicellariae* on the surface of sea urchins, some of which

can inject a toxin to paralyze prey. A sea urchin's mouth contains five teeth that form a chewing structure called *Aristotle's lantern* because of its resemblance to ancient Greek lanterns and in honor of an early student of echinoderm biology.

Sea urchins and their relatives are benthic organisms that can be found from shallow water to the ocean's depths. Sea urchins tend to occupy solid surfaces such as rocks, whereas sand dollars and heart urchins prefer to bury themselves in sandy bottoms.

Sea urchins are omnivorous but primarily feed on algae. Sand dollars and heart urchins feed on organic material that they find on the bottom.

Sea Cucumbers

Sea cucumbers (class Holothuroidea) (Figure 9–32) have elongated bodies, and, in most species, the body wall is leathery. Because of their body shape, sea cucumbers usually lie on one side. They move slowly using their ventral tube feet and the muscular contractions of their body wall. Located around the mouth of a sea cucumber are 10 to 30 modified tube feet called *oral tentacles.* Most sea cucumbers feed on small particles of food that they trap with these tentacles.

Some species use the ring of tentacles to strain plankton from the surrounding water. The tentacles are coated with a sticky mucus, and any small organism that comes into contact with them becomes stuck. Periodically sea cucumbers retract their tentacles into their mouths to remove

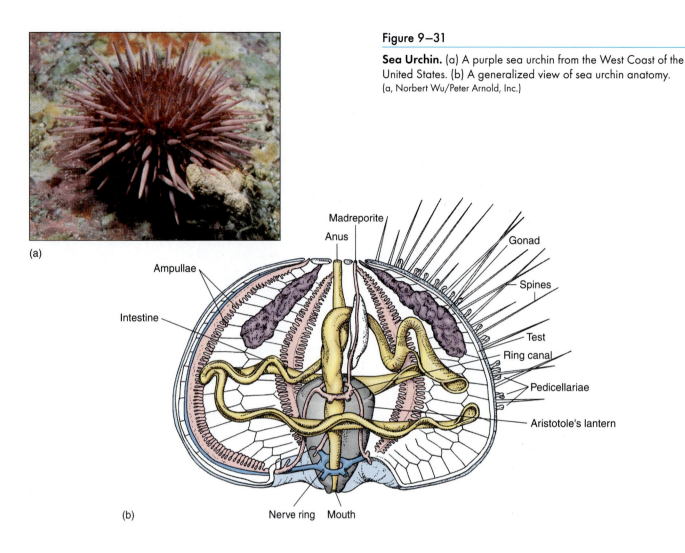

(a)

(b)

Figure 9–31

Sea Urchin. (a) A purple sea urchin from the West Coast of the United States. (b) A generalized view of sea urchin anatomy. (a, Norbert Wu/Peter Arnold, Inc.)

Madreporite

Anus

Ampullae

Intestine

Gonad

Spines

Test

Ring canal

Pedicellariae

Aristotole's lantern

Nerve ring   Mouth

Figure 9–32

Sea Cucumber. The anatomy of a sea cucumber.

Intestine   Gonad   Ring canal

Stomach

Madreporite

Anus

Tentacle

Ambulacrum

Pharynx

Respiratory tree

the food and then project them out to capture more. Other species use their tentacles to shovel sand and bottom sediments into their mouths and then digest the organic material, leaving conspicuous mounds of sand or fecal-mud castings behind.

When disturbed, some species of sea cucumber release tubules (tubules of Cuvier) from their anus (Figure 9–33). The tangle of tubules resembles spaghetti, and, on contact with seawater, they become sticky. The tubules will stick tenaciously to a predator and distract its attention to cleaning itself while the sea cucumber escapes. The tubules are strong enough to ensnare and immobilize crustaceans, and they are distasteful to fishes. This behavior helps slow-moving sea cucumbers deter potential predators.

Other species will eviscerate, that is, release some of their internal organs through either the anus or the mouth.

Figure 9–34

A Crinoid. A crinoid (also known as *feather star* or *sea lily*) from the Red Sea attached to a coralline rock. (Jeffrey Rotman/Peter Arnold, Inc.)

In each instance, the animal will ultimately regenerate the missing body parts. In some species, the process of evisceration appears to be seasonal and may represent the release of a digestive tract that is full of wastes following a long feeding period. These waste-laden intestines can serve as an important source of food for detritus-feeders.

Crinoids

Crinoids (class Crinoidea) are commonly called *sea lilies* and *feather stars* (Figure 9–34). They are the most primitive of the echinoderms and have a long fossil history. Their bodies look very much like flowers, and they are attached to the end of long stalks that fasten them to the bottom. Although some species remain fixed in one spot, others can use their long branched arms for locomotion. Crinoids feed on small organisms that are filtered from the water by their tube feet and by mucus nets of the ambulacral grooves. The ciliated grooves carry the food to the mouth, where it then enters the digestive system.

TUNICATES

Tunicates (subphylum Urochordata) are mostly sedentary animals that are widely distributed in all seas. These animals are named for their body covering called the *tunic*, which is largely composed of a substance similar to cellulose. Tunicates known as *sea squirts* (class Ascidea; Figure 9–35a) get their name because many species, when irritated, will forcefully expel a stream of water from their excurrent siphon. These animals can be solitary, colonial, or compound organisms composed of several individuals that share a com-

Figure 9–33

Sea Cucumber Defense. This sea cucumber from the Red Sea is releasing tubules of Cuvier. This behavior distracts potential predators and gives the sea cucumber time to escape. (Jeffrey Rotman)

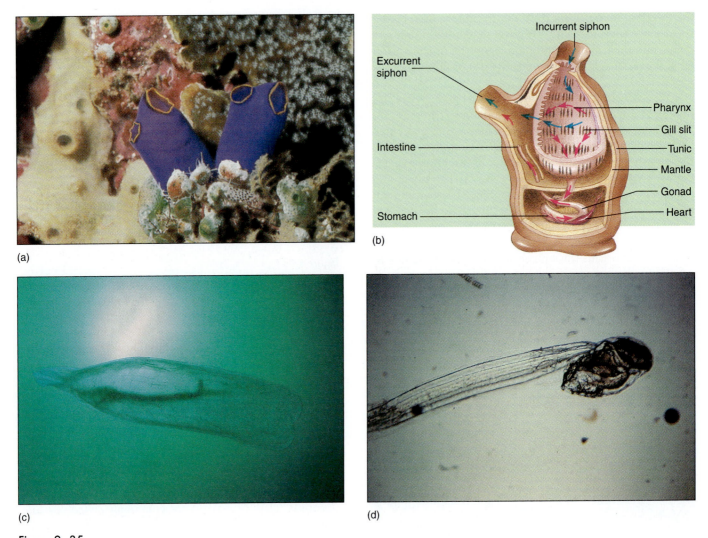

(a)

(b)

Incurrent siphon

Excurrent siphon

Pharynx

Gill slit

Tunic

Mantle

Gonad

Heart

Intestine

Stomach

(c)

(d)

Figure 9–35

Tunicates. (a) Two solitary blue and gold sea squirts on a reef in the Philippine Islands. (b) The anatomy of a sea squirt. The blue arrows represent the pattern of water flow through the animal's body, and the red arrows represent the movement of food. (c) Thaliaceans, also known as salps, are planktonic tunicates. (d) A larvacean. (a, F. Stuart Westmorland/Tom Stack and Associates; c, David Wrobel 1993/Biological Photo Service; d, William Jorgensen/Visuals Unlimited)

mon tunic. The bodies of sea squirts are generally round or cylindrical and have two tubes projecting from them: an incurrent siphon that brings in water and food, and an excurrent siphon that eliminates water and wastes. Food is trapped on a mucous net that is formed in the pharynx (Figure 9–35b).

Thaliaceans (tha-lee-AY-shunz; class Thaliacea; Figure 9–35c) are free-swimming tunicates and have their incurrent and excurrent siphons located on opposite ends of their barrel-shaped bodies. As they swim, they pump water through their bodies, extracting food and eliminating wastes. When food supplies are abundant, the populations

of thaliaceans increase dramatically. Some species are bioluminescent.

Larvaceans (class Larvacea) are another group of free-swimming tunicates (Figure 9–35d). They produce delicate enclosures made of mucus that are used in feeding. When these mucus networks are clogged (approximately every four hours) they shed them and produce another in a matter of minutes. Like thaliaceans, when food supplies are abundant, there are population explosions of larvaceans. SCUBA diving at these times among the shed mucous networks has been compared to swimming through a snowstorm.

CHAPTER SUMMARY

The phylum Mollusca includes chitons, tusk shells, gastropods, bivalves, and cephalopods. The generalized mollusc body plan consists of two parts: a head-foot and a visceral mass. Molluscs have soft bodies that, in most cases, are covered by a shell composed of calcium carbonate. The mollusc shell is formed by an organ called the *mantle*. In cephalopods, the mantle functions in locomotion. Some molluscan species rely on the mantle for gas exchange, though most marine molluscs exchange gases by means of gills. The radula, or ribbon tooth, is a structure unique to molluscs that is used for feeding.

Arthropods are the most successful group of animals. They have an exoskeleton, jointed appendages, and sophisticated sense organs. The arthropods known as *chelicerates* lack mouthparts, having instead appendages called *chelicerae* that are modified for feeding. Chelicerates are represented in the marine environment by sea spiders and horseshoe crabs. Crustaceans, which include lobsters, crabs, shrimp, barnacles, and numerous smaller forms, are widely distributed in marine environments. Crustaceans have two pairs of sensory antennae and mouthparts called *mandibles*. Crustaceans are important members of marine food webs.

Lophophorates are sessile animals that possess a feeding device called a *lophophore*. Lophophorates include phoronids, bryozoans, and brachiopods.

Echinoderms exhibit radial symmetry as adults, although their larvae exhibit bilateral symmetry, suggesting that they evolved from bilateral ancestors. Echinoderms have an internal skeleton and a unique water vascular system that functions in locomotion and food gathering. Echinoderms are represented by sea stars, brittle stars, sea urchins, sea cucumbers, feather stars, and sea lilies (crinoids).

Tunicates have bodies that are covered with a tunic composed of molecules similar to cellulose. Sea squirts are sedentary tunicates that filter their food from the water. They can be solitary, colonial, or compound animals. Thaliaceans and larvaceans are planktonic tunicates.

SELECTED KEY TERMS

mollusc, *p. 145*	veliger, *p. 150*	crustacean, *p. 158*	echinoderm, *p. 164*
radula, *p. 146*	bivalve, *p. 151*	parthenogenesis, *p. 158*	water vascular
chiton, *p. 147*	cephalopod, *p. 152*	krill, *p. 161*	system, *p. 164*
scaphopod, *p. 147*	arthropod, *p. 156*	lophophore, *p. 163*	crinoid, *p. 169*
gastropod, *p. 147*	exoskeleton, *p. 156*	bryozoan, *p. 163*	tunicate, *p. 169*
nudibranch, *p. 150*	molting, *p. 157*	brachiopod, *p. 163*	

QUESTIONS FOR REVIEW

MULTIPLE CHOICE

1. The molluscan shell is secreted by the
   a. foot
   b. skin
   c. gill
   d. kidney
   e. mantle

2. The _____ is a unique, toothlike structure found in many molluscs.
   a. mantle
   b. radula
   c. veliger
   d. trochophore
   e. exoskeleton

3. Molluscs that have shells composed of eight plates held together by a fleshy girdle are
   a. snails
   b. bivalves
   c. scaphopods
   d. cephalopods
   e. chitons

4. A type of bivalve that can damage wood is the
   a. scallop
   b. shipworm
   c. nudibranch
   d. oyster
   e. surf clam

5. Molluscs that have tentacles and a highly developed nervous system are
   a. gastropods
   b. bivalves
   c. scaphopods
   d. cephalopods
   e. chitons

6. Arthropod characteristics include
   a. a water vascular system
   b. a hard external shell composed of calcium carbonate
   c. jointed appendages
   d. soft segmented bodies with hydrostatic skeletons
   e. a tunic composed of cellulose

7. During the process of molting, arthropods
   a. grow extra appendages
   b. shed their old exoskeleton
   c. are well protected from predators
   d. shed excess appendages
   e. reproduce

8. While on a field trip to the seashore, you discover an animal with a spiny skin and a water vascular system. This animal is probably a(n)
   a. mollusc
   b. echinoderm
   c. arthropod
   d. lophophorate
   e. tunicate

9. Animals known as *brachiopods*
   a. possess an exoskeleton composed of chitin
   b. exhibit radial symmetry as adults
   c. produce a shell composed of two valves
   d. are all members of the plankton
   e. are also called *tunicates*

10. Which of the following adult animals is likely to be a member of the zooplankton?
    a. crab
    b. squid
    c. sea star
    d. larvacean
    e. sea cucumber

SHORT ANSWER

1. What adaptations allow squids to be successful predators?

2. What is a lophophore and how does it function in feeding?

3. Name four commercially important crustaceans.

4. Explain how the radula is modified in gastropods for different types of feeding.

5. Describe how a sea star uses its water vascular system to move.

6. Explain how slow-moving animals such as sea cucumbers avoid predation.

7. Describe how sea squirts feed.

8. Why are arthropods such a successful group of animals?

9. Distinguish between chelicerates and crustaceans.

THINKING CRITICALLY

1. What is the advantage of being able to alter the sex ratio in populations of the slipper limpet, *Crepidula fornicata?*

2. On a field trip to the ocean you discover a gastropod that you have never seen before. Since you are very interested in gastropod feeding, you want to know whether this animal is a herbivore or a carnivore. How could you determine with a fair amount of certainty which type of feeder this animal is?

3. What type of shell characteristics would you expect to observe in a gastropod that spends most of its life burrowing through soft sediments?

4. A sample of deep-sea dredgings contains an animal that is small, flat, and round. It has a ventrally located mouth, a stomach, but no intestines. To which of the groups of invertebrates introduced in this chapter does this animal probably belong? What other characteristics would the animal have to have in order to confirm your identification?

SUGGESTIONS FOR FURTHER READING

Abbott, R. T. 1974. *American Seashells,* 2nd ed. Van Nostrand Reinhold, New York.

Bavendam, F. 1996. Flowers of the Coral Seas: Feather Star Crinoids, *National Geographic* 190(6):118–130.

Bavendam, F. 1995. The Giant Cuttlefish: Chameleon of the Reef, *National Geographic* 188(3):94–107.

Bavendam, F. 1989. Even for Ethereal Phantasms, It's a Dog-Eat-Dog World, *Smithsonian* 20(5):94–101.

Bavendam, F. 1991. Eye to Eye with the Giant Octopus, *National Geographic* 179(3):86–97.

Bavendam, F. 1985. Sea Stars Deploy a Bag of Tricks in Marine Wars, *Smithsonian* 16(8):104–109.

Gosline, J. M., and M. E. DeMont 1985. Jet-Propelled Swimming in Squids, *Scientific American* 252(1):96–103.

Hadley, N. F. 1986. The Arthropod Cuticle, *Scientific American* 244(7):104–112.

Roper, C. F. E., and K. J. Boss. 1982. The Giant Squid, *Scientific American* 246(4):96–105.

Ruppert, E. E., and R. D. Barnes. 1994. *Invertebrate Zoology,* 6th ed. Saunders College Publishing, Philadelphia.

Chapter 10

The Fishes

Fishes, like amphibians, reptiles, birds, and mammals, are vertebrates. **Vertebrates** are animals that possess an internal skeletal rod (commonly called a backbone) composed of units called vertebrae. The vertebrae may be composed of cartilage, as in sharks, or bone, as in most other vertebrates. The backbone provides sites of attachment for muscles that can produce movements of the animal's trunk, thereby increasing and improving mobility.

The first fishes to evolve lacked paired fins and jaws. They probably spent their time scavenging food in the bottom sediments of early seas, a niche they filled for millions of years. About 425 million years ago, as a result of evolution, some fishes appeared that had paired fins and jaws. Fins allowed these fishes to maneuver better in the water, and they became active, agile swimmers. Jaws allowed them to feed on the slower moving fishes that lacked fins, as well as many types of invertebrate prey.

Fishes with paired fins and jaws could obtain more food than the fishes lacking these adaptations. Their ability to feed more efficiently allowed them to produce more offspring, and those offspring were more likely to survive because of the beneficial adaptations they had inherited from their parents. Ultimately, all but a few of the early jawless fishes were replaced by these successful newcomers. Modern fishes became such adept aquatic predators that they eventually replaced large arthropods, such as sea scorpions, from their niche as the highest-level consumers in the sea.

Today fishes can be found from the surface waters of the ocean to its deepest trenches. The size of some and bright colors and bizarre shapes of others inspire awe, delight, and sometimes dread in humans who have entered their realm. Fish species outnumber all other vertebrate species combined, and they display a wide array of adaptations that allow them to survive and reproduce in their aquatic world. Large segments of the world's human population depend on fish for food, and in certain areas, the impact of commercial fishing has endangered the survival of some species. **173**

KEY CONCEPTS

1. Hagfishes and lampreys are jawless fishes.

2. Sharks, skates, and rays have skeletons composed entirely of cartilage.

3. Sharks have streamlined bodies and highly developed senses that help them to be efficient predators.

4. Most marine fishes have skeletons composed primarily of bone.

5. The shape of a fish's body is primarily determined by the characteristics of its environment.

6. Many fishes exhibit coloration and color patterns that help them blend in with their environment.

7. Color in fishes functions in camouflage, species recognition, and communication.

8. Most bony fishes have a swim bladder that helps them maintain neutral buoyancy.

9. Most marine fishes are carnivorous, though herbivores, omnivores, and filter-feeders also exist.

10. Most marine fishes are oviparous and produce large numbers of eggs.

11. Fishes such as salmon and eels engage in lengthy migrations sometime during their life cycle.

AGNATHANS: THE JAWLESS FISHES

Agnathans (superclass Agnatha) are the most primitive fishes. The bodies of agnathans living today are slender, elongated, and eel-like, but they lack jaws and paired appendages. Their skeletons are composed of cartilage, and, unlike other fishes, their bodies are not covered by scales. They are represented in the marine environment by hagfishes and lampreys.

Hagfishes

Hagfishes (order Myxiniformes; Figure 10–1a) are scavengers feeding on dead and dying fish and soft-bodied invertebrates. They use their tongues, which are covered with teeth, to burrow into the carcasses of their prey. They then feed from the inside out. Hagfishes possess special glands positioned along the body that can produce large amounts of a milky fluid when the animal is disturbed.

(a)

Figure 10–1

Hagfish. (a) A hagfish from the Pacific Ocean. (b) Hagfish have flexible bodies that they can tie into overhand knots. They tie the knot at their tail, then slide it forward and over their head. This behavior is used to remove excess slime from their body, to escape predators, and to gain leverage when tearing flesh from their prey. (a, Tom Stack and Associates)

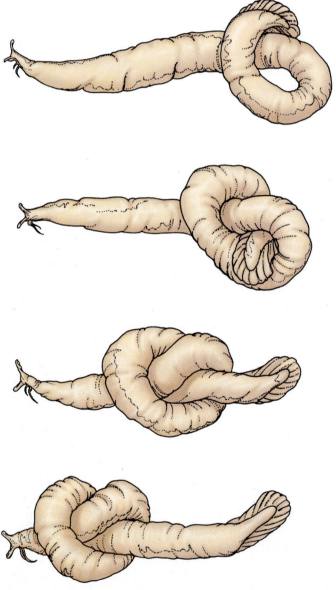

(b)

When the fluid comes in contact with seawater, it forms a slime so slippery that the hagfish is nearly impossible to grasp. The slime is probably used for physical protection. Hagfishes can remove the slime from their bodies by tying their tails in an overhand knot and sliding the knot forward and over the head (Figure 10–1b). The knot-tying maneuver is also used to escape predators and to pull off chunks of flesh when feeding.

Hagfishes have three or more hearts strategically placed throughout the body to help maintain a high enough blood pressure. Interestingly, only one of these hearts is controlled by the animal's nervous system. The others are regulated and coordinated by a chemical called eptatretin. Researchers are looking into the possibility of using eptatretin to treat individuals who suffer from certain types of irregular heartbeat. Although hagfishes contain both ovaries and testes, only one organ is functional in any one fish. Since they live in very deep water (as deep as 1000 meters), little is known about the reproductive behavior, development, and natural history of these animals.

Lampreys

Lampreys (order Petromyzontiformes) are parasitic fish that cause a tremendous amount of damage in the commercial fishing industry. Marine lampreys (*Petromyzon marinus*; Figure 10–2) spend their adult life in the ocean but return to freshwater to spawn. In North America, spawning takes place in the spring. Males migrate up rivers and build nests from stones in the shallow riffles of clear streams. The females arrive later and attach to one of the stones of the nest by their oral sucker. The male then attaches to the back of the female, and, as the eggs are shed, the male sheds his sperm. The fertilized eggs stick to stones in the nest and are ultimately covered by sand. The adults die shortly after reproducing.

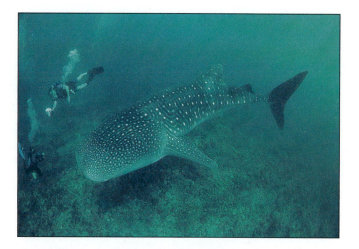

Figure 10–3

Whale Shark. A whale shark cruises a reef off Western Australia. (Jeffrey Rotman)

The larvae, known as ammocoetes (AM-oh-seets), hatch about two weeks later and migrate downstream, where they burrow into the river bottom with their mouths directed toward the current. The larvae are filter-feeders, and they remain burrowed for three to seven years before finally metamorphosing into adults and returning to the sea.

The parasitic adult attaches to its host by means of its oral sucker, and, using its sharp teeth, cuts through the host's flesh and sucks the blood. When the lamprey is finished feeding, it releases from its host, leaving an open wound that may ultimately prove fatal.

CARTILAGINOUS FISHES

Cartilaginous fishes (class Chondrichthyes), as well as all other vertebrates, possess jaws. They are represented by sharks, skates, rays, and chimaeras. The skeletons of these fishes are composed entirely of cartilage and do not contain any bone. The body of most species is streamlined, and they have well-developed sensory systems, making them some of nature's most sophisticated predators. With the exception of whales, sharks are the largest living vertebrate animals. The whale shark (*Rhincodon*; Figure 10–3), which may exceed 15 meters (49 feet) in length, is the largest species of fish.

Sharks

Sharks have streamlined bodies and are excellent swimmers (Figure 10–4). Using its massive trunk muscles, a shark swims with powerful, sideways sweeps of its tail, or **caudal**

Figure 10–2

Lamprey. The marine lamprey, *Petromyzon marinus*. (Zig Leszczynski/Animals Animals)

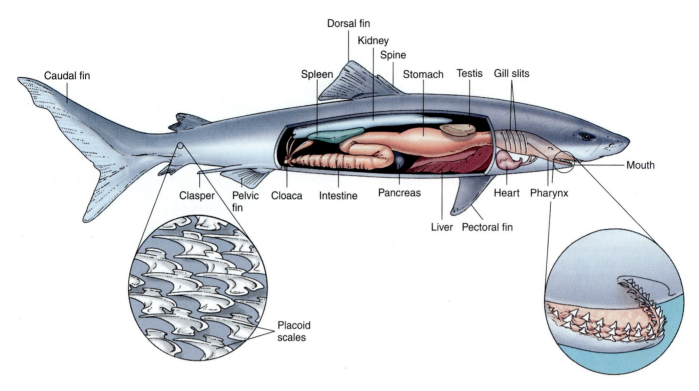

Figure 10—4

Shark Anatomy. The external and internal anatomy of a typical shark. Sharks have streamlined bodies and are well adapted for rapid swimming. Their skin is covered with placoid scales, which are similar to the teeth of other vertebrates. Modified placoid scales are found on the jaws and serve as teeth.

fin. The tail of many sharks is not symmetrical, and the vertebral column continues into the upper portion of the tail. This type of tail is called **heterocercal** (het-uh-roh-SIR-kuhl). The motion of the heterocercal tail lifts the rear of the animal up (tail lift) and forces the head down. The downward movement of the head is offset by a pair of large **pectoral fins** that provide lift for the front of the animal's body, allowing the shark to swim horizontally. In addition to providing lift, the pectoral fins also function in steering.

Additional fins include a pair of **pelvic fins** and one or two **dorsal fins.** A sharp spine may be associated with the dorsal fins, as in the spiny dogfish (*Squalus acanthias*). The pelvic fins of male sharks are partly modified to form claspers. The claspers function to transfer sperm from the male to the female during reproduction.

The skin of sharks is thick and leathery and is covered with a unique type of scale, placoid scales. Placoid scales resemble the teeth of other vertebrates. Sharks have several rows of teeth (actually modified placoid scales) in their jaws. If a tooth breaks or is pulled out, another is immediately available to take its place.

Sharks are slightly denser than water, and if they stop swimming, they will sink. Many sharks are able to compensate for this buoyancy problem with their large livers (in some species the liver may account for 20% of the shark's weight). The shark's liver produces large quantities of an oily material called squalene (SKWAY-leen). The squalene has a density less than saltwater (0.8 grams/cubic centimeter; the density of water is 1.0 gram/cubic centimeter), and this helps to offset the shark's high density.

Shark Sensory Systems

A shark's eyes lack eyelids, having instead a clear nictitating (NICK-ti-tay-ting) membrane that covers the eye and protects it. Color vision in many species is poor, but most have a large number of rod cells, specialized cells found in vertebrate eyes that are sensitive to low-intensity black and white light. The high rod content in their retinas allows them to see quite well under conditions of low light.

The shark's mouth is located ventrally, and a noselike structure, the **rostrum,** projects forward over the mouth and contains the nose and nostrils. Sharks have keen senses of taste and olfaction (smell) that allow them to monitor the chemical content of the water. Sharks actively pump water in and out of their nostrils, where sensory structures (chemoreceptors) respond to water-borne molecules. Almost two thirds of the cells in a shark's brain are involved in processing olfactory information. These animals can detect the presence of a drop of blood diluted in one million parts

MARINE BIOLOGY AND THE HUMAN CONNECTION

Sharks Aid in the Fight Against Cancer

When cells are crowded together and active, blood vessels invade the area to improve the supply of nutrients and oxygen. Cartilage, however, is an exception to this generalization. Cartilage cells prevent the invasion of blood vessels by secreting a chemical called *antiangio genesis factor* that blocks the growth of blood vessels. One reason that cancer cells grow so quickly is that blood vessels quickly grow into tumors to deliver the necessary materials for growth. Some researchers believe that the antiangiogenesis factor could be used to prevent blood vessels from supplying tumors, causing the tumor cells to die of starvation.

The problem with this possible treatment is that human cartilage cells produce very little of the antiangiogenesis factor, certainly not enough to use therapeutically. Since shark skeletons are composed entirely of cartilage, large amounts of the antiangiogenesis factor could be extracted from just a single animal. Sharks are currently being collected to supply quantities of this potentially valuable compound so that researchers can establish its effectiveness in the control of tumor growth.

of water and accurately find the source. It is no wonder that some biologists refer to sharks as "swimming noses."

The role that the sense of smell plays in locating prey may explain the odd shape of the hammerhead shark's head. This shark has a nostril at the tip of each end of the "hammer," and, as it swims, it moves its head from one side to the other. When the strength of a smell is equal in both nostrils, the shark senses that its prey is straight ahead.

Another important sensory organ is the **lateral line system** that consists of canals running the length of the animal's body and over the head (Figure 10–5). At regular intervals, the canals are open to the outside and there is a free movement of water in and out of them. Within the canals are sense organs called *neuromasts* that can detect vibrations in the fluid that fills the canals. Even the slightest movements in the water around the shark cause vibrations in the lateral line, where they are sensed by the neuromasts. The shark uses its lateral line system to locate prey and potential predators. Some researchers have suggested that the vibrations produced by a swimming human may be similar to those produced by a seal or an injured fish and may account for some sharks being attracted to swimmers.

Sharks and their relatives also have the ability to sense electricity with organs called the **ampullae of Lorenzini** (am-POOL-ee of loh-ren-ZEE-nee). These organs are found scattered over the top and sides of the animal's head, where they open to the outside by way of small pores in the skin. It has been suggested that the broad head of the hammerhead sharks (*Sphyrna*) is an adaptation to increase the field of the ampullae of Lorenzini. The ampullae can sense changes as little as 0.1 microvolt in the electrical fields in the water. Marine biologists have speculated that sharks may use these organs to sense injured animals and thus find food.

In a series of experiments using sharks and rays found in the Red Sea, marine biologists trained animals to eat in an

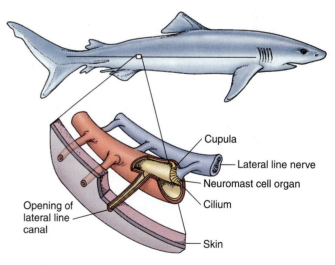

Figure 10–5

The Lateral Line System. The lateral line system consists of two canals, one on each side, that run the length of the animal's body. The fluid in the canals freely communicates with the water surrounding the animal. Even tiny vibrations in the surrounding water cause the fluid in the canals to move, moving the cupula of the neuromast cells and sending a signal to the animal's brain.

area directly over a pair of electrodes that had been buried in the sand. When the fish were fed, the current was turned on, producing 0.4 microvolt. Eventually the researchers discontinued putting out food, but when the electrodes were turned on, the animals swarmed over the area and, in some instances, dug the electrodes from the sand and began to snap at them. The experiments indicated that the animals could sense the minute electrical charges and trace them to their source and could associate the current with potential food.

Another series of experiments showed that even sharks that had not been previously conditioned could find injured fish that were buried in the sand on the basis of minute electrical changes produced by the prey. There is also some evidence that sharks may be able to detect the electrical impulses of normal swimming prey.

Digestion and Excretion in Sharks

A shark's mouth contains several rows of sharp, triangular teeth that are used for grasping their prey. When feeding, sharks generally swallow their food whole or tear off large chunks that are then swallowed whole. If hungry, sharks will usually eat anything that will fit into their mouths, and during frenzy feeding will eat nonfood items. Some of the unusual items that have been found in the stomachs and intestines of captured sharks include:

several legs of lamb	parts of a reindeer
horseflesh	a 30-foot roll of roofing
part of a ham	paper
the front half of a	telephone books
bulldog	pots and pans
the rear half of a pig	a bottle of Madeira wine
tin cans	a keg of nails
coal	sail canvas
false teeth	automobile license plates
34 leather jackets	a kerosene lantern

After the food is swallowed, it passes to the stomach and then on to a relatively short intestine that contains a structure called a *spiral valve*. The spiral valve forms a shape similar to a spiral staircase and is arranged so that the narrow part of the valve is anterior and the wider part is posterior. This arrangement slows the movement of food through the short intestine so that it can be fully digested. The spiral valve also increases the surface area of the intestine for the absorption of nutrients.

Characteristics of the shark excretory system suggest that, like bony fishes, cartilaginous fishes evolved in freshwater and later migrated to the sea after having thoroughly adapted to a freshwater environment. The move to a marine environment posed the problem of losing body water by osmosis to the animal's saltier surroundings. In sharks, this problem was solved by retaining nitrogenous wastes in the blood. This, in turn, maintains a blood osmotic pressure that is slightly higher than seawater and thus prevents the loss of body water to the environment. A prominent gland called the *rectal gland* empties into the rectum. This gland supplements the kidney by excreting a fluid with a high salt concentration.

Reproduction in Sharks

In males, a pair of ducts called *Wolffian ducts* carry sperm from the testes to the claspers, which are used to deposit sperm into the female reproductive tract (Figure 10–6). The female reproductive system consists of a single ovary and a pair of oviducts. The oviducts carry eggs from the ovary to a portion of the oviduct that is modified to function as a uterus. Fertilization is internal in sharks, and, in most species, the young develop in the uterus.

In some species, a primitive placenta actually attaches the young to the uterus. When the young are fully developed, they are born. This type of reproduction, in which the young develop from embryos attached to the mother, is called **viviparous** (vy-VIP-uh-ruhs) **reproduction.** In some species, the young develop from eggs that are simply retained within the mother's body until they hatch. This mode of reproduction is called **ovoviviparous** (oh-voh-vy-VIP-uh-ruhs) **reproduction.** Some species of shark and skates lay their eggs in leathery egg cases, which are frequently attached to seaweed. Development occurs within the egg case, and the young animal hatches when develop-

Figure 10–6

Shark Reproduction. Two nurse sharks copulating. The male holds the pectoral fin of the female in his mouth as he uses his claspers to introduce sperm into the female's genital opening. (Nick Caloyianis/ National Geographic Society)

MARINE ADAPTATION

Megamouth

In 1976 a new species of shark was discovered off the coast of California. It was named *megamouth* (*Megachasma pelagios*) because of its enormous jaws, which measure about one meter (3 to 4 feet) wide (see figure). These animals have large, blubbery lips and over 200 rows of tiny teeth that match fossils 26 million years old. Megamouth is one of only three known plankton-feeding sharks, and it may use its small teeth to strain food from large gulps of water. The silvery lining of the animal's large mouth is bioluminescent and may aid in attracting prey like euphausids (shrimp-like animals such as krill, see Chapter 9) that live at depths of 150 to 600 meters (500 to 2000 feet). The specimens of megamouth captured so far have ranged from 4.5 to 5 meters (15 to 16 feet) in length and weighed as much as one-half ton.

Figure 10–A
Megamouth. *Unknown to science until 1976, this unusual shark feeds on plankton.*
(Bruce Rosner/Jeffrey Rotman Photography)

Four more specimens have been captured since the first one in 1976. The most recent specimen was captured just south of Los Angeles, California, in a fisherman's gill net and was still alive. When no facility could be found that was large enough to keep the shark, Don Nelson at California State University, Long Beach,

was asked to track the animal upon its release. Dr. Nelson and his coworkers implanted two transmitters in the shark and were able to follow it at sea for two days. It was already known that megamouth was a plankton feeder, but Nelson and his team discovered that it was also a vertical migrator. On the night that the shark was released, it remained in 15-meter water (50 feet). At dawn, the shark descended to 151 meters (500 feet). The following evening, the animal ascended again to shallow water.

It may be surprising that an animal as large as megamouth could go unnoticed for so long. Part of the reason may be that the animal spends its days in the remote ocean depths. Another reason may be that, since it is a filter-feeder, it would not strike at baited hooks like other sharks, and thus would go unnoticed by fishermen.

---

ment is complete. Animals that lay their eggs outside of the body are called **oviparous** (oh-VIP-uh-ruhs).

Shark Attacks on Humans

Although sharks as a group have a rather bad reputation (due in part to the publication in 1974 of Peter Benchley's novel, *Jaws,* and the subsequent movie), they are generally timid and cautious animals. Some species, including the great white shark (*Charcharodon carcharias*), tiger shark (*Galeocerdo cuvieri*), mako shark (*Isurus oxyrhynchus*), and hammerhead sharks (*Sphyrna*), do pose a potential threat to humans (Figure 10–7). There are documented cases of each of these species causing human fatalities.

Most shark attacks occur in the region of the Southern Pacific Ocean and parts of the Caribbean Sea. Attacks by great white sharks are now most common along the West Coast of the United States. Between 1970 and 1991, there

have been 33 documented attacks by great white sharks along the California coast, of which four were fatal. Of these attacks, only one was on a swimmer. The remaining attacks were made on snorkelers (11), surfers (10), divers (9), kayakers (1), and paddleboarders (1). These data may support the hypothesis that many attacks by great white sharks involve mistaken identity. The shark mistakes a human for a seal or sea lion, its normal prey in California waters. More research will be needed, however, to arrive at a definitive explanation for shark attacks on humans.

Although there is no worldwide organization that keeps track of shark attacks, the U.S. Navy's Office of Information reports that the number of shark attacks that occur on a global basis averages between 40 to 50 annually. In 1961, the short-lived (1958–1970) Shark Research Panel of the American Institute of Biological Sciences issued a report in the journal *Science* that analyzed 790 shark attacks that occurred between the years 1580 and 1960. The panel concluded that 599 (76%) of the attacks were unprovoked, 32

(a)

(b)

Figure 10–7

Dangerous Sharks. Several species of shark are known to attack humans without provocation. They include (a) the great white shark and (b) the hammerhead shark. (a and b, Jeffrey Rotman)

(4%) were provoked, and 30 (3.8%) followed air and sea disasters. Fifty-three (6.7%) of the attacks may not have involved sharks, and in 76 (9.6%) cases, the attacks were on boats rather than persons.

Most shark attacks on boats involve the great white shark, and many of the recorded incidents have occurred in Canadian waters. On July 9, 1953, two men in a 14-foot dory were attacked by a large fish off Cape Breton Island, Nova Scotia. The fish hit with such force as to punch a 20-centimeter (8-inch) hole in the boat's hull and knock both men into the water, where one drowned. A tooth found embedded in the wood of the boat was examined by Dr. William Schroeder of Harvard and Woods Hole Oceanographic Institution. He identified the fish as a great white shark that must have been at least 3.6 meters (12 feet) long.

The chances of being attacked and killed by a shark are about the same as being struck and killed by lightning. In Australia, where shark encounters and attacks are more common than in the United States, approximately 92 swimmers die annually of drowning, 8 die in SCUBA-diving accidents, but only 1 dies of a shark attack. Many shark attacks can be avoided by not entering water known to have sharks or engaging in activities that attract the animals' attention. Large sharks regularly feed on animals approximately the size of humans. The sharks attack humans who are swimming or surfing, apparently mistaking them for seals or large fish.

Another common source of shark attacks involves overly confident divers who minimize the potential danger from smaller sharks, those around 1 meter (3.3 feet) in length. Even these animals are quite powerful and can inflict severe bites. Today, shark attacks are being recorded in colder water since humans increasingly use wetsuits to enter cold areas regularly.

Skates and Rays

Skates and rays are cartilaginous fishes that have greatly enlarged pectoral fins and much reduced dorsal and caudal fins (Figure 10–8a). The broad, flattened pectoral fins are attached to the head and give many species a characteristic leaf-shaped look. Skates and rays are more specialized than their shark cousins, and they differ from sharks, not only in their general body appearance but also in their mode of swimming. Skates and rays, with the exception of sawfishes (*Pristis*) and guitarfishes (*Rhinobatos*), swim with their pectoral fins.

One major difference between skates and rays is their mode of locomotion. Rays swim by moving their fins up and down as a bird moves its wings while flying. Skates, on the other hand, swim by creating a wave that begins at the forward edge of the fin and sweeps down the edge to the back of the fin, allowing the animal to glide easily along the bottom as its fins ripple. Skates can also be distinguished from rays by their lack of stinging spines on the tail and the lack of large electric organs between the pectoral fin and skull.

Skates and rays are adapted to life on the bottom. Their gill slits are on the underside of their bodies, and an opening called the **spiracle** is located on the dorsal surface. Since the mouth and gills are frequently in contact with the sandy bottom, water is drawn in through the spiracles and passed out over the gills. It then moves out through the animal's ventral surface. This arrangement helps to prevent the delicate gill filaments from becoming clogged with sand or debris. Skates and rays have powerful jaws modified for crushing their prey. They feed primarily on crustaceans and molluscs that they find along the bottom.

Skates and rays have evolved a variety of defenses to protect them from predators. Electric rays (*Torpedo*) have a pair of electric organs in their head. When the cells in these organs are simultaneously discharged, a relatively low-voltage, high-amperage current of several kilowatts passes into the surrounding water. In addition to defense, the electric rays use their electric charge to stun prey. It has been reported that the ancient Egyptians used electric rays in treating diseases such as arthritis and gout.

Stingrays (*Dasyatis*; Figure 10–8b) have hollow barbs connected to poison glands. These modified dorsal fin spines may be distributed along the tail, or there may be a single barb at the base of the tail. When disturbed, stingrays whip their tails around. If the barb punctures the skin, it injects a venom that causes swelling, cramping, and excruci-

ating pain. A common treatment for stingray injuries is to place the injured area into hot water. The heat from the water breaks down the protein toxin. Wounds from stingrays heal very slowly and are prone to bacterial infections that may be more serious than the wound.

Like most rays, stingrays are ovoviviparous. At birth, their barbs are flexible and covered with a sheath to prevent injury to the mother during the birthing process. Most skates are oviparous, releasing their eggs in a leathery, rectangular egg case called a *mermaid's purse* (Figure 10–8c).

Sawfishes (Figure 10–8d) have a series of barbs along their pointed rostrums. When disturbed or when feeding, they shake their heads sideways, using the sharp points of

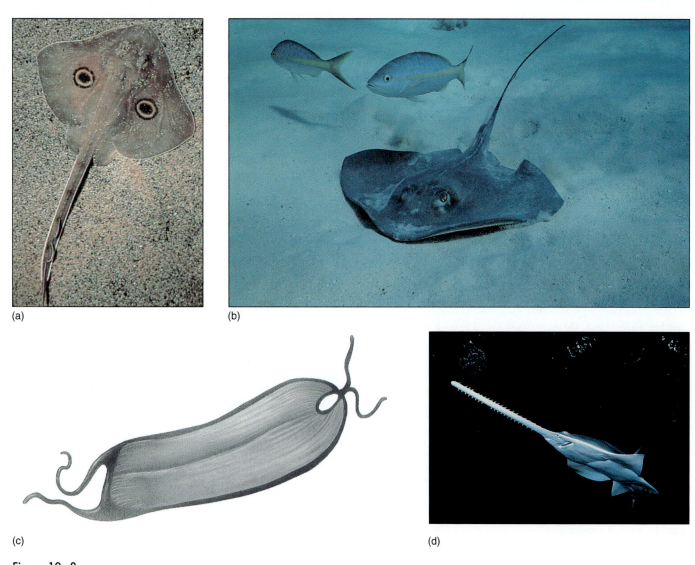

(a)　　　　　　(b)

(c)　　　　　　(d)

Figure 10–8

Skates and Rays. Both skates and rays have enlarged pectoral fins. (a) Skates lack a stinging spine associated with the tail, whereas (b) rays, like this stingray, have a spine that can inflict a painful injury. Rays are viviparous, whereas skates are oviparous. (c) Skates release their eggs in a leathery egg case called a *mermaid's purse*. (d) Sawfishes are related to skates and rays. (a, Charles Seaborn; b, Copyright David Wrobel 1993/Biological Photo Service; d, Norbert Wu)

Figure 10–9

Chimaeras. Relatives of the sharks, chimaeras like this female ratfish from Vancouver Island, Canada, are bottom dwellers that feed on a variety of fishes and invertebrates. (Copyright David Wrobel 1994/Biological Photo Service)

the "saw" to inflict injury. Like the sting of stingrays, the toothed saw of the baby sawfish is flexible and covered by a sheath that protects the mother when the young are born.

Chimaeras

Another group of cartilaginous fishes is the **chimaeras** (ky-MEER-uhz) (subclass Holocephali; Figure 10–9). These relatives of the sharks have bodies with unusual shapes and are

given common names such as *ratfish, rabbitfish,* and *spook-fish.* Instead of teeth, chimaeras have flat plates that they use to crush their prey. They feed on a wide variety of foods such as seaweed, crustaceans, molluscs, echinoderms, and fish. They are generally bottom dwellers, and their strangely shaped and colorful bodies help to camouflage them.

BONY FISHES

Modern bony fishes (class Osteichthyes) originally evolved in freshwater lakes and streams and then colonized the marine environment. Some of the most primitive bony fishes in the marine environment are the coelacanth (SEE-luh-kanth) (*Latimeria chalumnae*) and marine sturgeons (*Acipenser*).

The coelacanth (Figure 10–10a) belongs to a group of fishes known as *lobe-finned fishes,* which are named for the fleshy muscular base of the pelvic and pectoral fins. They are found in the deep waters of the Indian Ocean around the Comoro Islands. For years coelacanths were thought to be extinct, and the discovery of a live specimen in 1938 created quite a stir in the scientific community. Before this time, fishes like the coelacanth were known only from fossils. Biologists speculate that a fish similar to the coelacanth is a good candidate for the ancestor of amphibians. The discovery of a live coelacanth gave biologists a chance to test these theories.

Marine sturgeons (Figure 10–10b) are one of the most primitive bony fishes, exhibiting such characteristics as a heterocercal tail and ganoid scales. Ganoid scales (Figure 10–11) are very thick and heavy, giving the fish an armored appearance. More modern bony fishes have homocercal

(a)

(b)

Figure 10–10

Primitive Bony Fishes. (a) A coelacanth. Notice the muscular bundles at the base of the fins. (b) The scales of this marine sturgeon are modified into bony plates that give the fish an armored appearance. (a, Dr. J. Metzner/Peter Arnold, Inc.; b, Kenneth Lucas/Biological Photo Service)

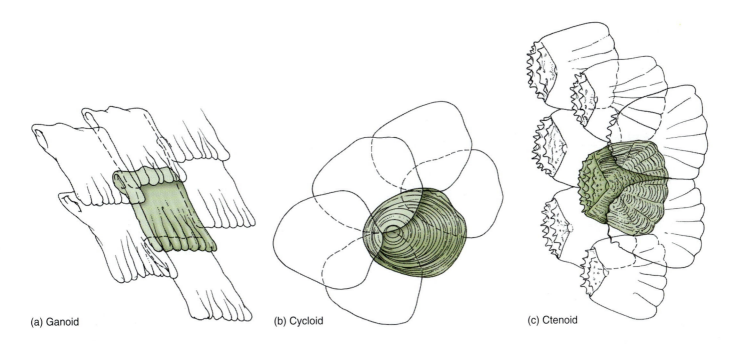

(a) Ganoid (b) Cycloid (c) Ctenoid

Figure 10–11

Fish Scales. (a) Thick, heavy, ganoid scales are characteristic of primitive bony fishes. More modern bony fishes have lighter, more flexible scales like (b) cycloid scales and (c) ctenoid scales.

tails, cycloid or ctenoid scales, and more maneuverable fins. **Homocercal** (hoh-moh-SIR-kuhl) **tails** have nearly equal dorsal and ventral lobes, and the vertebral column usually does not continue into the tail. Cycloid (SY-kloyd) and ctenoid (TEE-noyd) scales are lighter and more flexible than the heavier ganoid scales and are less cumbersome for active swimmers. The structure of their fins gives bony fishes better control of their movements.

Bony fishes possess both unpaired median fins and other paired fins (Figure 10–12). The median fins consist of one or more dorsal fins, a caudal or tail fin, and usually one anal fin. Median fins help fishes to maintain stability while swimming. The paired fins consist of pectoral and pelvic fins, both of which are used in steering. Pectoral fins also help to stabilize the fish.

Shape and Color in Fishes

The shape of a fish's body is mainly determined by the characteristics of its environment. Water is not very compressible. In order for a fish to move through water, it must displace the water—that is, literally push the water aside. The evolution of a streamlined body that is pointed at the front, bulkier in the forward half of the middle, and then tapered

toward the tail allows fish to move efficiently through their watery environment. This body plan allows water to flow easily along the sides of the fish, and in some species, the water may even give the fish an extra forward push as it passes the tail.

Fishes that are very active swimmers, such as the tuna (*Thunnus*) and marlin (*Makaira*), have a fusiform body (Figure 10–13a) with a very high and narrow tail. Fishes that live in seagrass or on coral reefs, like the butterflyfish and angelfish (*Pomacanthus*), have a laterally compressed or deep body that helps them to navigate more efficiently through their complex environment (Figure 10–13b).

Bottom-dwelling fishes, such as the sole (Pleuronectidae) and flounder (Bothidae), have depressed or flattened bodies (Figure 10–13c). In species like the flounder, the body is flattened sideways, but the animal swims horizontally rather than vertically. These fishes begin life looking like normal fish, but early in the juvenile stage, they begin side-swimming, and an eye migrates from what will become the bottom side to the upper side.

Fishes such as toadfish (order Batrachoidiformes), scorpionfish (*Scorpaena*), and anglerfish (*Antennarius*), which exhibit a more sedentary lifestyle, have globular bodies, and their pectoral fins are usually enlarged to help support the body (Figure 10–13d). Burrowing fishes and fishes that live

Figure 10–12

Bony Fish Anatomy. (a) The general internal and external anatomy of a bony fish. (b) In more primitive species, like the herring (Clupeidae), the pectoral fins are lower on the body and the pelvic fins are more posterior. (c) In more advanced species, like butterfly-fishes (*Chaetodon*), the pelvic fins are closer to the throat and the pectoral fins are higher on the body and more vertical.

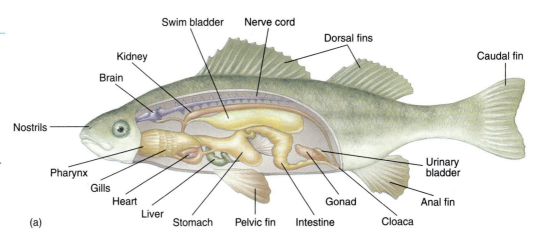

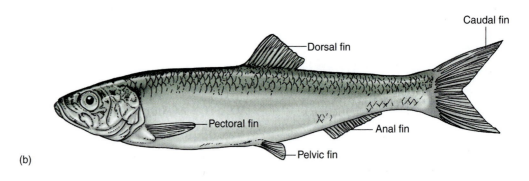

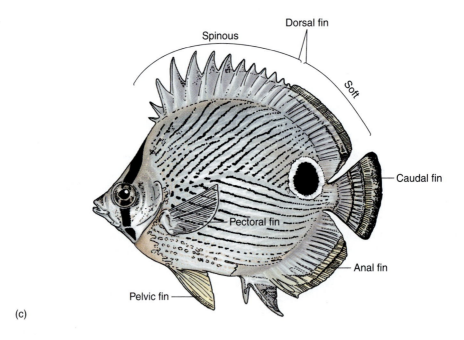

in tight spaces, such as moray eels (*Gymnothorax*), have long, snakelike bodies, and they lack or have reduced pelvic and pectoral fins (Figure 10–13e).

Not only does the environment play a key role in shaping the body of a fish, but it is also important in determining a fish's color. In fishes, color functions in species recognition, as an aid in behavioral displays, and in concealment.

The color of most fishes is located in a layer of cells called **chromatophores** (kro-MAT-uh-fohrz), which is located just beneath the layer of transparent scales.

Fishes that live in the open ocean, such as tuna, marlin, and the swordfish (*Xiphius gladius*), display a type of coloration known as **obliterative countershading.** In obliterative countershading, the fish's back (dorsum) is a dark

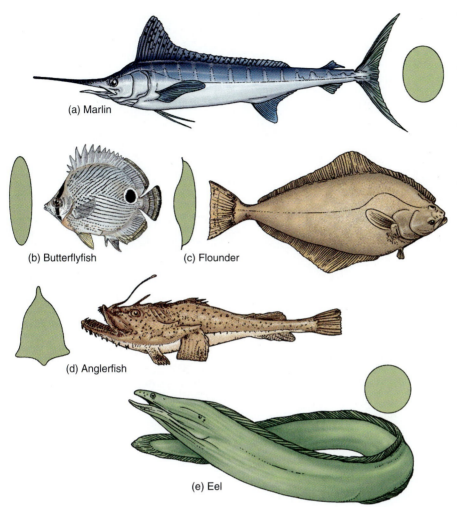

Figure 10–13

Fish Shapes. (a) Fishes that are active swimmers like this marlin have a fusiform body. (b) Reef fishes, such as this butterflyfish, that swim among the corals have laterally compressed or deep bodies. (c) Bottom dwellers like this flounder have horizontally compressed or depressed bodies. (d) Sedentary fishes such as this anglerfish have globular bodies. (e) Burrowing fishes and fishes, like this moray eel, that live in tight crevices have snakelike bodies.

(a) Marlin

(b) Butterflyfish   (c) Flounder

(d) Anglerfish

(e) Eel

green, dark blue, or grey. The shades become less dark on the sides, and the color of the belly is pure white. When looking down from above, the dark back blends in with the surrounding dark water. When looking up from below, the white belly blends in with the brightly lit surface. This effectively camouflages the animal even though it is in open water. This same type of coloring is also found in many inshore fish species, sharks, many marine mammals, and penguins.

Many species of coral reef fish exhibit **disruptive coloration** (Figure 10–14). In disruptive coloration, the background color of the body is broken up by lines that usually run vertically. This helps to break up the pattern of the fish's body and to make it more difficult for predators to see an individual. Often, one of the lines will pass through the eye, making it more difficult to be seen, and a dark dot, or eye spot, will be in the area of the tail. Many aquatic predators use the eye to determine which end of an animal is the head. The eye spot on the tail and the line through the eye draw a predator's attention to the wrong end of the fish, making it more likely that the prey will survive an attack.

Figure 10–14

Disruptive Coloration. The dark vertical bands on this butterflyfish from the Caribbean break up the background color, making it more difficult to see against the busy background of a coral reef. Notice the band that runs through the eye, making it more difficult for a predator to identify the fish's head. (Susan Blanchet/Copyright 1991 Dembinsky Photo Associates)

(b)

Figure 10–15

Camouflage. (a) This pipefish from the Solomon Islands in the Pacific Ocean resembles a piece of seaweed. (b) A scorpionfish from the Red Sea mimics its environment. (a, Norbert Wu; b, Jeffrey Rotman)

(a)

Some fishes, such as the pipefish (*Syngnathus*; Figure 10–15a), mimic their environment, blending in with the surrounding eelgrass. Others, such as the scorpionfish (Scorpaenidae; Figure 10–15b), are masters of camouflage. Their irregularly shaped bodies and cryptic coloration make them almost impossible to see when they are not moving. This not only helps them avoid predators but also helps them to be efficient predators. Smaller fishes on which they prey do not see the hidden scorpionfish and on swimming too close, become an easy meal.

Locomotion in Bony Fishes

Bony fishes use their trunk muscles to propel themselves through the water. The muscles are arranged as a series of muscle bands, and each band looks like a letter W lying on its side (Figure 10–16). Movement results when the bands of muscles contract alternately from one side of the body to the other. The muscle contractions originate at the anterior end of the fish and move toward the tail. Since the posterior portion of a fish is more flexible than the anterior, the wave of contraction increases in amplitude as it passes toward the tail. As a result of this pattern of contraction, the fish's body

bends. The body pushes against the water, and since the water cannot be compressed, the fish is pushed forward at an angle.

Fishes that swim slowly, such as the eel, tend to undulate their entire bodies, whereas fast-swimming fishes, such as marlin and swordfish, primarily use their tails. Increased use of the tail helps to decrease drag in the water and to maximize speed. Marlin have a series of interlocking projections down the sides of the spine that serve to spring the tail back toward the centerline after a thrust, thus reducing the effort involved in long-distance swimming. Not all fishes rely on their trunk muscles for swimming. Triggerfishes (Balistidae) swim by undulating their dorsal and ventral fins, and wrasses (Labridae) swim by undulating their pectoral fins.

The Swim Bladder

Most bony fishes, with the exception of some pelagic species, bottom dwellers, and deep-sea fishes, have a gas-filled sac called a **swim bladder** (see Figure 10–12a) to help them offset the density of their bodies and maintain a neutral buoyancy. Specialized networks of blood vessels located in the swim bladder can add or remove gases such as oxygen and carbon dioxide. By adjusting the amount of gas in the swim bladder, a fish can remain indefinitely at a given depth without any muscular movement and with expenditure of minimal energy.

When the fish descends in depth, more gas must be added to the swim bladder, or else the bladder will become compressed, and the fish will become denser and sink. On the other hand, as the fish ascends, it must remove gas from the swim bladder, or else the gas will expand, and the fish will become increasingly lighter and be forced to the surface of the water.

Pelagic fishes that are active swimmers, such as mackerels (*Scomber*) and skipjacks (*Katsuwonis pelamis*), do

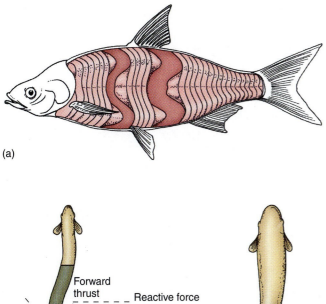

(a)

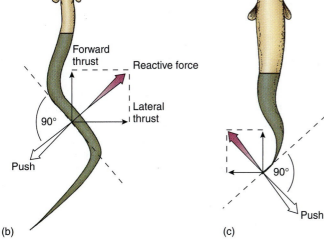

(b)   (c)

Figure 10–16

Fish Locomotion. (a) The trunk muscles of fishes are arranged as a series of W-shaped bands. These muscles contract in sequence from anterior to posterior and alternately from one side of the body to the other. By pushing the body against the water, the fish is propelled forward. (b) Fishes such as eels undulate their entire body as they swim, whereas fishes with shorter bodies (c) use mostly the posterior portion of the body as they swim.

not have swim bladders, probably because they frequently make rapid depth changes and would not be able to regulate the gas in a swim bladder as quickly as they can change depth. Like their relatives, the sharks, these animals must keep swimming or sink. Bottom dwellers, such as scorpionfishes, lack a swim bladder because they have no need to maintain neutral buoyancy. Many fishes that live in the deep ocean also lack a swim bladder. We will discuss these fishes and their adaptations in more detail in Chapter 18.

Respiration and Excretion

The evolution of gills allowed fish to extract oxygen efficiently from the water and to eliminate the carbon dioxide that is produced during metabolism. A fish's gill (Figure 10–17) is composed of thin, rod-like structures called **gill filaments,** and it has a very rich blood supply that runs in an opposite direction to the flow of water. This pattern allows for the most efficient exchange of gases between the gills and water.

The gills are covered by a moveable flap of tissue called the **operculum.** The operculum helps to protect the delicate gills from damage while maintaining the streamlined body necessary for efficient swimming. It also functions in pulling water in through the mouth and over the gills.

Water must be continuously moved past the gills to keep the blood properly oxygenated. Very active fishes with high oxygen demands, such as mackerel, can supply enough oxygen to the blood only by continuously swimming forward to force more water past the gills. Fish such as these would die of suffocation in an aquarium that did not permit enough room to swim, even if the water were saturated with oxygen.

The salt concentration in the blood of marine fishes is much lower than the concentration of salts in seawater. As a result, marine fishes tend to lose water through their gills and inner tissues of the mouth. This loss of water to the environment by osmosis poses the risk of drying out, much the same as for animals that live in desert environments. To compensate for the loss of water, marine fish drink seawater. This behavior, however, brings more salt into the fish's body.

Marine fish have evolved three mechanisms for dealing with this problem of excess salt. First, the salts that are most abundant in seawater, such as sodium chloride and potassium chloride, are carried by the blood to the gills, where they are actively secreted by specialized salt-secretory cells (Figure 10–18). Second, the remaining salts, such as magnesium, calcium, sulfates, and phosphates, are eliminated with the feces. Finally, ions that are not removed by the intestine or salt-secretory cells are excreted by the kidneys. Fishes that are adapted to seawater produce negligible amounts of urine, since they need to retain as much water as possible.

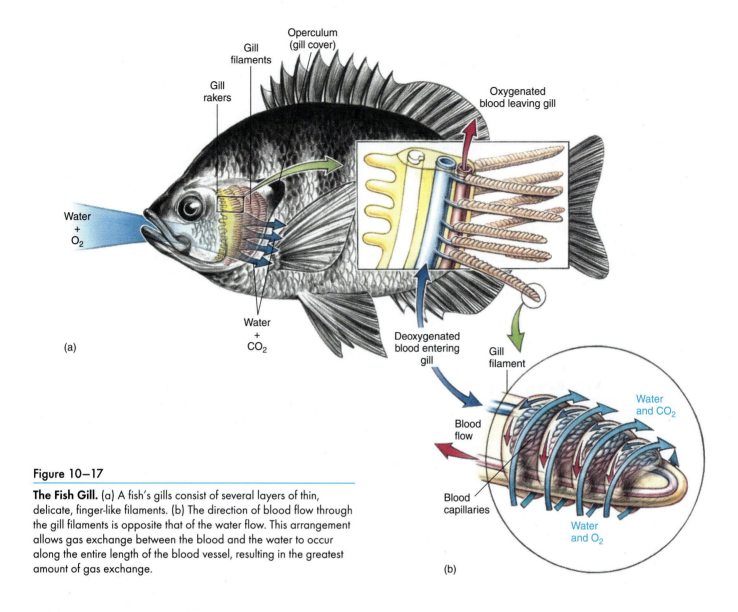

Figure 10–17

The Fish Gill. (a) A fish's gills consist of several layers of thin, delicate, finger-like filaments. (b) The direction of blood flow through the gill filaments is opposite that of the water flow. This arrangement allows gas exchange between the blood and the water to occur along the entire length of the blood vessel, resulting in the greatest amount of gas exchange.

Figure 10–18

Osmoregulation in Bony Marine Fishes. Marine fishes tend to lose water to their surroundings, especially from the gills. In order to replace the water lost, they drink seawater. For this process to be useful, the fish must retain as much water as possible while eliminating the excess salt. Excess salt is eliminated by the gills, the intestine, and the kidneys.

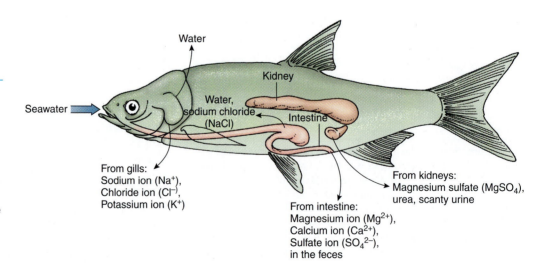

The Nervous System and Senses

Bony fishes, like their cartilaginous cousins, have a keen sense of smell. Their nostrils generally open into a blind sac that contains the olfactory (sense of smell) receptors. In some species, such as eels, the tubular sacs can project from the nostrils to smell the surrounding water. Bony fishes also have taste receptors that may be located on the surface of the head, jaws, tongue, and in the mouth and throat. Like sharks, bony fishes have a lateral line system, which helps them to detect the slightest movements in the water, and an inner ear. The inner ear of bony fishes is capable of detecting sounds in the range of 200 to 13,000 hertz (cycles per second).

Most bony fishes have very good vision. Their eyes lack eyelids, and they generally do not adjust the size of the pupil, since the quantity of light in water is frequently low. Because visibility in water is usually quite limited, fishes do not need to adjust their eyesight for a wide range of distances. As a result, their lenses are round and rigid, an adaptation for seeing things that are close. If the fish needs to adjust its vision to see more distant objects, the entire lens is moved backward, much like focusing a camera. Another reason for the round shape of the fish's lens has to do with the physics of light. The density of the lens is not much different than the density of the water surrounding it, and, as a result, the light does not bend or refract appreciably. In order to focus an image on the retina, the fish must have a sharply curved lens, and the sharpest curve possible is that presented by a round lens.

The eyes of most fishes are set on the sides of the head instead of the front of the head to keep the body streamlined for swimming. It is believed that a fish sees only a narrow field directly in front of it with each eye. For the most part, fishes have monocular vision, with each eye seeing its own independent field. In addition to rod cells that function in black and white vision, fish eyes also contain cone cells, which permit color vision.

Feeding Types

The evolution of jaws allowed fishes to diversify from their bottom-feeding ancestors and explore a variety of other feeding modes. The ability to be active predators placed increased selective advantage on adaptations that improved swimming, agility, feeding, and the senses, and these factors contributed greatly to shaping the body form and characteristics of modern fishes.

Most marine fishes are carnivores, using their sharp teeth to catch their prey rather than chew it. Since to spend time chewing their food would block the flow of water past the gills, most carnivorous fishes swallow their food whole. Some species, such as pufferfish (*Sphoeroides*) and the boxfish (*Ostracion tuberculata*), crush their prey with powerful jaws. Groupers (*Epinephelus*) have large mouths with small teeth. They lie in wait in their lairs until their prey comes along. When a mullet (*Mullus*), grunt (*Haemulon*), or large crustacean comes by, the grouper will open its huge mouth, creating a suction that draws in prey. If the prey is too large to fit completely into its mouth, a grouper will hold the prey's tail with its small teeth, while pharyngeal teeth in the grouper's throat crush the victim's head. Flounders lie camouflaged on the bottom. When a meal in the form of a crustacean, worm, or small fish comes along, the flounder springs up and grabs the unsuspecting prey.

Active swimmers, such as tuna, require large amounts of energy. To supply their energy needs, tuna feed almost constantly, consuming as much as one tenth of their body weight per day. The constant flexing of their muscles, coupled with a countercurrent heat-exchange system (Figure 10–19), produces a body temperature 2 to 10°C above that of the surrounding water. This increased body temperature and their modified respiratory and circulatory systems allow them to metabolize faster to support the energy needs of their very active lifestyle.

Some marine species are herbivores feeding on a variety of plants and algae. Surgeonfishes (*Acanthurus*) feed on the algae that grow on rocks and coral on coral reefs. In most species, the teeth are broad and flat with a sharp edge

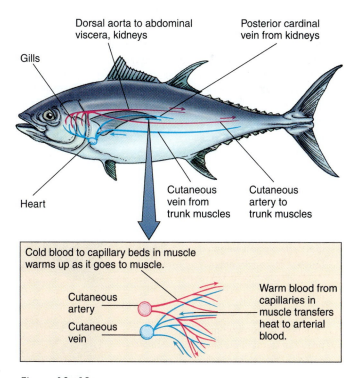

Figure 10–19

Countercurrent Heat Exchange in Tuna. By trapping some of the heat generated by the action of muscles, tuna are able to maintain an internal body temperature several degrees higher than their environment and to increase their level of metabolism.

MARINE ADAPTATION

Surviving in Near-Freezing Water

Of the approximately 20,000 species of fish on earth, only about 300 occur in the Antarctic Sea, and most of these are endemic. The greatest challenge to fishes living in these waters is to survive the subfreezing temperatures. Because water expands when it freezes, if the cells of animals like fishes were to freeze, they would rupture.

Marine invertebrates that inhabit the bottom of Antarctica's continental shelf regularly encounter temperatures as low as −2.7°C. They don't freeze because their bodies contain the same concentration of salts and minerals as the surrounding seawater, as well as a variety of organic compounds, which depresses the freezing point of their body fluids. As long as they stay away from the ice, they avoid freezing, and since ice floats, they simply remain near the ocean floor.

The blood and other body fluids of marine fishes, however, are more dilute than seawater, and their tissues freeze around −0.2°C. At this temperature, if an ice crystal were even to brush a fish's skin, it would quickly propagate and pierce the skin like a spear. Although ice is virtually everywhere during the winter, most Antarctic fishes encounter it without any problems. It was not until the 1960s that physiologists began to study how fishes in this environment can survive under such conditions. What they discovered was a relatively simple solution to the problem. The fishes have molecules in their blood and other body fluids composed of sugars and amino acids called glycopeptide chains that act as antifreeze. The glycopeptides depress the freezing point of water 200 to 300 times

more than would be predicted from their physical properties, thus protecting the fishes from the freezing water and contact with ice.

Another problem faced by fishes in the Antarctic Sea concerns the molecule hemoglobin that is found in the red blood cells of vertebrates. The hemoglobin molecule is responsible for the red blood cell's ability to carry oxygen, and at low temperatures it functions very inefficiently. To transport enough oxygen at the low temperatures of the Antarctic waters would require large amounts of hemoglobin and large numbers of red blood cells. Unfortunately, this would make the blood very viscous, and at the cold environmental temperatures at which these fishes live, the amount of energy necessary to move the blood and the amount of nutrients necessary to supply the energy would be extreme.

Because of these problems, Antarctic fishes gradually became more dependent upon oxygen dissolved in their blood plasma, thus eliminating some of the hemoglobin and red blood cells. The most extreme examples of this adapta-

tion can be seen in the 16 species of fishes known as *ice fishes* (Salangidae). Ice fishes have been known for over 100 years. Whalers visiting Antarctic waters used to call them *white crocodile fishes* because of their pale color and large mouths with many sharp teeth. Not only does the fish's body lack color, but even their gills are white and their blood is transparent. It was not until 1950 that biologists discovered that these fishes lacked hemoglobin and had very few red blood cells. To compensate for the anemia that results from this condition, these fish have a large heart and wide blood vessels, creating a circulatory system that can handle a large volume of blood under low pressure. This allows the fish to expend minimal amounts of energy to pump the large volumes of blood necessary to carry oxygen in the absence of hemoglobin. These fishes also have the ability to build up high oxygen debts that can be repaid later when the fish is less active—another energy-saving adaptation for life in a near-freezing environment.

Figure 10–B
Icefish. This icefish from Antarctica lacks the oxygen-carrying protein hemoglobin. The oxygen that its cells need is carried dissolved in the relatively large volume of blood that circulates through its body. (Norbert Wu)

Figure 10–20

Parrotfish. Parrotfish feed on the symbiotic algae of corals. Their beaklike mouthparts allow them to crush the coral to extract the algae. (Jeffrey Rotman)

(like a shovel), making them ideal tools for scraping food from the rocks. Parrotfishes (*Scarus*) (Figure 10–20) have teeth that are fused to form a beaklike structure. They feed mostly on coralline algae and, to a small extent, on the symbiotic algae that live in the tissues of coral polyps. Using its strong beak, the parrotfish bites off and ingests pieces of coralline algae or hard coral along with the coral's inhabitants. As the material passes through the fish's digestive tract, it is pulverized. The algae are extracted and digested, and a fine white sand is passed as part of the undigestible wastes. Since they extract so little organic material from each mouthful of food, parrotfishes feed almost constantly. A large parrotfish may weigh as much as 27 kilograms (59 pounds) and produce as much as two or three tons of sand per year. Parrotfishes have contributed sand to many of the white sand beaches of the Florida Gulf coast, as well as beaches in other parts of the world.

Many pelagic fishes, such as anchovies (*Engraulis*), are filter-feeders, feeding on the abundant plankton found in the sea. Many fish larvae are also filter-feeders. Filter-feeders use projections from the gill arches called **gill rakers** (see Figure 10–17a) to filter both phytoplankton and zooplankton from seawater. Most filter-feeding fishes travel in large schools and are an important source of food for larger carnivores.

Adaptations to Avoid Predation

Just as fishes have evolved a variety of feeding styles, they have also evolved many clever strategies to avoid being eaten. Although many marine fishes exhibit elaborate camouflage that obscures their presence, others have evolved more direct methods of avoiding predation. The pufferfishes and porcupinefish (*Diodon hystrix*), for instance, can swallow large amounts of air or water and inflate their bodies to a size that deters potential predators (Figure 10–21a). The rapid change in size frightens some potential predators, while others find the newly enlarged fish too large to fit conveniently into their mouths. In the case of the porcupinefish, or spiny boxfish, not only does the fish enlarge itself, but also in the process spines that normally lie flat against the body stick straight out, and thus add an extra measure of protection.

Flyingfishes (*Cypselurus*) have enlarged pectoral fins that they use to glide through the air (Figure 10–21b). When frightened or disturbed, these fishes swim forward

Figure 10–21

Avoiding Predation. (a) A porcupinefish inflates itself to become a large, prickly mouthful. (b) The large pectoral fins of the flyingfish help it to avoid predators by allowing it to leave the water and glide through the air. (a, Dave B. Fleetham/Tom Stack and Associates; b, Norbert Wu/Peter Arnold, Inc.)

(a)   (b)

very quickly and leap out of the water, spreading their large pectoral fins at the same time. This behavior allows the flying fish to glide out of the range of many predators.

The tiny pearlfish (*Carapus acus*) avoids being eaten by slipping into the bodies of other animals such as sea cucumbers or bivalve molluscs.

The surgeonfish is so named because of a pair of razor-sharp spines located on the sides of its tail. When this normally peaceable, herbivorous fish is disturbed, the spines snap out like a switchblade knife, and cut or stab the intruder. Many fishermen have injured their hands by unwittingly grabbing a surgeonfish.

Clingfishes (*Gobiesox*) have a powerful sucker (formed from modified pectoral fins) that it uses to secure its body to rocks in tidal pools, where it feeds on crustaceans that are washed into its jaws. Predators have a difficult time dislodging the clingfish and thus tend to leave it alone.

The many species of triggerfishes (*Balistes*) have dorsal fins that have the first three spines modified. These spines can be pulled flat against the body or can be projected up. When a predator tries to swallow a triggerfish, it projects the spines. This causes the triggerfish to become lodged in the predator's mouth and be rejected as a possible meal. The spines also help the triggerfish to wedge itself into the tight nooks and crannies of coral reefs, making it nearly impossible to be dislodged by a predator.

Some fishes have venomous spines that protect them from predators. Scorpionfishes generally have venom glands located in grooves associated with spines of the dorsal, anal, and pelvic fins. When these spines puncture the skin of a predator or an unsuspecting human, a painful wound is inflicted. Some species in the genus *Pterois* (lionfish, turkeyfish) have venom that is potent enough to kill a

Figure 10–22

Venomous Fishes. Some species, like this lionfish, have venom glands associated with their fin spines. These fishes can inflict a painful injury on an unwary antagonist. (Copyright David Wrobel 1995/Biological Photo Service)

human (Figure 10–22). Many of these venomous species are brightly colored, exhibiting warning coloration (see Chapter 9).

Stonefishes (*Synanceia*) of the tropical Indo-Pacific region have large venom glands associated with the dorsal, anal, and pelvic spines. Ducts carry the venom down a groove in the spine to the tip, where it can enter any wound that the spine makes. Stonefish venom is extremely toxic, and since the animals are so well camouflaged, there are many recorded instances of people accidentally stepping on them and being severely injured or killed.

Some species of fishes such as weevers and toadfishes have venomous opercular spines. Opercular spines are located along the edge of the operculum that covers the gills of these fishes. Weevers (*Trachinus*) are found in the North Sea and the Mediterranean Sea. When disturbed they are reported to attack aggressively, striking at their antagonist with their bladelike opercular spine. Toadfishes are found in shallow coastal areas worldwide. They have hollow opercular and fin spines that inject venom into the wounds the spines produce. Although their delivery system is more efficient than that of some other venomous fishes, their venom is not as dangerous.

Reproduction in Bony Fishes

Bony fishes exhibit a variety of reproductive styles and strategies. Since most bony fishes are oviparous, we will examine this mode of reproduction in more detail. Many marine fishes, including such commercially important species as cod and sardines, produce vast quantities of eggs that are released into the water column and fertilized by the males, who release sperm into the water without any preliminary courtship. The fertilized eggs drift with the currents, and only a small percentage of the zygotes from these eggs survive to maturity. Pelagic marine fishes, such as swordfishes, usually produce small, transparent eggs that are buoyant, so they remain in the plankton. Species that live closer to shore, such as blennies (Blenniidae) and clingfishes, produce eggs that are generally larger, with a large quantity of yolk. These eggs are usually nonbuoyant and are attached to some fixed object such as vegetation or rocks. In areas where currents or wave action can sweep the eggs into an unfavorable habitat, fishes such as grunions (*Leuresthes tenuis*) will bury their eggs or lay them in protected nests. In some species, the male will guard the nest to prevent intruders from feeding on the eggs.

The jawfish (*Ophistognathus macrognathus*; Figure 10–23a) lays her eggs in the mouth of the male, who incubates them in his mouth until they hatch. When the male must temporarily leave the eggs to feed, they are deposited in his protected burrow. Female seahorses (*Hippocampus*) lay eggs in a special pouch on the male's abdomen (Figure 10–23b). After the eggs are deposited, the male fertilizes

(a)

(b)

Figure 10–23

Males Incubating Eggs. In some fish species it is the job of the male to incubate and care for the eggs. (a) The male jawfish incubates the eggs in its mouth until they hatch. (b) Male seahorses have a special abdominal pouch in which the eggs are incubated. (a, Copyright Fred McConnaughey 1995/Photo Researchers, Inc.; b, Copyright 1994 Mark Conlin/Mo Yung Productions/Norbert Wu)

them and then carries and incubates them until the young hatch.

The larval stage of many fish species may look quite different from the adults. At first after hatching, larval fish still have the yolk sac attached to their abdomen (Figure 10–24),

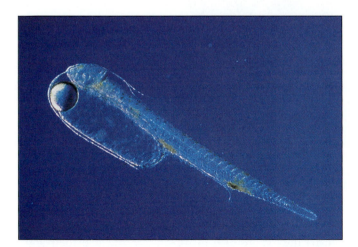

Figure 10–24

Fish Larva. After hatching, many fishes still have a yolk sac containing nutrients attached to their abdomen. The size of the yolk sac impairs the larva's ability to swim, and it remains a member of the plankton until the mouth and digestive tract are fully formed and the yolk sac is absorbed. (Copyright 1995 Peter Parks/Mo Yung Productions/Norbert Wu)

and the yolk sac is usually so large that it impairs the larva's movement. The larva is carried by the currents as a member of the plankton. As the larva continues to develop, a mouth and digestive tract form, and ultimately the yolk sac is absorbed. When the larva has transformed enough to leave the planktonic community, it becomes a juvenile and begins to feed as an adult. Unlike birds and mammals that cease to grow or grow very little after achieving maturity, fishes generally continue to grow for as long as they live.

Fish Migrations

Several species of fish engage in large migrational movements that may cover hundreds of miles of ocean. One of the most spectacular migrations is the spawning migration of freshwater eels (*Anguilla*). Eels are **catadromous** (kuh-TAD-ruh-muhs) fishes, which means that they breed in the ocean but live their adult lives in fresh water. Adult eels of both Europe and North America migrate down coastal rivers to the sea during the fall. They continue to swim for as long as two months until they reach the Sargasso Sea, which lies southeast of Bermuda (Figure 10–25). Here the eels spawn at depths of 300 meters (990 feet) or more, and the adults die after reproducing.

After the young hatch, the larvae begin their migration back to the rivers of Europe and North America. It takes approximately three years for the larvae to reach Europe and

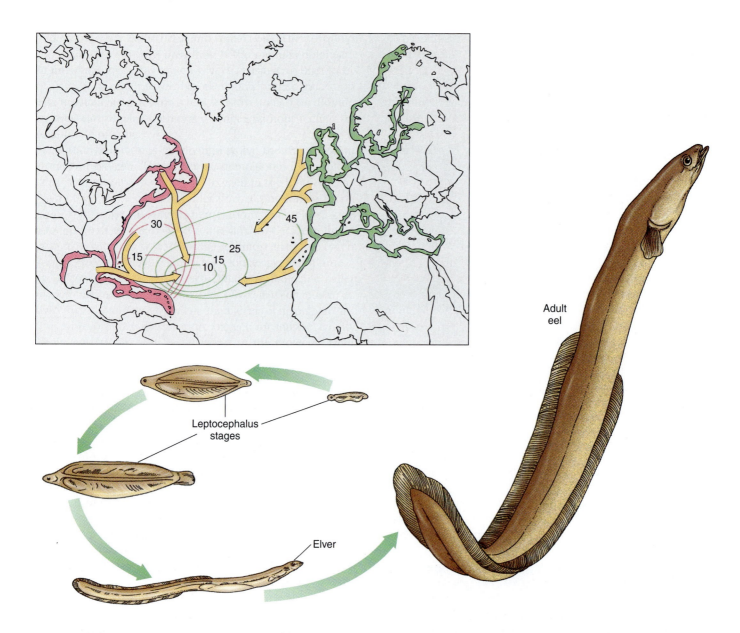

Adult
eel

Leptocephalus
stages

Elver

Figure 10–25

Life Cycle of the American and European Eels. Both the American eel (*Anguilla rostrata*) and the European eel (*Anguilla anguilla*) breed in the Sargasso Sea. Yellow arrows indicate the migratory paths the eels follow to the Sargasso Sea. Curved lines and numbers in the ocean indicate larval distribution and size of the larva in millimeters. Colored coastal areas show where the elvers enter rivers.

about eight months for the North American species to complete the trip. When they arrive at the coastal rivers, they undergo a metamorphosis, becoming juvenile eels called *elvers.* The males remain in the brackish water of coastal estuaries and rivers, while the females continue to migrate upstream. Depending on the species, it takes 8 to 15 years to reach sexual maturity. During the fall season following sexual maturity, the females migrate downstream and join the males for their return to the breeding grounds of the Sargasso Sea.

The Atlantic and Pacific species of salmon (Figure 10–26) are **anadromous** (uh-NAD-ruh-muhs) fishes, which means they reproduce in freshwater and live their adult lives in the sea. The single Atlantic species of salmon (*Salmo salar*) returns to the streams where it was born to spawn year after year. The six species of Pacific salmon (*Onchorhyncus*) return to their spawning grounds only once and then die after reproducing.

Young sockeye salmon (*Onchorhyncus nerca*), called *smolt,* migrate downstream and into the sea, where for the

Figure 10–26

Salmon. Salmon are anadromous fishes. They reproduce and mature in fresh water but spend their adult lives in the marine environment. This Pacific salmon has just returned to its native river to breed. (Jon L. Hawker)

next four years they range over hundreds of miles of ocean, feeding and growing. The adults return to the headwaters of the same streams that they left in order to go upstream to spawn and complete their life cycle. Experiments have shown that salmon are guided upstream by the characteristic odor of their stream. The odor is thought to be due to chemicals from the soil and a variety of plants that are found along the stream's edge. Odors of other streams that are passed along the trip downstream are used in reverse order as guides to finding the correct headwaters.

There is some controversy as to how the salmon locate the mouth of their particular coastal river from the open sea. Some suggest that the fish use cues such as the sun and the earth's magnetic field to find the coast. Others suggest that currents, temperature gradients, and food supplies ultimately bring a salmon close enough to the mouth of its river for it to use its odor map.

CHAPTER SUMMARY

Hagfishes and lampreys are descendants of early jawless fishes. Hagfishes are deep-sea scavengers, and very little is known about their natural history. Lampreys are generally parasitic and cause considerable damage in the commercial fishing industry.

Cartilaginous fishes are represented by sharks, skates, rays, and chimaeras. These animals have skeletons composed entirely of cartilage. Sharks have streamlined bodies and are efficient swimmers. They have many highly developed senses and are successful predators. Despite their bad reputation, most species are not aggressive.

Skates and rays have greatly enlarged pectoral fins and reduced dorsal and caudal fins. They differ from sharks not only in their appearance but also in their mode of swimming. Skates and rays are bottom dwellers that feed primarily on molluscs and crustaceans. Some species of ray are able to generate electricity, which they use for protection and to stun prey. Others have sharp spines associated with their tails. Some of these spines are hollow and connected to poison glands and are capable of inflicting serious injury.

Chimaeras have bodies with unusual shapes. Instead of teeth they have flat plates that are used for crushing their prey. Chimaeras are primarily bottom dwellers.

In bony fishes, most of the cartilage in the skeleton is replaced by bone during early growth. Coelacanths and sturgeons are examples of primitive bony fishes. The more advanced bony fishes have homocercal tails, cycloid or ctenoid scales, and more maneuverable fins than their cartilaginous cousins.

The shape of a fish's body is mainly determined by the characteristics of its environment. Pelagic fishes have streamlined bodies. Coral reef fishes frequently have a laterally compressed body. Bottom dwellers may have bodies that are dorsoventrally compressed, and burrowing fishes tend to have snakelike bodies. The environment also plays a major role in determining a fish's color. Pelagic fishes usually exhibit countershading. Coral reef fishes tend to exhibit disruptive coloration. Many fishes have body shapes and colors that camouflage them, helping them to avoid predation or get closer to their prey. Color also functions in identifying individuals of a particular sex and in communication.

Bony fishes use their strong trunk muscles to propel themselves through the water. Most bony fishes, with the exception of some pelagic fishes, bottom dwellers, and deep-sea fishes, have a swim bladder that helps them maintain neutral buoyancy.

Bony fishes have a keen sense of smell, and most species have very good vision. They also possess sense organs for hearing and sensing vibrations in the water.

Most marine fishes are carnivorous. Some species, however, are herbivores, omnivores, scavengers, or filter-feeders. Marine fishes have evolved a variety of adaptations for avoiding predation. These adaptations include sudden changes in size or color; sharp, sometimes poisonous spines; special behaviors, such as rapidly changing size or darting into crevices; and eluding predators by speed or gliding.

Most marine fishes are oviparous and produce large quantities of eggs. Many species have evolved specialized ways of incubating or protecting their eggs until they hatch.

Some species of fish engage in large migrational movements. Perhaps the best known are the several species of salmon that live their adult lives in the sea but return to the freshwater streams where they hatched in order to spawn. Freshwater eels also make large migrations to the Sargasso Sea in the Atlantic Ocean to reproduce.

SELECTED KEY TERMS

caudal fin, *p. 175*

pectoral fins, *p. 176*

pelvic fins, *p. 176*

dorsal fin, *p. 176*

heterocercal tail, *p. 176*

lateral line system, *p. 177*

ampullae of Lorenzini, *p. 177*

viviparous reproduction, *p. 178*

ovoviviparous reproduction, *p. 178*

oviparous reproduction, *p. 179*

chimaera, *p. 182*

homocercal tail, *p. 183*

chromatophore, *p. 184*

obliterative counter-shading, *p. 184*

disruptive coloration, *p. 185*

swim bladder, *p. 186*

catadromous, *p. 193*

anadromous, *p. 194*

QUESTIONS FOR REVIEW

MULTIPLE CHOICE

1. Lampreys and hagfishes lack
   a . heads
   b. mouths
   c. jaws
   d. tails
   e. all of these

2. The skeletons of sharks and rays are composed of
   a. bone
   b. cartilage
   c. soft tissue
   d. fluid
   e. cellulose

3. Shark's teeth are actually modified
   a. cartilage
   b. fins
   c. ctenoid scales
   d. placoid scales
   e. gill supports

4. A sense organ that allows fishes to detect even the slightest movements in the water is the
   a. lateral line
   b. ampulla of Lorenzini
   c. statocyst
   d. olfactory organ
   e. gill arch

5. The presence of _____ in the intestine of the shark slows the passage of food through this organ.
   a. stones
   b. chambers
   c. spiral valves
   d. gill rakers
   e. squalene

6. The type of caudal fin associated with most advanced fishes is
   a. homocercal
   b. heterocercal
   c. pelvic
   d. pectoral
   e. dorsal

7. Bony fishes can maintain neutral buoyancy by regulating the gas content of their
   a. blood
   b. gills
   c. intestines
   d. swim bladder
   e. liver

8. Many pelagic fishes have a dark dorsum and a light-colored belly. This type of coloration is known as
   a. cryptic
   b. disruptive
   c. warning
   d. countershading
   e. mimicry

9. Fishes that live in seagrass or on coral reefs have bodies that are
   a. fusiform
   b. globular
   c. snakelike
   d. round
   e. laterally compressed

10. An anadromous fish spawns in _____ and lives its adult life in _____.
    a. freshwater; saltwater
    b. saltwater; saltwater
    c. freshwater; brackish water
    d. saltwater; freshwater
    e. freshwater; freshwater

SHORT ANSWER

1. What are two adaptations that help prevent fishes from sinking due to their relatively high density?

2. Explain what is meant by "disruptive coloration" and give an example.

3. Describe how a fish uses its trunk muscles to swim.

4. Describe how fish gills are ideally suited for gas exchange in water.

5. Explain how the characteristics of the environment influence the shape of a fish's body.

6. Describe how reproduction in an oviparous fish differs from reproduction in an ovoviviparous fish.

7. Explain how you can distinguish a skate from a ray.

8. Why do most carnivorous fishes swallow their prey whole?

THINKING CRITICALLY

1. The jawfish and seahorse depend on the males to take care of the eggs. What is the advantage of this reproductive strategy?

2. While on a marine biology field trip you discover a fish that is new to you. Aside from analysis of the skeleton, how could you determine whether the fish was a cartilaginous fish or a bony fish?

3. What characteristics would you expect to observe in a fish that was adapted to a sedentary life hiding among rocks and coral on a coral reef?

4. Several species of marine fish display different colors as juveniles and as adults. What might be the benefit of such an arrangement?

SUGGESTIONS FOR FURTHER READING

Clark, E. 1992. Gentle Monsters of the Deep: Whale Sharks, *National Geographic* 182(6):123–138.

Curtsinger, B. 1995. Close Encounters with the Gray Reef Shark, *National Geographic* 187(1): 44–67.

Dorit, R. L., W. F. Walker, Jr., and R. D. Barnes. 1991. *Zoology.* Saunders College Publishing, Philadelphia.

Doubilet, D. 1997. Beneath the Tasman Sea, *National Geographic* 191(1):82–100.

Doubilet, D. 1987. Scorpionfish: Danger in Disguise, *National Geographic* 172(5):634–643.

Fricke, H. 1988. Coelacanths: The Fish That Time Forgot, *National Geographic* 173(6): 825–838.

Levine, J. S. 1990. For Fish in Schools Togetherness is the Only Way to Go, *Smithsonian* 21(4):88–93.

Oceanus, 1982. Special issue on sharks. 25(4).

Shapiro, D. Y. 1987. Differentiation and Evolution of Sex Change in Fishes, *Bioscience* 37(7):490–497.

Van Dyk, J. 1990. Long Journey of the Pacific Salmon, *National Geographic* 178(1):2–37.

Vincent, A. 1994. The Improbable Sea Horse, *National Geographic* 186(4):126–140.

Webb, P. W. 1984. Form and Function in Fish Swimming, *Scientific American* 251(1):72–82.

Chapter 11

Reptiles and Birds

Reptiles and birds generally occupy the higher trophic levels in oceanic food chains. Many are tertiary or higher order consumers. Their larger size allows them to feed on a wider variety of organisms while at the same time reducing the number of predators that can feed on them. They are well-adapted to moving freely about in search of food and have highly developed senses to supply them with a constant flow of information about their environment. A relatively complex nervous system processes this information, and their proportionately large brains allow them a greater capacity for learning. The ancestors of these marine animals were terrestrial, and the move back to the sea led to changes in body form, locomotion, insulation, osmotic balance, and feeding. In this chapter, we will examine how reptiles and birds have met the demands of the marine environment to become successful in this habitat.

MARINE REPTILES

Reptiles have been successful in both terrestrial and marine environments. The same characteristics that allowed the ancestors of reptiles to conquer land were also useful in allowing their descendants to return to the sea. A major reason for the success of reptiles was the evolution of an **amniotic egg** (Figure 11–1). An amniotic egg is covered by a protective shell and contains a liquid-filled sac called the **amnion** (AM-nee-uhn). The egg also contains a supply of food in the form of yolk, and an additional sac, the **allantois** (eh-LAN-toys), for the disposal of waste. A membrane called the **chorion** (KOR-ee-uhn) lines the inside of the shell and provides a surface for gas exchange during development.

The evolution of an amniotic egg about 340 million years ago gave reptiles a great advantage. It allowed for development within the protective egg to continue longer, eliminating the need for a free-swimming larval stage that

---

KEY CONCEPTS

1. The evolution of the amniotic egg gave reptiles a great reproductive advantage.

2. The Asian saltwater crocodile lives in estuaries and is adapted to life in the marine environment.

3. Sea turtles have streamlined bodies and appendages modified into flippers.

4. The marine iguana of the Galapagos Islands is the only marine lizard.

5. Several species of venomous sea snake live in the marine environment.

6. Shorebirds have long legs for wading and thin sharp bills for finding food in shallow water and sand.

7. A variety of bird species, including gulls, pelicans, and tubenoses, are adapted to feeding on marine organisms.

8. Penguins are the birds most adapted to life in the sea.

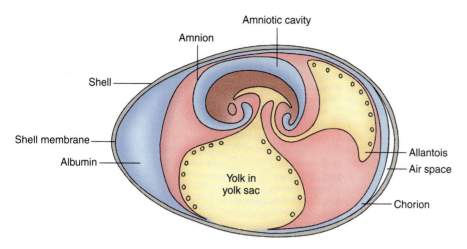

Amniotic cavity

Amnion

Shell

Shell membrane

Albumin

Yolk in yolk sac

Allantois

Air space

Chorion

Figure 11–1

Amniotic Egg. An amniotic egg is covered by a protective shell and contains a watery interior (amniotic fluid) in which the embryo develops. The yolk sac contains a supply of nutrients for the developing embryo. The allantois serves as a waste receptacle for products of metabolism. The membrane known as the *chorion* functions in gas exchange.

would be highly vulnerable to predation. It also allowed the eggs to be laid in a dry place out of the reach of aquatic predators.

To produce an egg with a shell required a means of fertilizing the ovum before the shell was added. The evolution of copulatory organs by reptiles allowed for increased efficiency of internal fertilization prior to the ovum being encased by a shell and being laid by the female.

Several other adaptations have helped reptiles survive better both on land and in the ocean. The circulatory system of reptiles is more advanced than that of fishes. The circulation through the lungs and the circulation through the rest of the body is nearly completely separated (in crocodilians the two circulatory paths are for the most part completely separated). This pattern of circulation results in a more efficient method of supplying oxygen to the animals' tissues and helps to support their active lifestyles. Their kid-

neys are very efficient in the elimination of wastes and conservation of water, allowing them to inhabit dry regions and the salty environment of the ocean. Reptiles have a skin that is covered with scales and that generally lacks glands. This adaptation allows them to resist losing body water in marine environments.

The most prominent of the first marine reptiles were ichthyosaurs (IK-thee-uh-sohrz) and plesiosaurs (PLEE-zee-uh-sohrz) (Figure 11–2). Ichthyosaurs were fast and agile swimmers that bore a striking resemblance to both sharks and dolphins. Fossils containing embryonic ichthyosaurs indicate that these reptiles did not have to come on land to reproduce. The female retained the eggs within her body until they hatched (ovoviviparous), and the young appeared live and swimming. Plesiosaurs were less streamlined than the ichthyosaurs and had bodies similar to that of a turtle. Their limbs were paddle-like, and they were probably slow

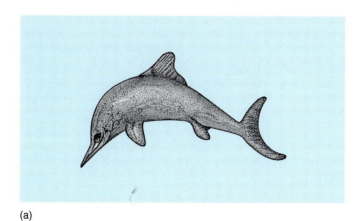

(a)

(b)

Figure 11–2

Extinct Marine Reptiles. (a) Ichthyosaurs resembled modern dolphins and porpoises. (b) Plesiosaurs had bodies that were similar to turtles. Like marine turtles, they used their pectoral and pelvic flippers for swimming.

swimmers. They had long, snakelike necks that may have compensated for their lack of speed by allowing them to dart quickly in any direction and snatch prey.

The ancestors of modern reptiles first began to appear around 100 million years ago. Modern-day reptiles include crocodilians, turtles, lizards, and snakes, all of which are represented in the marine environment.

Marine Crocodiles

Several species of crocodile, including the American crocodile (*Crocodylus acutus*) and the Nile crocodile (*Crocodylus niloticus*), venture into the marine environment to feed. The best adapted to the marine environment, however, is the Asian saltwater crocodile (*Crocodylus porosus*; Figure 11–3), which inhabits estuaries from India and Southeast Asia to northern Australia. This animal sometimes makes oceanic migrations of several hundred kilometers. One specimen is known to have traveled more than 1100 kilometers (683 miles) from Malaysia to the Cocos Islands in the Indian Ocean.

Saltwater crocodiles are large animals growing to lengths of 6 meters (20 feet). They feed mainly on fishes, and some individuals have been known to attack and kill large sharks that are close to their own size. Saltwater crocodiles are very aggressive and have been known to attack and kill humans in several parts of their range. Like other marine reptiles, saltwater crocodiles drink saltwater, eliminating the excess salt through **salt glands** on their tongues. As with other crocodiles, this animal lives in burrows along the shore, where it makes its nest and lays its eggs.

Figure 11–3

Saltwater Crocodile. Saltwater crocodiles inhabit estuaries along the Indian Ocean. (Jon L. Hawker)

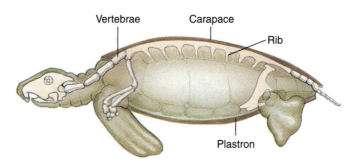

Figure 11–4

Sea Turtle Anatomy. The limbs of sea turtles are modified to form flippers. The turtle's shell is composed of an outer layer of hard protein and an inner layer of bone that is directly attached to the animal's backbone.

Sea Turtles

Turtles have changed very little since they first evolved (Figure 11–4). Their bodies are covered by protective shells that are fused to the skeleton, filling in the spaces between the vertebrae and ribs. The shell is composed of two layers: an outer layer consisting of keratin, similar to typical reptilian scales, and an inner layer composed of bone. The shell of sea turtles is reduced in both size and weight compared to that of their terrestrial cousins, an adaptation for buoyancy and swimming.

The front limbs of sea turtles are modified into large flippers, while their back limbs are paddle-shaped and used for steering and digging nests. On land their movement is awkward, but in the ocean they move with great grace and speed.

Turtles lack teeth but have a beaklike structure that they use to secure food. Most species of sea turtle are omnivorous, although the flatback sea turtle (*Chelonia depressa*) and two species of ridley sea turtle (*Lepidochelys*) are predominantly carnivorous. Table 11–1 summarizes the food preferences of some of the major species of sea turtle.

Marine turtles generally mate at sea. The females then come to shore to dig nests in the sand and lay their eggs (Figure 11–5). The eggs incubate in the sand, and after hatching, the hatchlings must make their way to the sea and relative safety. As they try to find their way to the water, turtle hatchlings are easy prey for birds. Once in the sea, hatchlings are still not completely safe, as they may become prey for large fishes (see the box on the Voyage of the Green Sea Turtle).

Most species of sea turtle are endangered as a result of real estate developments destroying their nesting sites and increased hunting of the turtles and their eggs for human consumption. Humans are not the only predators of turtle eggs; animals like raccoons also dig them up and feed on them. Many coastal U.S. states, including Florida, and several international nesting sites have strict laws protecting

Voyage of the Green Sea Turtle

The migrating and breeding behaviors of green sea turtles (*Chelonia mydas*) have been studied extensively by Dr. Archie Carr of the University of Florida. Green sea turtles feed on the turtle grass and manatee grass found in warm, shallow continental waters but breed on remote islands that can be thousands of kilometers away. One group of turtles studied by Dr. Carr feeds on manatee grass along the coast of Brazil, but migrates 2330 kilometers (1400 miles) to Ascension Island, a small island between South America and Africa, to breed.

Tagging studies showed that not all of these turtles migrate every year; some breed on a two- or three-year cycle. The voyage begins in December and ends in February to April. Arriving at their destination, females come ashore several times (four or more) at 12-day intervals. Each time, they dig a nest in the sand and lay about 100 golf ball–sized eggs. Between trips to land, females will mate with the males that remain in the water past the surf line. The sperm that the female receives are used to fertilize eggs that will be laid on a future trip two or three years later. Food supplies are scarce around Ascension Island, so after the females have finished laying their clutches of eggs, they return to the Brazilian coast to feed.

Warmed by the sun, the eggs hatch about two months later, and the young turtles dig their way out of the nest. The hatchlings, no larger than a silver dollar, immediately crawl to the water and enter the sea. It is assumed that the young turtles feed on plankton as they drift in the ocean currents that will eventually bring them to their Brazilian feeding grounds. After maturing, the young turtles will eventually return to their Ascension Island hatchery when it is their turn to reproduce.

Why do the turtles migrate such a long distance to reproduce, when suitable beaches for reproduction are located much closer to their feeding grounds? Dr. Carr suggests that at the time the turtles were evolving their migration patterns, the continents of South America and Africa were closer together. The turtles may have fed along the coast of Brazil and laid their eggs along the coast of Africa. As the continents continued to separate as a result of continental drift, the distance the turtles had to travel became longer. It is possible that they used Ascension Island as a resting place before proceeding to Africa, and then ultimately stopped there instead to reproduce.

There are many hypotheses as to how the green sea turtle, as well as other species, is able to navigate over such long distances. The senses of smell and taste may play an important role, as well as audible signals produced by other animals. Some researchers suggest that the turtles follow the earth's magnetic field or the forces produced by the earth spinning on its axis. Others, like Dr. Carr, believe the turtles may be able to navigate using the sun to determine their position. Studies to answer these and other questions concerning the natural history of sea turtles are currently being conducted.

Table 11–1
Food Preferences of Marine Turtles

Species	Feeding Type	Food
Green sea turtle (*Chelonia mydas*)	omnivore	green, brown, and red algae, mangrove roots and leaves, sponges, jellyfishes, molluscs, and crustaceans
Flatback sea turtle (*Chelonia depressa*)	carnivore	invertebrates: mainly sea cucumbers and prawns
Hawksbill sea turtle (*Eretmochelys imbricata*)	omnivore	sponges, cnidarians, ectoprocts, molluscs, sea urchins, tunicates, crustaceans (especially barnacles), fishes, sargassum, and red mangrove
Loggerhead sea turtle (*Caretta caretta*)	omnivore	sponges, cnidarians, molluscs, crustaceans, sea urchins, tunicates, fishes, seaweed, sargassum, and turtle grass
Kemp's ridley sea turtle (*Lepidochelys kempii*)	carnivore	crabs, snails, clams, jellyfishes, and fishes
Olive ridley sea turtle (*Lepidochelys olivacea*)	carnivore	fishes, crabs, snails, oysters, sea urchins, and jellyfishes
Leatherback sea turtle (*Dermochelys coriacea*)	omnivore	prefers jellyfishes; will also eat sea urchins, squids, tunicates, fishes, and floating seaweeds

sea turtle eggs and hatchlings. In some Florida coastal communities there are rules that require turning off bright lights along the beaches during the seasons when the eggs are laid so that the females and hatchlings will not become confused and lost.

The leatherback turtle (*Dermochelys coriacea*), the largest of the marine turtles, lacks a shell. Its body is covered by a thick hide that contains small bony plates. Leatherback turtles can attain lengths of as much as 2 meters (7 feet), a flipper span of 4 meters (13 feet), and a weight of 680 kilograms (1500 pounds). The smallest sea turtle is the Kemp's ridley (*Lepidochelys kempii*), which measures 0.6 meters (2 feet) long as an adult. Other species of marine turtle include the green sea turtle (*Chelonia mydas*), loggerhead (*Caretta caretta*), hawksbill (*Eretmochelys imbricata*), flatback (*Chelonia depressa*), and olive ridley sea turtle (*Lepidochelys olivacea*) (Figure 11–6).

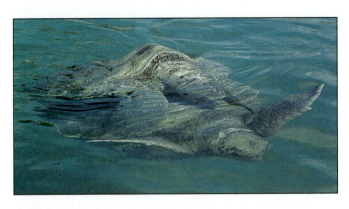

Figure 11–5

Sea Turtle Reproduction. Sea turtles, like these green sea turtles, generally mate at sea. The female then comes ashore to lay her eggs in shallow nests in the sand. (Copyright David Wrobel 1993/Biological Photo Service)

(a) Leatherback

(c) Kemp's ridley

(e) Loggerhead

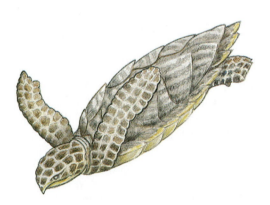

(b) Hawksbill

(d) Green

Figure 11–6

Sea Turtle Species. (a) Leatherback sea turtle. (b) Hawksbill sea turtle. (c) Kemp's ridley sea turtle. (d) Green sea turtle. (e) Loggerhead sea turtle.

The Marine Iguana

The only marine lizard is the marine iguana (*Amblyrhynchus subcristatus*) of the Galapagos Islands off the coast of Ecuador (Figure 11–7). These large lizards are descended from the green, vegetarian iguanas that still inhabit the tropical forests of the mainland. At some point in the distant past, chunks of riverbank from Central America may have broken loose and been carried out to sea, along with any plants and animals that were living there. At sea, some of these rafts of vegetation may have been caught up in ocean currents that carried them to the Galapagos Islands. If this happened many times over thousands of years, not only iguanas but also other animals and plants may have been brought to the Galapagos Islands in the same manner.

Any of the forest iguanas that survived the trip would have moved onto the islands to search for food. Some probably wandered inland and up the slopes where there was vegetation on which to feed, and they retained the lifestyle of a forest iguana. The Galapagos ecosystem would have been dramatically different from the forests that they left behind but would still have provided adequate resources for survival.

Any iguanas that remained near the shore on the barren lava rocks faced a more precarious situation. The only available food they could reach would have been the algae that were exposed at low tide. Although it was a foreign food source, it was nutritious and the animal's food preference may have changed out of necessity. As we find in all natural communities, individual differences that favor survival tend to persist and increase in frequency in subsequent generations. This would account for the unusual lifestyle of the marine iguana.

Figure 11–7

Marine Iguana. The marine iguana, the only marine lizard, is native to the Galapagos Islands. The scar on this animal's shoulder and side is the result of a shark bite. (Jon L. Hawker)

The marine iguana today is quite different from its relatives on the mainland. Some of these meter-long (3 feet) lizards are entirely black, while others are mottled red and black, showing some hint of green during the breeding season. It is thought that the dark coloration allows these lizards to absorb more heat energy to raise their body temperatures so they can swim and feed in the cold Pacific waters.

Instead of having a long snout like the forest iguana, the marine iguana has a short, heavy snout, similar to that of a bulldog, that is better suited for grazing on the dense mats of seaweed that hug the rocks. Marine iguanas feed by biting with one side of the mouth first and then the other. As the animal tears at the tough algae, it clings to the slippery rocks with powerful clawed toes. In order to feed better under water, the marine iguana swallows small stones to make its body less buoyant.

Whether feeding at low tide or underwater, the animal consumes large amounts of saltwater. The excess salt from the water is extracted and excreted by specialized tear and nasal glands. Salt that is extracted by nasal salt glands is periodically expelled by nasal spraying, a disgusting but effective behavior.

Marine iguanas are good swimmers, using lateral undulations of their body and tail to propel them through the water. They avoid heavy surf and rarely venture more than 10 meters (33 feet) from shore. When leaving the water, they tend to ride in with the swell, then swiftly crawl up on the rocks. If they do not find their territory immediately, they touch the rocks with their tongues that, like a snake's, carry scent to a receptor in the roof of the mouth. When they locate their own scent, they follow it to their territory, where they rest on the rocks, lying almost motionless above the high tide.

Each male occupies a small territory on the rocks, usually in the company of one or two females. If an intruder wanders by, even by accident, he is immediately attacked and driven off. Most fights between male marine iguanas are the result of deliberate challenges by an intruder. Combat begins with a great deal of posturing, threatening, and bluffing. Each male rises high on its legs and exposes the bright red interior of its mouth. They may also spray each other with a stream of moisture from their nostrils, and, as the battle escalates, the combatants butt heads and try to push each other away. In the end, the loser, often the challenger, lies down in a submissive posture, and the winner stands high and nods vigorously. Since such encounters rarely result in serious injury, the survival of the population remains unaffected.

Sea Snakes

Snakes are descendants of lizards that have lost their limbs as an adaptation to a burrowing lifestyle, a lifestyle that was later abandoned by many species. Although most species of snake are terrestrial or arboreal, there are about 50 species that live in the marine environment, each one bearing venomous fangs (Figure 11–8). Sea snakes are less tied to the land than other marine reptiles, with only about half of the species coming onto land to lay their eggs. Like the ancient ichthyosaurs, females of the remaining species are ovoviviparous, retaining their eggs within their body until they hatch. The young emerge able to swim and feed immediately.

Although all sea snakes breathe air, some species can remain submerged for several hours. The animal's single lung reaches almost to its tail, and its trachea (windpipe) has become modified to absorb oxygen, thus acting as an accessory lung. Sea snakes can also exchange gases through their skin when they are under water. These adaptations allow the sea snake to absorb large amounts of oxygen in a very efficient manner. Sea snakes are able to lower their metabolic rate so that they consume less oxygen when submerged, and some species may even be able to extract oxygen from swallowed water. Specialized valves in the snake's nostrils prevent water from entering when they are submerged.

Most sea snakes remain close to shore, but the yellow-bellied sea snake (*Palamis platurus*) is pelagic, feeding on surface fish. On several occasions, it has been sighted hundreds of miles from land. This species has migrated east and west from the coast of Asia and can be found off the east coast of Africa and the west coast of tropical America. Most of the other species are found in warm coastal waters from the Persian Gulf to Japan and east to Samoa. Sea snakes will sometimes congregate in enormous numbers, presumably for the purpose of mating. In 1932, a group of sea snakes that was 3 meters (10 feet) wide and 97 kilometers (60 miles) long was reported near Indonesia. There are many other reports of smaller aggregations.

Figure 11–8

Sea Snake. An olive sea snake patrols the bottom in search of eels. (Jon L. Hawker)

Sea snakes feed on fishes, mainly eels. They lie quietly at the surface or on the sea floor until the prey is within easy striking distance. The snake then strikes with amazing speed and agility. Like their land-dwelling cousins, sea snakes can swallow prey more than twice their own diameter. Sometimes two snakes try to feed on the same prey. When this happens, the smaller of the two is usually seen disappearing into the larger one. Some of the fishes that sea snakes feed upon have spines, and instead of digesting or voiding the spines, the sea snakes push them out through their body wall.

Although sea snake toxin is specifically adapted to killing fish, it is also quite toxic to warm-blooded animals. Sea snakes, however, are timid by nature and rarely bite humans. There are no accounts of sea snakes' attacking swimmers or divers, and when sea snakes are caught in fishing nets, they are handled with indifference. In those recorded instances in which sea snakes have caused human deaths, the snakes were inadvertently grabbed and were biting in self-defense.

In many parts of their range, sea snakes are eaten by humans. In Japan the consumption of sea snakes is so large that it supports a major fishery. The hunting of sea snakes for their skins has led to their near extinction in some areas.

SEABIRDS

Birds are homeothermic animals whose bodies are covered with feathers. Feathers help insulate the animal's body, allowing it to maintain a higher temperature than its surroundings. This ability allows the animal to metabolize more efficiently. The higher level of metabolism supplies the large amounts of energy needed for active flight and to maintain a rapidly functioning nervous system for flight control. In addition, the light weight and strength of the feather make it the perfect airfoil.

Adaptations such as strong muscles, quick responses, and a great deal of coordination greatly aid birds in the process of flight. The evolution of an advanced respiratory system and a circulatory system that includes a four-chambered heart allows them to provide more oxygen to the active muscles to support the high level of metabolism that flying demands. Their senses, especially sight and hearing, are keener than those of their reptilian ancestors, and they have larger brains relative to body weight that allow them to process more sensory information more effectively.

Only about 250 of the approximately 8500 bird species are adapted for life in and around the ocean. Although less varied than terrestrial and freshwater species, they are quite numerous. Seabirds feed in the sea and sometimes spend months out of the sight of land, but they must return to land to breed.

Like marine reptiles, seabirds consume large amounts of salt with the food they eat and salt water that they drink.

The seawater that these birds consume is about three times saltier than their body tissues. Their kidneys are not able to concentrate such high levels of salt in the urine; therefore, like marine turtles and marine iguanas, seabirds have salt glands (Figure 11–9). These are located one above each eye, and they are capable of producing a solution of sodium chloride that is twice the concentration of seawater. The salt solution that is produced by these glands is released by way of the internal or external nostrils so that seabirds, like gulls and petrels, appear to have constantly runny noses.

Shorebirds

Shorebirds, or waders, have long legs and thin, sharp bills that they use to feed on the abundance of marine life in the intertidal zone. They range in size from as small as a sparrow to larger than a chicken, and they exhibit countershading with dark backs and white belly surfaces. The group includes oystercatchers, tattlers, curlews, godwits, turnstones, sandpipers, jacanas, surfbirds, phalaropes, herons, and even inland species such as snipes and woodcocks.

Oystercatchers (Haematopodidae) have long, blunt, orange bills that are vertically flattened (Figure 11–10). They use their bills to slice through the adductor muscles of partially opened clams, mussels, and oysters so they can feed on the soft flesh. They can also use their bills to pry limpets off rocks, crush crabs, and probe mud and sand for worms and crustaceans.

The long-billed curlew (*Numenius americanus*) uses its 20-centimeter (8-inch) bill like a forceps to extract shell-

Figure 11–10

Oystercatcher. The oystercatcher uses its sharp bill to slice through the adductor muscle of bivalves. (Jon L. Hawker)

fish from their burrows. Turnstones (*Arenaria*) are heavyset birds with short necks and slightly upturned bills. They use their bills as crowbars for turning over stones, sticks, and beach debris in search of food.

Plovers (Charadriidae) are shorebirds with worldwide distribution. They have short, plump bodies, bills that resemble a pigeon's, and they are considerably shorter than other waders. These birds can be found on beaches, mudflats, and grassy fields. It is common to see flocks of hun-

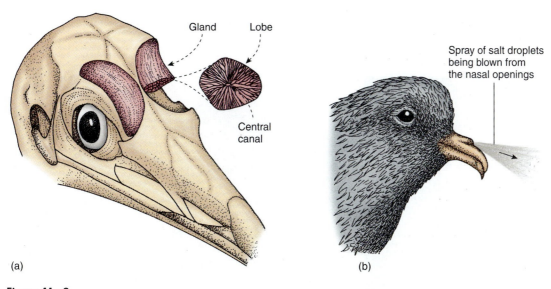

(a)   (b)

Figure 11–9

Salt Glands. (a) Salt glands located above the eyes of seabirds help to excrete the salt the birds ingest when drinking seawater and feeding on marine animals. The glands drain into the bird's nasal passageways, which explains why many marine birds appear to have a constant runny nose. (b) A petrel blowing salt droplets from the nasal openings.

dreds of plovers following the receding waves and plucking small animals from the damp sand. One of the most intriguing plovers is the wrybill (*Anarhynchus frontalis*) of New Zealand (Figure 11–11). It is the only bird in the world with a beak that is bent laterally to the right. The adaptive significance, if any, of this feature is not known.

The nests of plovers are characteristic of waders in general. Most nests are built on the ground in depressions or hollows. The females lay four pear-shaped eggs that fit together like the pieces of an orange, making it easier for a single bird to incubate the clutch. The chicks hatch in about three weeks and are covered with a camouflage down. They are active as soon as they hatch and leave the nest within hours to follow their parents and search for their own food.

Avocets (*Recurvirostra*) and stilts (*Himantopus*) have very long legs, elongated necks, and slender, graceful bodies. Avocets feed by wading through shallow water and moving their partially opened beak from side to side through the water. Stilts have legs that are longer in proportion to their body size than any other birds except flamingos, and they use their straight bills to probe the mud for insects, crustaceans, and other small animals. Both stilts and avocets are found in warm climates and breed colonially in marshes or at the edge of lagoons.

The American avocet (*Recurvirostra americana*) is anything but friendly during the breeding season. When another bird comes too close to their nesting sites, the birds will angrily charge the intruder. When there are no eggs or young to defend, these birds are quite indifferent to intruders, including humans. This characteristic combined with destruction of their habitats in the eastern United States has made them rare in this part of the country.

Sandpipers (Scolopacidae; Figure 11–12) are relatives of the avocets. As the surf and tide retreats, these birds scurry across the sand to feed on small crustaceans and mol-

Figure 11–12

Sandpipers. Sandpipers are a common sight along beaches, where they search for food in the sand. (M. Graybill/J. Hodder/Biological Photo Service)

luscs. Sandpipers are sociable birds that share the beaches with many other species of shorebird. They are found in most parts of the world and make annual migrations that take some species as far as 10,000 kilometers (6000 miles).

Herons (Ciconiiformes), which include egrets and bitterns, are one of the most widespread families of wading bird and are represented on each major continent. Although they may appear rather ungainly, their skinny legs and long necks are aids for hunting. The great blue heron (*Ardea herodias*), for instance, is a stalker that uses its bill as pincers to capture small fish and crustaceans (Figure 11–13). Its appetite for fish is enormous, and it will catch all sizes, even those that appear too big for the bird to swallow.

Figure 11–11

Wrybill. The wrybill is a plover from New Zealand. Notice the unique shape of this bird's bill. (Steven Holt/Aigrette)

Figure 11–13

Great Blue Heron. This great blue heron is about to land and feed in the shallow water. Herons are wading birds that feed on fishes. Jon L. Hawker)

Most herons feed by standing still and waiting for their prey, a small fish or crustacean, to come into range. An exception to this mode of feeding, however, is the little egret (*Egretta garzetta*). This impatient bird stalks its prey along mudflats. Sighting a fish, it will move cautiously until within range, then with a lightning jab of the beak, seize its prey. It will also vibrate its feet as it moves through shallow water to scare bottom-dwelling prey up to the surface. The snowy egret (*Leucophoyx thula*) also feeds by stirring the bottom with its feet and then rushing about grabbing the small fishes and crabs that it has frightened into motion.

Gulls and Their Relatives

Gulls (Charadriiformes) are noisy, aggressive birds that are efficient predators and scavengers. They are aggressively carnivorous, eating the eggs and young of other birds, stealing prey from other birds, and occasionally, when food is scarce, even eating the young of other gulls. In some areas they are so highly regarded for their cleaning of garbage from beaches and coastal waterways that they are protected by law.

Gulls are probably the best known seabirds because they are found mostly along shores and seaports, with some species venturing far inland. They have a worldwide distribution and are found everywhere the land and ocean meet, including polar regions. Although gulls have webbed feet and oil glands to waterproof their feathers, most species are not true ocean-going birds and so do not stray far from land. They have enormous appetites and are not very selective eaters. Many species will land on the water and feed on floating debris.

Gulls exhibit interesting behaviors when it comes to feeding. Some have been observed to pick up clams, mussels, oysters, or sea urchins from shallow water and then fly over rocks, dropping their food repeatedly until the shell breaks. They have also been observed using parking lots as shell-dropping areas. Gulls are so successful at getting food and surviving that in some areas they have reached nuisance proportions, menacing other birds, colliding with airplanes, and fouling buildings and sidewalks with excretory droppings.

Gulls are highly gregarious and gather in colonies of hundreds of thousands of individuals. They are not picky about nesting sites or nesting materials, using any material that is available. The female lays two to three eggs, and both sexes help in the incubation. The eggs hatch in three or four weeks, and the female has the capacity to lay another clutch of eggs immediately if the original clutch is lost. This survival mechanism can be repeated as many as three or four times a year. The newly hatched chicks remain in the nest until they are almost fully grown, camouflaged by their speckled down. Not only predators but also other gulls will feed on the young hatchlings, and it is not uncommon for

Figure 11–14

Sea Gulls. Herring gulls wait on a pier for fishing boats. (L. L. T. Rhodes/Animals Animals)

only one out of every five to survive the eight weeks after hatching until they are able to fly.

Herring gulls (*Larus argentatus*) are the most widespread and best known of the gull family (Figure 11–14). These vocal gray and white birds can be found along the shores of both North America and Europe, covering the waterfronts of large seaports and small fishing villages and ranging inland as well. They always travel in large groups. Hordes of these birds can be seen following fishing boats and garbage scows or migrating in groups of tens of thousands of individuals.

Related to gulls are the terns (Laridae), small, graceful birds with brightly colored and delicately sculptured bills. Because they have forked tails, they are sometimes referred to as *sea swallows*. Although a few species are freshwater birds, most terns live along the shoreline, where they hunt for small fish and invertebrates by plunging into the water but rarely landing on the water. They are generally not as avid hunters as the gulls, but they do steal food from other birds. Terns are gregarious nesters; often millions of birds swarm together. Not all terns, however, are so gregarious. The arctic tern (*Sterna paradisaea*) makes solo flights from north of the Arctic Circle to Antarctica. These seasonal migrations allow the bird to spend ten months of the year in continuous daylight, more than any other living creature.

Some very aggressive relatives of the gulls are the skuas and jaegers (Stercorariidae). Skuas (*Catharacta*) are found in both hemispheres, particularly around the poles, where they breed. One might consider these birds to be the hawks, falcons, and even vultures of the sea. They are omnivorous and keen predators, feeding on garbage, berries, other birds, eggs (especially penguin eggs), lemmings and other small mammals, insects, carrion, and even unprotected chicks of their own species. Jaegers (*Stercoraria*) are predators that steal fish from other birds. They can fre-

Figure 11—15

Skimmer. Skimmers fly along the water with their lower mandible just beneath the surface. When the mandible strikes a fish or shrimp, it snaps shut, capturing the prey. (Steven David Miller/Animals Animals)

Figure 11—16

Puffin. Although this puffin looks more like a relative of penguins, it is actually more closely related to the gulls. (Jon L. Hawker)

quently be seen pursuing terns and other marine birds and robbing them of their prey.

Skimmers or scissorbills (Rynchopidae; Figure 11–15) are perhaps the most interesting of the gull's relatives. They are small birds that grow to a length of 50 centimeters (20 inches) and have two unusual features: pupils that are vertical slits similar to those of cats and a lower jaw that is flexible and protrudes much farther than the upper bill. These birds fish along coastal rivers and streams, flying along with their lower bill just beneath the surface of the water and creating a thin ripple that attracts small fish to the surface. The skimmer then reverses direction of flight, and, with its bill, it follows the same line that it previously made in the water. When the mandible strikes a small fish or shrimp, it immediately snaps shut. The birds feed at dusk, dawn, and during the night. They race upstream and then turn to race downstream, occasionally pausing on the bank to rest and eat their catch.

Although auks, puffins, and murres (Alcidae; Figure 11–16) look more like penguins, they are actually related to gulls. Alcids are countershaded with black dorsums and white ventral surfaces. Like penguins, these birds are awkward on land. In the water, they are remarkably agile, using their stout swimming wings to fly through the water as they pursue their prey of fishes, squid, and shrimp. Alcids and penguins are examples of **ecological equivalents,** two different groups of animal that have evolved independently along the same lines in similar habitats and therefore display similar adaptations.

A major difference between all of the modern alcids and penguins is that alcids can fly. Ironically, an extinct alcid, the great auk (*Pinguinus impennis;* Figure 11–17), was the largest flightless bird in the Northern Hemisphere.

Standing 750 centimeters (30 inches) high, it was an easy mark for hunters and feather collectors and was hunted to extinction in the middle of the last century.

Alcids spend their winters in offshore waters and then gather in dense, noisy colonies in the cliffs along the northern Atlantic and Pacific oceans in early spring. Auks (*Alcae*) and murres (*Uria*) nest on ledges and among boulders, while puffins (*Fratercula*) prefer crevices and burrows on

Figure 11—17

Giant Auk. The giant auk is now extinct, the victim of hunters interested in its meat and feathers.

higher bluffs. The females lay a single egg that is cared for by both parents. Because these nesting sites are rather precarious, eggs of the ledge dwellers are pear-shaped, an adaptation that makes them less likely to roll over the edge. As one might expect, both egg and hatchling mortality rates are high. Adults jostling with each other on the ledges cause many eggs and chicks to fall, while others are preyed upon by the ever-present gulls.

Young murres that survive leave the ledges in July and literally plunge into the ocean, since they have not yet molted into their flight feathers. Each youngster's parents encourage it to the brink of the ledge where, after a moment of hesitation, it leaps into the ocean. The youngster is quickly joined by one or both parents, and they swim out to sea where they will spend the winter months.

Like other young alcids, chicks of the common puffin (*Fratercula arctica*) have enormous appetites. Both parents spend most of their time gathering food for their hungry chicks, sometimes carrying as many as 18 fish in a single load. The parents provide the chick with enough fish, mussels, and sea urchins per day to equal its body weight. After six weeks of constant care, the adult puffins abruptly leave their chicks and head to the open ocean. The young puffin spends about another week in the burrow or crevice by itself before venturing out to the water. It learns to swim, fly, and dive without the guidance of any adult birds.

Pelicans and Their Relatives

Pelicans (Pelecaniformes) are members of a group of birds that includes gannets, boobies, cormorants, darters, frigatebirds, and tropicbirds. Members of this group have webs between all four toes, and, in pelicans, cormorants, and frigatebirds, the upper mandible is hooked. Many species are brightly colored, and some, such as the darters (Anhingidae), have conspicuous head adornments, while others, such as the boobies (Sulidae), have brightly colored faces and feet. Members of this group enjoy worldwide distribution but are largely found in the tropics and warm temperate regions. Although some species can be seen in the mid ocean, most are found in coastal areas and some (pelicans, cormorants, and darters) in inland waters.

Pelicans (Pelecanidae) prefer warm latitudes and estuaries, coastal, and inland waters. They are large animals; some species weigh as much as 11 kilograms (24 pounds). Because they are so large and nest in sizable colonies, the waters where they live must support a large fish population to feed them. Sometimes pelicans will have to fly some distance from their nesting sites in order to find waters with enough food. After breeding, the birds will disperse or migrate, sometimes crossing deserts. Their numbers have declined in Europe, Africa, and Asia because of their dependence on inland waters that tend to be drained or to become polluted.

Figure 11–18

Pelican. A brown pelican with a fish in its gular pouch. (Steve Kaufman/Peter Arnold, Inc.)

Pelicans feed just under the surface of the water, using their large gular (throat) pouches as nets. The **gular pouch** is a sac of skin that hangs between the flexible bones of the bird's lower mandible and is characteristic of pelicans. The brown pelican (*Pelecanus occidentalis*) of the Americas is the most fully marine of the pelicans (Figure 11–18). It patrols about 9 meters (30 feet) above the water, and when it sights a school of small fishes, it folds its wings back and then crashes into the water with its gular pouch gaping. The impact stuns the fish, allowing the pelican to scoop up large numbers of them, as well as 8 to 12 liters (2 to 3 gallons) of water, with its pouch. The pelican then returns to the surface with its catch, buoyed by hundreds of air sacs located just beneath the skin (subcutaneous air sacs). It must then let the water drain from its pouch before it can become airborne again. An adult brown pelican will consume an average of 4.5 to 6.5 kilograms (10 to 14 pounds) of fish per day.

The six other species of pelican hunt in groups in shallow water by forming semicircles and dipping their bills to flush and trap fish. Occasionally they will beat their wings on the water to drive the fish together, making them easier to catch.

Pelicans are not the only birds in this order that exhibit interesting fishing techniques. Boobies and gannets dive into the sea from heights of 18 to 30 meters (59 to 98 feet), and they can catch flying fishes just before they reenter the water.

The red-footed booby (*Sula sula*; Figure 11–19a) is the smallest of the boobies and nests in trees. Since its nest is relatively safe from predators, the female lays only one egg. Other species of booby that are ground nesters lay several eggs since their nests are more accessible to predators. This strategy helps to make more likely that at least a few chicks

(a)

(b)

Figure 11–19

Boobies. (a) Red-footed booby. (b) Blue-footed booby. (a and b, Jon L. Hawker)

will survive. The largest of the boobies is the blue-footed booby (*Sula nebouxii*; Figure 11–19b). The female of this species lays two or three eggs, and both parents take turns incubating them. Even with such good care, usually only one nestling will survive to become a fledgling.

Cormorants (Phalacrocoracidae) are some of the most adept avian fishers but lack oil glands to waterproof their feathers and so must dry their wings on land periodically. Swimming along the surface, they scan the water for fish and then plunge to spectacular depths in pursuit of their prey. These birds have been found tangled in fishermen's nets at depths of 21 to 30 meters (69 to 98 feet). Cormorants swallow their prey headfirst so that the fish scales and spines will not catch in their throats. In Japan and some other Asian countries, cormorants are used by fishers to help catch fish (Figure 11–20). The birds are released with a rope or a steel ring around their throats to prevent them from swallowing the fish that they catch.

Most cormorants are strong fliers, but the flightless cormorant (*Phalacrocorax harrisi*) of the Galapagos Islands has completely lost the ability to fly. This bird has been isolated on two remote islands of the Galapagos group for millions of years. The waters around the islands contain abundant food, and the islands themselves have plenty of nesting material and nesting sites. There are also no natural enemies of the cormorant. A strong swimmer, this cormorant uses its webbed feet for propulsion and keeps its wings folded close to its sides to be more streamlined.

The guano cormorant (*Phalacrocorax bougainvillei*) of the coast of Peru is one of the world's most valuable birds because it produces a phosphate-rich manure called **guano,** which is used in making fertilizer. The high phosphate content of this material is the result of the birds' anchovy-rich diet. Commercial harvesters of guano have constructed large wooden platforms in sealed-off peninsulas where the birds can roost safely, isolated from predators.

Frigatebirds (Fregatidae; Figure 11–21) have lightweight bodies and wingspans that approach 2 meters (7 feet). Their large wingspan allows them to utilize air currents to soar for hours before returning to land to rest. Frigatebirds are another species with no oil glands in their skin to waterproof their feathers. If they are forced to settle on the ocean's surface, there is a good chance that they will drown. Because of this, frigatebirds feed without getting much more than their bills wet. They skim along just above the water, picking up jellyfish, squids, fishes, young sea turtles, and bits of carrion. They are particularly fond of flying fish and can catch them in midflight.

Frigatebirds are also pirates and will frequently steal fish from other birds. They have been observed attacking boobies even after the booby has swallowed its catch. The frigatebird beats and jostles its victim in flight until the

Figure 11–20

Cormorant. In China, cormorants are used for fishing. A rope made of grass is tied around the bird's neck to prevent it from swallowing its catch. This young boy is about to remove a fish from the cormorant's mouth. (Copyright 1995 Ted Wood)

booby regurgitates the fish. The frigatebird then catches the meal before it drops back into the water. A frigatebird will sometimes perch on the upper bill of a feeding pelican and steal fish while the pelican is emptying water from its gular pouch.

Tubenoses

Petrels, albatrosses, and shearwaters (Procellariiformes) all have a distinctive tubenose that may function in detecting odors, eliminating excess salts (osmoregulation), and discerning the strength of air currents. Members of this family can also be distinguished from other birds by the structure of their stomachs. Their stomach contains a large gland that produces an oil composed of liquified fat and vitamin A. This yellow oil is used to feed newly hatched young, and, since it has a strong, unpleasant odor, adult birds use it for defense by regurgitating it through the mouth and nostrils when they are disturbed.

Albatrosses are superb gliders with wings nearly 3.5 meters (11 feet) long, the largest of any living bird (Figure 11–22). The wandering albatross (*Diomedea exulans*) is the largest of all seabirds and master of the skies. It is capable of cruising the currents of turbulent ocean air for hours at a time with barely the flick of a wing. With their very large wings, albatrosses are able to take advantage of reliable sea winds to keep them aloft. If the wind suddenly dies down, the birds are forced into a labored, flapping flight, and in calm seas they are virtually incapable of leaving the water. For this reason, most species of albatross are restricted in their range to the Southern Hemisphere, where there is little land mass to block the ocean winds. These birds are not likely to move northward because the doldrums, areas of rising air at the equator (see Chapter 4), form an impassable barrier for soaring birds, especially in the Atlantic Ocean. In the Pacific Ocean, however, 3 of the 13 species of albatross have moved northward. Interestingly, these birds still migrate to the Southern Hemisphere to mate and breed at the same time as the southern species do, even though it is the middle of winter in the Northern Hemisphere.

Figure 11–21

Frigatebird. A male frigatebird inflates its red throat to attract mates. (Jon L. Hawker)

Figure 11–22

Albatross. A male and female albatross clap bills during a mating ritual. (Jon L. Hawker)

Figure 11—23

Storm Petrel. These Wilson's storm petrels exhibit a characteristic flapping type of flight. Early mariners thought the storm petrel could walk on water and named this bird after St. Peter. (Jon L. Hawker)

As a rule, albatrosses come to land only to breed, and some young birds may spend the first two years of their lives at sea before they return to nest. During the mating season, males court the females with elaborate displays that include gyrations, wing flapping, and bill snapping (see Figure 11—22). After mating, the female normally lays a single egg on a nest that is shaped like a volcano, and both parents take turns at incubating the egg. The young hatch 60 to 80 days later, usually in March, and are first fed on stomach oil and then regurgitated fish until they are able to feed themselves. Because it takes so long for the parents to raise a single chick, most species mate every other year rather than annually.

Storm petrels (Hydrobatidae) are small birds with long legs. They exhibit a characteristic, erratic type of flight that looks more like a fluttering moth than a flying bird. Wilson's storm petrels (*Oceanites oceanicus*) are the most abundant seabirds. They feed with their legs extended and their feet paddling rapidly just below the surface of the water so that the birds appear to be walking on water (Figure 11—23). For small birds, storm petrels have an extremely long life span (20 years or more). They usually form long-term pair bonds and return to the same nesting sites every year. They do not breed until they are four or five years old, which is late for a bird of this size. They breed seasonally and, like other species in this group, lay only a single egg.

Although diving petrels (Pelecanoididae) have the nostrils, beak, and stomach glands that are characteristic of tubenoses, in appearance they look more like the auks of the Northern Hemisphere. Diving petrels are found exclusively in the Southern Hemisphere. They require cold water throughout the year and are found where skies are gray, seas are rough, and precipitation is high, generally not straying far from their colonies. Diving petrels have the ability to spot small crustaceans and fish from the air. Sighting its prey, the petrel performs a headlong dive and continues the chase by "flying" under water. Grasping its meal in its beak, the bird re-emerges from the water and enters the air again without missing a wing beat.

Penguins

The birds that are most highly adapted to life in the sea are penguins. Although penguins are usually associated with Antarctica, only two species, the emperor penguin (*Aptenodytes forsteri*) and the Adelie penguin (*Pygoscelis adeliae*), live there. Fourteen other species of penguin are found living on barren, rocky islands in the Antarctic Sea and along the shores of South America. One species, the Galapagos penguin (*Spheniscus mendiculus*), is found as far north as the equator. It is able to survive in this atypical region because of the cold, food-rich waters of the Humboldt Current that bathe the western coast of South America.

On land, penguins are somewhat awkward, moving by hopping, waddling on short legs, or tobogganing on their bellies as they push on the ice with their flipper-like wings.

Figure 11—24

Penguins. A male emperor penguin incubating an egg. (Bruno P. Zehnder/Peter Arnold, Inc.)

In the sea, they are as swift and agile as many fishes, literally flying through the water by flapping their wings. They can swim at speeds of 25 kilometers (15 miles) per hour for long distances and can achieve speeds twice as fast while actively pursuing prey. Their torpedo-shaped bodies are streamlined to offer less resistance to movement through the water, and their flat, webbed feet, which are stretched out behind them, are used for steering. While swimming, penguins will break the surface periodically to breathe. They arch through the air like a dolphin and disappear into the water again with scarcely a ripple.

Penguins feed on the fishes, squid, and krill that abound in the cold southern seas. In turn, penguins are preyed upon by leopard seals and killer whales. On land, adult penguins have virtually no enemies, but gull-like skuas (discussed previously in this chapter) that live on the outskirts of penguin colonies take a heavy toll on eggs and chicks.

Adelie and emperor penguins breed on the Antarctic continent. Adelies lay their eggs during the summer when the temperatures are above freezing, but the emperor penguins lay their eggs in the middle of the Antarctic winter, when temperatures drop to −63°C (−80°F) and blizzard winds blow at 130 to 170 kilometers (81 to 106 miles) per hour. This strategy may help to ensure that the young will be mature enough by summer for life in the sea.

The female emperor penguin lays one egg and then leaves for her feeding grounds, a trip that may take her over vast stretches of ice. While she is gone, the egg is incubated by the male (Figure 11–24). For two months he stands on his feet, while a fold of skin on his lower abdomen covers the egg and keeps the developing embryo warm. During the incubation period, the male fasts, living off his reserves of fat. If the egg hatches before the mother returns, the chick is fed a secretion from the father's **crop,** a digestive organ that stores food before it is processed. When the mother returns, her crop is filled with fish, and she feeds the chick. The male is now free to travel to open water to feed. When the chick is about six weeks old, it requires both parents to provide it with enough food, but by the time summer arrives, the young penguins are capable of feeding themselves. The young birds do not have to travel as far as the parents did earlier because the seasonal breakup of the ice floes has reduced the distance between the hatchery and the water.

CHAPTER SUMMARY

The evolution of an amniotic egg allowed reptiles to sever completely all ties with their aquatic environment, giving them a great reproductive advantage. Other adaptations, such as more efficient respiratory, circulatory, and excretory systems, as well as a highly developed nervous system, helped reptiles survive on land and, later, in the marine environment. The first marine aquatic reptiles were probably similar to ichthyosaurs and plesiosaurs.

Several species of crocodiles will enter the ocean to feed, but the best adapted is the Asian saltwater crocodile. Although usually associated with estuaries, this species will sometimes make long ocean migrations. They feed mainly on fish, and in some parts of their range they have been known to attack and kill humans.

Marine turtles have changed very little since they first evolved. Their bodies are streamlined, and their appendages are modified into flippers. Turtles lack teeth, having instead a beaklike structure that is used in securing food. Most sea turtles mate at sea. The female then comes ashore to dig a nest and deposit her eggs. Sea turtles are currently endangered because of damage to or loss of nesting sites.

The only marine lizard is the marine iguana of the Galapagos Islands. This relative of terrestrial iguanas feeds on seaweeds and exhibits several adaptations that allow it to feed in the cold marine waters of its habitat. Snakes evolved from the lizard line of reptiles, and several species are found in tropical marine environments. Sea snakes are venomous and feed primarily on fish, mainly eels. Only about one half of the species come to land to lay their eggs; the others are ovoviviparous.

Instead of a body covering of scales, birds have feathers. Feathers are light and provide good insulation, two important factors for flight. In addition to feathers, the evolution of powerful muscles, homeothermy, and a highly developed nervous system that can coordinate quick responses helped these animals adapt to a flying lifestyle. Seabirds feed in the sea and sometimes spend months out of sight of land. They must return to land, however, to breed.

Shorebirds, or waders, have long legs for wading and thin, sharp bills for finding food in the shallow water and sand. Some well-known shorebirds are oystercatchers, turnstones, plovers, sandpipers, surfbirds, and herons.

Gulls are efficient predators and scavengers. Since gulls are found mostly along shores, they are probably the best known seabirds. Gulls are highly gregarious and gather in colonies of 100,000 or more individuals. Terns, skuas, skimmers, puffins, and murres are all related to the gulls. Terns are found along the shoreline, where they hunt for small fishes and invertebrates. Skuas are aggressively carnivorous and will even eat the young of their own species. Skimmers are small birds that fly along the surface of the water, scooping up food with their enlarged mandible. Puffins and murres resemble penguins, but, unlike penguins, these birds can still fly. Penguins and puffins are examples of ecological equivalents. Puffins and murres are found in the

Northern Hemisphere. They breed on bluffs and cliffs and feed on fish.

Pelicans are large-bodied birds that feed by scooping up large amounts of water with their gular pouch. Related to pelicans are boobies, cormorants, and frigatebirds.

Birds known as *tubenoses* are represented by albatrosses and petrels. Albatrosses are superb gliders, having the largest wings of any living bird. Storm petrels are small birds that exhibit a characteristic, erratic flight. As a group, tubenoses feed primarily on fish.

Penguins are the birds most adapted to life at sea. Their streamlined bodies and flippers instead of wings allow them to be excellent swimmers. Most penguin species are found in the Antarctic, where they feed on fish, squid, and krill. In turn, penguins are preyed upon by leopard seals and killer whales.

SELECTED KEY TERMS

amniotic egg, *p. 198*

amnion, *p. 198*

allantois, *p. 198*

chorion, *p. 198*

salt glands, *p. 200*

ecological equivalent, *p. 208*

gular pouch, *p. 209*

guano, *p. 210*

crop, *p. 213*

QUESTIONS FOR REVIEW

MULTIPLE CHOICE

1. The key to the evolutionary success of reptiles is
   a. a skin covered with scales
   b. a four-chambered heart
   c. the amniotic egg
   d. an efficient excretory system
   e. a large brain

2. Most marine reptiles must still return to land to
   a. sleep
   b. reproduce
   c. feed
   d. molt
   e. die

3. Marine iguanas feed on
   a. grass
   b. crustaceans
   c. fishes
   d. seaweed
   e. molluscs

4. Marine crocodiles drink saltwater and eliminate the excess salt by way of
   a. their urine
   b. their feces
   c. skin glands
   d. their gills
   e. specialized glands on the tongue

5. Birds that exhibit soaring flight have
   a. small bodies
   b. few feathers
   c. long wings
   d. large heads
   e. small feet

6. _____ use their gular pouch as a net to catch fish.
   a. Petrels
   b. Pelicans
   c. Albatrosses
   d. Gulls
   e. Penguins

7. The seabird that plays an important role in keeping beaches clean is the
   a. albatross
   b. pelican
   c. petrel
   d. tern
   e. gull

8. Two types of seabird that are ecological equivalents are
   a. gulls and skuas
   b. petrels and albatrosses
   c. penguins and puffins
   d. herons and sandpipers
   e. boobies and frigatebirds

9. Skuas
   a. are related to pelicans
   b. feed on krill
   c. are most frequently found along tropical shores
   d. are predators of penguin chicks
   e. nest on ledges along the arctic shore

10. Shorebirds generally have
    a. short legs and long beaks
    b. short necks and webbed feet
    c. long legs and flat beaks
    d. long legs and long beaks
    e. short legs and webbed feet

SHORT ANSWER

1. What similarities are shared by most species of shorebird?

2. Describe how marine turtles are adapted to life in the sea.

3. Compare marine iguanas with their more terrestrial cousins.

4. Explain how sea snakes that do not lay their eggs on land reproduce.

5. Explain how birds are well adapted for flight.

6. Describe how sea birds maintain their osmotic balance while drinking seawater and eating marine animals.

7. Why are most soaring birds like albatrosses primarily confined to the Southern Hemisphere?

8. Why do the birds known as *alcids* resemble penguins more than the gulls and terns they are more closely related to?

THINKING CRITICALLY

1. Why do seabirds that nest in trees or other protected areas tend to lay fewer eggs than seabirds that nest on the ground?

2. Why is it important for marine reptiles and birds to be able to efficiently eliminate salt from their bodies?

3. Which of the following would you expect to suffer more from predation: albatross adults or chicks? Why?

SUGGESTIONS FOR FURTHER READING

Alderton, D. 1988. *Turtles and Tortoises of the World.* Facts on File, Inc., New York.

Doubilet, D. 1996. Australia's Saltwater Crocodiles, *National Geographic* 189(6):34–47.

Jackson, D. D. 1989. The Bad and the Beautiful: Gulls Remind Us of Us, *Smithsonian* 20(7):72–85.

Oeland, G. 1996. Emperors of the Ice, *National Geographic* 189(3):52–71.

Rudloe, J. and A. Rudloe, 1989. Shrimpers and Lawmakers Collide Over a Move to Save the Sea Turtles, *Smithsonian* 20(9):44–55.

Taylor, K. 1996. Puffins, *National Geographic* 189(1):112–131.

Trivelpiece, S. G. and W. Z. Trivelpiece. 1989. Antarctica's Well-Bred Penguins, *Natural History* December, 28–37.

Marine Mammals

Marine mammals live mostly or entirely in the water, depend completely on food taken from the sea, and display anatomical features, such as fins, that adapt them for their aquatic lifestyle. Most mammals, whether terrestrial or aquatic, have an insulating body covering of hair, and all maintain a constant, warm body temperature (homeothermic). Mothers feed their young with milk, a secretion produced by special glands in the female called **mammary glands.** It is this characteristic that gives the class Mammalia its name.

Marine mammals are placental mammals. Placental mammals retain their young inside their body until they are ready to be born. Within the mother they are sustained by her systems through a remarkable organ, the **placenta,** which is present only during pregnancy. Although mammals produce fewer offspring than many other animals, more of the young survive to adulthood because they receive a great deal of parental care.

The same characteristics that allowed mammals to become so well adapted to terrestrial environments also allowed them to function well in the sea. The hair (fur) of some and the layers of blubber in others effectively decrease the loss of body heat to the surrounding water. Like other warm-blooded animals, mammals expend about ten times as much energy as similarly sized reptiles and they need more food than a fish of comparable size to support their high metabolic rate. Being homeothermic allows marine mammals to be active feeders around the clock and to adapt to a wide range of habitats. Some, such as baleen whales, feed closer to the base of the food chain, while others, such as sea otters and toothed whales, are second order or higher consumers.

Most marine mammals are quite intelligent compared to other marine animals. This trait, combined with their generally friendly nature, makes them very popular with the public. Unfortunately, marine mammals share another common characteristic: the bodies of many contain materials that are commercially valuable to humans. As a result, they have been hunted in large numbers over the centuries. During the last 50 years, international conservation measures have helped to reverse the decline in many populations, and the numbers of some species are now gradually increasing.

SEA OTTERS

Sea otters (*Enhydra lutris;* order Carnivora) are found along the coast of California and as far north as the Aleutian Islands of Alaska. Instead of having a thick layer of blubber like other marine mammals, their skin is covered by a thick fur with an underlying air layer that protects the animal from the cold. They have short, erect ears, five-fingered forelimbs that they use with great dexterity, and well-defined hindlimbs with finlike feet.

Sea otters seldom venture more than a mile from shore. They favor the areas around coastal reefs and kelp beds, where they spend most of their time floating lazily on their backs (Figure 12–1). They do not come ashore very often except during storms, although some individuals appear to prefer to sleep on shore, possibly to avoid their primary predators, sharks and killer whales. Females normally give birth on shoreline rocks to one pup, and the offspring soon follows its mother into the sea.

Sea otters have a large appetite and consume nearly 25% of their body weight in food per day. Their diet consists

Figure 12–1

Sea Otter. Sea otters bring their food to the surface and eat it while floating on their backs. This otter is preparing to feed on a sea urchin. (Jon L. Hawker)

almost entirely of sea urchins; molluscs, especially abalone; crustaceans; and a few species of fish. They pick their prey from the seabed and carry it to the surface, where, floating on their backs, they use their chests as eating surfaces. They have been observed bringing a stone up along with a mollusc and using it as a tool to smash open the shell. They also have extraordinarily powerful cheek teeth (molars) to help them break through the hard shells of their prey.

Sea otters are diurnal and gregarious animals and can often be observed playing. They are also quite vocal. Pups will cry when hungry, and females will coo affectionately to their offspring and mates. In distress, both sexes will scream loudly, and they produce a companionship call that is a high-pitched squeal that from a distance sounds like a whistle.

Because of the value of their fur, sea otters were nearly hunted to extinction. At one time sea otters ranged from Baja, California, all along the western coast of North America, and west along the Aleutian Islands to Japan. After 170 years of unrestricted killing, the sea otter was almost wiped out. In 1911 only about 1000 animals were left in the world. Since that time, international treaties have protected the animal, and currently 130,000 animals now occupy about 20% of their original range. Their survival today is one of the successes of the marine mammal conservation movement.

PINNIPEDS: SEALS, SEA LIONS, AND WALRUSES

General Characteristics

The term **pinniped** means "featherfooted" and is the name applied to the group of marine mammals (order Pinnipedia) that includes seals, elephant seals, sea lions, and walruses. Although pinnipeds are more at home in the water than on land, they still retain the four limbs that are characteristic of terrestrial mammals. All pinnipeds come ashore to give birth and to molt. Most species also mate on shore, and some species prefer to sleep on land or ice floes where they are safe from sharks and killer whales. Pinnipeds can be found in all oceans, although most species prefer the colder waters of the Northern and Southern hemispheres. They feed on fish and larger invertebrates and a few, notably the leopard seal (*Hydrurga leponyx*), feed on warm-blooded animals such as penguins and other seals. The natural predators of pinnipeds are sharks, killer whales, and humans. Although hunted by humans for over 200 years, the world's pinniped populations are estimated to number around 25 million, though some species are still endangered.

Pinnipeds are divided into three major groups (families): eared seals (Otaridae), true seals (Phocidae), and walruses (Odobenidae). As their name suggests, eared seals, which include sea lions and fur seals, have visible but small external ears (Figure 12–2a). True seals, or phocids, and walruses lack external ears and are thus more streamlined for swimming under water (Figure 12–12b). Eared seals and phocids also differ in their manner of swimming. The main propulsive force for swimming in eared seals is produced by the forelimbs, and they appear to be flying underwater. Their hindlimbs remain nearly motionless during swimming and are used for steering. Phocids, on the other hand, propel themselves through the water with a sculling movement of their hind flippers. Walruses use a combination of the two methods, relying primarily on their forelimbs to move their bulky bodies.

The body of pinnipeds is spindle-shaped, and many species have several thick layers of subcutaneous fat. The head is round and is carried on a distinct neck so that it can be moved independently of the rest of the body. They have large brains and well developed senses. The two sets of pinniped limbs are modified into flippers. The arm bones are shorter than those of whales so that movement occurs only at the shoulder. Their hindlimbs, which are set well back, overlap the stumpy, vestigial tail. In phocids, the hindlimbs serve only for swimming and must be dragged about on land. Eared seals have hindlimbs that can be rotated at right angles to the body and act as legs on land.

Pinnipeds are fast swimmers and can achieve speeds of 25 to 30 kilometers per hour (15 to 18 miles per hour) for short distances. They are also expert divers, remaining under water for as long as 45 minutes without coming up for a breath. Pinnipeds exhale before diving to decrease the amount of air in their lungs and thus make themselves less buoyant. While under water, the animal's metabolism slows by 20%, and the heart rate decreases in order to conserve oxygen. During the dive, blood is redistributed so that vital organs, such as the brain and heart, receive sufficient amounts of oxygen to maintain proper levels of function. These physiological adaptations allow pinnipeds to feed on animals in deeper water as well as at the surface.

Some pinnipeds are capable of deep diving. A Weddell seal (*Leptonychotoes weddelli*), which is found in the cold waters of Antarctica, has carried a depth recorder as deep as 596 meters (1968 feet) and remained there for as long as 70 minutes. The record, however, belongs to the northern elephant seal (*Mirounga angustirostris*). In 1988, a female of this species carried a recorder to a depth of over 1250 meters (4125 feet). Other members of this species were found

(a)　　　　　　　　　　　　　　　　　　　(b)

Figure 12–2

Pinnipeds. Pinnipeds are well adapted to the marine environment, with torpedo-shaped bodies and limbs modified to form flippers. (a) Eared seals, like this fur seal, have visible external ears and long necks. They also have front flippers that rotate backward and rear flippers that can be moved forward to better support the animals' body weight as they move on land. (b) True seals or phocids lack external ears, have short necks, and have front and rear flippers that cannot be rotated and, thus, make movement on land awkward. (a, Patti Murray/Animals Animals; b, David J. Boyle/Animals Animals)

to regularly dive below 697 meters (over 2300 feet), making elephant seals the deepest diving pinnipeds.

Most pinnipeds leave the water during breeding season and congregate on well-established breeding beaches. The bulls arrive first to establish territories and to await the arrival of females. Some species, such as elephant seals (*Mirounga*) and fur seals (*Callorhinus*), are **polygynous** (pah-LIJ-eh-nehs; *poly* meaning "many" and *gyn* meaning "female") with bulls establishing harems of fifteen or more females. The females arrive ready to give birth to pups conceived the previous year. They then mate again almost immediately after giving birth. Although some species mate every two years, most mate annually.

Gestation varies between 9 and 12 months, depending on the species, and typically one or two pups are born. The length of time that the pups nurse (lactation period) depends upon the species and the habitat. Pinnipeds that breed in the coldest habitats, mainly phocids, have the shortest lactation periods, from a few days to two to three weeks, whereas the sea lions of temperate waters will generally nurse their young for as long as six months. One reason for the short lactation period of many phocids is that females fast or feed very little during lactation, placing a severe physiological stress on their body. There is also a limit to the amount of weight that the mother can lose before she will be too weak to survive. Another reason that pinnipeds in cold habitats have short lactation periods has to do with the nature of the habitat. Many species that live in cold environments breed on pack ice. The ice is unstable in many cases, and, as it breaks up, there is the possibility that shifting ice will crush and injure or kill the pup. The shorter the lactation period, the faster a pup develops insulation, and the sooner the pup can enter the water. Also, because newborn pups have very little blubber, they need to develop insulation as quickly as possible in order to survive the cold climate.

Eared Seals

There are two groups of eared seals: sea lions and fur seals. Sea lions have a coarse coat of nothing but hairs, while fur seals have thick dense underfur beneath the stiff, outer guard hairs. The best known of the pinnipeds is the eared seal known as the California sea lion (*Zalophus californianus*). They are the trained seals of zoos and circuses and are intelligent animals that display an eagerness to learn tricks. These sleek pinnipeds are naturally playful and chase each other in the water while vocalizing with a variety of honking and barking sounds. Before becoming popular attractions at animal parks, sea lions were hunted for their blubber.

Sea lions are highly social animals and usually congregate in groups when they come ashore. During the mating season, breeding beaches become battlegrounds as bulls compete with each other for territory and females. Harems consist of as many as 12 females protected by a dominant bull that can be 2.5 meters long (8 feet) and weigh as much as one ton. The bull aggressively guards his territory and harem.

Fur seals can be distinguished from the short-haired sea lions by their thick, woolly undercoats. This thick fur is prized by hunters and is quite valuable on the fur market, a fact that has resulted in massive fur seal slaughters in the past. They are smaller than sea lions, with bulls averaging 2 meters (7 feet) and weighing around 270 kilograms (600 pounds). There are eight species of fur seal found in the Southern Hemisphere, but only one is found in the Northern Hemisphere. Both northern and southern species exhibit similar habits, although the northern species has been studied more extensively.

Of the eight species, the northern fur seal (*Callorhinus ursinus*) is the most abundant. At one time it was hunted relentlessly for its valuable fur, and by 1914 only 200,000 animals remained. This animal is now protected by international treaties, and only a specified number of young males are allowed to be killed each year. As a result of these protective measures, the number of northern fur seals is estimated to be 1.5 million animals. Similar conservation measures have been taken to protect the southern species (*Arctocephalus*) of fur seal, such as those that breed on the islands off the coast of South America.

Phocids or True Seals

The forelimbs of true seals, or phocids, are set closer to the head and are smaller than the hindlimbs. As a result, they are less well adapted to life on land than other pinnipeds. On ice floes and land, they move by dragging their bodies and take every opportunity to slide or roll instead of crawl. Most phocids congregate during the breeding season, but, instead of forming a harem, a male, after establishing his territory on the land, mates with a single female, and the couple remains together for the entire breeding season.

The most abundant pinniped is probably the crabeater seal (*Lobodon carcinophagus;* Figure 12–3a), with a population estimated at over 30 million animals. When these animals were first described in 1842, biologists erroneously thought they fed on crabs, thus the name. The crabeater seal actually feeds on plankton, primarily krill, that it strains from the water with its teeth.

Some of the most familiar species of seal are the harbor seals (*Phoca*) found along cooler coasts of the Northern Hemisphere. Although individuals of most species are marine, a few inhabit freshwater lakes in Canada and Finland. All harbor seals spend some time on land, inching along with a caterpillar-like locomotion. In water they use alternating strokes of their flippers to propel themselves through the water. Harbor seals are generally solitary feeders that eat almost any type of fish, crustacean, or mollusc.

Figure 12–3

Seals. (a) Crabeater seal. (b) Harp seal with pup. (c) Leopard seal. (d) Elephant seal with trunk inflated.

A close relative of harbor seals is the harp seal (*Phoca groenlandica;* Figure 12–3b). They are found in the cold waters of the Atlantic and Arctic oceans. Newborn harp seals have long, silky, yellowish white fur that turns pure white a few days after birth. The white fur of the pups, known as "white coat" to fur merchants, is highly prized and expensive. During the mid-1960s conservationists drew worldwide attention to the cruelty with which harp seal pups were killed and skinned in the major hunting grounds of Canada. Deluged with complaints, the Canadian govern-

ment implemented stricter hunting laws and stronger control over the killing and skinning of these animals.

The leopard seal (*Hydrurga leptonyx;* Figure 12–3c) is the only phocid that will eat warm-blooded prey. Although leopard seals will feed on fish and krill like other phocids, penguins, sea birds, and other seals make up the bulk of their diet. They are powerful swimmers and can be found lurking patiently along the edges of ice floes in the Antarctic Sea, where they wait for their unwary prey to enter the water. They are less dangerous on land since they must laboriously drag their bodies across the ice floes and cannot move quickly enough to catch their prey. Their primary predators are killer whales.

The giants among the pinnipeds are elephant seals (*Mirounga;* Figure 12–3d). The common name for these animals comes not only from their large size but also from the unique proboscis found on the males. This structure is actually an enlarged and inflatable nasal cavity. It is usually limp and deflated, but, during the mating season, it can be inflated to form a trunk 50 centimeters (20 inches) long. The inflated trunk acts as a resonating chamber that amplifies the bull's roar, warning rival bulls from his territory. The inflated proboscis also plays a role in attracting mates.

There are two species of elephant seal. The largest of the two is the southern elephant seal (*Mirounga leonina*), which will grow to a length of 6 meters (20 feet) and will weigh as much as four tons. It is found on islands at the tip of South America and in the Antarctic. The northern elephant seal (*Mirounga angustirostris*) rarely exceeds 5 meters (16 feet) and is estimated to weigh as much as 3 tons. This closely related species can be found on the Pacific coast of North America, from Vancouver Island in Canada to Baja California. Northern elephant seals breed during the winter months from December to mid-March. This allows the young pups to first enter the sea at an optimal time during the spring upwelling. During summer and fall, the adults and pups remain at sea, where they feed primarily on squid.

Walruses

Walruses (*Odobenus rosmarus*) lack external ears and have a distinct neck, as well as hind limbs that can be used for walking on land (Figure 12–4). They average 3 to 5 meters (10 to 16 feet) in length and weigh up to 1364 kilograms (3000 pounds). The canine teeth of their upper jaw have developed into tusks in the males, which are used for fighting with other males. Most bulls carry the scars of battles during the mating season. The tusks can also function as a pick ax, helping the animal to hoist its massive body onto ice floes. Tusks first appear when the animals are five years old and continue to grow for the rest of the animal's life.

The typical family group consists of a large, dominant bull that presides over a harem of as many as three females and half a dozen calves of various ages. One or two calves are born at the end of an 11-month gestation period, and they will stay with their mother until they are between four and five years old. Initially, young walruses have a yellow-brown fur. As they age, the fur gets paler and eventually disappears. Although females are devoted mothers, old bulls are known to kill the young, and the remains of infant walruses have been found in the stomachs of adult males.

Walruses are found in the arctic region of the Northern Hemisphere, where they feed on fishes, crustaceans, molluscs, and echinoderms. At one time, walruses were killed by hunters for their ivory tusks and by Eskimos and other northern peoples for their succulent flesh. As a result, the walrus populations declined to such low levels as to make the species endangered. Protective laws have virtually eliminated walrus hunting for sport and for ivory. However, native peoples who subsist on walrus meat are exempt from the restrictions and continue to take the animals for food.

SIRENS: MANATEES AND DUGONGS

Manatees and dugongs are collectively referred to as **sirenians** (order Sirenia). They get their name from the mythical *sirens* of Homer's *Odyssey*. These sweet-voiced creatures, who lured Ulysses and his men with temptations, were

Figure 12–4

Walrus. Two male walruses fight for a female. (Jon L. Hawker)

probably the forerunners of the mythical mermaids. Mermaids have been a part of the mythology of the sea for over 2500 years. Pliny the Elder, a respected Roman naturalist, gave detailed descriptions of them, and Christopher Columbus wrote in his log: "Today I saw three mermaids, but they were not as beautiful as they are painted." No doubt what Columbus and other seafarers saw and mistook for mermaids were manatees and their relatives, although it is difficult to imagine how anyone could mistake these large, gray animals with wrinkled skin and whiskers for the alluring maidens of legend.

At one time sirenians enjoyed a wide distribution. Now they are confined to coastal areas and estuaries of tropical seas. They share many similarities with whales. Both have streamlined, practically hairless bodies, forelimbs that form flippers, a vestigial pelvis with no hindlimbs, and tail flukes. Sirenians are completely aquatic animals and are helpless on land, not even able to crawl as do the pinnipeds. There are two families of sirenians represented by the manatees (Trichechidae) of the Atlantic Ocean and Caribbean Sea, and the dugongs (Dugongidae) of the Indian Ocean.

The primary differences between manatees and dugongs are in anatomy and habitat. Manatees inhabit both the sea and inland rivers and lakes. Dugongs are strictly marine mammals living in coastal areas, where they can feed on shallow-water grasses. The head of the dugong is larger and the flippers are shorter than those of the manatee. The dugong's tail is notched, whereas the manatee's tail is rounded.

There is only a single species of dugong (*Dugong dugong;* Figure 12–5a). Large numbers of this animal were once found along coastal areas of the Indian Ocean, Red Sea, and South China Sea. It has been hunted extensively for its tasty flesh and for oil from its blubber. Today the species is endangered and found only in isolated areas of its once large range.

There are three species of manatee. The northern manatee (*Trichechis manatus;* Figure 12–5b) is found from the southeastern United States to northern South America. The Brazilian manatee (*Trichechis inunguis*) is a freshwater species endemic to the Amazon and Orinoco rivers. The African manatee (*Trichechis senegalensis*) is found in coastal habitats, rivers, and lakes of western Africa. All three species of manatee have been extensively hunted, and the northern species is now protected by law in Florida.

Manatees mate and give birth under water, and the male remains with his mate even after the breeding season. The female usually gives birth to a single calf after an 11-month gestation period. Manatees are strict vegetarians and consume large amounts of shallow-water plants. A single manatee will consume at least 27 kilograms (60 pounds) of aquatic plants per day. When manatees eat, they guide the water plants to their mouths with their flippers. This observation may account for the association with mermaids. Be-

cause of their voracious appetites, they provide a service by keeping coastal waterways free of non-native plant species, such as the water hyacinth, that tend to take over their new habitats. In Guyana, manatees have been introduced to clear weeds from the canals that transport irrigation water to sugar cane fields.

(a)

(b)

Figure 12–5

Sirenians. (a) Dugongs are found along coastal areas of India and Asia. (b) The scars on the back of this northern manatee were caused by a boat propeller. (a, David B. Fleetham/Oxford Scientific Films/Animals Animals; b, Norbert Wu)

Neither the manatee nor dugong are found in Arctic waters. The only sirenian from this area, Steller's sea cow (*Rhytina gigas*), is extinct. During 1741–1742, Georg Wilhelm Steller, a German, was the physician and naturalist for Russia's expedition to the northern Pacific Ocean led by Captain Vitus Bering. Steller discovered the enormous mammal when his ship was wrecked off one of the Commander Islands. He made many notes on the physical characteristics and behavior of this animal and correctly identified it as a previously unknown species of Sirenia. Desperate for food, Steller and his companions killed and ate several of the creatures. When their ship was repaired, they returned to Russia with the furs of sea otters and arctic foxes that they had also killed for food. Word spread quickly that a fortune in furs was available for the taking on the Commander Islands, and hundreds of hunters descended on them. The hunters found the meat of Steller's sea cow tasty and easy to obtain, and 27 years after its initial discovery, the species became extinct.

Sirenians are by nature gentle animals that become quite tame and trusting in captivity and in their natural habitat. There are several preserves in Florida where divers and snorkelers can swim with manatees and even touch them and rub their bellies. Only in those areas of the world, like India, where they are hunted nearly to extinction, have sirenians become shy and elusive.

The greatest threat to northern manatees are the propellers of motorboats. Many of these relatively slow moving animals are mauled or killed by boats (Figure 12–5b). Although there are strict laws governing the speed and use of motorboats in areas where the manatees congregate, there are few regulations governing the canals and waterways that connect these freshwater bodies to the sea. As the manatees migrate from the freshwater areas to the ocean, they must travel these waterways, and it is then that they are in the greatest danger.

CETACEANS: WHALES AND THEIR RELATIVES

General Characteristics

Of all marine mammals, **cetaceans** (seh-TAY-shenz; whales, dolphins, and porpoises) are the ones most extensively adapted to a marine environment. Humans have been fascinated and awed by these animals for centuries. Ancient Greeks and Phoenicians worshiped the dolphin or "sacred fish," as did many Polynesian cultures. In Australia more than 5000 years ago, aborigines painted on stones recognizable pictures of dolphins. The biblical story of Jonah being swallowed by a whale, the appearance of marine mammals in the art of the Middle Ages, and the Hermann Melville novel *Moby Dick* all attest to our fascination with these marine mammals.

The body of cetaceans closely resembles those of fishes, and early naturalists even included them both in the same group of animals. Their exact ancestry is not known, but it is generally agreed that they evolved from an ancient group of land-dwelling carnivorous mammals. Recent discoveries in Pakistan, Egypt, and the Arabian desert of fossil whales with fore and hind limbs and teeth similar to carnivores support this hypothesis. Other evidence for the terrestrial origins of cetaceans can be seen in developing fetuses. Whale fetuses are remarkably similar to those of land mammals. Early in development they have four limbs. The rear appendages disappear externally before birth, although there are a vestigial pelvis and leg bones in some species (Figure 12–6). The front appendages persist and develop into flippers with the bone structure of a five-fingered hand. Initially the animal's nostrils are located at the end of its snout, but before birth they migrate to the top of the head to form the characteristic **blowhole.** The location of the blowhole allows these animals to surface and breathe with a minimum of effort.

The cetacean body is streamlined, and beneath the skin is a uniformly thick layer of fat called *blubber.* The fat provides insulation to conserve heat and acts as an energy reserve as well as a source of water when the fat is metabolized. In becoming streamlined, the neck has disappeared, and the head has become continuous with the rest of the body. The seven cervical vertebrae of cetaceans are very compressed, and in some species they are fused into a single structure. The result of these modifications is that, in many cetaceans, the head cannot be moved relative to the body.

Cetaceans have no external ears or projecting nostrils to hamper their movements through the water. The whale ear consists of a small opening on the side of the head that leads to an eardrum. The opening is plugged with wax to prevent water from entering and damaging the eardrum. The inner ear is contained in a bone that floats freely in the soft tissue surrounding the skull.

The cetacean body is essentially devoid of hair, with the exception of a few hairs on the rostrum. The skin of cetaceans lacks the sweat glands that are characteristic of other mammals, since they have no need of cooling themselves. The absence of these glands also helps them to conserve water in their salty environment.

The forelimbs of cetaceans have become modified into flippers that can move only up and down and twist a bit. They function as stabilizers. The cetacean tail consists of flat, cartilaginous **flukes** and acts as the main propulsion organ. The tail flukes not only supply the force of propulsion but also act to regulate vertical movement. Most cetaceans also have a dorsal fin that apparently helps them to control roll.

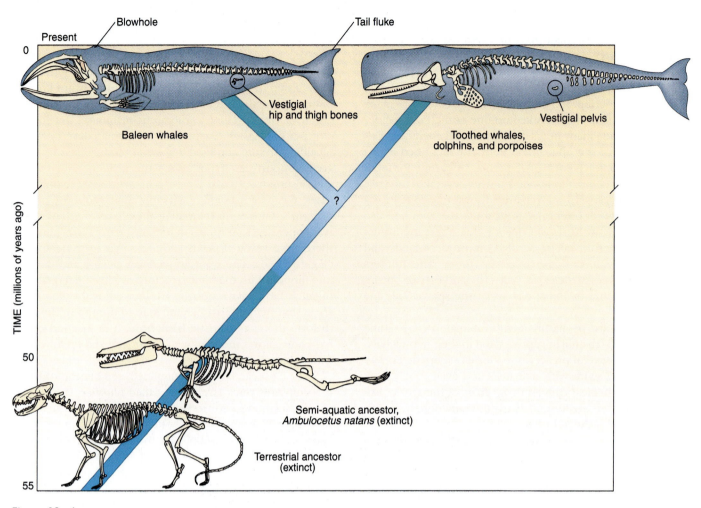

Figure 12–6

Cetacean Evolution. Whales evolved from four-legged, terrestrial carnivores. Their ancestors may have moved about on land in addition to swimming in shallow seas. Modern cetaceans are well adapted to life in the sea, with streamlined bodies, appendages modified to form flippers, and a large tail flipper or fluke.

The flippers (as well as the tail flukes) also serve an important role in retaining body heat (Figure 12–7). The arteries that carry blood to the flippers are surrounded by veins that carry blood back to the heart, an example of countercurrent flow (see also the discussion of tuna in Chapter 10 and Figure 10–19). As warm blood in the arteries moves out into the flipper, it transfers heat to the cold venous blood that is returning to the core of the body. Most of the heat is returned to the body, and cold blood is carried to the uninsulated flippers and flukes.

Adaptations for Diving

Perhaps the most striking adaptations of cetaceans involve those of the circulatory and respiratory systems that allow them to dive to great depths and remain submerged for rel-atively long periods of time (Figure 12–8). Before diving, a whale takes an enormous breath of air and then blows it out. During the inhalation, oxygen is rapidly transferred to the blood. Expelling the air from the lungs then helps the animal become less buoyant. Cetacean lungs are proportionately large for the body size, and they contain a large number of small air sacs, or alveoli. This feature increases the internal lung surface area that is exposed to blood vessels and allows for a more efficient diffusion of gases into and out of the blood. The lungs and rib cage of whales are structured so that they collapse easily upon descent. As a result, the lungs contain little air during a dive. Since gas volumes change dramatically with changes in pressure, this adaptation helps the animal to avoid the problems of compression and decompression while diving and surfacing.

While diving, the animal's metabolism and heart rate decrease, and blood is preferentially shunted to vital organs

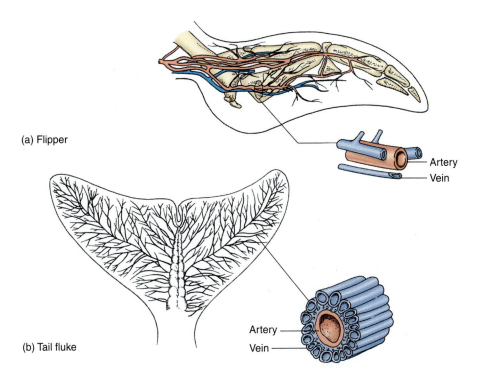

(a) Flipper

Artery

Vein

(b) Tail fluke

Artery

Vein

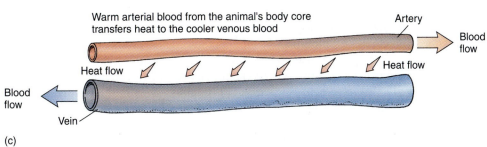

Warm arterial blood from the animal's body core transfers heat to the cooler venous blood

Artery

Blood flow

Heat flow

Heat flow

Blood flow

Vein

(c)

Figure 12–7

Countercurrent Blood Flow in Cetacean Flippers and Tail Flukes. The blood vessels in a cetacean's (a) flippers and (b) tail fluke are arranged so that (c) some of the heat carried by arterial blood is transferred to the cooler venous blood returning to the body core, thus conserving body heat.

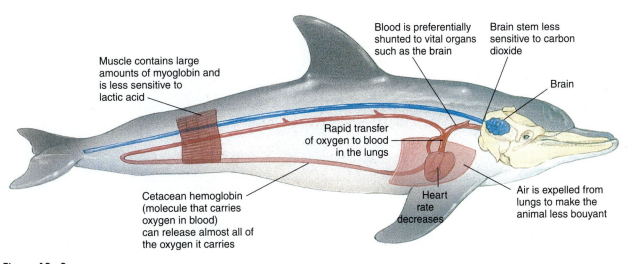

Muscle contains large amounts of myoglobin and is less sensitive to lactic acid

Blood is preferentially shunted to vital organs such as the brain

Brain stem less sensitive to carbon dioxide

Brain

Rapid transfer of oxygen to blood in the lungs

Cetacean hemoglobin (molecule that carries oxygen in blood) can release almost all of the oxygen it carries

Heart rate decreases

Air is expelled from lungs to make the animal less bouyant

Figure 12–8

Diving Physiology. Physiological changes in the respiratory system, nervous system, circulatory system, and muscles allow whales to dive to great depths and remain submerged for over 1 hour.

and tissues such as the brain and spinal cord. The portion of the brain that controls breathing is the medulla oblongata. In most mammals this region is very sensitive to high levels of carbon dioxide in the blood, which trigger inhalation. The medulla of cetaceans is less sensitive to carbon dioxide, so they can remain submerged longer without feeling the urgency to breathe.

Other adaptations for diving involve the transport and storage of oxygen. The molecule hemoglobin in red blood cells is responsible for carrying oxygen in the blood and for giving blood its red color. Under normal physiological conditions in most mammals, hemoglobin will release only about 30% to 40% of the oxygen it carries. Cetacean hemoglobin, however, is able to discharge almost all of its oxygen, supplying the needs of tissues for a longer period of time.

Mammalian muscle tissue contains a molecule called *myoglobin* that acts as a reservoir of oxygen for muscle activity. Cetacean muscles contain high levels of myoglobin and thus store proportionately more oxygen for muscle activity. Their muscles are also less sensitive to the molecule lactic acid. Lactic acid is a waste produced during vigorous or extended muscle activity when there is not enough oxygen. In humans and terrestrial mammals, moderate amounts of lactic acid can impair muscle function, but cetaceans can tolerate much higher accumulations of lactic acid with no ill effects.

When a cetacean surfaces, the top of its head emerges first, thus exposing the blowhole to air. From this opening the animal emits a spout that consists of vapor condensing in the cool air as well as mucous droplets. The mucous droplets may play a role in eliminating nitrogen gas from inhaled air by absorbing it before it reaches the lungs, thus preventing a condition known as "the bends." The bends occur when nitrogen gas that is dissolved in the blood comes out of solution, and forms gas bubbles as the pressure decreases during an ascent. These bubbles can interfere with the circulation of blood to body tissues and can cause tissue damage or even death. As a cetacean descends, the increased pressure of the water causes the alveoli or gas sacs in the lungs to collapse. Any residual air remaining in the lungs is forced into airways, where gas exchange with the blood cannot occur. This prevents nitrogen gas from entering the blood and helps the animal to avoid the bends.

Cetaceans also exhibit adaptations for preventing water from entering their respiratory passages. The larynx, or voice box, does not open into the back of the throat as it does in some mammals but rather into the nasal chambers. This allows cetaceans to open their mouths under water without the danger of food or water entering their respiratory passageways.

Reproduction and Development

Many cetaceans travel in groups called **pods** that are composed of both adults and young. Normally cetaceans bear only one offspring at a time, rarely two. The young are fed extremely rich milk, containing 40% to 50% fat and 10% to 12% protein, as compared with cow's milk, which contains 2% to 4% fat and 1% to 3% protein. The high caloric content of whale's milk allows the infant whale to grow at an exceedingly fast rate and to produce enough body heat to resist the cold until it develops its thick layer of blubber. Baby blue whales, for instance, double their birth weight in the first week and then gain some 200 pounds a day thereafter.

Baleen Whales

Baleen whales (suborder Mysteceti) have enormous mouths to accommodate plates of **baleen,** which take the place of teeth (Figure 12–9). Each plate of baleen has an elongated triangular shape and is anchored at its base to the gum of the upper jaw. The baleen plate is composed of keratin (a tough protein found in mammalian hair and nails) fibers that are fused together, except at the inner edge, where they form a fringe. Hundreds of these plates, each one 1 to 4 meters (3.3 to 13 feet) long, form a tight mesh on the sides of the mouth. When not actively feeding, the baleen is covered by the underlip, which projects above the lower jaw.

The diet of baleen whales consists mainly of plankton, especially krill, or fish. A baleen whale feeds by opening its mouth and swimming into dense groups of krill or schools of fish. When it has filled its mouth, it closes it and expels the water through the baleen plates, straining out the prey in the process. The retained food is then moved by the tongue to the back of the mouth and swallowed. The process is repeated until, at the end of a feeding, a large whale has as much as 10 tons of krill or fish in its stomach. Humpback whales (*Megaptera novaeangliae*) sometimes

Figure 12–9

Baleen. This piece of baleen is from a humpback whale. Baleen whales use these structures to strain plankton from the water column. (Jon L. Hawker)

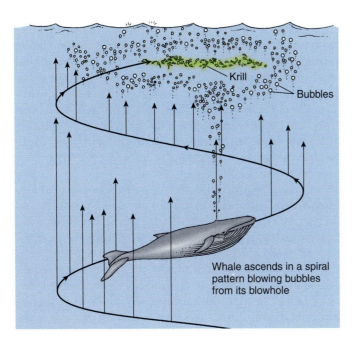

Figure 12–10

Bubble Net. When feeding, humpback whales will frequently blow rings of bubbles as they rise to the surface. Small prey hesitate to cross the bubble barrier and become easy prey for the whales.

capture their food by blowing a ring of bubbles called a **bubble net** that traps krill near the surface (Figure 12–10).

Baleen whales include the right whales, rorquals, and gray whales. Right whales (Balaenidae) can be distinguished from other cetaceans by the lack of a dorsal fin and the absence of grooves on the throat and chest. Little is known about most species of right whale, and they are considered rare. Right whales received their name from early whalers. When lookouts sighted a whale, if it was the "correct" whale for hunting, they would yell: "It's the right whale!" thus the origin of the name. Right whales were the favorite catch of early whalers because they were relatively slow and easily harpooned, floated when killed, and could be sliced up with relative ease. As the result of overhunting, three of the four species of right whale were nearly extinct by the end of the last century, and one species, the Greenland, or bowhead, whale (*Balaena mysticetis*), is the rarest of all whales.

Rorquals (Balaenopteridae) have the dorsal fin and ventral grooves that the right whales lack. The ventral grooves or pleats under the mouth allow the throat to expand while the animal is feeding. Rorquals are generally slender and streamlined and are capable of relatively fast swimming. Some of the better known rorqual species are the blue whale, humpback whale, sei whale, and common rorqual. The blue whale (*Balaenoptera musculus*) has the distinction of being the largest of the whales and may be the largest animal that has ever lived (Figure 12–11). They range in size from 24 to 30 meters (80 to 100 feet) and

weigh in excess of 100 tons (as much as 1600 humans or 25 elephants!). Because of excessive hunting, however, large individuals are rarely seen anymore. Whaling records indicate that the weight of a blue whale consisted of 50 tons of muscle; 8 tons of blood; 1 ton of lungs; and 60 tons of skin, bones, and internal organs (other than the lungs). The heart alone weighed 591 kilograms (1300 pounds) and measured about 1 meter (3.3 feet) in each direction.

Until modern whaling techniques were introduced, rorquals like the blue whale were relatively safe from human hunters. They were too large to approach in small boats and fast enough to escape from hand-hurled harpoons. On the few occasions when a rorqual was killed, it usually frustrated the whalers by sinking or swamping the whaleboats. All of this changed in this century with the introduction of harpoon guns, motorized boats, factory ships, and techniques for injecting air into the whale carcass so

Figure 12–11

Blue Whale. The blue whale, which may be the largest animal that has ever lived, feeds on plankton, especially krill. (Copyright 1995 Peter Howorth/Mo Yung Productions/Norbert Wu)

Figure 12–12

Whaling Ship. The development of modern whaling vessels allowed whalers to kill more whales, pushing many species to the brink of extinction. (Mitsuaki Iwago/Minden Pictures)

that it would float and could be towed for great distances (Figure 12–12). Since the average blue whale could furnish 120 barrels of high-grade oil, they became the primary targets of the whaling industry, and their numbers plummeted. In the 1930–1931 whaling season, the worldwide catch of blue whales was 29,649. By the 1964 to 1965 season, whalers could find only 372 blue whales to kill. In 1966, the International Whaling Commission gave the blue whale worldwide protection, and their numbers have started to increase again.

The humpback whale (*Megaptera novaeangliae*) can be distinguished by the low hump on its back; the large bumps, called *bosses,* on its snout; and the very long flippers that may be one third the animal's length, or 5 meters (15 feet). Humpbacks are slow-moving, heavyset creatures that may be up to 15 meters (50 feet) long and 12 meters (40 feet) around. It is one of the slowest swimming whales, averaging 3 to 8 kilometers (2 to 5 miles) per hour. Humpbacks are mainly coastal inhabitants and will frequently enter harbors and even venture up the mouths of large rivers. These characteristics have made the humpback a very vulnerable species. Although it has been protected by the International Whaling Commission since 1966, it is still considered to be an endangered species. There are three distinct populations of humpback whale: in the North Atlantic, the North Pacific, and the Southern Hemisphere. Of these three, only the North Atlantic population is showing signs of recovery from the impact of whaling.

Humpbacks spend most of their time in polar seas feeding on krill and other plankton that are abundant there, but they come to warmer waters to mate and bear their young. They are a popular sight for tourists during the month of March in the waters off of Maui, Hawaii, where the north-ern Pacific population comes to breed. During the mating season, males will literally jump out of the water and come crashing back down to the surface with a loud noise and large splash (Figure 12–13). This behavior, called **breeching,** is thought by some to be part of the mating ritual. Since humpbacks are coastal species, they are more accessible than other whale species and have been the subjects of many studies. Much research has focused on their vocalizations or "songs" that may play a role in mating and intraspecies communication.

At one time, three different populations of gray whales (*Eschrichtius gibbosus*) existed in the western and eastern Pacific Ocean and the North Atlantic Ocean. Today, only the eastern Pacific population survives. The other two populations were decimated by the whaling industry. The eastern Pacific population, which initially contained an estimated 24,000 individuals, was on the brink of extinction in 1946. The number of individuals in the population at that time was estimated to be 250. In that year, strict laws were enacted to protect the remaining gray whales, and today the population consists of more than 15,000 individuals, allowing them to be removed from the list of endangered species in 1994.

Gray whales migrate from their summer feeding grounds in the Bering Sea to the waters off Baja California to mate and give birth. Females aggressively defend their

Figure 12–13

Humpback Breeching. Humpback whales migrate to the warm Hawaiian waters off the island of Maui in March to breed. During this time males can be seen jumping out of the water and crashing back down, a behavior known as *breeching.* This photo shows a rare occurrence: two humpbacks breeching at the same time. Breeching is thought to play a role in mating. (Animals Animals Copyright 1995 James D. Watt)

Figure 12–14

Narwhal. The upper incisor (usually the left) of male narwhals develops into a large tusk that projects straight out from the head. (Flip Nicklin/Minden Pictures)

young, a trait that prompted whalers to refer to them as "devilfish." Gray whales carry large accumulations of barnacles on the skin, more than any other cetacean, and for this reason are sometimes referred to as *mossback whales.* Although the gray whale has been recovering from the impact of whaling, its popularity with tourists is now causing problems. Large numbers of whale watchers and their motorboats disturb the whales while they attempt to mate and give birth, resulting in fewer offspring.

Toothed Whales

Toothed whales (suborder Odontoceti) include sperm whales, dolphins, porpoises, killer whales, and narwhals. As with other aquatic mammals, the teeth of these whales tend to be simplified. Dolphins and porpoises have between 100 and 200 identical teeth that are fused to the jaws. They function to grasp the fish on which these animals feed. Sperm whales have functional teeth only in the lower jaw, which fit into sockets in the upper jaw. It is believed that this is an adaptation for feeding on the slimy cephalopods, which are their primary food. Beaked whales have only a few teeth, sometimes only two in each jaw, and the narwhal is peculiar in that one of its teeth forms a tusk that projects straight out from its head (Figure 12–14).

Echolocation

Even on a clear day, toothed whales cannot see much farther than 30 meters (100 feet) in the water. To compensate for this, they have ears that are modified to receive a wide range of underwater vibrations. This adaptation not only improves the animal's hearing under water but refines it for the purpose of **echolocation** (Figure 12–15), a process that, like sonar, allows the animal to distinguish and home in on objects from distances of several hundred meters. Arthur F. McBride, the first curator of Marineland in Florida, was one of the first to discover echolocation in cetaceans in the early 1950s. He found that every time he tried to capture dolphins for an exhibit by driving them toward a net,

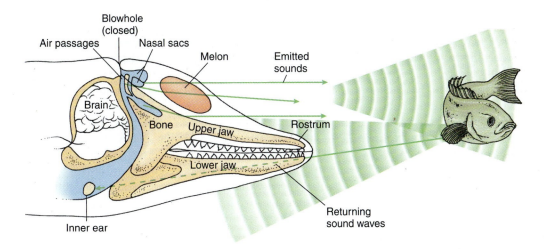

Figure 12–15

Echolocation. Toothed whales use sound to locate objects, determine shapes and distance, and navigate. Sounds produced by the animal's larynx (voicebox) are focused by a structure in the head called the *melon.* When the sounds strike a target, they are reflected as echoes (returning sound waves). The echoes are picked up by the animal's lower jaw and transmitted to the inner ear. The vibrations received by the inner ear are then converted into signals that can be carried by nerves to the brain for interpretation.

the animals would stop short of the net and swim away. It didn't matter whether he tried during day or night, and even in murky water where they could not see the net they would still avoid it. McBride hypothesized that the animals might somehow use sound to sense the presence of the nets in the water.

Later in the 1950s, Winthrop N. Kellogg of Florida State University demonstrated that dolphins use a series of clicking sounds in the same way that humans use sonar. Dolphins can emit a wide spectrum of sounds, lasting from a fraction of a millisecond to as long as 25 milliseconds. To obtain information about their environment, dolphins emit sounds with frequencies that range from less than 2000 to greater than 100,000 hertz. Only a portion of these sounds can be perceived by the human ear, and these we hear as clicks. The low-frequency clicks are orientation clicks that give the animal a general idea of its surroundings. The high-frequency clicks are discrimination clicks that give the animal a precise picture of a particular object. The clicks can be produced as a single sound or as a series of sounds strung together.

The dolphin's larynx does not have vocal cords but instead has a ring of muscles that acts as a valve, enabling the animal to control the air flow through the organ. Air under pressure circulates through the animal's nasal passages, producing the clicks and other characteristic sounds. The sounds are directed by being focused in the **melon,** an oval mass of fatty, waxy material that is located between the blowhole and the end of the head. The sounds are directed toward objects in front of the animal, but not on objects that are below a line level with its jaws. This fact may explain why some toothed whales cannot sense gradually sloping bottoms and sometimes run aground. The clicking sounds travel through the water and bounce off anything solid. The reflected sounds (echoes) are picked up by sensitive areas of the lower jaw that transmit the vibrations to the inner ear. The animal's brain then processes the information to produce a mental image of the target object.

When the sounds bounce off a target, they produce echoes that provide at least four types of information: the direction from which the echo is coming, the change in frequency, the amplitude, and the time elapsed before the emitted sound returns. With this information, dolphins, as well as other toothed whales, can determine not only the range and bearing of an object, but also its size, shape, texture, and density. In tests, blindfolded dolphins have been able to discriminate between two very small objects that are about 2 inches in diameter and have different shapes at a distance of 3 meters (10 feet).

When a dolphin travels, it usually moves its head from side to side and up and down. This behavior allows the animal to scan a broad path ahead of it. The scanning motions become fast and jerky when the dolphin becomes interested in a small target, such as a fish. As the dolphin scans,

it can determine the direction from which echoes are returning and thus determine the bearing of the target. Changes in the frequency of the returning sound give the dolphin information about the size and shape of the object. The amplitude of the echo and time elapsed before it returns help the dolphin determine distance.

Sperm Whales, Beluga Whales, and Narwhals

The giant of the toothed whales is the sperm whale or cachalot (KASH-uh-loh; *Physeter catodon*), the third largest animal on earth behind the blue whale and fin whale. Sperm whales (Figure 12–16a) are sometimes confused with right whales and rorquals, but they have teeth instead of baleen and a massive, blunt snout that projects forward beyond the mouth. At one time, the sperm whale was abundant in most seas and for almost a century and a half it was the favorite catch of American whalers. Sperm whales were challenging targets for whalers of the 18th and 19th centuries because of their large size (up to 18 meters or 60 feet), toothed jaw, and powerful tail. Whalers judged the risks well worth taking because the animal's commercial value was so great. During the 20th century, modern whaling techniques decreased the risks, while profits remained high. As a result, the once large populations have decreased significantly, and, although protected by international treaty, the sperm whale is still considered to be a rare and endangered species.

The sperm whale received its name from an oily, wax-like substance that is found in the animal's head. The snout contains a cavity that is developed from one of the nasal chambers. This cavity contains a thin, colorless, transparent oil that forms a waxy material when it comes into contact with air. Early whalers erroneously thought it to be an enormous reserve of semen and named it **spermaceti** (spur-meh-SEH-tee; Figure 12–16b). At the height of the whaling industry, spermaceti was one of the most prized products of the sperm whale catch because it was found to be a high-grade wax. It was used in the manufacture of ointments, face creams, luxury candles, and lubricants. Now synthetics have taken its place. It is generally believed that the spermaceti plays a role similar to that of the melon in dolphins and thus is part of the whale's echolocation system.

The sperm whales' fondness for squid and cuttlefish generates a digestive by-product called *ambergris,* which in the past was even more valuable than spermaceti. The ambergris is a secretion of the whale's digestive tract and may function in protecting the enormous digestive system from undigested squid beaks and cuttlefish's cuttlebone. When this secretion is exposed to air, it forms a solid, gray, waxy material with a disagreeable odor, but over time it acquires properties that make it highly prized as a base for the most expensive perfumes. Even though synthetic substitutes are

(a)

Figure 12–16

Sperm Whale. (a) A sperm whale nears the surface of the water in the North Atlantic Ocean. (b) Spermaceti is believed to function in echolocation and was prized by whalers as a high-quality wax. (a, Doug Perrine/Jeffrey Rotman Photography)

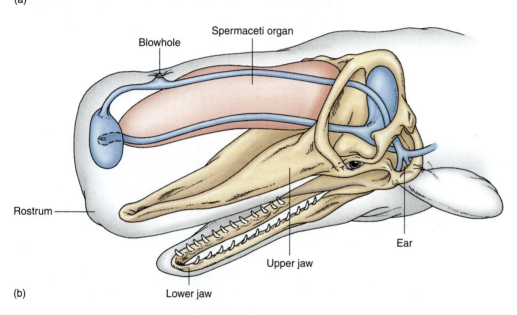

(b)

available, ambergris is still very much in demand. Lumps of ambergris that are evacuated by the animals sometimes wash up on shore. A 418-kilogram (920 pound) lump of ambergris that washed up on an Australian beach in 1953 sold for $120,000.

Sperm whales have no real dorsal fin but instead have a series of humps on the rear third of their body. Unlike most other species of whale, sperm whales are aggressive, attacking squid, fishes, and, on occasion, whalers in small boats. They are polygynous, and males are always accompanied by several females. Although mainly found in tropical and temperate waters, they do range occasionally into the polar seas.

The beluga whale (*Delphinapterus leucas*) is unique among whales for its white color. In fact, the name is derived from the Russian word for *white*. The beluga lives in the northern polar seas. They usually travel in small family groups and feed on crabs, cuttlefish, flounder, and halibut. On occasion they have been observed to enter and swim up arctic rivers.

The narwhal (*Monodon monoceris;* see Figure 12–14) is a close relative of the beluga whale and also lives in arctic waters. Although the fetal animals have tooth buds, by birth these have all disappeared, with the exception of the two upper incisors. These teeth remain undeveloped in the skull of the females, but in males a tusk as long as 3 meters (10 feet) develops from one of the two teeth, almost always the left. In some individuals, the right tooth also develops a tusk, giving an already odd-looking creature an even more bizarre appearance.

During the Middle Ages, the narwhal tusk was passed off as the magical horn of a unicorn. Since the tusk was literally worth more than its weight in gold, the narwhal was hunted relentlessly. In later times, the narwhal was killed incidentally by whalers in the Arctic Ocean who were searching for bowhead whales. Since the demise of the bowhead fishery, the narwhal population has recovered, and the animals survive relatively undisturbed. Their only remaining predator appears to be Eskimos who kill the animal for its nutritious skin.

Dolphins and Porpoises

Since most people think of whales as being giant animals, few are aware that the playful, acrobatic dolphins and porpoises are also toothed whales. In fact dolphins are the most numerous cetaceans. Dolphins and porpoises (superfamily Delphinoidea) are collectively referred to as **delphinids.** The name is derived from ancient Greek mythology. According to legend, Apollo rose from the sea in the shape of a dolphin to lead settlers to Delphi, home of the Delphic oracle. The word *porpoise* comes from a contraction of the Latin word *porcopiscis* (meaning "pigfish"), apparently referring to the animal's stocky body. The Greek naturalist Aristotle observed that the porpoise resembled a small dolphin but differed by being broader across the back. Modern biologists point out that the most obvious difference between the dolphin and the porpoise is that the dolphin has a beak, while the porpoise's head is rounded off.

One of the smallest cetaceans is the harbor porpoise (*Phocoena phocoena;* Figure 12–17a), which measures only 1 to 2 meters (3 to 7 feet) and weighs nearly 45 kilograms (100 pounds) when fully grown. This animal is widely distributed in the North Atlantic Ocean and elsewhere and is a familiar sight to boaters and beachcombers on both sides of the Atlantic. It is also widely distributed along the Pacific coast of North America. The harbor porpoise is one of the most intelligent cetaceans, and its capacity for learning is surpassed only by that of the bottlenose dolphins, killer whales, and pilot whales. It feeds on a variety of schooling fish such as herring, sardines, and mackerel, sometimes feeding on fish that have already been caught in nets. A related species, the vaquita or "little cow" (*Phocoena sinus*), is endemic to the upper Gulf of California and is in danger of extinction as a result of fishers' depleting the stocks of fish and squids on which the animals feed.

(a)

(b)

Figure 12–17

Comparison of Dolphin and Porpoise. Dolphins and porpoises are both delphinids. (a) The head of a porpoise is rounded, whereas (b) dolphins can be distinguished by their beaks. (a, Michael Sacca/ Animals Animals; b, Norbert Wu)

The Dall's porpoise (*Phocaenoides dalli*) is another porpoise without a prominent beak. For years Europeans considered this animal a delicacy, and it was quite rare. The rarity was probably more a factor that the porpoise eluded its human predators than the fact that the numbers were few. This animal is perhaps the first to be protected by law. In Normandy as far back as 1098, laws were passed regulating the size and number of the Dall's porpoise harvest. Dall's porpoises feed on fish, preferring herring, whiting, and sole. These animals have been observed swimming up rivers as far as 58 kilometers (36 miles), and one specimen was observed in the Maas River in the Netherlands, 323 kilometers (200 miles) from the sea.

The best known delphinids are the common dolphin (*Delphinus delphis*) and bottlenose dolphin (*Tursiops truncatus;* Figure 12–17b). The common dolphin has a definite beak that is separated from the snout by a groove. This animal is small in comparison with other cetaceans; adults grow to no more than 3 meters (10 feet). They can be found in all temperate oceans and are among the swiftest cetacean swimmers. They feed on schooling fish like herrings and sardines that they devour in enormous quantities. The common dolphin is frequently seen escorting ships. When common dolphins find a ship, they encircle it. Some of the dolphins will lead at the bow, while others follow along at the sides. They will leap completely out of the water in a graceful arc that covers several meters and will repeat this behavior for hours and hundreds of kilometers without ever seeming to tire.

The most abundant dolphin species along the eastern coast of the United States and the Caribbean is the bottlenose dolphin. These dolphins also inhabit the Mediterranean, Baltic, and Caspian seas, and a subspecies is found in the temperate waters of the Pacific Ocean. These are the dolphins that are most frequently used for scientific studies of cetacean intelligence. Bottlenose dolphins are often seen as performers at aquariums or as stars of television and movies. These intelligent animals are quite playful and appear to like the company of others of their species. They have been observed to aid disabled members of their species, pushing them to the surface so they can breathe or helping them to shallow water. There are also many stories, some authenticated, about bottlenose dolphins coming to the aid of injured or drowning humans.

The largest of the dolphins is the killer whale or orca (*Orcinus orca;* Figure 12–18). Males sometimes reach lengths of 10 meters (33 feet), while females are only half as large. These animals received their common name because they are the only cetaceans that feed on warm-blooded animals. Killer whales have a high dorsal fin (as much as 1.8 meters or 6 feet) and broad, rounded flippers. Their coloration is particularly distinctive. They are primarily black with a white patch over the eye and a white ventral stripe. These powerful animals are quite agile and are the fastest

Figure 12–18

Killer Whale. Killer whales are the only cetaceans known to feed on warm-blooded prey. (Copyright Brandon Cole/Norbert Wu)

cetacean swimmers, reaching speeds of 48 kilometers per hour (30 miles per hour). They can be found in all seas, but are most common in the cold waters of the Arctic and Antarctic oceans.

Both the upper and lower jaws of killer whales contain large, conical teeth. Although they tend to eat more fish than anything else, they will also prey upon seals, sea lions, baby walruses, penguins and other sea birds, as well as other cetaceans. The stomach of one captured specimen contained the remains of 13 porpoises and 14 seals. Pods of killer whales have also been known to attack larger whales.

There are no authenticated reports of killer whales killing or injuring humans in the wild, although there was an incident at the Vancouver oceanarium in which a trainer was killed by an orca. There have been a few incidents in which killer whales have rammed boats, usually after the inhabitants of the boat have antagonized the whale. There are also reports of killer whales ramming ice flows on which humans are standing, but this appears to be a case of mistaken identity. In captivity, killer whales are quite docile, and their high level of intelligence and ability to learn makes them popular performers in aquarium and marine park shows. A relative, the false killer whale (*Pseudorca crassidens*), has a similar appearance but is much smaller and feeds on squids.

Pilot whales (*Globicephala*) are also related to the killer whales. These animals have a globular head, projecting forehead, and a muzzle that forms a small beak. Pilot whales are inoffensive animals that live in pods of several hundred individuals that appear to follow a single leader. They feed mainly on cuttlefish and, when attacked, group closely together. Pilot whales exhibit an odd behavior of following individual members that stray from the pod. Whalers exploit this behavior by harpooning one or two pilot whales, bringing them to shallow water, and then waiting for the rest of the pod to follow to mass slaughter. These animals are not protected by any international agreement and are still hunted for dog food and oil. Large numbers of the North Atlantic species are caught each year.

Pilot whales will frequently beach themselves, sometimes in large numbers. There are several theories on these mass strandings. One theory suggests that the cause of the beachings may be a parasite that attacks the inner ear and interferes with the animal's ability to navigate by echolocation.

Dolphin Intelligence

Although dolphins have large brains, brain size alone is not a good indicator of intelligence. Generally, large animals have large brains and small animals have small ones. A better indicator of intelligence is something called the *encephalization quotient,* or EQ, which is a ratio of actual brain mass to the expected brain mass for a defined body size. The EQ allows researchers to make comparisons between species. Large whales, like baleen whales and sperm whales, have low EQs, whereas the small toothed whales, like the bottlenose dolphin, have high EQs. Primates are the only other group of mammals known to have high EQs.

Another way to compare brains is on the basis of the thickness of the cerebral cortex and the degree of folding (convolutions) on the surface. The cerebral cortex is the part of the brain that is responsible for higher brain functions such as intelligence and memory. The surface of the cerebral cortex of the bottlenose dolphin is more convoluted than any other mammal, including humans. Its cortex, however, is only about one-half as thick as a human's but thicker than a chimpanzee's.

Aside from the characteristics of their brains, bottlenose dolphins appear to be quite clever, sometimes devising ingenious solutions to problems. One example was observed in two captive dolphins trying to extract a moray eel from a rock crevice in their tank. One of the dolphins captured a scorpionfish and, with the fish in its mouth, nudged the moray eel from behind with the spine of the fish. As the eel fled from its crevice, it was captured by the second dolphin.

Studies conducted in Hawaii by researcher Lou Herman have greatly increased our knowledge of dolphin learning abilities. A female dolphin named Ake (short for Akeakamai) and her mate, Phoenix, learned that they would receive rewards when they mimicked sounds that were generated by a computer. They also were able to make associations. When taught the hand signal to go through the gate in their tank, they also responded correctly to a similar hand signal that meant go through a hoop. They were able to learn that ball not only meant the small, red ball that they were first trained with but any other ball regardless of size, color, or its ability to sink or float.

Though originally taught to respond to hand signals given by their trainers, Ake and Phoenix could also recognize and respond to the same signals when they saw them on a television screen. In a test, dolphins outscored new trainers in interpreting televised gestural commands.

Many animals, including humans, communicate by way of body language. Dolphins are no exception. For instance, a loud opening and closing of the jaws, known as a jaw clap, or the slapping of tail flukes on the surface of water can indicate threat or displeasure.

Although dolphins probably communicate with each other in a number of ways, the most important are auditory signals. Bottlenose dolphins produce two types of vocalizations: whistles and pulsed sounds (which include echolocation clicks). It was discovered in the 1960s that each dolphin produced its own unique whistle, called a *signature whistle.* The signature whistle identifies the dolphin that is vocalizing, gives its location, and may relay other information as well. Current research indicates that perhaps as many as 90% of the whistles made by captive dolphins are signature whistles. It appears that female offspring have whistles that are quite different from their mothers', whereas male offspring have whistles that are very similar to their mothers'. Since females generally stay with their mothers when they mature, it may be important for them to have distinct signature whistles to avoid confusion. Males usually leave their mothers' group and thus do not have to differ as much in their signature whistle. Calves develop signature whistles between the age of two months and one year. The whistles remain unchanged for at least 12 years and possibly for the animal's life.

While scientists are just beginning to learn the meaning of signature whistles, little is known of the pulsed sounds the dolphins make. One type of pulsed sound is the echolocation click that is used in navigation. Other pulsed sounds may be used for communication.

Bottlenose dolphins live their lives in carefully structured, complex social groups. Studies of dolphins near Sarasota, Florida, indicate that females associate with other females that are in a similar reproductive state. For instance, mothers with calves swim with other mothers and their calves. Calves remain with their mothers for three to six years before joining juvenile groups. During this time, a calf learns to identify the other dolphins in the group, it learns its relationship to the other members, it learns how to find food and capture it, and it learns where there is danger and where it can find safety. As they mature, females frequently return to their mothers' group, but males travel alone or with one or two other adult males with whom they have grown up.

Dolphin feeding behaviors also exhibit signs of learned behaviors. Some dolphins in southwestern Florida beat their prey to death with their tail flukes. This appears to be a local behavior that the young learn from their mothers. In the Gulf of Mexico, dolphins have learned to follow shrimp boats for a free meal. This also appears to be a learned behavior that is passed from a mother to her offspring.

Occasionally, dolphins will socialize with other species, including humans. In a coastal region of Brazil, female bottlenose dolphins and their young have assisted fishers for almost 150 years. The fishers cast hand-thrown nets from the shore and wait for the dolphins to chase fish, usually mullet, into them. The dolphins feed on the fish that escape the nets.

As a result of the many long-term studies on dolphins, researchers know much more about bottlenose dolphin behavior. We now know that dolphins are strongly social, exhibit problem-solving skills, have long periods to mature with many learning experiences, and are capable of intraspecies and interspecies cooperation. Yet these observations do not amount to a direct measure of dolphin intelligence, and a precise understanding of dolphin intelligence is an ongoing pursuit.

CHAPTER SUMMARY

Mammals can be distinguished from other animals by their body covering of hair, constant warm body temperature, and mammary glands that produce a secretion called *milk* that nourishes the young.

Sea otters inhabit the northern Pacific Ocean. Instead of thick layers of blubber, this animal has a thick coat of fur to keep it warm. Sea otters mainly stay close to shore, favoring areas around coastal reefs and kelp beds. Their diet consists mainly of sea urchins, crustaceans, and molluscs (especially abalone). Nearly hunted to extinction because of their valuable fur, they are now making a comeback after international protective measures were enacted.

Pinnipeds have four limbs that are modified into flippers. This group includes seals, sea lions, elephant seals, and walruses. Although they are more at home in the water, they can and do come onto land, mainly to mate, give birth, and molt. Their bodies are spindle-shaped, and many species have several layers of fat under the skin to provide insulation. Eared seals, which include fur seals and sea lions, have visible external ears. True seals and walruses lack external ears. Eared seals primarily use their front limbs to propel themselves through the water. True seals use their hind limbs, and walruses use a combination of both. The hind limbs of eared seals can be rotated at right angles to the body axis and act as legs on land. Sea lions have a coarse coat of nothing but hairs, while fur seals have a thick coat of fur. Fur seals were at one time relentlessly hunted for their coats, but now they are protected by international law.

Walruses are restricted to the arctic seas where they feed on fishes, crustaceans, echinoderms, and molluscs. At one time these animals were slaughtered for their ivory tusks, but now they are protected by law.

Sirenians are represented by manatees and dugongs. Although at one time these animals enjoyed a wide distribution, they are currently confined to coastal areas and estuaries of the tropics. Sirenians are vegetarians feeding on shallow-water grasses and a variety of water plants.

The mammals that are most suited to life in the sea are the cetaceans. The forelimbs of cetaceans are modified into flippers, and the hind limbs are absent. Their tail forms a large horizontal fluke that is used in swimming. Beneath the skin is a thick layer of fat that helps to insulate the body. Some of the most striking cetacean adaptations are those involving the respiratory and circulatory systems that allow them to dive to great depths and remain submerged for over 1 hour.

Cetaceans known as baleen whales have enormous mouths that contain plates of baleen instead of teeth. The baleen consists of keratin fibers that are fused together and functions to strain the plankton, mainly krill, on which the animals feed, from the water. Baleen whales include right whales, rorquals, and grey whales. Many species of baleen whale are endangered because of the impact of whaling.

Toothed whales include sperm whales, dolphins, porpoises, killer whales, and narwhals. The largest of the group is the sperm whale, which feeds primarily on squid. Perhaps the best known of the toothed whales are the dolphins that perform in shows at numerous aquariums and oceanariums. The harbor porpoise is one of the smallest cetaceans and is a familiar site to both boaters and beachcombers. The largest of the dolphins is the killer whale or orca. These are the only cetaceans to feed on warm-blooded animals, mainly seals and penguins.

The ears of toothed whales are modified to receive a wide range of water vibrations. This adaptation improves the animal's ability to hear underwater and refines its hearing for the purpose of echolocation. Echolocation is similar to sonar. It allows toothed whales to distinguish and home in on objects from distances of several hundred meters.

Dolphins are strongly social animals. They exhibit problem-solving skills, have long periods to mature with many learning experiences, and are capable of intraspecies and interspecies cooperation.

SELECTED KEY TERMS

mammary glands, *p. 216*	sirenian, *p. 221*	pod, *p. 226*	echolocation, *p. 229*
placenta, *p. 216*	cetacean, *p. 223*	baleen, *p. 226*	melon, *p. 230*
pinniped, *p. 217*	blowhole, *p. 223*	bubble net, *p. 227*	delphinids, *p. 232*
polygynous, *p. 219*	fluke, *p. 223*	breeching, *p. 228*	

QUESTIONS FOR REVIEW

MULTIPLE CHOICE

1. Mammals feed their young on secretions produced by the mother's
   a. digestive system
   b. oil glands
   c. mammary glands
   d. placenta
   e. salivary glands

2. Phocid seals lack
   a. tails
   b. hind limbs
   c. forelimbs
   d. external ears
   e. nostrils

3. Whales in the suborder Mysticeti have _____ instead of teeth.
   a. tusks
   b. bony plates
   c. baleen
   d. strainer nets
   e. suckers

4. Sperm whales feed primarily on
   a. squid
   b. krill
   c. penguins
   d. clams
   e. sea urchins

5. The primary food of baleen whales is
   a. squid
   b. fish
   c. krill
   d. crabs
   e. other whales

6. The spermaceti in the head of a sperm whale is thought to play a role in
   a. digestion
   b. reproduction
   c. excretion of salt
   d. echolocation
   e. swimming

7. The largest dolphin is the
   a. beluga whale
   b. sperm whale
   c. narwhal
   d. killer whale
   e. blue whale

8. The cetacean most often seen in captivity is the
   a. California sea lion
   b. common dolphin
   c. bottlenose dolphin
   d. beluga whale
   e. walrus

9. The only pinniped to feed on warm-blooded animals is the
   a. California sea lion
   b. killer whale
   c. elephant seal
   d. walrus
   e. leopard seal

10. Sea otters are protected from the cold by
    a. a thick skin
    b. a thick fur
    c. a layer of blubber
    d. a counter-current exchange mechanism
    e. high metabolism

SHORT ANSWER

1. What factor contributed to the near extinction of the sea otter?

2. List three characteristics that would help you to distinguish a fur seal from a "true" seal.

3. Explain the function of the large proboscis in the male elephant seal.

4. Explain how the arrangement of blood vessels in the flippers and tail flukes of cetaceans help them to retain body heat.

5. Describe the changes that occur in a cetacean's body when it dives for long periods of time.

6. Explain how toothed whales use echolocation to navigate.

7. Describe the effects of whaling over the past several centuries on cetacean populations.

8. Describe how baleen whales feed.

9. How could you distinguish between a manatee and a dugong?

10. Why do mammals produce fewer offspring than most other animal groups?

THINKING CRITICALLY

1. Some marine mammals are polygynous. What is the advantage of this lifestyle?

2. From an ecological standpoint, why aren't more of the large marine mammals predators of birds and other mammals?

3. Toothed whales use echolocation to find their prey, but baleen whales do not. Why?

4. If diving mammals were not able to expel most of the air from their lungs before diving, how would the length and depth of their dive be affected?

SUGGESTIONS FOR FURTHER READING

Bonner, W. N. 1989. *Whales of the World.* Facts on File, Inc., New York.

Bruemmer, F. 1990. Survival of the Fattest, *Natural History* 99(7), 26–33.

Darling, J. 1988. Whales: An Era of Discovery, *National Geographic* 174(6):872–909.

Gentry, R. L. 1987. Seals and Their Kin, *National Geographic* 171(4):474–501.

LeBoeuf, B. J. 1989. Incredible Diving Machines, *Natural History* 98(2), 34–41.

Marten, K., K. Shariff, S. Psarakos, and D. J. White. 1996. Ring Bubbles of Dolphins, *Scientific American* 275(2):82–87.

Nicklin, F. 1995. Bowhead Whales: Leviathans of Icy Seas, *National Geographic* 188(2):114–129.

Norris, K. S. 1994. White Whale of the North: Beluga, *National Geographic* 185(6):2–31.

Norris, K. S. 1992. Dolphins in Crisis, *National Geographic* 182(3):2–35.

O'Shea, T. J. 1994. Manatees, *Scientific American* 271(1): 66–72.

Trefil, J. 1991. One of Evolution's Great Missing Links Turns Up in the Egyptian Desert, *Discover* 12(5):44–48.

Whitehead, H. 1985. Why Whales Leap, *Scientific American* 252(3):84–93.

Whitehead, H. 1995. The Realm of the Elusive Sperm Whale, *National Geographic* 188(5):56–73.

Wiley, J. P. 1987. Manatees, Like Their Siren Namesakes, Lure Us to the Deep, *Smithsonian* 18(6):92–97.

Wolkomir, R. 1995. The Fragile Recovery of California Sea Otters, *National Geographic* 187(6):42–61.

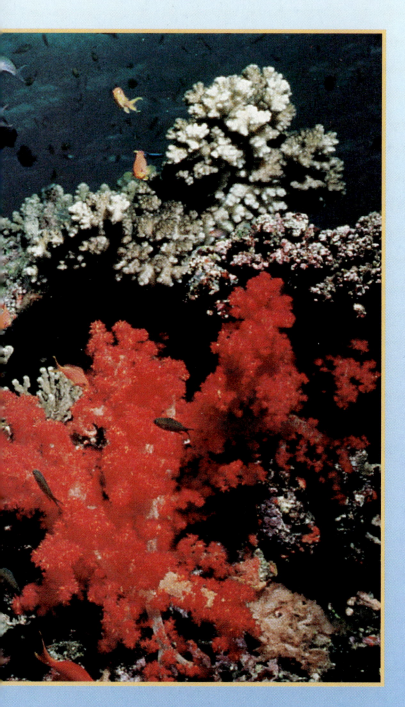

Marine Ecosystems

All of the living organisms found in a particular area along with their physical environment, such as this Pacific coral reef, make up an ecosystem. In this unit we will study the world's major marine ecosystems. Coral reefs are covered in Chapter 15. (Copyright 1992 Norbert Wu)

Chapter 13

Estuaries

Estuaries form where rivers meet the sea. These are regions of constant environmental change, and although many organisms cannot survive under these conditions, some have adapted quite well and thrive. Because of the relatively harsh conditions, estuaries contain fewer resident species than either the marine or freshwater ecosystems nearby, resulting in less competition for food and space. Since there is less competition, many estuarine species tend to be generalists. That is, they are able to feed on a variety of foods depending on what is available. The nutrients available in fresh water and salt water complement each other and, when mixed in the relatively shallow, sunlit waters of estuaries, promote high levels of primary production. Animals that can survive the changes in salinity and temperature and the occasional exposure to air that occurs in estuaries can take advantage of this high productivity, grow rapidly, and produce large populations.

Estuaries are important to marine ecosystems for a number of reasons. Much of the organic material that is produced there is exported to, and enriches, adjacent ocean waters. Many fishes and shellfishes of coastal waters spend a part of their lives in estuaries, and the juveniles of many species seek protection from predators in them.

Estuaries also support many commercially important animals that humans rely on for food. Many species of oyster, crab, scallop, and shrimp spend a part or all of their lives in estuaries. Estuaries are nurseries for a number of fish species including flounder, fluke, bluefish, tarpon, striped bass, and several species of herring. The economic importance of estuaries cannot be overstated. About 85% of the fishes and shellfishes that are sold in the commercial markets of the world spend all or part of their lives in estuaries.

Unfortunately, these are also fragile habitats, and the consequences of their pollution are far reaching. Estuaries are also damaged by the building of dams that block the supply of fresh water. This is a major problem along the Gulf of Mexico and is altering the salinity of the estuaries and killing many estuarine organisms.

KEY CONCEPTS

1. Estuaries form in embayments where fresh water from rivers and streams mixes with sea water.

2. The salinity of water in estuaries varies both vertically and horizontally.

3. The mixing of nutrients from salt water and fresh water, combined with plentiful sunlight and relatively shallow water, makes estuaries very productive ecosystems.

4. Animals and plants that live in estuaries must be able to adapt to changing salinities.

5. The physical characteristics of estuaries tend to favor benthic organisms.

6. Many important commercial fishes and shellfishes spend a portion of their life cycle in estuaries.

7. Estuarine communities include oyster reefs, mud flats, seagrasses, salt marshes, and mangrove forests.

THE GEOLOGY OF ESTUARIES

Along coastlines there are areas where portions of the ocean are partially cut off from the rest of the sea. These coastal areas are called **embayments** (Figure 13–1). Into some of these embayments, rivers and streams carry freshwater runoff from the land. The fresh water mixes with salt water from the oceans and forms an area called an **estuary.** All estuaries are partially isolated from the sea by land and diluted by fresh water. These physical characteristics produce a unique habitat inhabited by a variety of hardy organisms that have adapted to the rigors of estuarine life.

The characteristics of estuaries, such as size, shape, and water flow, can vary greatly depending on the geology of the region where they occur. Coastal plain estuaries and drowned river valley estuaries (Figure 13–2a) are formed between glacial periods when water from melting glaciers causes the sea level to rise and flood coastal plains and low-lying rivers. There are several examples of coastal plain estuaries throughout the Gulf of Mexico and eastern Atlantic. The Chesapeake Bay and Long Island Sound are examples of drowned river valleys. Estuaries like the San Francisco Bay area were created when earthquakes caused the land to sink, allowing seawater to cover it. This type of estuary is called a *tectonic estuary* (Figure 13–2b). During the last glacial period, retreating glaciers cut deep valleys into some

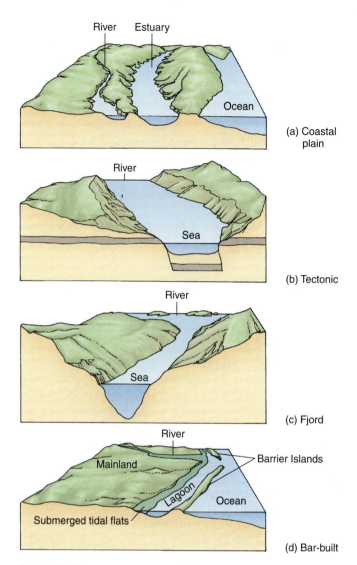

Figure 13–2

Types of Estuaries. (a) Coastal plain estuaries form when rising water levels from melting glaciers flood coastal plains. (b) Tectonic estuaries form when geological events, such as an earthquake, cause the land to sink below sea level, allowing seawater to cover it. (c) Fjords form when retreating glaciers carve large valleys in coastal areas. (d) Bar-built estuaries form when geographical barriers, such as an island, form a wall between fresh water and salt water.

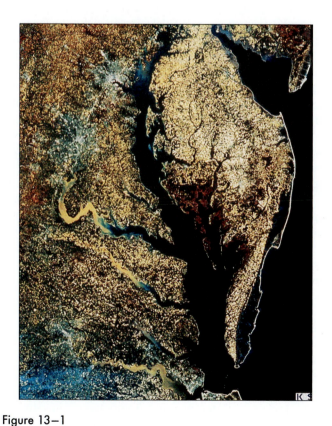

Figure 13–1

Embayment. Chesapeake Bay is an example of an embayment and an estuary. (Courtesy of Chesapeake Bay Foundation)

coasts. These valleys filled with water and formed a type of estuary known as a **fjord** (FYORD; Figure 13–2c). Spectacular fjords can be found in Alaska and along the coasts of Scandinavia.

As rivers and streams flow to the sea, they carry along sediments that are ultimately deposited at the mouth of the river. As these sediments accumulate, they form deltas in the upper part of the river mouth, shortening the estuary. **Tidal flats** develop when enough sediment accumulates to be exposed at low tide. The original channel of the estuary is then divided by these tidal flats. At the same time, currents and tides erode the coastal area and deposit sediment on the seaward side of the estuary. When more sediment is deposited than is carried away, barrier islands, beaches, and brackish water lagoons appear. These islands and other geographical barriers form a wall between the fresh water from the rivers and the salt water from the oceans, forming bar-built estuaries (Figure 13–2d). The Cape Hatteras region of North Carolina, the Texas Gulf coast, the Indian River complex of Florida's east coast, and the coast of northern Europe are all good examples of this type of estuary.

ENVIRONMENTAL CHARACTERISTICS OF ESTUARIES

Salinity and Mixing Patterns

As mentioned in Chapter 4, seawater has an average salinity of approximately 35‰. By comparison, the salinity of fresh water ranges from 0.065‰ to 0.30‰. The concentration of ions in river water also varies from one river drainage to the next and will affect the chemistry and salinity of the water in estuaries. Although the quantity of dissolved salt in an estuary is about the same as in seawater, it is distributed in a gradient from the fresh water to the ocean.

The salinity in estuaries varies both vertically and horizontally. The least salty waters are located near the mouth of the river where it joins the sea. As one moves farther out to sea, the salinity tends to increase. Salinity can be uniform, or it can be layered. Uniform salinity results when currents are strong enough to mix thoroughly the fresh water and salt water from top to bottom. In some estuaries, the vertical salinity is uniform during low tide, but at high tide the seawater at the surface moves upstream more quickly than the bottom water. The denser seawater at the surface tends to sink as the lighter fresh water beneath it rises, creating a mixing action from the surface to the bottom, a phenomenon called *tidal overmixing*. Strong winds can also mix fresh water and salt water. Normally, though, fresh water flows seaward over the seawater moving upstream.

The pattern of water circulation and vertical distribution of salinity are important characteristics of estuaries. **Salt-wedge estuaries** (Figure 13–3a) occur in the mouths of rivers that are flowing into salt water. At the surface, fresh water flows rapidly out to sea, while at the bottom the denser salt water flows upstream along the river bottom. The rapid flow of the river prevents salt water from entering and produces an angled boundary between the fresh water moving downstream and the seawater moving upstream, called a **salt wedge.** When the tide rises or the river flow decreases, the salt wedge moves upstream. When the tide falls or the river flow is increased, the salt wedge moves downstream. The fresh water normally moves rapidly enough to create a sharp boundary between the fresh water

and seawater. The moving fresh water takes salt water from the face of the salt wedge and mixes it upward with the river water. This action increases the salinity of the surface water that is moving out to sea. Salt water from the sea is constantly replacing the water removed from the salt wedge, so the salt wedge does not become smaller. Very little of the river water is mixed downward with the salt water; thus, the mixing of salt water with fresh water is essentially a one-way process. In this type of estuary, the circulation and mixing of the water is controlled by the flow rate of the river. By comparison, tidal currents play only a small role. Some examples of salt-wedge estuaries are found at the mouths of the Mississippi, Amazon, Congo, and other large rivers and at the mouth of the Sacramento River in San Francisco Bay.

Another type of estuary is the **well-mixed estuary** (Figure 13–3b). In well-mixed estuaries, river flow is low and tidal currents play a major role in the circulation of the water. The net result is a seaward flow of water and a uniform salinity at all depths. The salinity of the water decreases as it approaches the river. Lines of constant salinity move toward land when the tide rises or when river flow decreases. On the other hand, when the tide falls or river flow increases, the lines of salinity move toward the sea. Delaware and Chesapeake Bays are examples of well-mixed estuaries.

Estuaries that have a strong surface flow of fresh water and a strong influx of seawater are called **partially mixed estuaries** (Figure 13–3c). Tidal currents force the seawater upward, where it mixes with the surface water, producing a seaward flow of surface water. This system of circulation produces a rapid exchange of surface water between the estuary and the ocean. Salinity is increased by the influx of seawater. San Francisco Bay and Puget Sound are examples of partially mixed estuaries.

In fjords, river water remains at the surface and moves seaward, mixing little with the salt water beneath. Salinity increases slowly in the estuary with the slow influx of ocean water.

The mixing patterns of estuaries do not always fit neatly into one of the above categories. Some estuaries are intermediate between two of the types we have discussed. Others change from season to season as changes in rainfall, tides, and winds alter the volume of water and strength of currents as in Galveston Bay, Texas.

Temperature

Salinity is not the only environmental factor that varies in an estuary. Temperature is also important. Since estuaries are relatively shallow, the water temperature changes rapidly with changes in air temperature. Temperatures in estuaries can fluctuate dramatically on a seasonal and even a daily basis. For instance, in northern temperate regions, water tem-

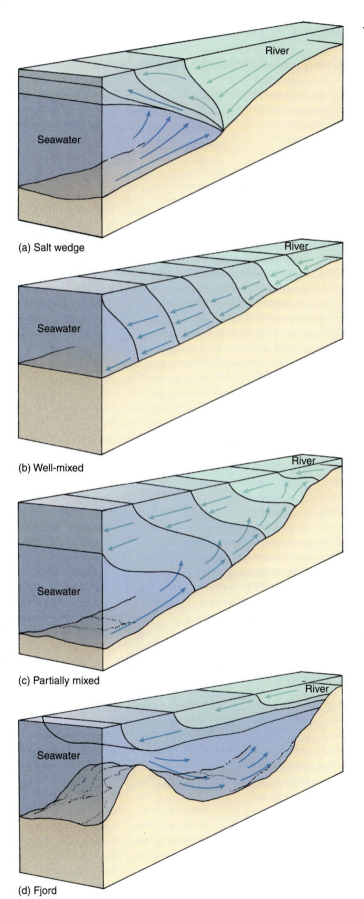

(a) Salt wedge

(b) Well-mixed

(c) Partially mixed

(d) Fjord

◀ Figure 13–3

Mixing Patterns in Estuaries. Green represents low-salinity freshwater, and blue represents high-salinity saltwater. The darker the blue color, the higher the salinity of the water. The arrows indicate the direction of water flow. (a) In salt-wedge estuaries, less dense fresh water from a river flows rapidly to sea over the denser, slower-moving salt water that is flowing upstream. This produces an angled boundary (salt wedge) between the two. (b) In well-mixed estuaries, river flow is low and tidal currents mix fresh water with salt water, producing uniform salinity at all depths. (c) Partially mixed estuaries have a strong surface flow of fresh water and a strong influx of seawater. (d) In a fjord the river water remains at the surface and mixes very little with the seawater beneath it.

perature may range from almost −2°C to 30°C, and portions of estuaries may freeze during the winter.

Like other bodies of water, estuaries are heated by the sun. Depending on the season, tidal currents will warm or cool an estuary. In some estuaries, the surface water is cooler in the winter and warmer in the summer than deeper water, as a result of more or less solar energy reaching the surface. This situation produces a winter turnover in which cooler (denser) surface water sinks and is replaced by warmer (less dense) deeper water. This mixing of water during the winter provides an important means of trapping nutrients. Without vertical mixing, important nutrients would be swept back out to sea with the next tide. The vertical mixing that occurs as a result of the turnover acts to circulate nutrients vertically between the water and bottom sediments.

PRODUCTIVITY IN ESTUARIES

Fishers and those who depend on the sea for food have long known how productive estuaries are. Part of the reason for their high productivity is the mixing of fresh water and salt water that occurs there. Freshwater runoff from the land often contains nitrogen, phosphorus, and silica. Water at the surface of the open sea often has less nitrogen and silica but more phosphorus and other nutrients. Individually, the two types of water can support only a limited variety of plant and algal growth due to the lack of one or more essential nutrients. When the two are combined, however, by strong mixing currents in the shallow, sunlit waters of an estuary, their nutrients complement each other and high levels of primary production can occur.

Once nutrients enter an estuary, several factors act to hold them there in forms usable by plants, algae, and animals. Rivers dump loads of silt and clay as they meet the sea. These particles readily absorb any excess nutrients from the surrounding water and release them back to the water when nutrients are in short supply. In this way, the silt and clay act as a sort of nutrient buffer, helping to maintain a more or less constant nutrient level in an estuary.

The activities of some estuarine animals also contribute to keeping nutrients in an estuary. When there are large amounts of phytoplankton in an estuary, filter-feeders, such as bivalves, remove more of these cells from the water than they are able to digest. The excess phytoplankton are not consumed but rather are eliminated in large, semisolid particles called **pseudofeces.** Pseudofeces contain substantial amounts of nutrients and are relatively large so they can be easily manipulated by other organisms. Thus the normal feeding behavior of bivalves acts to package and store nutrients for use by bottom feeders like gastropod molluscs and polychaete worms.

Species that can tolerate the salinity and temperature changes that occur in estuaries can exploit the area's high productivity, grow rapidly, and multiply into enormous populations. At one time in the rich waters of the Chesapeake Bay, for instance, large populations of oysters (Ostreidae) and blue crabs (*Callinectes sapidus*) literally carpeted the bottom. Today populations that large are seen infrequently due to pollution, disease, and overfishing. Great South Bay along the coast of Long Island contains enough clams to support New York's largest single fishery. In both instances the characteristics of the populations are typical of those encountered in estuaries. Estuarine populations contain large numbers of individuals belonging to relatively few species, and the dominant animals are those with relatively broad tolerances and ecological requirements.

LIFE IN AN ESTUARY

Living with Changing Salinities

Many marine animals have body fluids that contain about the same concentration of salts as seawater, and their body fluids are essentially **isosmotic** to the surrounding water. That is, the osmotic pressure of their body fluids is equal to the osmotic pressure of the seawater, and they neither gain water from nor lose water to their surroundings. Since the marine environment remains relatively constant, they do not have a problem maintaining water balance. Animals that live in estuaries, however, must have some physiological mechanism for dealing with the varying salinities. Otherwise their tissues and cells would absorb water and lose salts as they encountered an environment with lower salinity than the sea. Animals that live in estuaries can survive by having tissues and cells that can tolerate dilution (**osmoconformers**), or by maintaining an optimal salt concentration in their tissues, regardless of the salt content of their environment (**osmoregulators**).

Animals such as tunicates, jellyfishes, and sea anemones are unable to actively adjust the amount of water in their tissues. When their environment becomes less saline, their body fluid gains water and loses ions until it is isosmotic to the surroundings. These organisms are examples of osmoconformers (Figure 13–4a). The ability of osmoconformers to inhabit estuaries is limited by their tolerance for changes in their body fluid.

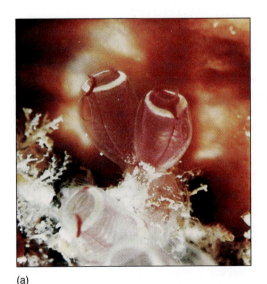

(a)

(b)

Figure 13–4

Osmoconformers and Osmoregulators. (a) These tunicates are osmoconformers, losing and gaining water and ions until their body fluid is isosmotic to the environment. (b) This hermit crab is an osmoregulator, actively adjusting the concentration of salts in its body fluid when the concentration in the surrounding water changes. (a, Steven Webster; b, Robin Lewis/Coastal Creations)

In contrast, osmoregulators employ a variety of strategies to maintain a constant salt concentration in their bodies. Osmoregulators that live in estuarine waters have the ability to concentrate salts in their body fluids when the concentration of salts in the surrounding water decreases. For instance, some crabs and fishes are able to regulate their salt content in less saline water by actively absorbing salt ions through the gills to compensate for salt ions lost from their body (Figure 13–4b). This helps them to maintain a relatively constant body fluid. Some animals can either concentrate salts when their environment is less saline or excrete salts when the environment is hypersaline. The latter are generally animals, such as some crustaceans, that are semiterrestrial or that live in areas such as salt marshes and mangrove swamps that may periodically receive large amounts of rain. Other animals, such as the blue crab and polychaete worms in the genus *Nereis,* are osmoregulators at lower environmental salinities and osmoconformers at higher environmental salinities.

One mechanism that is employed by some estuarine organisms to decrease water and salt exchange with their surroundings is to wall themselves off from their external environment. Many estuarine animals have a body surface that shows a decreased permeability compared to purely marine forms. This decreased permeability can be the result of increased amounts of calcium in the exoskeleton, as in arthropods, or increased numbers of mucous glands in the skin, as in some fishes. Structural adaptations, such as the operculum of a snail, can be used to isolate the body surface from the environment when necessary to prevent salt and water loss or gain.

Remaining Stationary in a Changing Environment

In addition to changes in salinity, the problem of remaining stationary in a changing environment also affects the distribution of organisms in estuaries. The more or less constant movement of water in an estuary makes it difficult for some organisms to remain stationary long enough to feed and carry on other vital functions. Because of this, the changing environment primarily selects those organisms that are benthic. Marine plants and algae that are found in estuaries have substantial root systems or holdfasts to prevent moving water from pulling them up and carrying them out to sea. Benthic animals live attached to the bottom either in the available spaces around other sedentary animals and plants or buried in the small crevices between sediment particles.

Of the nonbenthic animals, crustaceans and fishes, especially the young, are the most dominant. These animals usually spawn in the seawater offshore and then spend a portion of their development in the estuary, which is a more protected habitat than the open sea. These animals maintain their position in the estuary by actively swimming or by moving back and forth with the movement of the tides.

Estuaries as Nurseries

Although estuaries can be challenging habitats for many animals, they provide excellent habitats in which juveniles and young of many species can grow and develop. The high level of nutrients and lower number of predators allow juveniles and young to grow and develop to a size or stage where they will have a better chance of survival in the open sea.

The striped bass (*Morone saxatilis*), for instance, spawns at the border of fresh water and water of low salinity. As the larvae and young mature, they move downstream toward water with higher salinity. The shad (*Alosa*) is an anadromous species that spawns in fresh water but spends its adult life in the marine environment. Young shad spend their first summer in an estuary feeding and growing before moving out to the open ocean. Species such as the croaker (Sciaenidae) spawn at the mouth of an estuary, and then their young move upstream to feed in the plankton-rich, low-salinity water. In the eastern United States, young bluefish (*Pomatomus saltatrix*) come to estuaries to feed but spend the rest of their time in the ocean. Many species of shellfish, such as blue crabs and white shrimp (*Penaeus*), also spend a part of their life cycle developing in the relatively protected waters of estuaries.

ESTUARINE COMMUNITIES

Oyster Reefs

A prominent mollusc in the intertidal zone of temperate estuaries is the oyster (Ostreidae; Figure 13–5a). Like other bivalves, oysters have a free-swimming larval stage in their life cycle that allows these sessile animals to disperse to other areas (Figure 13–5b). Oyster larvae attach themselves to any solid surface and form extensive oyster beds, or large reefs composed of oysters growing on the shells of previous generations. Oyster reefs are usually oriented at right angles to tidal currents and occur generally at the point of lower salinities. The currents bring food to the oysters and carry away their waste. Tidal currents also play a role in clearing sediment from the oysters. If the sediments were allowed to accumulate they would suffocate the oysters.

Oyster reefs provide habitats for a variety of other organisms including algae, sponges, hydrozoans, bryozoans, polychaetes, molluscs, echinoderms, and barnacles. Many of these organisms depend on the oysters not only for protection and a surface for attachment, but also for food. The oyster drill snail (*Urosalpinx*) preys upon the sedentary

(a)

Figure 13–5

Oysters. (a) Oysters form extensive beds composed of thousands of individuals, providing a habitat for many other species. (b) Life cycle of the commercial oyster *Crassostrea virginica*. Adult oysters shed their sperm and eggs into the water column, where the eggs are fertilized. The fertilized egg undergoes cell division, forming an embryo that develops into a free-swimming trochophore larva. The trochophore becomes a veliger larva with a small, hinged shell. Within four weeks, the veliger develops a foot, settles to the bottom, and cements its left valve to a solid surface. The veliger then develops into an immature oyster called a *spat*. The spat grows and develops into an adult oyster. (a, Fred Whitehead/Animals Animals)

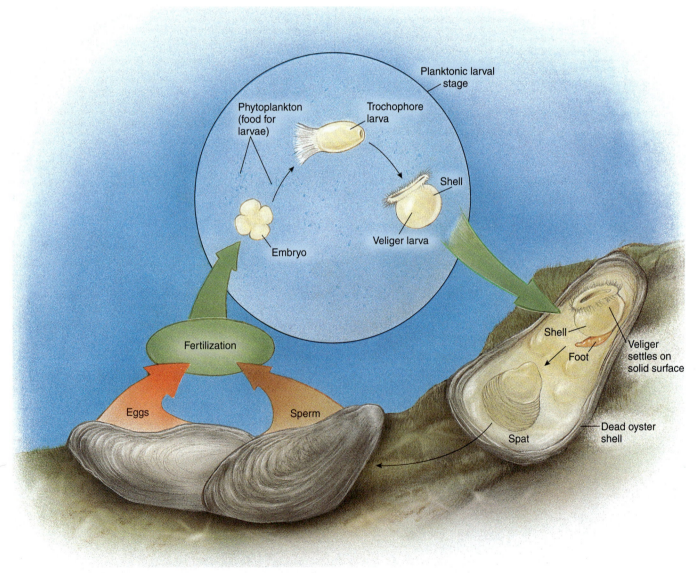

(b)

Figure 13–6

Oyster Drill. Oyster drill snails can cause extensive damage to oyster beds by feeding on the sedentary oysters. (Herb Segars/ Animals Animals)

oysters, drilling through their shells with its radula and feeding on the contents (Figure 13–6). The veliger larvae of oyster drills are quite sensitive to changes in salinity and are more affected by fresh water than adults. Short periods of low salinity kill predatory oyster drills and disease-causing organisms in the oyster beds. However, rapid changes in salinity as the result of prolonged rainfall or hurricanes kill off large numbers of both oysters and oyster drills as well as the other organisms that grow with them.

Mud Flats

Mud flats are found in bays and around the mouths of rivers wherever the land is protected from wave action. Mud flats contain rich deposits of organic material among small inorganic sediment grains. Detritus from nearby communities and nutrients carried in from the sea by tides contribute to the rich food reserves. Bacteria and other microorganisms thrive in the mud and produce a variety of sulfur-containing gases that give mud flats a characteristic odor of rotten eggs. Mud provides good mechanical support for animals of the flats, many of which have very thin shells or flabby bodies. Mud is also cohesive, permitting the construction of a permanent burrow. Sand is frequently mixed with the mud, making it softer and providing a much better bottom material for burrowing organisms.

Most inhabitants of mud flats are burrowing organisms (Figure 13–7). They live just beneath the surface, where they avoid predators and exposure to the drying air during low tides. Unlike sandy beaches, mud flats are not very porous. The silt packs closely together and interferes with the circulation of water that is necessary for carrying oxygen through the sediments. As a result, burrowing animals that exchange gases through their skins generally cannot survive in this habitat unless they circulate water through their burrows or tubes, or maintain a "snorkel" connection to the surface.

One common resident of the mud flats of North America is the soft-shelled clam (*Mya arenaria*). Like other bivalves, it is a filter-feeder, drawing water in through its incurrent siphon and removing from the water oxygen and planktonic food. Wastes are carried away by the water that is expelled from its excurrent siphon. When the tide retreats, the soft-shelled clam withdraws its siphon and metabolizes anaerobically. In laboratory experiments, soft-shelled clams have been found to survive as long as eight days without oxygen.

Lugworms (*Arenicola*) are common residents of mud flats worldwide. Concentrations as high as 820,000 individuals per acre have been reported from some areas. Another burrowing worm that is found in California mud flats is the innkeeper worm (*Urechis*). This worm may be 30 centimeters (12 inches) long with a fat, pink, sausage-shaped body. Like lugworms, it lives in a U-shaped burrow through which it continuously pumps water to supply oxygen and remove waste. It feeds by producing a mucous net that traps tiny food particles. When the net is full, it is drawn into the animal's mouth and digested and another net is produced to take its place. The innkeeper worm derives its name from the variety of organisms that live with it in its burrow. Some species, such as the small, red scaleworm (Polynoidae) are commensal symbionts in the burrows of innkeeper worms. The scaleworm maintains almost constant contact with the innkeeper worm and feeds on discarded food particles and sometimes on the mucous net. Another common symbiont in the innkeeper's burrow are tiny pea crabs (Pinnotheridae) that may compete with scaleworms for bits of food. Tiny fish called *gobies* (Gobiidae) also inhabit the burrows but do not compete with the other organisms for food. Instead, when pieces of food are too large for their own use, gobies have been observed bringing them to the pea crabs.

Innkeeper worms are not the only burrowing organisms on mud flats to take in boarders. The ghost shrimp (*Callianasa*) burrow is composed of a vertical shaft connected to a number of lateral passageways. The lateral tunnels branch repeatedly, sometimes widening to allow the animal to turn around. Within this burrow the shrimp may house small clams, pea crabs, several species of marine worms, and gobies. The ghost shrimp digs its tunnels by scooping up the sediments with its first pair of legs and storing the material in a pouch that is formed by the fleshy appendages around its mouth. When the pouch is full, the shrimp takes it to the opening of the burrow and dumps it.

On mud flats, the energy base is organic matter. It consists of decaying remains of local plants and animals as well as organic matter that is deposited during high tides. A great deal of this organic material becomes available to other organisms as the result of bacterial decomposition. The action

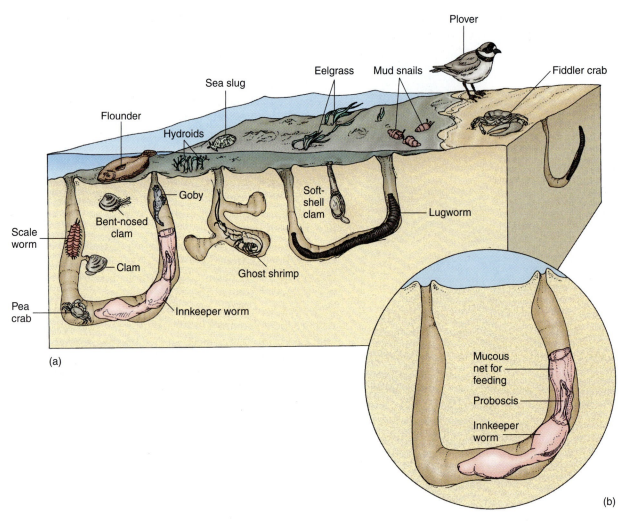

Figure 13—7

Mud Flats. (a) Most inhabitants of mud flats are burrowing animals that feed on detritus or the meiofauna. (b) Some, like the innkeeper worm, live in burrows that provide protection for many other

of bacteria is also important in recycling nutrients such as nitrogen and phosphate back to the sea (see Chapter 6). Bacteria in mudflats are not just decomposers but are also basic consumers that serve as a food source for many higher-level organisms. Many deposit-feeding organisms, such as nematodes, polychaetes, some gastropod molluscs, and arthropods, ingest organic material to feed on the bacteria that it contains.

The producers on mud flats consist of photosynthetic bacteria, chemosynthetic bacteria, phytoplankton such as diatoms and dinoflagellates, and, in some cases, large algae such as *Ulva*. A variety of small organisms, including the larvae and juveniles of many animal species, as well as filter-feeding animals, rely on these rich sources of food. The smaller organisms, in turn, provide food for larger inverte-

brates, fishes, birds, and even some terrestrial carnivores like raccoons.

Seagrass Communities

Stretching seaward from many mud flats along the coasts of North and South America, Europe, Asia, and Australia are vast expanses of seagrasses. These plants (see Chapter 7), thrive in protected waters from the low tide zone to a depth of about 6 meters (20 feet), where they are able to grow in sandy mud that is not able to support the holdfasts of large marine algae.

The surfaces of seagrasses provide a place of attachment for many tiny organisms. Older leaves may be completely covered by hydrozoans, tube worms, bryozoans, and tufts of red algae. Juvenile scallops (*Argopecten*) often attach to seagrass leaves by byssal threads. Those attached well above the bottom sediments tend to avoid predation by benthic predators such as crabs.

Most seagrasses are not consumed directly by herbivores because they are too tough. In temperate regions waterfowl are the major seagrass herbivores, but they often consume relatively little, less than 5% of the total seagrass production. In the Caribbean, green sea turtles, certain parrotfishes, some sea urchins, and manatees feed extensively on seagrasses. In the case of the green sea turtle and the bucktooth parrotfish (*Sparisoma radians*), about 90% of their diet consists of turtle grass (*Thalassia*). Although few animals feed directly on seagrasses, seagrasses provide a substantial food source for many organisms in the form of detritus.

The mud among the blades of seagrass is home to a wide variety of filter-feeders, such as adult scallops, jacknife clams (*Ensis*), and some species of sea cucumber. Seagrass rhizoids and root complexes provide sites for more permanent attachments by tube worms and mussels. Many species of snail, both herbivores such as conchs and carnivores such as whelks and tulip snails, thrive in Caribbean and temperate seagrass communities (Figure 13–8). As snails fall

prey to carnivores, their shells become available for a variety of hermit crab species that are mainly scavengers. During high tide, predators like eels, flounder, and other large fishes come to the seagrass beds to feed.

The high productivity of seagrass communities allows them to support a large and diverse group of organisms, including the larval and juvenile stages of many animal species. The combination of changing salinity, available hiding places, and shallower water allow small juveniles to be protected from larger predators. In the seagrass nurseries these animals feed and grow without the heavy predation pressure that will face them when they enter the unprotected sea.

Salt Marsh Communities

Salt marsh communities are found on the shoreward side of mud flats in temperate and subarctic regions of the world. The dominant plant life in this community on many North American shores is cordgrass (*Spartina*; see Chapter 7) that has moved from land to the shallow intertidal area. The cordgrass *Spartina alterniflora* dominates the marshlands of the Atlantic coast, while *Spartina foliosa* is dominant along the coast of California. On the Gulf Coast of Florida,

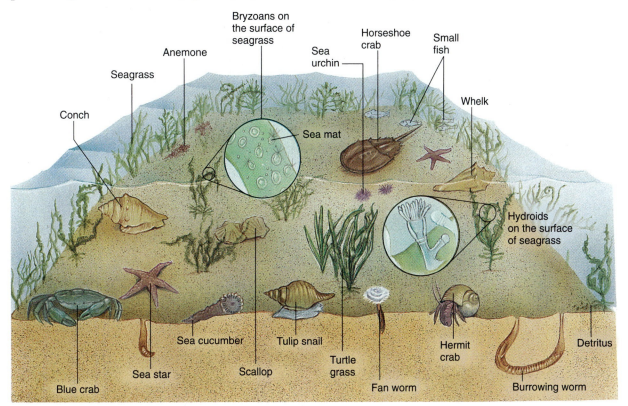

Figure 13–8

Seagrass Community. The seagrass blades provide a suitable surface for attachment for many tiny organisms. The sandy mud provides habitat to a wide variety of animals. Although the seagrass is too tough for most animals to eat, it is an important source of detritus.

the tallest and most common marsh grass is *Juncus*, which is found at higher elevations (higher up on shore). The species found most seaward is tall cordgrass, *Spartina alterniflora*. Tall cordgrass may grow as high as 3 meters (10 feet) and is found growing in the **low marsh,** the lower part of the intertidal zone that is covered by tidal water much of the day. In the **high marsh,** the region closer to shore that is covered briefly by saltwater each day, is a thick carpet of short, fine grasses that usually do not grow taller than 60 centimeters (2 feet). These may include salt meadow hay (*Spartina patens*) and spike grass (*Distichlis spicata*), as well as *Juncus*.

The low marsh that is dominated by the tall cordgrass is typically flushed by the tides twice each day, whereas the high marsh meadows are usually flooded by only the spring tides. In areas where the tall cordgrass grows thickly, leaves, stalks, and debris are flushed by tidal currents so that the water is very clear. The beds of short cordgrass are not flushed by tidal currents, and dead leaves and debris accumulate on the marsh surface, forming a moist mat that is an ideal habitat for many species.

Marsh grasses form the basis of a complex and distinct community (Figure 13–9). On the southern Atlantic coast, marsh periwinkle snails (*Littorina irrorata*) can be found

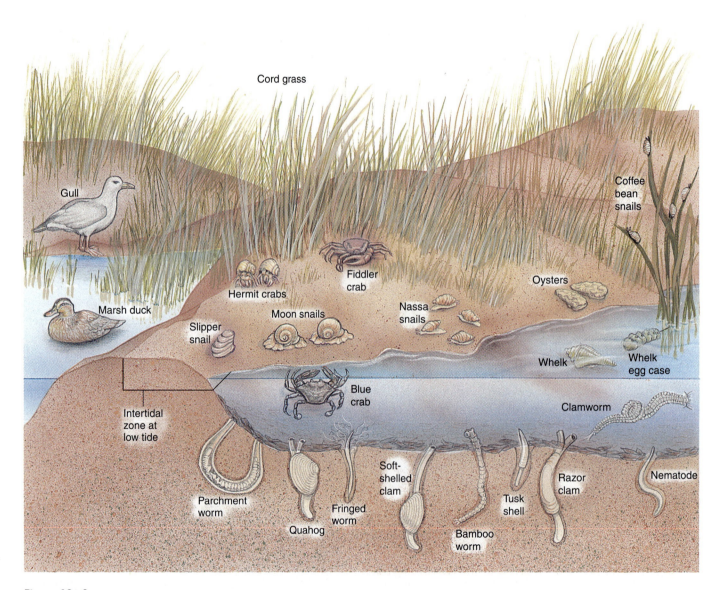

Figure 13–9

The Salt Marsh Community. Salt marshes provide habitat for a number of animals. Burrowing animals dominate the low marsh, whereas crabs, periwinkles, and marsh snails are common in the high marsh.

Figure 13–10

Fiddler Crab. Fiddler crabs are common inhabitants of salt marshes. The large left claw (cheliped) of the male is used to attract mates. (Courtesy of David Campbell)

on the marsh grass, where they feed on algae that grow on the surface of the leaves. The marsh periwinkle, as with most periwinkles, survives out of water because its mantle cavity can function as a type of lung, allowing the snail to absorb oxygen from the air. At low tide it is found on the underside of leaves of the marsh grass, where it is protected from the sun and predators. As the tide moves in, it crawls higher up the plant to avoid being submerged. When high spring tides submerge the entire plant and the snail as well, the snail can go without breathing for about an hour, long enough to survive until the tide begins to retreat. The periwinkles exhibit an interesting behavior in that they appear to be able to sense an impending flood tide and crawl to the highest parts of the plant even before the water begins to rise.

The tidal marsh snail (*Melampus bidentatus*) is the dominant gastropod mollusc on many Atlantic coast marshes, frequently attaining densities of more than 1,000 per square meter. This snail is somewhat unique in that, although it leads a largely terrestrial existence, it possesses an aquatic larval stage. The adults lay eggs in the water, and the larvae hatch and begin to develop in the water column during spring tides when at least part of the high marsh is flooded. The larvae later settle back onto the marsh to take up a more terrestrial lifestyle.

Burrowed in the mud of the low marsh along the Atlantic coast is the ribbed mussel (*Geukensia demissa*). At low tide, the mussel closes its valves to trap moisture and prevent desiccation. During high tide, it opens its valves to filter food particles from the water. Particles that are not acceptable for food are trapped in mucous ribbons and ejected from the mussel as pseudofeces. Another prominent bivalve in some salt marshes along the Atlantic coast is

the razor clam (*Solen*). Like ribbed mussels, these clams burrow into the mud and filter feed at high tide. It is not unusual to find large clamworms, sometimes as long as 19 cm (7.5 inches), feeding on the razor clams.

At low tide in more southern salt marshes, purple marsh crabs (*Sesarma reticulatum*) come out of their burrows to feed on the marsh grass. A relative of the marsh crab, the fiddler crab (*Uca;* Figure 13–10), is one of the most prominent inhabitants of salt marshes. At low tide, hundreds to thousands of these animals emerge from their burrows and scurry across the mud in search of food. Fiddler crabs are omnivores and feed on detritus, algae, and small animals. The fiddler crab is named for the one oversized claw (cheliped) of the male that is used for attracting a mate and defending his territory against other males. Fiddler crabs can exchange gas in both air and water and can survive several weeks without being immersed in water. The space beneath its carapace that is just above the legs forms a lung cavity. This space traps air and has a rich blood supply. As long as the lining of the lung cavity remains moist, the crab is able to exchange sufficient quantities of gas across it to survive out of the water.

Fiddler crabs dig their burrows at the base of tall cordgrass. As they excavate their burrows, they form the mud that is being removed into neatly packed pellets that they carry to the opening and stack around it. During high tide, the crab fashions a door from some of the mud pellets and seals the entrance, trapping air in the burrow. When the tide retreats, the crab reemerges to feed and court mates.

Burrowing animals, like the fiddler crab, play a vital role in the overall marsh ecology. Their burrowing activities constantly bring nutrient-rich mud from deeper down to the surface where the nutrients have been depleted, and their burrowing allows oxygen to penetrate into deeper sediments.

Amphipods called sandhoppers (*Orchestia*) are also abundant in some salt marshes. These small arthropods are common around the base of the cordgrass and under debris. They feed on detritus and are an important source of food for several species of marsh bird. Another prominent arthropod is the grass shrimp (*Palaemonetes*), which is common in the water associated with both salt marshes and seagrass beds. Seasonally, edible shrimp of the genus *Penaeus* are also very common. These animals feed on detritus and plankton and are important sources of food for fish and birds.

A variety of predators feed in the salt marsh with the tides. During high tide, aquatic predators such as fish and squid swim around the cordgrass in search of food. Blue crabs also come to the flooded marsh to feed. At low tide along the Atlantic coast, terrestrial predators such as birds, diamondback turtles (*Malaclemys terrapin*), raccoons (*Procyon*), otters (*Lutra*), and mink (*Mustela vison*) enter salt marshes in search of prey. Many insect and spider species are residents of salt marshes. These animals not only

The Role of Nutrients in the Growth of Marsh Grass

Ivan Valiela is an ecologist who has spent many years studying salt marshes. While conducting studies on salt marshes in New England, he observed that some grasses seemed to grow better in areas that appeared to have higher levels of nutrients. His observations led him to hypothesize that the growth of marsh grass was limited by the availability of nitrogen, an important nutrient that is necessary for plant growth. Based on this hypothesis, he predicted that if nitrogen were added to marsh grass, it would grow larger or faster or both.

In order to test his hypothesis, Valiela picked several plots of salt marsh that were as identical as possible in several respects. In each plot the types of plants, plant density, soil type, freshwater input, and height above the average tide level were as similar as possible. To his experimental plots, Valiela added nitrogen-containing fertilizer. The control plots received no fertilizer but otherwise were subjected to the same environmental conditions as experimental plots.

All through the growing season, Valiela monitored the plots, measuring the growth rate of the plants, chemical composition of the leaves, and rate at which the leaves decayed when they were removed from the plant. He analyzed his data statistically and, among other things, found that several important species of marsh grasses grew taller and larger than the controls when they were supplemented with nitrogen fertilizer. Further tests of the hypothesis continued to support these findings, thus leading to the conclusion that a major limiting factor in the growth of salt marsh plants is the availability of nitrogen.

serve as predators but are also food to many species of reptile, bird, and mammal. Herbivorous animals from the land, such as swamp rabbits (*Sylvilagus aquaticus*), white-footed mice (*Peromyscus*), and sometimes deer (*Cervus*), also come at low tide to browse on the marsh plants.

Birds are especially plentiful in salt marsh communities. Marsh wrens (*Telmatodytes* and *Cistothorus*), clapper rails (*Rallus longirostris*), red-winged blackbirds (*Agelaius phoenicius*), and seaside sparrows (*Ammospiza maritima*) all nest in the tall marsh grass. Marsh hawks (*Circus cyaneus*) feed on the many rodents found in this area.

Salt marshes can be the first stage in a succession process that will eventually produce more land. The roots of marsh plants act as a sediment trap, holding sediments carried down by rivers and preventing their removal by tidal currents out to sea. Over time, the area becomes built up with sand and silt, and this in turn becomes mud as it becomes enriched with decaying material from dead plants and animals. Eventually small islands of mud appear, and, as the cordgrass traps more sediment, the size of these islands increases, and they begin to merge with each other. As land masses build, high tide covers less and less of them. Tall cordgrass is ultimately replaced by short cordgrass, and that, in turn, is replaced by rushes. At this point, land plants begin to establish themselves, and this is followed by the influx of terrestrial animals.

Mangrove Communities

Mangrove forests, or **mangals,** replace salt marshes in tropical regions, where they can cover as much as 75% of the coastline. Mangrove forests appear where there is little wave action and where sediments accumulate and muddy sediments lack oxygen. The most highly developed mangals supporting the greatest number of species are found along the coasts of Malaysia and Indonesia. Mangals in this region may contain as many as 40 species or more and may exhibit a definite pattern of zonation.

Since the mud in which mangroves grow is soft and oxygen-poor, their roots are adapted to anchor the trees firmly as well as to supply sufficient oxygen to those parts of the plant buried in the mud. To help anchor the plants, mangroves have root systems that are shallow and widely spread. Red mangroves gain extra support from aboveground roots called **prop roots** (Figure 13–11), which grow from the trunk and branches. Prop roots also aid in gas exchange for the roots that are buried in the mud. Black mangroves have many erect, aerial roots called **pneumatophores,** which branch from horizontal roots beneath the mud. The pneumatophores not only help to support the mangrove but also exchange gases for the buried roots. Without prop roots and pneumatophores, the buried roots of these plants would suffocate, and the plants would die.

Prop roots and pneumatophores form a tangle that collects various sediments and organic material. This, in turn, slows the movement of tidal waters, allowing more sediment to build up. Eventually the area becomes a terrestrial habitat, as the colonizing mangroves continue their growth toward the sea.

In the Americas, red mangroves are frequently the pioneering species and are usually found closest to the edge of the water. These plants occupy regions of the mangal that experience the greatest degree of tidal flooding. The red mangrove produces seeds that germinate while still attached to the parent plant. When the seeds finally drop off, some take root next to the parent, while others are carried by tidal currents to other areas to take root (see Chapter 7). Seedlings grow into mature plants that range in size from small shrubs to trees that can reach heights of 30 meters (100 feet).

Shoreward are the black mangroves, occupying areas that receive only shallow flooding during high tide. Closest

Figure 13–11

Mangrove Roots. Prop roots give red mangroves extra support and aid in supplying oxygen to the roots underground. Black mangroves have pneumatophores that project upward through the mud from the roots and aid in supplying oxygen to the roots. (Jon L. Hawker)

to land are white mangroves and buttonwoods. Buttonwoods are not true mangroves and represent a transition to terrestrial vegetation. The various types of mangroves usually remain separated by their ability to tolerate flooding by salt water during high tide and their different tolerances to soil salinity.

The prop roots of red mangroves and pneumatophores of black mangroves provide a habitat for a variety of animals. In mangrove forests of the west coast of Florida, prop roots and pneumatophores eventually become encrusted with a purple oyster called the coon oyster (*Lopha frons*). This bivalve gets its name because it is a favorite food of raccoons. Barnacles and mussels that filter feed when the tide is high compete with coon oysters for space on the roots of red and black mangroves. On the east coast of Florida, mussels and acorn barnacles are dominant. Periwinkle (*Littorina*) snails that are related to the marsh periwinkle graze on algae that grow on the stems and prop roots of the mangroves as well as on the shells of the sedentary organisms attached to them. The king's crown conch (*Melongena corona*), a carnivorous snail, feeds on oysters by prying open their valves with its strong muscular foot and then digesting the contents. Mangrove crabs (*Aratus*) go through their larval stages in the water beneath the mangroves. When mature, they crawl up on the mangroves and feed on the leaves.

The roots of mangroves provide a habitat for many of the same organisms that are found on mud flats and in salt marshes (Figure 13–12). These include fiddler crabs, hermit crabs, marsh crabs, marsh snails, and ghost shrimp. Some animals that cannot tolerate the varying salinity of the salt marsh and mud flats can survive in the more stable environment of the mangrove forest. These include sea stars, brittle stars, and sea squirts.

In the mangals of Malaysia and Indonesia, fish known as *mud skippers* (*Periopthalmus chrysospilos*) live burrowed in the mud (Figure 13–13). These fish come out of their burrows at low tide and scoot around on the surface of the moist mud, behaving more like amphibians than fish. The sheltered waters around mangrove roots are also an ideal nursery for crab larvae, shrimp, and fishes. On Florida's southern coasts, the role of mangroves as a nursery is equal to that of seagrass communities.

The rich food supplies of mangrove forests along the Gulf Coast of the United States attract a variety of predators including clapper rails, killifishes (Cyprinodontidae), diamondback turtles, water moccasins (*Agkistrodon*), and raccoons. Birds such as pelicans, herons, egrets, roseate spoonbills (*Ajaia ajaia*), and wood ibises (*Mycteria americana*) also find the upper branches of mangroves an ideal nesting site. Until much of this habitat was destroyed by drainage projects and pollution, mangals along the west coast of the Everglades provided major nesting sites or rookeries for many bird species.

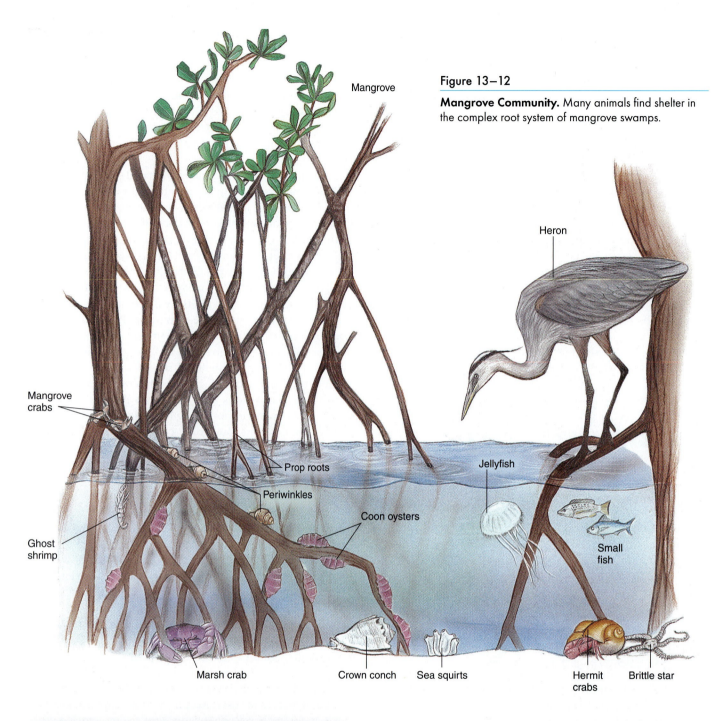

Figure 13-12

Mangrove Community. Many animals find shelter in the complex root system of mangrove swamps.

Mangrove

Heron

Mangrove crabs

Prop roots

Jellyfish

Periwinkles

Coon oysters

Small fish

Ghost shrimp

Marsh crab

Crown conch

Sea squirts

Hermit crabs

Brittle star

In tropical regions, burrowing and climbing crabs feed on mangrove leaves. Falling leaves and other detritus from the mangrove forests are removed by tidal currents to the surrounding water and become the basis of a detritus food web (Figure 13–14). This very productive food web supports a variety of important commercial fishes and shellfishes such as blue crab, shrimp, mullet (*Mullus*), spotted sea trout (*Cynoscion nebulosus*), and red drum (*Sciaenops occellatus*).

◄ Figure 13–13

Mud Skipper. Mud skippers, residents of mangrove swamps in Asia, can spend time out of water by exchanging gases at their moist gills. (Norbert Wu)

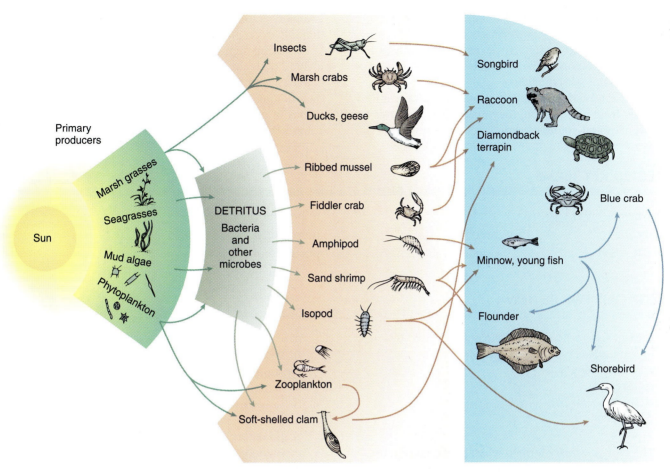

Figure 13–14

Detritus Food Web. Detritus from falling leaves and other decaying plant material forms the basis for a productive food web in estuarine communities.

CHAPTER SUMMARY

Estuaries form in embayments where fresh water from rivers and streams mixes with seawater. The characteristics of estuaries vary with the geology of the regions in which they occur. The mixing of salt water and fresh water and the distribution of salinity are important characteristics of estuaries. In salt-wedge estuaries there is a sharp boundary between fresh water and salt water. In well-mixed estuaries there is uniform salinity at all depths and a gradient of decreasing salinity as one proceeds toward the source of fresh water. Partially mixed estuaries occur where there is a strong surface flow of fresh water and a strong influx of seawater. In fjords, there is very little mixing between surface fresh water and deeper salt water. Many estuaries may exhibit a combination of mixing patterns, or mixing patterns may vary with season.

The salinity of estuaries varies both vertically and horizontally. In some, a phenomenon known as *tidal overmix-*

ing occurs in which denser seawater at the surface will sink and be replaced by less dense fresh water. Estuaries also exhibit fluctuations in temperature. In some, seasonal temperature change promotes a turnover that acts to circulate nutrients.

The mixing of nutrients from fresh water and seawater that occurs in estuaries, along with plentiful sunlight, make these areas some of the most productive in the marine environment. Nutrients are trapped in estuaries by silt and clay that is brought in by rivers and streams and by the action of some organisms, for instance, the formation of pseudofeces by bivalves.

Animals that live in estuaries must be able to adapt to changing salinities. Osmoregulators use a variety of physiological mechanisms to maintain optimum salt concentrations in their tissues regardless of the salinity of their surroundings. Osmoconformers have tissues and cells that can tolerate changes in salinity. Remaining stationary is another challenge that is faced by organisms in an estuary. The characteristics of estuaries tend to favor benthic organisms. Motile organisms must actively work to maintain position or move in and out with the tides.

Because estuaries are highly productive and relatively protected from wave action, they make good nurseries for the juveniles and young of many species. Many important commercial fishes and shellfishes spend at least a portion of their life cycle in the protected waters of an estuary.

Estuaries support a variety of distinct communities that include oyster reefs, mud flats, seagrass communities, salt marshes, and mangrove forests. Each community has its own unique characteristics and supports distinctive populations of producer and consumer organisms that have become adapted to life in each habitat.

SELECTED KEY TERMS

embayments, *p. 241*

estuary, *p. 241*

tidal flat, *p. 241*

salt wedge, *p. 242*

pseudofeces, *p. 244*

osmoregulator, *p. 244*

osmoconformer, *p. 244*

prop root, *p. 252*

pneumatophore, *p. 252*

QUESTIONS FOR REVIEW

1. Coastal areas where part of the ocean is partially cut off from the rest of the sea is called a(n)
   a. embayment
   b. estuary
   c. inlet
   d. tidal flat
   e. delta

2. The type of estuary formed when earthquakes cause land to sink, allowing seawater to enter, is a
   a. fjord
   b. tectonic estuary
   c. drowned river valley estuary
   d. coastal plain estuary
   e. delta

3. A deep estuary that has high river input but little tidal mixing is a
   a. fjord
   b. tectonic estuary
   c. drowned river valley estuary
   d. coastal plain estuary
   e. delta

4. In a well-mixed estuary, vertical salinity is
   a. stratified
   b. uniform

5. The environmental factor that affects the types of organisms found in estuaries more than anything else is
   a. sunlight
   b. temperature
   c. salinity
   d. nutrient density
   e. type of sediment

6. If the concentration of salts in an animal's body tissues varies with the salinity of the environment, the animal would be an

a. osmoregulator
b. osmoconformer

7. Most organisms that inhabit mud flats are
   a. active swimmers
   b. terrestrial
   c. burrowers
   d. predators
   e. immobile

8. The mangrove species that is usually found closest to the water in American mangrove forests is the
   a. white mangrove
   b. black mangrove
   c. red mangrove
   d. green mangrove
   e. buttonwood

9. An example of a deposit-feeder found on mud flats would be the
   a. jacknife clam
   b. fiddler crab
   c. lugworm
   d. soft-shelled clam
   e. innkeeper worm

10. An animal that feeds directly on seagrasses is the
    a. queen conch
    b. green sea turtle
    c. sea cucumber
    d. fiddler crab
    e. mullet

SHORT ANSWER

1. What are the distinguishing characteristics of an estuary?

2. What is a salt-wedge estuary, and how does it differ from other types of estuaries?

3. What factor(s) contributes to the high productivity of estuaries?

4. What adaptations have evolved in mangroves that help them survive in their habitat?

5. Explain how organisms that are osmoconformers survive in estuaries.

6. Explain how fiddler crabs are well adapted to life in the salt marsh?

7. Describe the process of succession in a salt marsh.

8. Sketch a chart that traces energy flow in a mud flat.

THINKING CRITICALLY

1. Why is it difficult for burrowing animals that exchange gas through their skin to survive in mud flats?

2. Predict what effect agricultural runoff would have on a neighboring estuary.

3. In order to control flooding, a series of dams is constructed along a river that feeds a large estuary. What effect will the dams have on the estuary's productivity?

SUGGESTIONS FOR FURTHER READING

Boicourt, W. C. 1993. Estuaries: Where the River Meets the Sea, *Oceanus* 36(2):29–37.

Lippson, A. J. and R. L. Lippson. 1984. *Life in the Chesapeake Bay.* Johns Hopkins University Press, Baltimore, MD.

Rutzler, K. and C. Fuller. 1987. Mangrove Swamp Communities, *Oceanus* 30(4):16–24.

Teal, J. M. 1996. Salt Marshes: They Offer Diversity of Habitat, *Oceanus* 39(1):13–15.

Valiela, I., and J. Teal. 1979. The Nitrogen Budget of a Salt Marsh Ecosystem, *Nature* 280:652–656

Chapter 14

Intertidal Communities

Bordering the water where land meets ocean, we find the realm of the seashore. Sandy beaches and rocky shores offer us places to ponder the wonders of nature or to relax and play in the surf. Although rocky and sandy shores are quite different, they share one feature in common. They are alternately submerged and exposed by the tides and pounded by the action of waves.

Imagine living in a world where radical environmental change was the norm. Twice a day, every day, from burning, desert-like heat you would be plunged into icy cold water. It is unlikely that a human could withstand such brutal external pressures, let alone carry on with daily activities. Yet, this is the type of habitat in which intertidal organisms not only survive, but thrive. At low tide they are baked by the sun, dried by the air, and pounded relentlessly by waves. At high tide, these organisms are submerged and become part of the aquatic world. This is the world of the **intertidal zone, the area between the tides. To survive in this habitat of extremes, organisms must be particularly hardy and adaptable.** In this chapter we will explore how marine organisms survive and make a living in such a seemingly hostile environment.

CHARACTERISTICS OF THE INTERTIDAL ZONE

The world's coastlines host a variety of living organisms. Whether the shoreline is rocky, sandy, or muddy, the interaction of wind, waves, sunlight and other physical factors creates a complex environment. **Organisms that live in this area must be able to tolerate radical changes in temperature, salinity, and moisture, as well as the potentially crushing force of waves, in order to survive.**

The intertidal zone can be composed of sandy beaches, rocky shores, tidepools, sand flats, mud flats, salt marshes, mangrove swamps, or a combination of these. In this chapter we focus on rocky shores and sandy beaches. The major

---

KEY CONCEPTS

1. The intertidal zone is the coastal area alternately exposed and submerged by tides.

2. Organisms that inhabit intertidal zones must be able to tolerate radical changes in temperature, salinity, and moisture and also be able to withstand wave shock.

3. Organisms on rocky shores tend to be found in definite bands or zones on the rocks.

4. Rocky shores provide a relatively stable surface for organisms to attach to.

5. Subtidal zones tend to be more stable and thus support a greater diversity of organisms than the supratidal or intertidal zones.

6. Tidepool organisms must be able to adjust to abrupt changes in temperature, salinity, pH, and oxygen levels.

7. The intertidal zone of sandy shores is inhabited mostly by burrowing organisms.

characteristics of the other habitats were described in Chapter 13. Regardless of the type of habitat, the inhabitants of the intertidal zone experience daily fluctuations in their environment. During high tide, when the area is submerged by seawater, the inhabitants are most active, foraging for food, finding mates, and reproducing. The water contains food for filter-feeders and oxygen for those organisms that use gills to exchange gases.

As the tide retreats, the organisms become exposed to air. Those with gills must protect their respiratory structures from drying out and collapsing. Animals that are filter-feeders withdraw into protective coverings. The pace of life slows greatly. In summer the sun bakes the exposed area,

causing temperatures to rise and tissues to lose water. During winter months on temperate beaches there is a danger of freezing. Most animals survive during low tide by sealing themselves up in shells or burrows or retreating into cracks and crevices until the high tide returns.

ROCKY SHORES

As the tide retreats from a rocky shore, the higher regions of the coast become exposed to air. Each region is composed of a specific group of organisms that form definite horizontal bands or zones on the rocks (Figure 14–1). This separa-

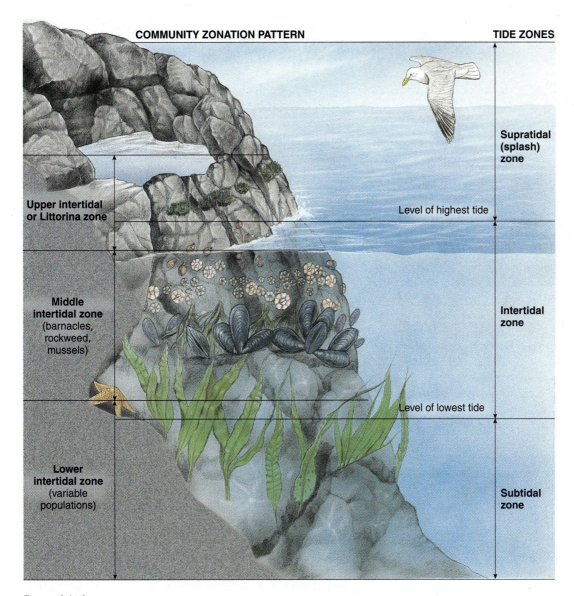

Figure 14–1

A Rocky Shore. Since rock provides a relatively stable surface on which organisms can attach, rocky shores exhibit definite bands or zones, each inhabited by organisms adapted to the special conditions of the environment.

tion of organisms into such definite bands is called **zonation.** Nowhere else is zonation more striking than on rocky shores. Unlike shores of sand or mud, rocks provide a relatively stable surface to which organisms can attach, as well as a variety of hiding places. As the tide retreats, organisms in the upper regions are exposed to air, changing temperatures, solar radiation, and desiccation for prolonged periods of time. The lower regions, on the other hand, are exposed only for a short period of time before the tide returns to cover them. The pattern of zonation differs from one rocky shore to the next, depending on such factors as wave action, tidal cycle, climate, length of exposure, amount of light, shape of the shore, size and shape of the ocean basin, and type of rock.

The uppermost area of a rocky shore, which is covered only by the highest (spring) tides and is usually just dampened by the spray of crashing waves, is the **supratidal** or **splash zone.** The **intertidal zone** lies between the supratidal zone and the subtidal zone and is regularly exposed during low tides and covered during high tides. The intertidal zone can be divided into three parts: upper, middle, and lower intertidal zones. The **subtidal zone** is the region of the shore that is covered by water, even during low tide. The distinctions between zones are not always clear, but each rocky shore exhibits some variation of this plan.

The Supratidal (Splash) Zone of Rocky Shores

The supratidal, or splash, zone receives very little moisture. It is exposed to the drying heat of the sun in the summer and to extreme low temperatures in winter (less than or equal to −20°C in northern temperate regions). As a result of these harsh conditions, the supratidal zone supports only a few hardy organisms. Gray and orange lichens composed of fungi and algae are common in this zone (Figure 14–2). The fungi in these symbiotic relationships trap moisture for both themselves and their algal partners, while the alga produces nutrients for the pair by photosynthesis. On the North Atlantic coasts, tarlike patches of cyanobacteria, mostly of the genus *Calothrix,* survive by producing a gelatinous covering that traps and stores moisture.

Sea hair (*Ulothrix*), a filamentous green alga also found on rocks of the North Atlantic coasts, is capable of surviving on the moisture provided by sea spray from waves. During winter months, it is found growing lower on the intertidal rocks than where it grows during the summer. At this time of year, the reproductive spores the alga produces can survive only where the ocean provides the necessary warmth. The adult algae growing higher on the rocks gradually die out as the air temperature changes.

On the Atlantic coasts of North America and Europe, rough periwinkles (*Littorina saxatalis*) graze on various types of algae growing at the lower edge of the splash zone

Figure 14–2

The Supratidal Zone. The supratidal zone, or splash zone, receives very little moisture and is inhabited only by a few hardy organisms, such as these yellow lichens and cyanobacteria. (Laurie Campbell/ Natural History Photographic Agency)

(Figure 14–3a). Although basically marine animals, these snails are well adapted to life out of water. Even though they lack both functional gills and lungs, their mantle cavity is very vascular and acts as a surface for gas exchange. These snails are so well adapted to breathing air that they would drown if submerged in water for several hours. To avoid the problem of desiccation, rough periwinkles hide in the cracks and crevices of rocks or seal the opening of their shell with mucus, thus trapping moisture in their mantle cavity until temperatures moderate. In order to prevent her eggs from being damaged by exposure, the female rough periwinkle retains them inside of her mantle cavity, where they can be kept moist and oxygenated until they hatch. Other species of periwinkle, such as the common periwinkle (*Littorina littorea*), deal with the problem by producing planktonic eggs inside jelly coats. Others, such as the north-

Figure 14–3

Animals of the Supratidal Zone. (a) Rough periwinkles graze on the algae that cover the rocks.
(b) These isopods of the genus *Ligia* are scavengers feeding on the available organic material.
(a, Jeffrey Rotman; b, A. Kerstitch/Visuals Unlimited)

ern yellow, or smooth, periwinkle (*Littorina obtusata*) attach gelatinous egg masses to large algae.

Crustaceans known as isopods (*Ligia;* Figure 14–3b) are also common inhabitants of the supratidal zone. They are scavengers and feed on available organic material. Like the rough periwinkles, isopods are adapted to exchanging gas with the air and will drown if submerged under water. Female isopods have thoracic pouches in which they carry their eggs to prevent them from drying out.

The Intertidal Zone of Rocky Shores

When the tide retreats, the intertidal zone becomes exposed. As the water recedes, intertidal animals must face the problems of gas exchange, desiccation, temperature extremes, and feeding. Tissues that are involved in gas exchange must be kept moist, and animals must keep from losing too much body water to the environment as they are exposed to air and the radiant energy of the sun. Filter-feeders and organisms that feed on aquatic prey must temporarily suspend feeding and conserve their energy reserves. The solutions to these problems are as varied as the organisms that inhabit the intertidal zone.

In addition to dealing with the previously mentioned problems, organisms that live in the intertidal zone must be adapted to withstand the force of the waves as they crash against the rocks during low tide. This force is called **wave shock.** Not only do rock-dwelling organisms have to deal with the crushing force of a wave as it strikes the rocks but they also must deal with the drag that is created as the water moves back out to sea. Animals that live in this zone ex-

hibit compressed or dorsally flattened bodies or shells to dissipate the force of waves. These animals have also adapted mechanisms for adhering tightly to the surface of the rocks to prevent them from being dislodged and washed away by the waves (Figure 14–4).

The most evident organisms in the upper portion of the rocky intertidal zone are acorn barnacles (*Balanus*) and rock barnacles (*Semibalanus*), which form a line at and below the high tide mark. On some rocky shores, the barnacle population may be as dense as 9,000 individuals per square meter. Barnacles cement themselves permanently to solid surfaces. They are particularly evident on beaches that are pounded by heavy surf where other, less well-adapted, organisms would not be able to survive. During low tide, the barnacle closes the opening of its shell with calcareous plates, trapping enough water inside to support the animal until covered again by water at high tide. If the temperature becomes too warm, the barnacle will open its shell just a little to allow some of the trapped water to evaporate. This evaporative cooling helps to keep the animal from overheating. When the tide is high, barnacles open their shells and strain food from the surrounding water, using modified thoracic appendages.

In the middle and low intertidal zones oysters (*Ostrea*), mussels (*Mytilus*), limpets (*Patella* and *Acmaea*), and periwinkles (*Littorina* species) dominate. Like barnacles, oysters and mussels fasten themselves to the rocks. Oysters cement the lower valve of their bivalve shell to a solid surface, while mussels attach themselves to the rocks by means of strong protein fibers called **byssal threads** that are secreted by the animal (see Figure 14–4c). At low tide, these bivalves close their shells, trapping enough water inside to

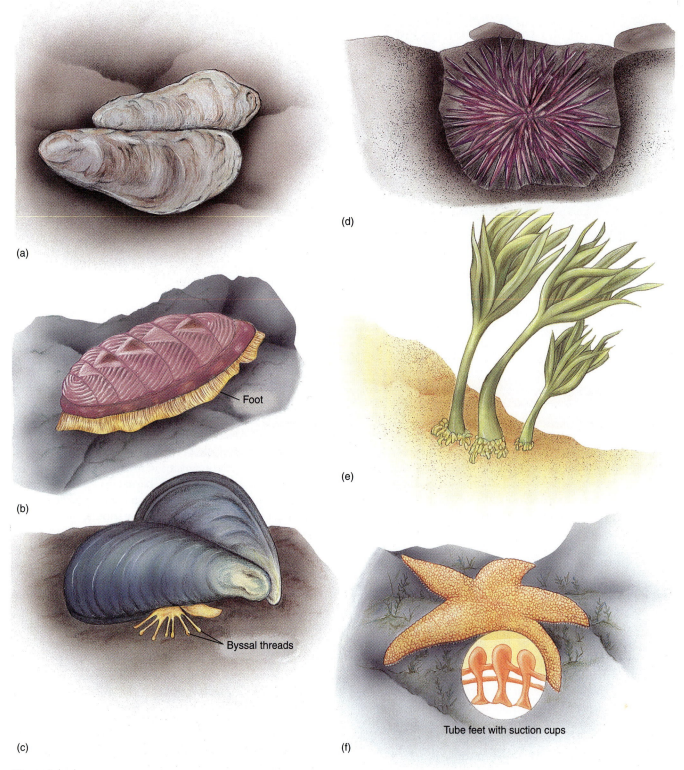

(a)

(b)

Foot

(c)

Byssal threads

(d)

(e)

Tube feet with suction cups

(f)

Figure 14–4

Adaptations to Wave Shock. Organisms that live on rocky shores have evolved a number of different strategies for dealing with the crushing force of waves. (a) Some rock dwellers, such as barnacles and oysters, cement themselves to the rock's surface. (b) Chitons and limpets hold tight with their powerful, muscular feet. These animals also present a low body profile, exposing less surface area to the oncoming waves. (c) Mussels attach themselves with tough byssal threads, and (d) rock urchins hollow out cavities in the rocks. (e) Seaweeds have flexible bodies that bend with the flow of water that passes over them. (f) Sea stars use suction cups on the ends of their tube feet to cling to the surface of rocks.

ECOLOGY AND THE MARINE ENVIRONMENT

Connell's Barnacles

As far back as Darwin, biologists have recognized that the organisms that live on rocky coasts are arranged in horizontal layers. These layers are arranged in an orderly and predictable fashion from the highest splash zone, which receives moisture from waves that crash against the rocks, to the wet zone beneath the low tide mark. An explanation for this arrangement of organisms was proposed by a biologist named J. H. Connell from the University of California.

Connell performed experiments on the rocky coasts of Scotland, where he noticed that two different genera of barnacles, *Balanus* and *Chthamalus,* live next to each other in separate layers on the rocks. Members of the genus *Chthamalus* live in the layer of organisms just above the high tide line. In the zone just beneath this, the population of *Chthamalus* is replaced by populations of the genus *Balanus.* Barnacles are sedentary animals, and the species that Connell worked with are relatively small and live in large, dense populations. Be-

cause of these characteristics, Connell could perform experiments that would have been impossible with larger, more mobile organisms.

On one section of rocky shore, Connell observed the normal growth, reproduction, and daily activities of the two types of barnacles. This section of shore acted as the control for his experiment. On other sections he performed the following experiments. In one experiment, he transplanted barnacles to different zones on the rocks. In another experiment, he completely cleared the barnacles from areas of rock in order to see which species would colonize.

Connell determined from these experiments that individuals of *Chthamalus* were better adapted to living above the high tide mark. Members of this genus had a much greater tolerance for the higher temperature and desiccation that occurs in this zone compared to the members of the genus *Balanus.* The members of the genus *Balanus,* however, grow faster than those of *Chthalamus.*

When larvae of both species settle on bare rock in the lower zone, the members of the genus *Balanus* compete better for the vital resource of space. The *Balanus* barnacles would either overgrow the *Chthamalus* barnacles, causing them to starve, or undercut the slower growing *Chthamalus* barnacles, literally prying them off of the rocks. As a result the *Chthamalus* barnacles did not have the opportunity to become established in this zone. *Chthamalus* barnacles live in the upper zone because they are better adapted to the harsh environment in the splash zone. The *Balanus* lives in the lower zone where both species can survive but where *Balanus* is a superior competitor.

Experiments such as these show the importance of the interplay between the physical environment and biological factors, such as competition, in determining the distribution of organisms in an ecosystem.

provide for their needs until high tide. When the tide is in, they open their shells to filter feed on plankton that the incoming water brings.

Limpets are small, oval molluscs that graze at high tide on algae (see Chapter 9). At low tide, some limpet species return to an area of rock where they have hollowed out a shallow depression that perfectly conforms to the shape of their shells. They trap moisture in their mantle cavity to prevent desiccation and cling tightly to the rocks with their muscular foot to prevent waves from dislodging them. Another mollusc, the chiton (see Figure 14–4b), fills a niche similar to the limpet and in some areas competes with limpets for food on the rocks.

The common periwinkle is found on northern Atlantic shores of North America and Europe. Unlike its relative that inhabits the splash zone, this snail cannot breathe air and

must remain moist in order to exchange gases. During low tide it can be found buried in masses of seaweed that trap enough moisture for the snail to survive until the tide returns. At high tide, the common periwinkle feeds on algae that inhabit this area.

In some regions, echinoderms called *rock urchins* (*Arbacia* on the East Coast and *Strongylocentrotus* on the West Coast) use their jaws to hollow out a space in the rock. During low tide they wedge themselves firmly in place with their spines (see Figure 14–4d). This behavior, combined with their low profile, helps them to survive wave shock. Rock urchins feed mostly on algae that cling to the rocks. They also feed on drifting bits of seaweeds that they capture among their spines and tube feet. They actively feed during high tide and at low tide return to their depressions, where they are sheltered and less likely to desiccate.

MARINE BIOLOGY AND THE HUMAN CONNECTION

Mussels and Medicine

Many species of mussel from rocky shores around the world are being studied as possible sources of chemicals that might have medicinal value. The green-lipped mussel (*Perna viridis*) of New Zealand, for instance, is the source of a substance used to help arthritic patients. Many medications taken by people suffering from arthritis have very unpleasant side effects. Pernine, a substance isolated from the green-lipped mussel, especially the gonad, is added to some arthritis medications to lessen the severity of the side effects.

Many researchers are looking into possible uses of the mussel glue protein (the protein that forms byssal threads). This protein has tremendous adhesive properties and is used by the mussel to attach itself to rocks, where even powerful waves cannot dislodge it. In fact, the glue is so strong that it has been incorporated into a sealant for tiles on the U. S. space shuttles. The glue's adhesive properties come from modifications of the amino acid tyrosine, which is present in the protein. Some of the possible uses that are being investigated include use as a dental cement, as an adhesive for reconstructing the inner ear, as a replacement for sutures in corneal transplants, and as a glue to mend broken bones.

The most characteristic seaweeds in the intertidal zone on the northern Atlantic and northern Pacific coasts, especially in colder temperate areas, are the brown algae or rockweeds like *Fucus* (Figure 14–5). Rockweeds are usually small, less than 30 centimeters (1 foot) in length, and grow on rocks that do not have full exposure to the sea, as they do not tolerate large waves. In protected areas, such as bays, rockweeds may be as large as 1.8 meters (6 feet) and almost completely cover the rocks. Some species of rockweed possess gas-filled chambers called *air bladders* that help to buoy them during high tide.

In most rocky intertidal habitats, rockweeds must compete with barnacles for space on the rocks. The large blades of the algae move across the rocky surface in a sweeping action that is powered by waves. This action prevents the cyprid larvae of barnacles (see Figure 9–25c) from settling down and gaining a foothold, thus providing the algae with more space for growth. Another problem faced by rockweeds is grazing by herbivores. Some species produce chemicals that deter herbivores from feeding on them. *Fucus,* for instance, produces a chemical that discourages some herbivores, such as some molluscs, from eating them. Studies have revealed that when herbivorous molluscs begin to graze on *Fucus,* the alga produces more of the bad-tasting chemical in the area where the molluscs are feeding. The increased amount of chemical usually succeeds in deterring any further grazing.

To prevent desiccation, rockweeds produce a gelatinous covering that retards water loss (see Chapter 7). Their holdfasts anchor them tightly to the rock's surface, and, rather than resisting the force of the waves, the blades and stipes of the algae bend gracefully with the wave action. As the tide goes out, the seaweeds form large mats that trap water and provide a haven for many species of animal, such as juvenile sea stars, brittle stars, sea urchins and bryozoans. The shelter provided by large algae, such as rockweeds, may significantly reduce the stress of increased temperature

Figure 14–5

Rockweeds. Rockweeds such as *Fucus* are common along rocky coasts of the North Atlantic and North Pacific oceans. (Richard Thom/ Visuals Unlimited)

Figure 14–6

A Tidepool. Organisms that inhabit tidepools must be able to adjust to rapid changes in temperature, pH, salinity, and the oxygen content of the water. (B. Kent/Animals Animals)

and desiccation on these associated animals during low tide.

Not all areas of the rocky intertidal are exposed when the tide retreats. Depressions in the rocks continue to hold water, forming areas called **tidepools** (Figure 14–6). Tidepools prevent organisms within them from being exposed to air, but in turn they present their own unique set of problems. As the small body of water heats up in the sun, it loses oxygen. This can lead to the suffocation of some inhabitants. During heavy rains, the salinity of tidepools decreases drastically as rainwater accumulates and dilutes the seawater. On the other hand, on a hot sunny day, so much water can evaporate that the salinity will increase to dangerous levels. If the tidepool contains a large amount of algae, the oxygen content of the pool will be high during the day while sunlight powers photosynthesis. At night, however, the level of oxygen declines and the level of carbon dioxide rises, resulting in a lower pH. Large tidepools near the low tide mark usually show the least amount of fluctuation, while smaller ones and those near the high tide line that are exposed the longest exhibit the widest range of fluctuation. Regardless of their position on the shore, most tidepools return abruptly to marine conditions as the tide rises and seawater floods the pool. This return to the ocean produces almost instantaneous changes in salinity, temperature, and pH. The organisms that inhabit tidepools have evolved a variety of mechanisms, similar to those found in salt marsh organisms, that allow them to survive in this changeable environment (see Chapter 13).

In cooler coastal areas, common tidepool inhabitants include sea stars, mussels, anemones, tube worms, and hermit crabs. In warmer coastal areas, anemones, tube worms, hermit crabs, and a variety of mollusc species dominate. Many of the tidepool organisms are filter-feeders, feeding on phytoplankton and zooplankton. Anemones and tube worms expose a large surface area while feeding. If disturbed, they will retract their tentacles into their bodies or tubes to protect these delicate appendages from damage. This behavior is also an efficient way to avoid desiccation.

The Subtidal Zone of Rocky Shores

The subtidal zone is covered by water even at low tide (Figure 14–7). Only during the lowest spring tides does this area of rocky shore become exposed, and even then, only partly. Since this zone is more stable than the other two, it exhibits a greater diversity of marine life. Along coasts where the water is cold, kelp such as *Laminaria* form a dense forest with smaller algal species and animals living among them. In these cooler waters a variety of mussel species can be found, as well as young sea stars and brittle stars. In more temperate regions and the tropics, brown algae give way to various species of red algae. Attached to many of the algae are lacey colonies of bryozoans. Hydrozoans can be found attached to both rocks and the broad blades of algae. On the West Coast of the United States, sea urchins graze on algae and bits of decaying material that they find, while spider crabs (*Libinia*) and Jonah crabs (*Cancer borealis*) scavenge for their food. At high tide on the Pacific coast, where kelp is prominent, rock eels (*Xiphister*), rock bass (*Centropristus philodelphica*), gobies, and a variety of other fishes visit the subtidal zone to feed.

Ecology of the Rocky Shore

Life on the rocky shore is heavily influenced by biotic factors. Grazing, predation, and competition all play an important role in determining the distribution of organisms in this community (Figure 14–8). For instance, in New England, where the wave action can be heavy, mussels and barnacles are most abundant, and there are fewer periwinkles. With fewer of these grazing snails, the algal growth is greater, and species of algae like *Ulva* are more abundant. In areas where there are many periwinkles, their persistent grazing removes the *Ulva* and allows *Fucus,* which they don't eat, to become well established.

(a)

(b)

Figure 14–7

The Rocky Subtidal Zone. Rocky subtidal zone of the (a) West and (b) East coasts of the United States. Since the subtidal zone of rocky shores offers a more stable environment than the intertidal zone, it exhibits greater diversity.

Along the New England rocky coast, mussels win the competition for space on the rocks against the barnacles and algae. Predation on mussels by sea stars and the carnivorous dogwinkle snail (*Nucella*), however, prevents the mussels from becoming completely dominant. A notable exception are rocks that are so wave beaten that sea stars and dogwinkles are not prevalent. On these rocks, mussels dominate completely in the middle and lower intertidal zones. Rock barnacles frequently dominate in the upper intertidal

zone since they are more tolerant of desiccation and temperature extremes than are mussels. Physical factors, such as temperature, frequently set the upper limit of organism distribution in the intertidal zone.

On the Pacific coasts, barnacles compete with algae for available space, and mussels displace barnacles by growing over them. Ochre sea stars (*Pisaster*) maintain the balance by eating enough mussels to prevent them from completely overtaking the barnacles.

Figure 14–8

Rocky Shore Food Web. Plankton, algae, and detritus form the basis of food webs on rocky shores.

THE SANDY SHORE

Sandy beaches account for much of temperate seashores and many tropical shores. The nature of a sandy beach, the porosity of its sediments, and the ability of animals to burrow into these sediments are all influenced by the size of the sand particles that form it. Waves play a significant role in determining the types of sediment on a sandy beach. Heavy wave action carries off much of the finer sediment, leaving behind only coarser material. Thus beaches with fine sand are found along coasts with little wave action or in areas protected from heavy surf. Sandy beaches with steep slopes usually receive more wave action. Beaches that receive less wave action are usually flat. On all sandy beaches, a cushion of water separates the grains of sand below a certain depth. This is especially true of beaches with fine sand where capillary action (see Chapter 4) is the greatest.

The most important factor in determining the distribution of life on sandy beaches is wave action. The amount of moisture and oxygen that is available to organisms on a sandy beach depends on waves. Although wave action provides organisms with more moisture, oxygen, and plankton, it also removes large amounts of organic material from the beach, leaving little for scavengers to consume. For an organism to survive in this hostile environment, the wave action must be able to meet the organism's specific needs for food, oxygen, and waste removal.

Sandy beaches can be classified into four groups based on the amount and type of wave action they receive.

Exposed beaches receive the greatest amount of wave action.

Semi-exposed beaches are only partly exposed to the surf and receive less wave action.

Table 14–1
Classification of Sandy Beaches by Exposure

	Exposed (Reflective)	Semi-Exposed (Dissipated)	Protected	Very Protected
Wave Action	high	moderate	low	none
Wave Types	plunging, surging	spilling	–	–
Slope	steep	gradual	gradual	flat
Width	narrow	wide	various	extensive
Oxygen	highest	–	–	lowest
Moisture	lowest	–	–	highest
Particle Size	coarsest	finer	finer	varies
Organic Matter	lowest	–	–	highest
Permanent Burrows	no	no	some	frequently

Protected beaches receive very little wave action because of barriers such as reefs, rocks, or coastal geographical features that absorb most of the wave energy.

Very protected beaches are exposed to no wave action at all.

Table 14–1 summarizes the characteristics of these four types of sandy beaches.

In sharp contrast to the rocky shore, sandy beaches appear barren and devoid of life at first. Sandy beaches do not exhibit the obvious pattern of zonation that is so characteristic of rocky shores. Since the shifting sands offer no solid

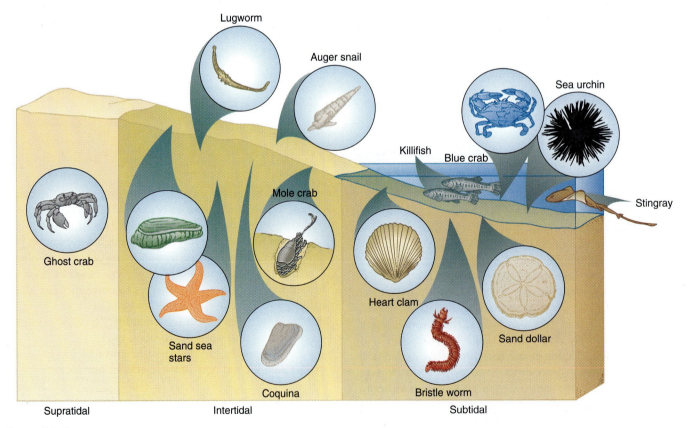

Figure 14–9

A Sandy Shore. Sandy shores do not exhibit the obvious zonation of rocky shores. Most of the organisms inhabiting the intertidal zone are burrowers that reveal their presence only during high tides.

Figure 14–10

Sand Dunes. Dunes such as these form along the coast when sand is blown away from the beach. They are most common along the Atlantic and Gulf coasts. (G. Karleskint)

footing, organisms that live here are either mobile or they burrow into the sand. If we were able to see beneath the surface of the sand, we would find a variety of organisms just waiting for the return of the tide to resume the business of life (Figure 14–9).

Like rocky shores, sandy beaches and sand flats can be divided into three zones: **supratidal, intertidal,** and **subtidal.** In addition, below the surface of the sandy beach,

there is also vertical zonation in the sand itself. These zones will be examined in more detail in the following sections.

Life above the High Tide Line

The supratidal zone of a sandy shore stretches from the high tide line to the point where terrestrial vegetation begins. In some areas, sand dunes (Figure 14–10) are found at the uppermost extent of this zone. This portion of the sandy beach at first appears to be barren. The sun bakes the surface sand, raising its temperature and making it unsuitable for habitation. In this zone most of the living organisms are buried in the sand (**infauna**), where they can survive dry periods and the intense heat of the sun. Most of the infauna live in permanent or semipermanent tubes or burrows, or they are able to quickly burrow into the sand. Multicellular forms obtain oxygen either through their skin or through gills that are sometimes bathed in supratidal water drawn in by elaborate siphons.

By far the most numerous inhabitants of the sandy beach are small organisms between 60 micrometers and 0.5 millimeters (less than 0.02 inches) in length (Figure 14–11). These microscopic organisms that inhabit the spaces between sediment particles make up the **meiofauna** of the sandy beach. The meiofauna includes nematodes, gastrotrichs, small arthropods, and many other organisms that live between the grains of sand. These organisms are generally elongated with few lateral projections

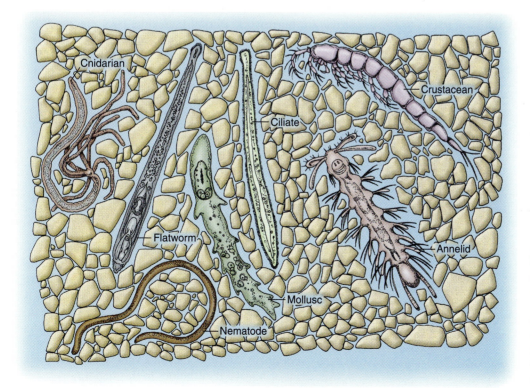

Figure 14–11

Meiofauna. Meiofauna are tiny organisms that inhabit the spaces between sediment particles. Members of the meiofauna that inhabit sandy shores include tiny protozoans, nematodes, cnidarians, flatworms, annelids, crustaceans, and molluscs.

of any type, such as setae or spines, that would interfere with their ability to move freely and quickly in the sand. Since there is very little water available in this zone, the majority of these animals have no pelagic larval form. They feed mostly on algae, bacteria, and detritus. The sizes of meiofauna populations exhibit seasonal variations, reaching a peak during summer months. Beaches that are protected from wave action have the greatest abundance of meiofauna.

Within the supratidal zone there is definite vertical zonation, with the two different zones separated by their moisture content. The uppermost layer is the **zone of dry sand.** This zone contains no moisture, and the heat during the day makes it impossible for living organisms to survive. Just below this level is the **zone of drying sand.** Moisture reaches this zone during the highest tides and gradually evaporates over time. In temperate regions, several species of amphipod crustaceans are abundant in this zone, as are many insect species.

In the supratidal zone of temperate regions and the tropics, the dominant organisms are ghost crabs (*Ocypode*) (Figure 14–12). The ghost crab gets its name from its light color and its habit of coming out at night to forage for food. During the day, it spends most of its time in its burrow to avoid the heat and predatory birds. The ghost crab is not as aquatic as many of its relatives. Its legs are not adapted for swimming but are useful for scurrying sideways along the beach. When the crab must enter the water, for instance while being pursued by predators, it moves by running along the bottom. Female ghost crabs deposit their eggs in the ocean, and the young develop in the water. Young crabs rely on currents to carry them about and eventually to deposit them on a beach, where they can take up an adult existence. Once the crabs arrive on land, they return to the

sea only occasionally to moisten their gills or to avoid predators. In either case, their visit is short, and they quickly return to the shore. Like its aquatic relatives, the ghost crab extracts oxygen from water through its gills. A special chamber surrounding the gills traps enough seawater to supply the crab's needs until it can enter the water and replenish its supply.

Adult ghost crabs live in burrows that consist of a deep shaft with a chamber at the end. Sometimes another shaft branches from the chamber and serves as an alternate escape route. The crabs emerge at night to scavenge in the beach wrack for food. They spend the early morning hours making necessary repairs to their burrows, and, by midday, when the temperatures are reaching a peak, they enter their burrows and seal the opening to escape the high temperatures and dessication. They will emerge again after the sun sets to resume their foraging.

In areas like the New Jersey coast, ghost crabs will prepare for winter by moving their burrows farther up on the beach, well out of the reach of high tide. They fashion a larger chamber at the end of the shaft that they stock with food for the winter. As the temperatures begin to fall, the crabs retire to their burrows, seal the openings, and live their subterranean existence until the warmth of spring brings them out for another season.

Life in the Sandy Shore Intertidal Zone

Like its rocky shore counterpart, the sandy beach intertidal zone is exposed at low tide and submerged at high tide. As in the supratidal zone, most inhabitants of the intertidal zone are burrowers. During high tide, they emerge from their burrows to look for food, find mates and reproduce, or extend specialized appendages to filter the water for food and extract oxygen. At low tide, they burrow into the moist sand and retract their appendages to avoid desiccation and exposure to land predators.

Like the supratidal zone, the intertidal zone also exhibits vertical zonation (Figure 14–13). The zone of dry sand and the zone of drying sand are not as thick as in the supratidal zone. Just beneath the zone of drying sand in the high intertidal region is the **zone of retention.** This region retains moisture at low tide because of the capillary action of water and is inhabited worldwide by several species of isopod crustaceans. In the mid and low intertidal regions are the **zone of resurgence** and the **zone of saturation.** In the zone of resurgence, water is retained at low tide and supports a variety of assorted crustaceans and polychaete worms. Farther down is the zone of saturation. This zone is constantly moist and supports the greatest diversity of organisms. Polychaetes are found burrowing in this area along with bivalves like *Donax*, tellins (Tellinidae), and Venus clams (Veneridae). Amphipods and a variety of other small crustaceans also make their homes in this region.

Figure 14–12

Ghost Crab. Ghost crabs dominate the supratidal zone of sandy shores in temperate and tropical regions. They spend most of the daytime in their burrows, coming out at night to scavenge for food. (M. Graybill/Biological Photo Service)

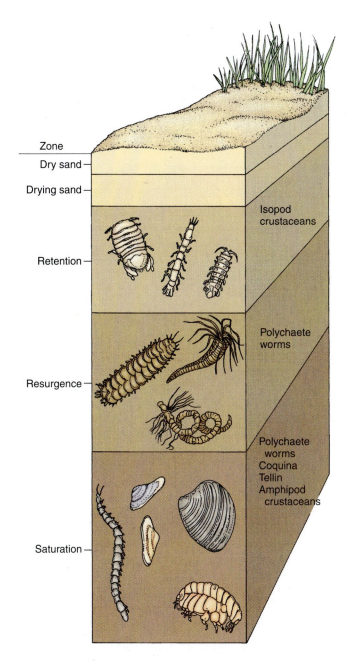

Figure 14—13

Vertical Zonation of the Sandy Intertidal Zone. The sandy intertidal zone exhibits vertical zonation, with each zone characterized by its moisture content. The zones of retention, resurgence, and saturation are home to a variety of burrowing organisms.

At low tide, a careful examination of the sandy bottom will usually reveal telltale signs of the presence of burrowing organisms (Figure 14—14). Flat, ribbon-like trails are signs of echinoderms such as sand dollars or sea stars. The sand dollars use tiny spines and their tube feet to pick detritus out of the sand. In Florida and the Caribbean, sand sea stars (*Luidia*) feed mostly on small molluscs and crustaceans that they find burrowed in the sand. These sea stars

Figure 14—14

Animal Trails. This trail was produced by a gastropod mollusc known as an auger (*Terebra*). Tracks in the sand such as these indicate the presence of burrowing animals, frequently molluscs, crustaceans, or echinoderms. (G. Karleskint)

swallow their prey whole and then regurgitate the shells and hard parts. Unlike the sea stars that inhabit rocky shores, these animals lack suckers at the ends of their tube feet so they are not able to cling to rocks. They are, however, very efficient burrowers and can cover themselves with sand in a matter of seconds.

Many snails leave trails that resemble the mounds of elevated earth made by moles. Moon snails (*Polinices*) and olive snails (*Oliva*) are especially common in this habitat and leave hills of sand that are raised by the animal's foot as it crawls along the bottom. Moon snails feed on bivalves by using their radula to drill a neat hole into the bivalve shell (Figure 14—15). Once the shell is pierced, the moon snail inserts its proboscis through the opening and rasps out the flesh from inside the shell. The average moon snail will consume enough clams each week to equal one third of its body weight. Olive snails are also carnivorous. They feed on a variety of organisms but especially small bivalves.

Figure 14—15

Moon Snail. Moon snails (*Polinices; Lunatia*) are carnivores feeding mostly on bivalves. (H. W. Pratt/Biological Photo Service)

(a)

(b)

Figure 14—16

Lugworm. (a) A common burrowing organism in the sandy intertidal zone is the lugworm. (b) The worm ingests sand as it constructs its burrow and deposits it in the form of fecal castings that are stacked in a cone-shaped pile outside the burrow entrance. (a, J. & M. Bain/Natural History Photographic Agency; b, Claudia Mills/Biological Photo Service)

Another prominent inhabitant of the intertidal zone is the lugworm (Figure 14–16a). Lugworms are deposit-feeders. They ingest sand as they dig their burrows and digest the organic material that it contains. The sand passes through the worm's digestive system and is deposited during defecation at the entrance to the worm's burrow. The coiled, cone-shaped casts are prominent on sandy beaches at low tide (Figure 14–16b).

The activity of intertidal organisms is keyed to the movement of the tides. As the tide moves in, the pace of life begins to quicken. During high tide, bivalves, such as cockles (*Cardium*), tellins, and surf clams (*Spisula*), project their siphons from their burrows and begin to filter the water for food and bathe their gills with oxygen. Carnivorous

snails, called *whelks* (Buccinidae), glide along the bottom in search of bivalves, which they locate by sensing the currents produced by the clam's siphons. Sea stars and sand dollars also become more apparent as they come out of the sand in search of food.

Two filter-feeding organisms that move up and down the beach with the movement of the tide are mole crabs

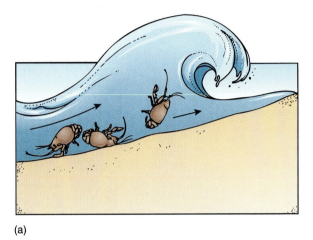

(a)

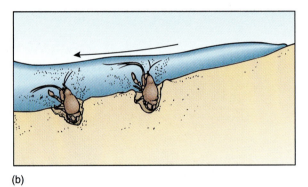

(b)

(c)

Figure 14—17

Feeding Behavior of Mole Crabs. Mole crabs are filter-feeders that move up and down the beach with the tides. (a) Incoming waves carry the mole crabs up the beach. (b) As the wave retreats, the crabs burrow into the sand. (c) They then extend their antennae and filter food from the water as the wave retreats.

(*Emerita*) and coquina clams (*Donax*). As the tide moves in, large numbers of mole crabs emerge from the sand (Figure 14–17). Incoming waves carry these creatures up the sand, and, as the waves retreat, the animals force their way backwards into the sand again, using their large antennae to strain food from the water as it returns to the sea. When the tide retreats, the mole crabs repeat the process in the opposite direction.

Coquina clams move along with the tide in a similar fashion (Figure 14–18). As the water covers the sand, thousands of these tiny bivalves come out of their burrows, and the waves carry them up the beach. Before the backwash can carry them back into the water, they burrow into the sand and put out their siphons to feed. As the tide retreats they follow the water back, always staying at the edge of the water.

The incoming tide brings not only food and oxygen but also predators (Figure 14–19). Blue crabs and green crabs

Figure 14–18

Coquinas. Like mole crabs, coquinas emerge in advance of a wave so the wave will carry them up the beach. As the wave retreats, they burrow into the sand, straining the receding water for food. (Jim Doran/Animals Animals)

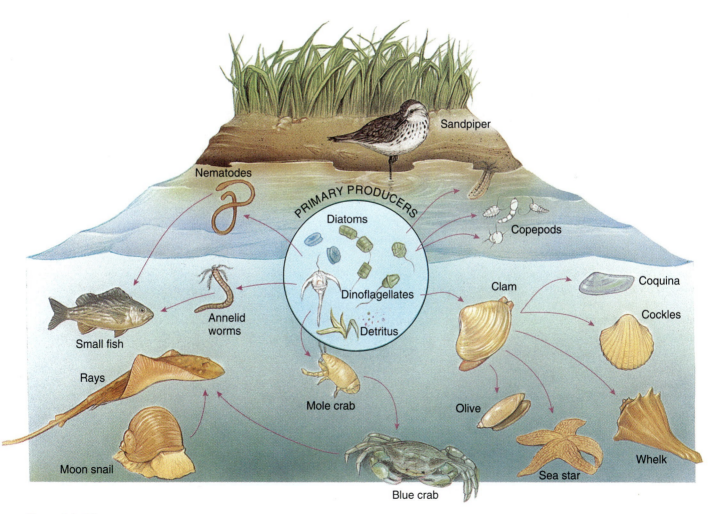

Figure 14–19

Sandy Shore Food Web. Detritus and single-celled algae provide food for numerous burrowing animals. These burrowing animals are a source of food for many larger animals including crabs, fishes, and birds.

(Portunidae) move in to feed on smaller crustaceans and clams. Small fish, like the killifish (Cyprinodontidae) and silversides (Atherinidae), feed on crustaceans, worms, and a variety of larval forms. Skates and rays cruise the bottom, preying on crustaceans and molluscs. When the tide recedes again, these predators will be replaced by gulls and a variety of shorebirds that scurry across the intertidal zone in search of food.

Life below the Low Tide Line

The subtidal zone is a truly marine environment that is exposed only during the lowest spring tides. The variety and distribution of organisms in this zone is primarily influenced by characteristics of the bottom sediments. Where the sediments are bare sand, the inhabitants are predominantly burrowing organisms similar to those inhabiting the intertidal zone. Several species of bivalves extend their siphons to filter food from the water. Tube worms and other polychaetes either trap their food by filter- and deposit-feeding or search the sand particles for detritus. Along the southeastern coast of the United States and along Caribbean shores, heart urchins (*Moira*) move slowly along the sandy bottom, feeding on detritus that is mixed with the grains of sand. As they travel just beneath the surface of the sand, they angle their spines backward to avoid resistance. This causes them to leave a V-shaped trail as they move along the bottom, half-buried in sand. A variety of molluscan species feed on algae, detritus, or each other.

Along some coasts, fields of seagrass are found in the subtidal zone. In addition to arthropods and molluscs, the seagrass beds host sea urchins, sea stars, brittle stars, sea cucumbers, and anemones. The addition of rocks and broken pieces of coral increases the complexity of the bottom and the diversity of life. Since the subtidal zone is rarely exposed, many species of fish are found here, primarily species that move into the intertidal zone at high tide to feed there. The pace of life in the subtidal region is relatively constant, and only during the lowest ebb tides do these organisms temporarily suspend their activity.

CHAPTER SUMMARY

Intertidal organisms must survive in complex habitats that are alternately exposed and submerged by tides. Organisms that inhabit these areas must be able to tolerate radical changes in temperature, salinity, and moisture, as well as the crushing force of waves, in order to survive.

Rocky shores are generally inhabited by organisms that tend to form bands or zones on the rocks. This occurs because the rocks provide a relatively stable surface for attachment. The supratidal, or splash, zone of a rocky shore receives only the little moisture delivered by sea spray. The intertidal zone is the area that is alternately exposed and submerged by tides. Organisms that live in the intertidal zone display specific adaptations to deal with the problems of desiccation and wave shock. The subtidal zone is generally covered by water even at low tide. Since this zone is usually more stable, it supports a greater diversity of organisms than the others. In addition to physical factors, biotic factors such as grazing, predation, and competition all play an important role in determining the distribution of organisms on the rocky shore.

Tidepools are areas that trap water when the tide recedes, preventing organisms that inhabit them from becoming fully exposed to the air at low tide. However, life in tidepools presents a unique set of problems. Organisms that inhabit tidepools must be able to tolerate abrupt changes in temperature, salinity, pH, and oxygen content of the water.

The distribution and types of organisms found on sandy shores are greatly influenced by wave action. Sandy shores can be divided into zones based on the amount of tidal coverage they receive. In addition to horizontal zonation, vertical zonation is also exhibited by sandy shores, based on the amount of water the sand retains.

In the supratidal zone of sandy shores, most of the living organisms are buried in the sand. The most numerous are the microscopic organisms that comprise the meiofauna. The intertidal zone is the area of shore that is exposed at low tide and submerged at high tide. Organisms that live in this region are predominantly burrowers. During low tide, they remain hidden in their burrows. At high tide, they emerge and forage for food. Like rocky shores, the subtidal zone of sandy shores is exposed only during spring tides and is truly a marine environment. Where the bottom is bare sand, burrowing organisms predominate. In other areas where the bottom is covered with seagrass, rocks, or coral rubble, a greater diversity of organisms can be found.

SELECTED KEY TERMS

zonation, *p. 260*

splash zone, *p. 260*

supratidal zone, *p. 260*

intertidal zone, *p. 260*

subtidal zone, *p. 260*

byssal threads, *p. 261*

wave shock, *p. 261*

tidepool, *p. 265*

infauna, *p. 269*

meiofauna, *p. 269*

QUESTIONS FOR REVIEW

MULTIPLE CHOICE

1. A prominent herbivore that can be found grazing on algae at the lower edge of the supratidal zone of rocky shores is the
   a. mussel
   b. barnacle
   c. periwinkle
   d. ochre sea star
   e. rock crab

2. A major competitor of barnacles for space on rocks is the
   a. oyster
   b. mussel
   c. periwinkle
   d. rockweed
   e. rock crab

3. The low body profile of animals that live on intertidal rocks is an adaptation that protects against
   a. sunlight
   b. high temperatures
   c. desiccation
   d. predation
   e. wave shock

4. The most common alga in the intertidal zone of temperate rocky shores is
   a. *Fucus*
   b. *Laminaria*
   c. *Sargassum*
   d. kelp
   e. red alga

5. The most important factor in determining the distribution of life on a sandy beach is
   a. temperature
   b. salinity
   c. pH
   d. wave action
   e. sediment characteristics

6. On temperate and tropical sandy beaches, the dominant animal in the supratidal zone is the
   a. periwinkle
   b. sea star
   c. sea urchin
   d. ghost crab
   e. mole crab

7. Most organisms that inhabit the intertidal zone of a sandy beach are
   a. predators
   b. grazers
   c. burrowers
   d. multicellular algae
   e. surface dwellers

8. The vertical zone of the sandy beach that supports the highest number of organisms is the zone of
   a. retention
   b. saturation
   c. resurgence
   d. dry sand
   e. drying sand

SHORT ANSWER

1. What characteristics are exhibited by organisms that live in the rocky intertidal zone?

2. What environmental challenges are encountered by organisms that live in a tidepool?

3. Explain how rough periwinkles avoid desiccation.

4. Describe some of the adaptations exhibited by organisms inhabiting rocky coasts that help them survive wave shock.

5. Describe the vertical zones found in the sandy beach.

THINKING CRITICALLY

1. Do you think pollutants entering from the ocean would have a greater effect on rocky shores or on sandy shores? Explain your answer.

2. What abiotic factor is probably most important in terms of influencing the number of organisms that can inhabit a rocky shore?

3. What kinds of organisms that inhabit sandy shores would be least affected by the recreational use of beaches?

SUGGESTIONS FOR FURTHER READING

Connell, J. H. 1961. The Influence of Interspecific Competition and Other Factors on the Distribution of the Barnacle *Chthamalus stellatus, Ecology* 42:710–723.

Gore, R. 1990. Between Monterey Tides, *National Geographic* 177(2):2–43.

Horn, M. H. and R. N. Gibson. 1988. Intertidal Fishes, *Scientific American* 258:64–70.

Koehl, M. A. R. 1982. The Interaction of Moving Water and Sessile Organisms, *Scientific American* 247:124–134

Moore, P. G. and R. Seed. 1986. *The Ecology of Rock Coasts,* Columbia University Press, New York.

Morell, V. 1995. Life on a Grain of Sand, *Discover* 16(4):78–86.

Peteron, C. H. 1991. Intertidal Zonation of Marine Invertebrates in Sand and Mud, *American Scientist* 79:236–248

Wolcott, T. G. and D. L. Wolcott. 1990. Wet Behind the Gills, *Natural History* 99(10):46–55.

Chapter 15

Coral Reefs

KEY CONCEPTS

1. Coral reefs are found in warm, clear water to depths of 60 meters.

2. The three major types of coral reef are fringing reefs, barrier reefs, and atolls.

3. The physical environment determines which species of coral will be found on a given reef or part of a reef.

4. Hard corals are responsible for the large colonial masses that make up the bulk of a coral reef.

5. Reef-forming corals rely on symbiotic dinoflagellates called *zooxanthellae* to supply nutrients and to produce an environment suitable for formation of the coral skeleton.

6. Coral reefs are constantly forming and breaking down.

7. The primary producers on coral reefs are mostly symbiotic zooxanthellae and algae.

8. Coral reefs provide food and habitat for a large number of species.

9. Inhabitants of coral reefs display many adaptations that help them to avoid predation or to be more efficient predators.

10. Coral reefs are huge, symbiotic complexes full of intricate interdependencies.

Located in some of the oceans' clear, blue, warm waters are complex communities teeming with life known as *coral reefs*. Like underwater cities, these communities are crowded with animals representing virtually every major animal phylum. Nowhere else on earth can you find living organisms with such spectacular colors or fantastic shapes as on coral reefs. Corals are both living members of their ecosystem and vital components of its structure. In fact, it is the complexity of the reef's physical and biological structure that provides the abundance of habitats and niches that results in the enormous diversity of animal life.

Interestingly, the ocean water surrounding reefs contains little in the way of nutrients. It is, in fact, this lack of nutrients that gives the water its blue color. Yet on the reefs beneath the water's surface there is an abundance of life, most notably consumers. For years biologists were puzzled as to how so many consumer organisms could be supported in a community that appeared to contain so few primary producers. The solution to the problem can be found in the magnificent corals that make up this intricate world, for it is the corals themselves that make possible the abundance of life found on reefs.

The link between coral animals and primary production is a symbiotic relationship involving the coral animals and dinoflagellate species known as *zooxanthellae*. Together with benthic algae they form the basis of a community upon which all the other reef animals depend for food, shelter, or both. Far from being barren wastelands, coral reefs are one of the most productive ecosystems on earth owing to the intricate network of symbiotic relationships and specializations that occur there.

THE DISTRIBUTION OF CORAL REEFS

Coral reefs are some of the most complex biological communities on earth. The diversity in these communities is rivaled only by that in tropical rain forests. Like rain forests, coral reefs are tropical in distribution. The coral animals responsible for these massive structures are anthozoan cnidarians that exhibit the polyp body form (see Chapter 8). The physical and nutritional requirements of these animals and their symbionts restrict their distribution to a rather narrow range. Coral reefs are found in areas where the water is warm, generally not less than 20°C (annual average), and clear. These requirements restrict coral reefs to tropical and semitropical seas (Figure 15–1). Coral reefs are found in shallow water that barely covers the tops of the corals to depths of about 60 meters (200 feet). The coral polyps that form reefs maintain an important symbiotic relationship with dinoflagellates known as *zooxanthellae* that require sunlight for photosynthesis. The depths at which coral reefs can form is limited by the depth to which sunlight can penetrate to power the vital process of photosynthesis.

TYPES OF CORAL REEFS

Coral reefs can be divided into three categories on the basis of their structure and their relationship to the underlying geologic features. **Fringing reefs** (Figure 15–2a) are found close to and surrounding newer volcanic islands as well as bordering continental land masses. They are located directly offshore and project outward to the sea. Fringing reefs are the most common type of reef.

Another reef type is the **atoll** (Figure 15–2b). Atolls are usually somewhat circular in shape with a centrally located lagoon. They are formed on top of the cones of submerged volcanos or where islands with barrier reefs subside below sea level, leaving only the reef around a large lagoon. Alexander Agassiz originally proposed the former theory, while the latter was proposed by Darwin. The water in the lagoon is not isolated but is connected to the open sea by breaks in the reef. Over time, parts of the reef may become exposed and eroded by the action of wind and waves. The sand formed by these physical processes can be the basis for the formation of an island, and frequently islands are found surrounding the lagoon. More than 300 atolls are located in certain parts of the Indian and Pacific oceans, whereas only 10 atolls are found in the Atlantic Ocean. The Atlantic Ocean supports fewer reefs because the water is generally too cool and turbid.

Barrier reefs (Figure 15–2c) are separated by a lagoon from the land mass with which they are associated. Those surrounding volcanic islands are formed from subsiding islands with fringing reefs. The lagoon forms between the island and the reef. The largest barrier reef is the Great Barrier Reef of Australia, which runs over 2,000 kilometers (over 1,000 miles) along the northeastern coast of Australia to New Guinea (Figure 15–2d). This reef is so large that it is even visible to astronauts orbiting the earth in the space shuttle. The second largest barrier reef is located in the Caribbean Sea off the coast of Belize.

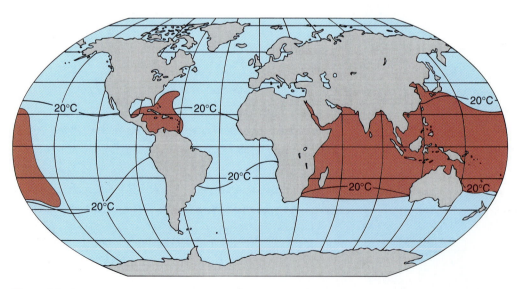

Figure 15–1

Distribution of Coral Reefs. Coral reefs (brown areas) are restricted to the tropics and subtropics, where the water is both warm and clear.

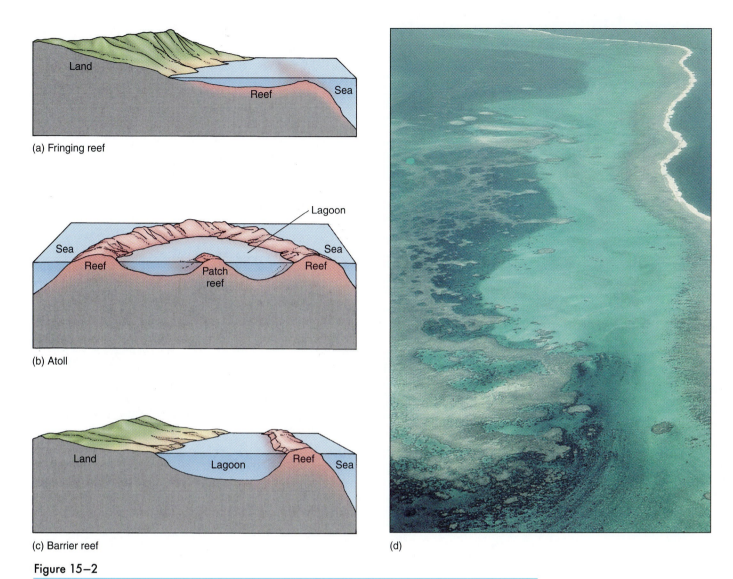

(a) Fringing reef

(b) Atoll

(c) Barrier reef

(d)

Figure 15–2

Types of Coral Reefs. (a) Fringing Reef. A fringing reef is connected directly to the shore. (b) Atoll. An atoll is a circular reef that encloses a lagoon. Atolls usually form on the cones of extinct volcanos. (c) Barrier Reef. A barrier reef is separated from the shore by a lagoon. (d) The largest barrier reef in the world, the Great Barrier Reef, runs along the northeastern coast of Australia. (d, Edward Hodgson/Visuals Unlimited)

In addition to these main reef types, there are several special reef arrangements. **Table reefs** are small reefs found in the open ocean that have no central islands or lagoons. **Patch reefs** are small patches of reef that are located in lagoons associated with atolls and barrier reefs. When patch reefs occur, they are usually quite numerous.

REEF CHARACTERISTICS

All three types of reef share several common characteristics. On the seaward side, the reef rises from the lower depths of the ocean to a level just at or just below the surface of the

water. This portion of the reef is called the **reef front** or **fore-reef** (Figure 15–3a). The slope of the reef front can be either gentle or quite steep. In some cases, the reef front forms a vertical wall referred to as a *dropoff.* The reef front does not generally form a solid sea wall; instead, finger-like projections of the reef protrude seaward. This arrangement, called a **spur and groove formation,** disperses wave energy and prevents damage to the reef and its inhabitants (Figure 15–3b). Between some of these projections (spurs) lie sand-filled pockets (grooves) that allow sediments to be channeled down and away from the living coral surface and provide a habitat for many species of burrowing organism.

The highest point on the reef is called the **reef crest,** and the area opposite the reef front levels off and is referred

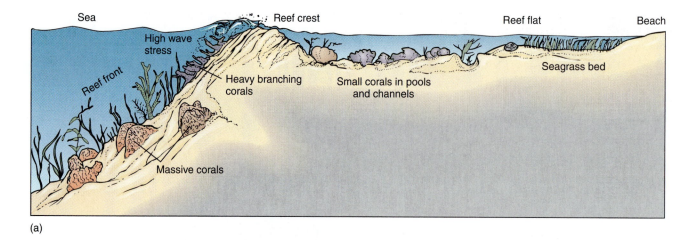

(a)

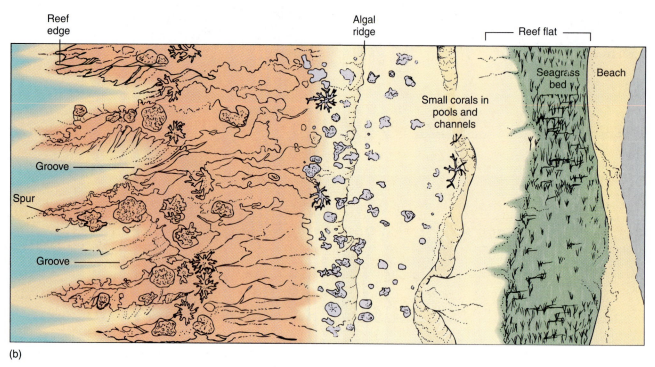

(b)

Figure 15–3

Reef Characteristics. (a) The reef front rises from the ocean's depths to the reef crest. The area behind the reef is the reef flat. (b) A view of a reef from above looking down shows the characteristic spur-and-groove formation.

to as a **reef flat** or **back reef.** This area exhibits a high degree of variability. The flat may be short or several hundred meters long. It may be shallow or cut through by channels several meters deep. The bottom of the flat may consist of rock, sand, coral rubble, or some combination of these. Seagrass beds are commonly found in the reef flat area. The reef flat of fringing reefs ends at the shoreline. The reef flat of atolls and barrier reefs descends into the lagoon.

Different areas of a reef support different species of coral as well as other organisms. Coral populations on the reef front are usually found at depths of 10 to 60 meters (33 to 200 feet). Massive, dome-shaped brain corals (*Diploria*) and columnar pillar corals (*Dendrogyra*) are found on intermediate slopes. Below this region, species that form

platelike formations, like lettuce leaf and elephant ear coral (*Pectinia, Pavona,* and *Agaricia* species; Figure 15–4a), predominate. Higher up on the reef, where wave stress is greatest, branching species of coral are found. Wave stress is one of the most important factors in determining what species of coral and other organisms can occupy the reef crest. In the Caribbean, this upper area is the habitat of elkhorn coral (*Acropora palmata;* Figure 15–4b). This coral's heavy, spreading branches project toward the sea, where they break the force of incoming waves.

In more protected areas behind the reef front, the deeper and less turbulent water supports more delicate species of coral. In the Caribbean, this is frequently staghorn coral (*Acropora cervicornis*). In the Indo-Pacific

(a)
(b)

— Elkhorn coral

— Brain coral

Figure 15–4

Types of Coral. (a) Corals that produce large, platelike formations like this lettuce leaf coral are located farther down the reef front, where wave action is minimal. (b) Areas of a reef that receive heavy wave action contain thick, branching corals such as this elkhorn coral, as well as large, solid corals such as brain coral. (a, Jon L. Hawker; b, David J. Wrobel/Biological Photo Service)

region, species such as staghorn, finger (*Stylophora*), cluster (*Pocillopora*), and lace corals (*Pocillopora damicornis*) are prominent. Farther from the reef front in shallow, calmer water are small species of coral such as rose (*Meandrina* and *Manicina*), flower (*Mussa* and *Eusmilia*), and star (*Montastraea*). Species of this type are found associated with both Caribbean and Indo-Pacific reefs. On some

reefs there are signs of zonation as certain species of coral gradually appear and disappear along the reef.

There are approximately 20 genera of coral represented on Caribbean reefs, and most Caribbean reefs are dominated by 10 species of hard coral and the hydrozoan coral *Millepora*, also known as *fire* or *stinging coral* (Figure 15–5a). By contrast, there are more than 80 genera of coral in the Indo-

(a)

(b)

Figure 15–5

View of Coral Reefs. Coral reefs are colorful, complex communities. (a) A panoramic view of a Caribbean reef. (b) A panoramic view of a Pacific reef. (a and b, Norbert Wu)

Pacific, and the number of species there is equally diverse (Figure 15–5b). For instance, in the Caribbean, there are only three species of the genus *Acropora,* whereas there are at least 150 species of this genus in the Indo-Pacific. The Great Barrier Reef of Australia is inhabited by over 200 species in the genus *Acropora.* Cooler water temperatures and higher turbidity are the primary reasons that there are fewer coral species in Caribbean waters. The Indo-Pacific region is also geologically older and covers a larger area than the Caribbean; therefore more species are to be expected.

THE MAKING OF A CORAL REEF

Corals

The principal reef-building organisms are **hard corals** (scleractinian corals). Colonies of hard corals begin when a planula larva settles down and attaches itself to some solid surface, frequently the dead remains of other coral colonies (Figure 15–6). Once attached, the planula develops into a coral polyp and secretes a cup of calcium carbonate around

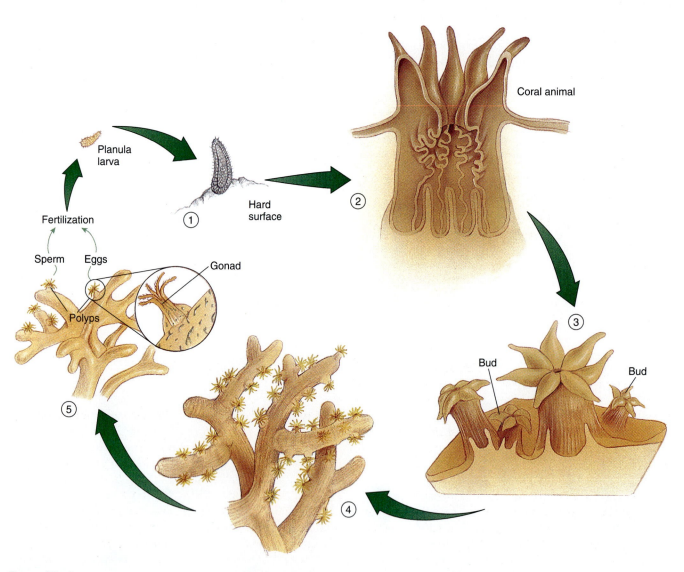

Figure 15–6

Formation of a Coral Colony. (1) A coral colony begins when a planula larva attaches to a hard surface. (2) The planula develops into a polyp, and the polyp secretes a cup of calcium carbonate. (3) The polyp reproduces asexually by budding, and each new polyp forms a calcareous cup around its body. (4) The colony continues to grow as the new polyps repeat the budding process. The shape of the colony is genetically determined. (5) As the colony continues to grow, some polyps develop gonads. The sperm and eggs produced by these polyps are shed into the surrounding water, where the eggs are fertilized. The fertilized eggs develop into new planula larvae that begin the process of colony formation again.

its body. The polyp reproduces by budding, and the new polyps that are formed by this process also form cups of calcium carbonate. In turn, these new polyps will form buds, and, as the process of asexual reproduction continues, the colony grows. Each polyp remains attached to its parent by a thin tissue that covers its limy skeleton. This tissue is usually quite colorful; red, pink, yellow, green, purple, or brown are common. The colony grows both upward and outward and assumes the shape that is genetically determined for the species, with some variation due to environmental stresses. As the colony grows, some polyps develop gonads and are capable of sexual reproduction. These polyps release their sperm and eggs into the surrounding seawater, where fertilization occurs.

With the possibility of several species of coral releasing eggs and sperm at the same time, how does the right sperm find the correct egg? At least two processes seem to be involved. Both sperm and egg cells have protein molecules on their surfaces that identify them as belonging to a particular species. As a result, most of the time a sperm cell will only attach to an egg cell of the same species. The fertilized egg ultimately develops into a planula larva that can establish a new coral colony. In some closely related species of coral, the sperm of one species may fertilize the egg of another. In this case, differences in chromosome number or other genetic differences usually prevent further development.

Fragmentation appears to be a common form of asexual reproduction for branching corals. Some branching corals, such as staghorn coral, are fragile, and water turbulence resulting from storms may cause many broken branches and toppled colonies. If the storms are not so severe that the fragments are killed, many of them can reattach and regrow, forming new colonies.

Like other cnidarian polyps, the soft, cylindrical body of a coral polyp has a ring of tentacles surrounding the mouth. The tentacles of coral polyps are strewn with stinging cells, and any small animal that brushes against them is immediately paralyzed and passed through the mouth into the digestive cavity. Some coral polyps have short tentacles that are covered with cilia. The cilia beat toward the mouth, creating a current that draws small organisms toward the tentacles and then into the mouth. When the polyp is not feeding, the cilia reverse the direction of their beat so that silt and other debris is moved away from the polyp. During periods of inactivity or when they are disturbed, polyps withdraw their tentacles and body wall into their calcareous cup. Since the polyps of many coral species are nocturnal feeders, day visitors to a reef rarely have the opportunity to view the living animal (Figure 15–7).

Living within the tissues of the coral polyp are symbiotic zooxanthellae. The zooxanthellae provide nutrients to the polyps in the form of carbohydrates, primarily glycerol, and they reduce levels of carbon dioxide, making conditions more suitable for the formation of the coral skeleton. The presence of zooxanthellae are so important to the life

Figure 15–7

Coral Polyp Feeding. At night coral polyps extend their tentacles and feed on plankton. (Biological Photo Service)

of the coral polyp that reef-building corals cannot grow any deeper than sunlight can penetrate to supply the photosynthetic needs of their symbionts. Experiments have shown that corals kept in the dark, or deprived of their symbionts, deposit calcium carbonate at a significantly slower rate than normal corals. For its part, the polyp provides a suitable habitat for the zooxanthellae and a variety of nutrients. We will examine this relationship in more detail later in this chapter when we discuss reef productivity.

Not all corals are reef builders. Some species of hard coral exist as single, solitary polyps, such as the mushroom and feather, or fungus, corals (*Fungia;* Figure 15–8). Some temperate species of coral form small colonies only. **Soft corals** lack the hard outer coverings of calcium carbonate associated with hard corals. Each of these species, however, still contributes in its own way to the formation of a reef, either by precipitating calcium from seawater or by binding smaller fragments of coral and shell together.

Figure 15—8

Solitary Hard Corals. Hard corals like this mushroom coral are formed by single polyps that do not form colonies. (Jeffrey Rotman)

Reef Formation

The process of reef formation involves both constructive and destructive phases. It is not a simple matter of new coral building on top of old existing coral but rather a complex series of events involving many animal and algal species. The larger hard corals form the basic material of

the reef, while the skeletons of other smaller species provide the "cement" to hold the pieces together.

The destructive phase of reef formation is usually underway well before the death of a coral colony. Any time the coral surface is exposed, it is quickly attacked by boring organisms such as boring clams or sponges (Figure 15–9). These organisms usually attack the underside of large coral stands, since these areas do not receive enough sunlight to support the growth of coral polyps and are relatively bare surfaces. As the boring organisms riddle the base with tunnels, the coral stand weakens, and a storm or heavy ocean surge will eventually topple the coral, crushing and killing many of the polyps in the process. The resulting coral fragments provide a surface that is quickly colonized by coralline algae and encrusting bryozoans. These organisms deposit more calcium carbonate onto the basement surface.

The length of the destructive phase is determined by the amount of time it takes for the fine sediment produced by the boring organisms to settle on the surface of the coral and the bottom. As the debris from boring accumulates, it eventually covers and smothers the boring organisms and the process of boring stops. If this occurs quickly, the chunks of coral left over from the process will be relatively large. If the process takes a longer time, the large corals will be broken up into much smaller pieces.

Along with the fine calcium sediment produced by boring, the calcareous shells and body coverings of a variety of organisms, including calcareous red (*Lithothamnion*,

---

MARINE BIOLOGY AND THE HUMAN CONNECTION

Using Coral to Repair Skeletons

Doctors are discovering that coral is an excellent material for replacing bone in several types of reconstructive surgery. Physicians at the University of South Florida in Tampa are currently using coral to repair fractured bones that require a graft. Plastic surgeons at other hospitals have successfully used dead coral as a replacement for portions of jaw bones that have been destroyed by cancer. Coral is becoming popular because it is quite compatible with bone and, once in place, fuses almost per-

fectly with human bone. Coral species belonging to the genera *Porites* and *Goniopora* are currently used because they have a physical structure consisting of numerous channels that is almost identical to bone. The maze of channels in the porous coral provides binding sites for projections of bone and spaces for blood vessels from the grafted bone. This arrangement allows for a strong fusion of the two materials and a permanent seal.

In traditional bone replacements, physicians use either bone from other

parts of a patient's body or bone taken from cadavers. However, only a limited amount of bone can be removed from other sites in an individual's body, and with cadaver bone there are the risks of rejection and of AIDS. Coral does not present any of these problems. Coral is widely available, and, because it is essentially inert, it cannot transmit disease or trigger a rejection reaction in the human body.

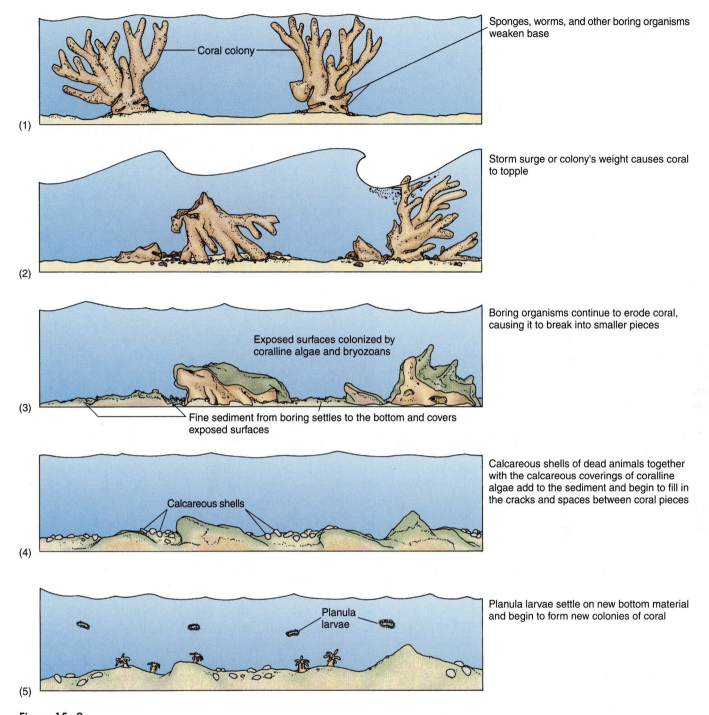

(1) Coral colony — Sponges, worms, and other boring organisms weaken base

(2) Storm surge or colony's weight causes coral to topple

(3) Exposed surfaces colonized by coralline algae and bryozoans — Boring organisms continue to erode coral, causing it to break into smaller pieces — Fine sediment from boring settles to the bottom and covers exposed surfaces

(4) Calcareous shells — Calcareous shells of dead animals together with the calcareous coverings of coralline algae add to the sediment and begin to fill in the cracks and spaces between coral pieces

(5) Planula larvae — Planula larvae settle on new bottom material and begin to form new colonies of coral

Figure 15—9

Reef Cycle. (1) The destructive phase of a reef cycle begins when boring animals burrow through the bases of large corals, weakening them. (2) The weight of the coral and/or storm surges cause the coral to topple, killing the polyps. (3) The coral is broken into smaller pieces, and bryozoans and coralline algae begin to colonize the exposed surfaces. (4) Shells of dead organisms contribute to filling in cracks and crevices. Coralline algae help to cement the pieces together to form a new, solid surface. (5) Planula larvae begin to settle and form new coral colonies.

Porolithon, Lithophyllum) and green (*Halimeda*) algae, are added to the sediment that settles into the cracks and crevices of the reef. Eventually this material fills in the spaces between the larger pieces of coral with a wet calcium material that acts as cement to hold the larger pieces in place. Sponges and calcareous algae also contribute to the cementing process that will produce large areas of solid calcium carbonate where planula larvae can settle and begin to produce new stands of coral. Calcareous red and green algae also contribute to the reef sands found in pockets among the coral, and they may dominate reef crests, especially in the Indo-Pacific region.

Reef Growth

Reefs grow both horizontally and vertically. Vertical growth, however, is limited by water depth and light. Core samples of most modern reefs indicate that the reef platforms of many are located below the photic zone. This observation can be explained by changes in sea level, or subsidence, that have occurred over the time that the reef formed. During the last glacial period, large amounts of seawater were tied up in glaciers, and the sea level was about 120 meters (396 feet) lower than its current level. As the glaciers melted, the sea level gradually rose again, and the reefs grew as well.

Reef growth in the vertical direction is the result not only of corals' following the rising sea levels, but also of wave action. Reefs that receive large amounts of wave energy grow more slowly and more compactly than those that receive less wave energy. Slow-growing reefs are usually dominated by coralline algae in the Indo-Pacific region and coralline algae and fire coral in the Caribbean.

Reefs with the thickest platforms result from subsidence of the bottom sediments. Subsidence has been an important factor in the formation of several atolls and barrier reefs, including the Great Barrier Reef of Australia. Typically, atolls form on the top of sunken volcanos (see Chapter 3), where the rate of subsidence of a volcano is matched by the upward growth of coral.

REEF PRODUCTIVITY

The water that surrounds and bathes coral reefs is not very rich in nutrients. The presence of suspended material and organisms in productive waters reduces the amount of light that can penetrate the water. As a result, important wavelengths of visible light disappear rapidly below the surface, and the water appears green. Since corals require clear water so that there will be enough light to support their photosynthetic symbionts, it is not surprising to find that coral reefs develop only in water that contains minimal amounts

Figure 15—10

Coral Polyp and Symbionts. Symbiotic dinoflagellates known as *zooxanthellae* (they appear as greenish-brown speckles in the polyp in this photo) live within the tissues of coral polyps and are responsible for the brownish color of many live corals. The zooxanthellae provide the polyps with carbohydrates and remove carbon dioxide. In return, the polyps supply the zooxanthellae with nitrogen compounds and carbon dioxide. (Animals Animals)

of plankton and is quite blue. Although the water that bathes the coral reef is nutrient poor, the reef itself is one of the most productive of all marine ecosystems.

The key to the high productivity of coral reefs is the symbiotic relationship between zooxanthellae and many of the reef's inhabitants, especially the hard coral polyps (Figure 15–10). Other reef animals, such as giant clams and their relatives (Tridacnidae), also support an enormous number of symbiotic zooxanthellae. Symbiotic zooxanthellae can be as much as three times more productive than the equivalent amount of phytoplankton. Coral animals and their symbionts are so intimately connected that they behave as a single organism. As previously mentioned, zooxanthellae provide the corals with nutrients essential for reef building. A large portion of the zooxanthellae's tremendous photosynthetic output is channeled back to their host, and much of the oxygen produced by the photosynthesis is consumed by the coral polyps. Zooxanthellae manufacture a variety of amino acids, sugars, and other organic compounds that are absorbed directly into their host's tissues. In turn, carbon dioxide and ammonia, wastes of the coral animal's metabolism, are perfect nutrients for zooxanthellae. The symbionts absorb these materials directly from the animal's tissues so efficiently that several species of coral release virtually no nitrogen wastes to the surrounding water. Some compounds may be passed back and forth between polyp and zooxanthellae several times, each time being altered or combined with other molecules. It is not clear what regulates the exchange of nutrients, considering the fact that many of the materials taken by the coral could just as easily serve as growth factors for the zooxanthellae. It ap-

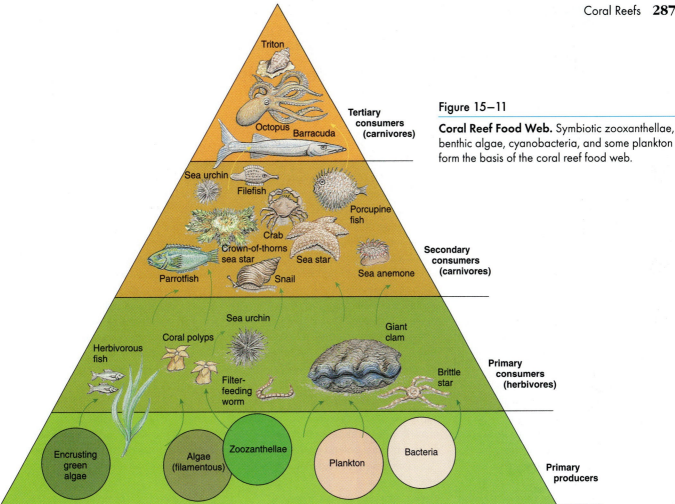

Figure 15–11

Coral Reef Food Web. Symbiotic zooxanthellae, benthic algae, cyanobacteria, and some plankton form the basis of the coral reef food web.

pears that the coral polyps somehow stimulate the zooxanthellae to release the nutrients, but the nature of this process is not clear.

Coral animals actively position their algal symbionts to receive maximum exposure to sunlight. During the day, the polyp's tissue layer containing the zooxanthellae is spread out like a carpet to catch as much sunlight as possible. At night, when photosynthesis is not carried out, the tissue containing the zooxanthellae is withdrawn, and the coral polyp extends its feeding tentacles and preys upon plankton. In shallow water, where sunlight is abundant, coral tissues contain several layers of zooxanthellae, packing in as many zooxanthellae per unit volume as possible. The colonies formed by these species are generally quite branched, allowing maximum numbers of zooxanthellae to be exposed to sunlight. In deeper water, where sunlight is less available, coral colonies tend to be flatter, forming large horizontal tables over which the polyps can spread their tissues containing the zooxanthellae.

When coral polyps are stressed, they expel the zooxanthellae from their tissues, a phenomenon known as **bleaching.** Lacking zooxanthellae that give coral its characteristic yellow-brown color, the coral colonies appear white, like the bleached specimens in souvenir shops, thus the term *bleaching.* Without symbiotic zooxanthellae, coral polyps will cease to grow and will die within a few months. It ap-

pears that slight increases in water temperature (heat stress) is one of the most important causes of bleaching. During the 1980s marine biologists recorded widespread coral bleaching throughout the world, a process that is unfortunately continuing today. Some researchers think that a combination of global warming due to increased carbon dioxide in the atmosphere and El Niño events are the primary culprits. Whatever the cause of bleaching, if environmental conditions return to a more normal range, the coral polyps will collect new zooxanthellae and continue to grow and thrive.

In addition to deriving nutrients from symbiotic zooxanthellae, some coral polyps use sticky strands or nets of mucus to trap bacteria, plankton, and detritus on which they feed. Soft corals frequently lack zooxanthellae and probably employ methods such as absorption and mucous nets to gather food. The coral mucus is an important source of food for many reef organisms. Other coral species that lack zooxanthellae are covered with numerous microscopic projections that probably allow them to absorb scarce, dissolved nutrients from the seawater.

Many organisms, including crabs, shrimp, molluscs, and fishes, depend on the abundant coral mucus for energy-rich triglycerides and fatty acids. Some species of benthic shrimp can live almost entirely on a diet of coral mucus. The corals themselves provide food for some echinoderms, molluscs, crabs, and fishes (Figure 15–11).

Why Are Florida's Reefs Disappearing?

The Florida Keys are a chain of islands that stretches from Key Biscayne, five miles off the coast of Miami, 135 miles away to Key West (see Figure 15–Aa). The waters that surround these islands contain the only coral reefs that are associated with the continental United States. Florida's reefs contain fewer species of coral and are not as complex as reefs farther south in the Caribbean or in the Indo-Pacific region. The reason for this is that they are located fairly far north, and during winter the water gets colder than most species of coral can tolerate. Historically, the southernmost reefs near Key West have been the most diverse and best developed, whereas those farther north showed less diversity and coral growth. In recent years, this pattern has changed dramatically.

Although all of Florida's reefs are protected by law, they are showing signs of damage. Researchers James Porter and Ouida Meier of the University of Georgia have discovered that the number of coral species on these reefs, as well as the area covered by live coral, is declining. During a study conducted between 1984 and 1989, they found that some reefs lost as much as 29% of their coral species, and five out of six reefs surveyed lost as much as 45% of their coverage. If this trend continues, the researchers estimate that some of Florida's reefs will be devoid of living coral by the year 2000.

Of the reefs studied, the ones most affected were those in the southern end of the chain, around Looe Key. In the past, these reefs were some of the best developed and fastest growing, and they are still a popular dive site. Surprisingly, of the six study sites examined, the only site that showed increases in live coral coverage was the northern site near Key Biscayne. Because of its proximity to metropolitan Miami, researchers expected

this reef to show the greatest amount of damage if the cause of the coral damage was some type of coastal pollution. Since coastal pollution did not seem to be the cause of the problem, Porter and his colleagues were puzzled.

A breakthrough in the study occurred when Porter and his co-workers combined their efforts with other researchers in the area. Rather than just concentrating on the immediate area around the reefs, the researchers began to look at all of southern Florida and to research environmental changes that had occurred in the area over several years prior to the time of the reef study. What they discovered was an interesting series of environmental changes that may be producing the problems on the reefs.

Florida Bay, located at the tip of the peninsula, was at one time a productive estuary with an average salinity of 18‰. Currently the salinity averages 42‰ and sometimes reaches as high as 70‰ (recall that normal seawater has salinities in the range of 32 to 36‰). As a result of the high salinities, almost 90% of the seagrass that carpeted the estuary has died. As the dead seagrass decomposed, nutrient levels in the bay increased and the water became cloudier. The increase in nutrient levels was accompanied by a decrease in dissolved oxygen (see Chapter 6). The decrease in oxygen caused many animals to die and others to move away. The commercial fisheries for shrimp, lobster, and stone crabs, an industry worth $250 million a year, has collapsed as a result of these changes.

How did these changes occur, and what, if anything, do they have to do with the changes in the coral reefs? To answer these questions, Porter and his associates have proposed the "Florida Bay Water Hypothesis." In the past, much of the rain that fell over central

and southern Florida collected in a large lake called Lake Okeechobee. From the air, the lake appears to be landlocked, but obviously all of the water that collected there had to go somewhere. Because the southern shore of the lake is lower than the other shores, the overflowing water once moved like a large, shallow river over and through the Everglades, a vast wetland that covers the southern end of the Florida peninsula. The water eventually reached Florida Bay, where it mixed with seawater to form a productive estuary (Figure 15–Aa). In 1961, in order to prevent flooding of low-lying wetlands and to make more land available for agricultural use, the Army Corps of Engineers began construction of an elaborate system of canals and levees. As a result of this action, the northern portion of the Everglades is now completely dry. The water that once drained into that area is now diverted to the east, where it empties into Biscayne Bay. In turn, Florida Bay's water supply has also been drastically reduced, which accounts for the increased salinity and the resultant damage to the estuary (Figure 15–Ab).

The winds and current patterns of the Gulf of Mexico cause water that accumulates in Florida Bay to spill out through the Keys and over the reefs. Before the freshwater was diverted, water in Florida Bay had a lower salinity than seawater and thus was less dense. As it moved out of the bay, it stayed on the surface and passed over the reefs, eventually mixing with the ocean water of the Atlantic. Since the water now has a higher salinity than seawater, it is more dense, and as it moves through the Keys, it sinks and flows directly over the corals. The combination of high salinity, low oxygen, high nutrient levels, and high turbidity kills some coral species while severely weakening others.

Recall that this is only a hypothesis and that it has not as yet been tested. If the hypothesis is true, diverting water from Lake Okeechobee back to Florida Bay could reverse the situation. Most of the diverted water is not used commercially; it is simply dumped into Biscayne Bay. What little is used could be replaced by other sources. If alternative methods of flood control could be implemented, then a great deal of water could be redirected back to the Everglades and, ultimately, Florida Bay. Such alternatives have been proposed but would have an estimated price tag of $1 billion. Thus, more testing of the hypothesis will have to be done before state and federal agencies will step in to stem the damage to the coral reef. These studies are now underway, and hopefully they will produce concrete results before Florida's reefs are gone forever.

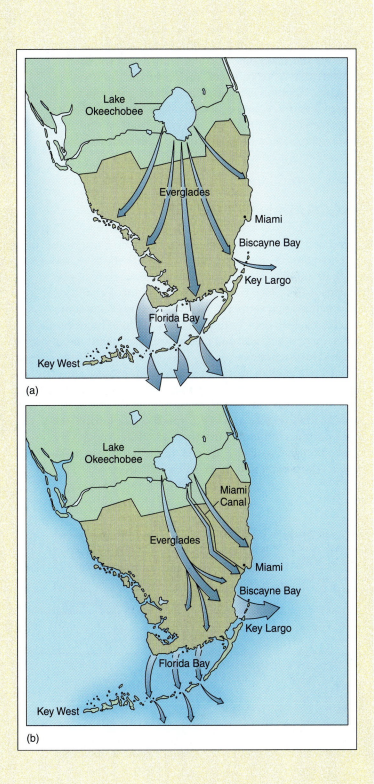

Figure 15–A
The Florida Bay Water Hypothesis.
(a) Before construction of drainage canals in the 1960s, fresh water flowed south from Lake Okeechobee through the Everglades and into Florida Bay, forming a very productive estuary (light blue color). The less dense water from the estuary then flowed through the channels in the Florida Keys out and over the coral reefs.
(b) Diversion of fresh water from the Everglades to Biscayne Bay has caused the water in Florida Bay to become hypersaline (darker blue color). As it flows through the Keys across the reefs, the dense water sinks and damages and kills the corals.

Growing attached to the reef are a variety of benthic algae, especially green algae, that grow in and on the dead portions of calcium carbonate structures and other available surfaces. These algae contribute significantly to primary productivity on the reef. The water surrounding coral reefs is not completely devoid of plankton, and the little plankton that is available represents an important food source for the many filter-feeding and suspension-feeding organisms that inhabit a reef, including the coral polyps themselves. Sponges are one of the largest components of the many benthic coral reef fauna. Many of these sponges have a symbiotic relationship with cyanobacteria similar to that between corals and zooxanthellae, thus contributing to the overall reef productivity.

THE CORAL REEF COMMUNITY

Where coral reefs form, hundreds of other organisms are found. Not only does coral provide a basis for the reef food web, their colonies also provide shelter for an array of resident organisms. Close up, the reefs are multicolored, complex structures that support a variety of beautiful and sometimes bizarre creatures. In this complex world of branches, boulders, caves, tunnels, and mazes swarms a multitude of fishes and invertebrates.

Many reef animals are either sessile or move very slowly as adults. Instead of actively pursuing their prey, these animals sift through the water currents and sediments in search of food (Figure 15–12). Large sponges inhabit the deeper portions of the reef front, while smaller species are found in the waters behind the reef. In the Caribbean, large sea fans (*Gorgonia*) and other soft coral species are con-

(a)

(b)

Figure 15–13

Soft Corals. (a) Sea fans like this yellow one are prominent members of the Caribbean reef community, while (b) delicate branching soft corals such as these are more prevalent on Pacific reefs. (a and b, Jon L. Hawker)

Figure 15–12

Reef Filter-Feeders. Reef-dwellers like these annelid Christmas tree worms filter their food from the surrounding water. (Jon L. Hawker)

spicuous occupants of the reef (Figure 15–13a). Although also found on Indo-Pacific reefs, they are not as conspicuous there. On these reefs, there are larger numbers of beautiful, branching soft corals (Figure 15–13b).

Sea anemones are another type of cnidarian found on reefs (see Chapter 8). Some species of these large polyps will achieve diameters of 1 meter (3 feet) on some coral reefs. Using the stinging cells located on their tentacles, sea anemones feed on plankton, small crustaceans, and small fishes. Sea anemones usually remain fixed in one place by firmly holding the bottom by means of their muscular, mucus-secreting basal disk. If conditions become unfavorable, however, some species will release themselves and, by undulating their basal disk, crawl in a snail-like fashion along the bottom at the rate of 10 centimeters (4 inches) per hour until they find a more suitable area. Other species may turn themselves upside-down and walk on the tips of their tentacles, while others lie on their sides and ripple their bodies, moving along like oversized inchworms.

Several species of beautifully colored featherduster worms and crinoids, or feather stars, also feed on available plankton. Featherduster worms, or fanworms, are represented on reefs by two families (see Chapter 8). The serpulid worms, which produce tubes of calcium carbonate, contribute to the formation of a reef. Sabellid worms form tubes of sand and small particles that are held together by mucus. They also contribute, in part, to a reef. Not all annelid worms found on a reef, however, contribute to its formation. The palolo worm (*Eunice viridis*) of the Pacific Ocean and its relatives in the Caribbean Sea burrow through the coral, weakening it. Other polychaetes feed on the coral itself, or on the dinoflagellates living in the coral polyps.

Feather stars are echinoderms that attach themselves to coral with structures called **cirri** (see Chapter 9). They feed by forming a basket of mucus with their lacy, cilia-covered tentacles and moving them in such a way as to create a current that carries the food to the animal's mouth. Though generally sedentary, the feather star can move by creeping along on its cirri or using its arms in a swimming movement.

One of the more interesting plankton feeders on the reef is the tiny gall crab (*Domecia*). The female crab goes through several free-swimming larval stages before settling on a section of reef. She then secretes a chemical substance that alters the growth pattern of the coral so that the coral forms a chamber, called a **gall,** in which the crab lives. The openings to the gall are quite small and prevent potential predators from entering. Currents produced by the movement of her gills bring water into the gall, from which the crab extracts food and oxygen. The male of the species leads a more precarious life on the open reef. When the time to reproduce occurs, the male, which is small enough to enter the gall, enters and mates with the female. After hatching, the tiny larvae exit the gall and repeat the life cycle.

Many reef invertebrates are active hunters like mantis and pistol shrimp. Mantis shrimp (Stomatopoda) have forward appendages that are shaped like the forelimbs of a praying mantis (Figure 15–14). Their forelimbs are so sharp that they can cut another shrimp in half when they strike. They hide in crevices and under coral, where they wait for their prey to come close enough to attack. Pistol shrimp (*Alpheus*) use their modified cheliped to produce snapping sounds involved in territorial defense. They sometimes use the sound to stun small fishes on which they feed.

Reefs are home to a wide variety of molluscs. Cowries (Cypraeidae; Figure 15–15a) are mainly nocturnal gastropods that come out at night to graze on algae, sponges, and coral polyps. They do not actually feed on the polyps themselves but on the algae growing on their surface. During the day, these animals hide in cracks and crevices and beneath coral formations. Large helmet snails (*Cassis*) are found partially buried in the sand pockets of the reef. At night they come out to feed on sea urchins and other echinoderms. On Indo-Pacific reefs, the Triton's trumpet snail (*Charonia tritonis*) searches for sea urchins, sea stars, and molluscs that make up its diet (Figure 15–15b). A similar species, *Charonia variegata,* fills the same niche on Caribbean reefs. Many species of cone snail (Conidae) are found on reefs. Some live in the cracks and crevices of a reef, while others are found under rocks and pieces of coral or buried in the sand. These venomous snails use their modified radula to sting potential prey, thus paralyzing it so the snail can devour it. A variety of nudibranchs, sporting amazing colors and patterns, feed on animals that most other predators would not consider (Figure 15–15c). Sponges, anemones, and jellyfish are the regular diet of these naked snails (see Chapter 9).

One of the largest inhabitants of Indo-Pacific reefs is the giant clam (*Tridacna gigas;* Figure 15–16). Growing as long as 1 meter (3 feet) and sometimes weighing more than 180 kilograms (400 pounds), these huge clams contribute a great deal of mass to the reef in the form of their shells. Like other clams, the giant clam is a filter-feeder, but the quantity of plankton available on a coral reef is not great enough to sustain such a large animal. To supplement its energy needs, the clam hosts symbiotic zooxanthellae in its mantle. In this relationship, the clam provides carbon dioxide and nitrogen to the zooxanthellae, which photosynthesize and return oxygen and carbohydrates to the clam. As needed, the giant clam will also "harvest" some of its symbionts to supplement its diet. In many areas of the Pacific Ocean, this magnificent clam is endangered, as it is being overfished as a food source. The giant clam is the largest member of the family Tridacnidae. Other species of this family of bivalves

Figure 15–14

Reef-Dwelling Crustaceans. The anterior appendages of this mantis shrimp are used to capture prey. (Peter Arnold Stock Workbook/Norbert Wu)

(a)

(b)

(c)

Figure 15–15

Reef-Dwelling Molluscs. (a) A tiger cowry from the tropical Pacific feeds on the algae that grow on the reef. (b) This Triton's trumpet snail is attacking a crown-of-thorns sea star, which feeds on corals. (c) Many colorful nudibranchs make their home on the reef.
(a, W. Gregory Brown/Animals Animals; b, Norbert Wu; c, Jeffrey Rotman)

ranging in size from 4 centimeters to 0.5 meter (2 inches to 1.5 feet) are also found on Indo-Pacific reefs. Like the giant clam, they too host zooxanthellae. These dinoflagellates, along with those found in the coral polyps, are an important basic energy source for the food webs of Pacific reefs.

Although seldom larger than 60 centimeters (2 feet) and exceedingly shy of human swimmers, the octopus is one of the coral reef's most formidable predators (Figure 15–17). It has keen eyesight, a well-developed nervous system, and the ability to rapidly change its color in response to its background or as a social signal (see Chapter 9). These characteristics make it a superb, active predator. Octopuses feed on fish, molluscs, and crustaceans, especially crabs. They stalk their prey slowly and, when within range, pounce with amazing speed, enveloping the prey with their tentacles. Drawing the prey to its mouth, the octopus bites it and injects a venom that paralyzes it. Once the prey ceases to fight, the octopus crushes it with its beaklike jaws and tears it into morsels small enough to fit easily into its mouth.

Octopuses are nocturnal hunters, returning by day to a den. This can be a crevice in the reef, a hollowed portion under the coral, or the deserted shell of a large mollusc. The opening to the den is quite small, not much larger than the diameter of one of the octopus's arms. Octopuses have an amazing ability to squeeze through small spaces where potential predators cannot gain access. Outside of the den, one usually finds piles of crab carapaces and mollusc shells, the leftovers of previous meals.

Although octopuses are skilled hunters, moray eels (Muraenidae) are the dominant nocturnal predators. They are found on coral reefs, where they live in cracks, crevices, and holes in the reef, sometimes competing with the octopus for space. At night they come out to feed on shrimp, crabs, octopuses, and other fish. During the day, the barracuda (*Sphyraena;* Figure 15–18) takes its place, patrolling reef waters for schools of fish. Barracudas have elongated, streamlined bodies and long, protruding jaws filled with thin, double-edged teeth. They rely primarily on their sense of sight to locate their prey. Adult barracudas are generally solitary hunters. They will swim into a school of fish and

Figure 15—17

Reef Octopus. This octopus is about to feed on the sand crab that it has just captured. (Fred Bavendam/Peter Arnold, Inc.)

Figure 15—16

Giant Clam. Giant clams like this one on the Great Barrier Reef contribute to the mass of coral reefs with their large, heavy shells. They also maintain a symbiotic relationship with zooxanthellae similar to that of corals, contributing to primary production on the reef. Notice how the opening of the clam is facing upward. This allows the animal to drape its mantle over the edges of its shell so as to give the symbiotic zooxanthellae maximum exposure to sunlight. (Jon L. Hawker)

Figure 15—18

Barracuda. A barracuda patrols the reef off Grand Cayman Island in the Caribbean for fish to feed on. (Biological Photo Service)

strike with lightning speed, swallowing small prey whole. Larger prey are first slashed to pieces by the razor-sharp teeth before they are consumed.

Many echinoderms live in the reef community. Sea urchins graze on the algae that cover some of the rock and coral. Brittle stars scavenge the bottom for food; sea stars feed on molluscs and even the coral polyps themselves. One of the most important coral-eating sea stars is the crown-of-thorns (*Acanthaster planci;* see Chapter 9). This large,

sixteen-rayed animal of the tropical Indo-Pacific feeds on a variety of coral polyps. During the late 1960s there was a large increase in the number of these animals, and between 1965 and 1969, 405 square kilometers (140 square miles) of the Great Barrier Reef were laid bare by the crown-of-thorns sea star. Drastic measures were taken at the time to deal with the potential problem, but the number of sea stars eventually declined naturally. The cause of the crown-of-thorns population explosion is still hotly debated.

EVOLUTIONARY ADAPTATIONS OF REEF DWELLERS

Protective Body Covering

Reef fishes display an amazing array of adaptations that help them to survive in this fiercely competitive community. Some avoid predation by having tough defensive exteriors. Trunkfishes (Ostraciidae; Figure 15–19), for instance, have a bony skin similar to the shell of a turtle. The armor is not without its drawbacks, since a certain amount of mobility and growth potential had to be sacrificed for this protective exterior. Trunkfishes swim slowly using their pectoral fins in a sort of paddling motion. Filefishes (Balistidae) are relatives of the trunkfish. Their skin is covered by small, prickly scales that give them the texture of sandpaper. Triggerfishes (Balistidae) have tough, leathery skins that protect them from some predators.

Protective Behaviors

Many mechanisms have evolved in reef fishes to avoid predation. When attacked, soapfishes (Grammistidae; Figure 15–20) produce a sudsy coating of mucus that is poisonous and has an unappealing taste. Trunkfishes will also produce a sudsy coating of poisonous mucus when disturbed. Brightly colored damselfishes (Pomacentridae) do not wander far from the branched coral, where they retreat immediately if disturbed. The pearly razorfish (*Hemipteronotus novacula*), which feeds on small molluscs on the floor of the reef, dives headfirst into the sand and buries itself when

Figure 15–20

Protective Behaviors. This soapfish produces a toxic mucus when it is disturbed. (Copyright 1993 by Joyce Bunek/Animals Animals)

disturbed. Pufferfishes (Tetraodontidae) inflate themselves to be too much of a mouthful for a potential predator.

A much higher proportion of predators are more active at night than during the daytime. As a result, many diurnal fishes, like parrotfishes (Scaridae) and wrasses (Labridae), seek places to hide as twilight approaches. These animals have well-developed eyesight that serves them well during the day, but at night they cannot see well enough to avoid predators. As night approaches, these fishes converge on the reefs and wedge themselves into crevices for the night. Some species of parrotfish take the extra precaution of secreting a mucous cocoon that surrounds their body and masks their scent from nocturnal predators.

The Role of Color in Reef Organisms

Coloration plays a vital role in survival on a coral reef. The barracuda is a good example of countershading, and butterflyfishes (*Chaetodon*) exhibit disruptive coloration (see Chapter 10). Most reef fishes are brilliantly colored. Although in the open water such colorful fishes would stand out, in the world of the coral reef they are protected. Bright colors help to break up the shape of the body, and the reef world itself is such a maze of brilliantly colored corals, algae, and other sedentary organisms that a brightly colored fish simply blends in well. Some fishes, like the Nassau grouper (*Epinephelus striatus*) of the Caribbean, conceal themselves by changing color to match their background. Fishes are not the only animals that have mastered the art of camouflage. Many invertebrates, including shrimp, crabs, molluscs, and worms, also have colors and shapes that allow them to blend with their environment.

Figure 15–19

Protective Body Coverings. The trunkfish from the Caribbean has a tough exterior similar to that of a turtle, making it unappealing to many predators. (W. Gregory Brown/Animals Animals)

Some fishes sport bright colors during the day but take on duller, less conspicuous colors at night. The Spanish grunt (*Haemulon macrostomum*), which is silver and yellow by day, becomes dull and blotched at night. At night, the large yellow dorsal fin of the porkfish (*Anisotremus virginicus*) turns black. It is interesting to note that these colors cannot be seen without flashlights. Even in bright moonlight, only a dull outline is visible. Why fishes change color at night is uncertain, but physiological responses rather than camouflage are thought to be involved.

Camouflage is not the only role of coloration in reef organisms. Fishes that have an unpleasant taste or that have specific defensive weapons, such as sharp, venomous spines, frequently display warning coloration. Lionfishes (*Pterois*) with their venomous dorsal spines are an example of a strikingly colored reef fish displaying this type of coloration (see Chapter 10).

Coloration may also play a role in defending territories and in mating rituals. Some species, such as the harlequin tusk wrasse (*Lienardella fasciatus*) of Australia (Figure 15–21), display bright colors as they vigorously defend a territory. In essence, the bright coloration is a "no trespassing" sign to others of the same species.

The bright colors of some territorial fishes provoke aggression. As they reach the time for courtship and mating however, these bright colors are temporarily lost and another set of mating colors takes their place. For example, the bright colors of adult parrotfishes indicate the sex of an individual and are important at mating time in order to attract mates. Male gobies also display bright colors at mating time. The brightly colored male attracts the attention of a female, and she is lured to his lair by an elaborate courtship dance.

Figure 15–21

Harlequin Tusk Wrasse. The harlequin tusk wrasse of the Great Barrier Reef displays bright colors while defending its territory. (Norbert Wu)

Not all reef fish change colors at mating time. In species such as the beaugregory (*Pomacentrus leucostictus*) and blue angelfish (*Holacanthus bermudensis*), the bright colors of both males and females are required to defend the territory. Both sexes maintain their bright colors at all times to ward off intruders.

Other Types of Camouflage

Color is not the only means of camouflage employed by reef organisms; body shape also plays an important role. The thin bodies of trumpetfishes (Aulostomidae) and pipefishes (Syngnathidae) help to conceal them as they hover head down among gorgonians and other soft corals that are swaying gently in the currents. In the Indo-Pacific the extremely venomous stonefish (*Synanceja horrida*) can be found half buried in the sand. Its irregular body resembles rock complete with algal growth. When a small fish swims too close, the stonefish opens its mouth and swallows its unwary prey.

Symbiotic Relationships on Coral Reefs

At one time, marine biologists were puzzled by gatherings of many fish species at particular sites on a reef. It was later recognized that these fishes were waiting their turn to be cleaned by one of the reef's many cleaner organisms, such as the cleaner wrasse (*Labroides dimidiatus*). Cleaner wrasses are small fish that feed on the parasites of larger fishes (Figure 15–22a). The wrasse sets up a territory on the coral reef known as a **cleaning station,** where clients will visit at regular intervals to have parasites and dead tissue removed. The cleaners benefit by obtaining a constant food supply and protection from predators while the clients are kept healthy by the occasional grooming. The wrasse attracts potential customers to its station with a series of movements referred to as a *dance.* When larger fishes come by, the wrasse scours their exterior, removing and feeding on parasites and dead skin. The cleaner wrasse's blue color and distinctive dance are a signal to large carnivores that the fish is a friend and not a meal. Even aggressive predators, such as the moray eel, that usually eat small fish like the wrasse, will submit to cleaning without harming the cleaner. Color changes or ritual postures by clients (also called hosts) may also be used to initiate and terminate the cleaning process. The Atlantic surgeonfish (*Acanthurus bahianus*), for instance, changes color from a purple-brown to an olive color as it nears the station of a bluehead wrasse (*Thalassoma bifasciatum*). This apparently communicates to the wrasse that the fish is coming to be cleaned and will not attack the wrasse. Several species of shrimp, including the peppermint shrimp (*Lysmata wurdemanni*), also engage in cleaning symbiosis.

Small fishes known as *blennies* (*Blennius*) frequently build burrows that they defend vigorously against intruders.

(a)

(b)

Figure 15–22

Symbiosis. (a) A cleaner wrasse removes parasites and dead skin from the surface of its "client," a coral trout. (b) The boxer crab uses anemones attached to its chelipeds as defensive weapons. (a, Norbert Wu; b, Rudie H. Kuiter/Oxford Scientific Films/Animals Animals)

Most species are carnivorous, feeding on small organisms that venture into their territory, but some are omnivorous. One species that lives on coral reefs resembles the cleaner wrasse, both in body shape and color. This blenny goes so far as to mimic the wrasse's dance that attracts other fish to its "cleaning station." When the unsuspecting fish comes near to be cleaned of parasites, the blenny viciously attacks, biting pieces of flesh from the unwary fish.

Cleaning is not the only example of symbiosis in the coral reef community. The reef is full of examples of organisms that have evolved special relationships with other species. Clownfishes, for instance, seek shelter from their enemies in the tentacles of large anemones (see Chapter 2). Pearlfishes (*Carapus*) live in the hindgut of sea cucumbers, coming out through the anal opening at night to feed. They may also supplement their diet by feeding on the easily regenerated intestines of their hosts. Pearlfishes of the Indo-Pacific can be found living in the shells of oysters and clams and in certain species of sea stars. The tiny conchfish (*Astrapogon stellatus*) is found only in the mantle cavity of the large queen conch (*Strombus gigas*). At night, the conchfish emerges to feed on shrimp, sea lice, and other small crustaceans. On Indo-Pacific reefs certain species of gobies are found at the entrance to burrows that are dug by pistol shrimp. The shrimp digs the burrow and continuously clears it of debris. When predators appear, the gobies hide in the shrimp's burrow. The adult shrimp may benefit from the warning given by the gobies, but, on the other hand, the gobies feed on young shrimp.

Hermit crabs rely on the shells of dead snails to provide them with a protective covering. As the crabs grow, they exchange smaller shells for larger ones. Some species of hermit crab will attach anemones to their shells for added protection. The anemones are carefully transferred to new shells when an exchange is made. The crab massages the base of the anemone until it releases its hold on the old shell. The crab then transfers the anemone to the new shell and holds it in place until the anemone attaches. The anemones benefit by being carried to better feeding grounds, and some species of hermit crabs even feed their anemones by dropping morsels of food into their mouths. The boxer crab (*Dromidia;* Figure 15–22b) attaches anemones to its claws, making them even more formidable weapons against potential predators.

Tiny shrimpfishes (Centriscidae) hide between the spines of sea urchins, where they hover head down with their striped bodies looking very much like the urchin's spines. A species of cardinalfish seeks refuge in the spines of urchins and in return cleans its host's body.

These are only a few examples of the thousands of symbiotic relationships that occur in the coral reef community. In a sense, the entire coral reef community can be thought of as a huge, symbiotic complex where all of the various organisms exist bound together by intricate interdependencies.

CHAPTER SUMMARY

Coral reefs are found in shallow water to depths of 60 meters in areas of ocean where the water is warm and clear. They can be divided into three major categories. Fringing reefs are found surrounding islands and bordering continental land masses. Barrier reefs are separated from the nearest land mass by a lagoon. Atolls are found on the tops of submerged volcanos and are generally circular with a centrally located lagoon. All three types of reef share some common physical characteristics. The reef front rises from the lower depths to a point just below the surface of the water. The highest point of the reef is the reef crest, and the area behind the reef is the back reef. Because of differences in the physical environment, each area supports different combinations of coral and other reef organisms.

Hard (scleractinian) corals are responsible for the large colonial masses that comprise the bulk of a reef. These animals rely on symbiotic dinoflagellates known as *zooxanthellae* to supply them with some nutrients and to produce an environment in which the coral polyp can deposit its calcareous cup. Most coral polyps feed at night in order to obtain nitrogen for their growth and the needs of their symbionts. Not all hard corals form large colonies; some are solitary polyps. These, along with the soft corals and many other skeleton-building and shell-building organisms, however, still contribute to the structure of the reef.

Coral reefs are constantly forming and breaking down. Boring organisms weaken the bases of large coral stands, causing them to topple, and tropical storms regularly cause breakdown and destruction of corals. Following the destructive phase of the reef cycle, new solid surfaces form and are ultimately colonized by new polyps. The shells of other animals and calcareous algae also contribute to the formation of a reef.

The water that surrounds coral reefs is, by necessity, quite clear and thus does not contain a great deal of plankton. The primary producers on a coral reef are mostly symbiotic zooxanthellae, benthic algae, and cyanobacteria.

Where coral reefs flourish, large numbers of other organisms can be found. The corals not only provide a basis for the food web, but they also provide shelter for resident animals.

Evolution has provided many reef animals with adaptations that help them to avoid predation or to be more efficient predators. Some animals are protected by special body coverings. Others produce toxic substances to deter predators or exhibit special behaviors that help them avoid predation. Color also plays a vital role in a reef animal's quest for survival. Some color patterns make animals more difficult to see, allowing them to avoid predation or sneak up on their prey. Color helps some species to identify members of the opposite sex or to distinguish juveniles from adults. Color combines with body shape in some species to disguise and camouflage.

The reef is a huge, symbiotic complex full of intricate interdependencies. These important symbiotic relationships are one of the reasons that so many organisms can live and thrive in these communities.

SELECTED KEY TERMS

fringing reef, *p. 278*

atoll, *p. 278*

barrier reef, *p. 278*

hard coral, *p. 282*

soft coral, *p. 283*

bleaching, *p. 287*

cleaning station, *p. 295*

QUESTIONS FOR REVIEW

MULTIPLE CHOICE

1. The type of reef that is separated from its associated land mass by a lagoon is called
   a. a barrier reef
   b. an atoll
   c. a fringing reef
   d. a table reef
   e. a patch reef

2. The portion of a reef that rises from the ocean's depths is the

   a. reef crest
   b. reef front
   c. back reef
   d. reef flat
   e. spur-and-groove formation

3. The corals that dominate areas of a reef that receive the most wave energy are
   a. fire corals
   b. elkhorn corals
   c. mushroom corals

   d. cluster corals

   e. flower corals

4. Corals supply their zooxanthellae with

   a. oxygen

   b. lipids

   c. sugars

   d. glycerol

   e. nitrogen wastes

5. Which of the following are major primary producers on coral reefs?

   a. kelps

   b. red algae

   c. sea grasses

   d. zooxanthellae

   e. phytoplankton

6. An echinoderm that feeds on coral polyps is the

   a. sand dollar

   b. sea cucumber

   c. crown-of-thorns sea star

   d. ochre sea star

   e. pencil urchin

7. Many reef organisms exhibit color patterns and body shapes that help them to

   a. blend in with their background

   b. defend a territory

   c. attract a mate

   d. be more efficient predators

   e. all of the above

8. An example of an animal that exhibits warning coloration is the

   a. stonefish

   b. lionfish

   c. cleaner wrasse

   d. harlequin tusk wrasse

   e. surgeonfish

9. An important nocturnal predator of the coral reef is the

   a. barracuda

   b. stingray

   c. moray eel

   d. parrotfish

   e. wrasse

10. An important predator of sea stars and sea urchins is the

   a. octopus

   b. squid

   c. shark

   d. Triton's trumpet snail

   e. sea cucumber

SHORT ANSWER

1. What role do coralline and calcareous algae play in reef formation?

2. Explain how the physical characteristics of the reef environment influence the species of corals that inhabit them.

3. Describe how a coral colony is formed.

4. Describe the process of reef formation.

5. If the water surrounding coral reefs is not very productive, how can the coral reef support large numbers of organisms?

6. Describe what is meant by *cleaning symbiosis.*

7. Why don't most coral species grow in the aphotic zone?

8. Why are coral reef fishes so brightly colored?

THINKING CRITICALLY

1. Although illegal, some fishers in the Pacific Ocean region use charges of dynamite to stun fish, causing them to float to the surface, where they can be readily harvested. What effect would this practice have on nearby reef communities? What impact would this practice have on future fish catches?

2. What impact would daily visits by hundreds of sport divers and snorkelers have on a reef community?

3. Why are coral reefs not found along warm coastal areas where large rivers such as the Mississippi, Amazon, and Congo empty into the sea?

SUGGESTIONS FOR FURTHER READING

Birkeland, C. 1989. The Faustian traits of the Crown-of-Thorns Starfish, *American Scientist* 77(2):154–163.

Brown, B. E. and J. C. Ogden 1993. Coral Bleaching, *Scientific American* 268(1):64–70.

Doubilet, D. 1996. The Desert Sea, *Natural History* 105(11):48–53.

Endean, R. 1983. *Australia's Great Barrier Reef.* The University of Queensland Press, New York.

Jackson, J. B. C. and T. P. Hughes. 1985. Adaptive Strategies of Coral Reef Invertebrates, *American Scientist* 73:265–274.

Kaplan, E. H. 1988. *A Field Guide to Coral Reefs of the Caribbean and Florida Including Bermuda and the Bahamas.* Houghton Mifflin, Boston.

Levine, J. S. 1990. Coral Reef Fishes Use Riotous Colors to Communicate, *Smithsonian* 21(8):98–103.

Ruppert, E. E. and R. D. Barnes 1994. *Invertebrate Zoology,* sixth edition. Saunders College Publishing, Philadelphia.

Ward, F. 1990. Florida's Coral Reefs Are Imperiled, *National Geographic Magazine* 179(3):115–132.

Chapter 16

Coastal Seas and Continental Shelves

KEY CONCEPTS

1. The coastal seas lie above the continental shelves.

2. Coastal seas receive high levels of nutrient input from rivers, coastal runoff, and upwellings.

3. Coastal seas support enormous amounts of phytoplankton.

4. The number and kinds of benthic organisms on continental shelves are influenced by sediment characteristics.

5. The high productivity of coastal seas supports large numbers of fishes, birds, and marine mammals.

6. In areas north and south of the tropics, kelps (a type of brown algae) dominate the subtidal zone of coastal seas, where the water is cold and the sediments are hard. They are important primary producers and provide habitats for many animals.

Do you like seafood? Maybe you like lobster, or scallops, or shrimp. Perhaps your taste runs more toward whitefish or flounder. Anyone who has enjoyed a seafood meal at a fine restaurant, a fish sandwich at a fast-food restaurant, or even a can of sardines has sampled the bounty of the coastal seas.

In the relatively shallow waters above the continental shelves, plentiful sunlight and abundant nutrients combine to support enormous numbers of primary producers. Grazing on these producers are equally impressive numbers of consumers. In the water column, huge schools of fish feast on the abundance of food. On the bottom, large numbers of filter-feeders, deposit-feeders, and scavengers feast on the detritus that rains down from the sunlit waters above. In turn, bottom fishes prey on the abundance of benthic organisms or scavenge for food. In shallower waters with hard bottoms, forests and thickets of sizeable algae provide food and shelter for many animals.

The productive and bountiful coastal waters are the source of most of the saltwater fish and shellfish that humans consume worldwide. Because of their high productivity, these areas are hotly contested by coastal countries, which vie for the exclusive rights to fish the waters. These areas are not infinitely productive, however, and overfishing in most parts of the world has started to impact on the productivity of coastal seas.

THE PRODUCTIVE COASTAL SEAS

The coastal seas that lie along the edges of the continents teem with life. These waters contain an enormous number of phytoplankton, the microscopic photosynthetic organisms that form the basis of aquatic food webs. Unlike the clear, blue waters around coral reefs, coastal seas are green, a sign of high productivity. Grazing on these pastures of phytoplankton are hordes of tiny animals, the zooplankton. The abundant plankton provides food for large schools of fish, and these, in turn, provide food for larger fishes, some marine mammals, seabirds, and humans (Figure 16–1).

Beneath the surface lie productive benthic (bottom) communities. In some areas where the bottom is primarily composed of rock, large forests of seaweed form a habitat for a variety of animals that rely on these large algae for food and shelter. Where the bottom is soft, burrowing organisms of all kinds dominate. One of the most important factors in determining the type of organism found in benthic communities is the stability of the environment. The benthic communities of the continental shelves are quite stable and not subject to the same changing forces that affect organisms living along the shore. On continental shelves, food supply is the major limiting factor. In most areas of a continental shelf, large amounts of dead and dying plant and animal material settle to the bottom to fertilize the sediments. This continuous food supply supports many filter-feeders, suspension-feeders, and detritus-feeders, which in turn supply food for other animals.

The coastal seas cover an area about the size of Asia and contain only 10% of the ocean's expanse, but they produce 90% of the world's annual harvest of fish and shellfish. By comparison, the open ocean is a barren desert. The density of populations in coastal seas and on continental shelves is astounding. A 10-milliliter sample of surface water may contain thousands of planktonic organisms. At one time, it was estimated that as many as 10,000 cod (Gadidae) lived above each acre of the continental shelf bottom of the Grand Banks off Nova Scotia, Newfoundland. As the result of overfishing, however, there are so few fishes now that the area will no longer support a major commercial fishery. Some areas off the coast of Great Britain support as many as 80 million brittle stars per square kilometer (0.36 square mile).

THE CONTINENTAL SHELVES

Most coastal seas lie over continental shelves (see Chapter 3). These shelves extend as little as 1.6 kilometers (1 mile) from the coastline of western North and South America to as far as 1,500 kilometers (900 miles) off the arctic coast of Siberia. Most continental shelves, however, average about 67 kilometers (40 miles) wide. They descend gradually from the shore, reaching depths of 130 meters (430 feet) at their most distant edge. At this point, the gently sloping bottom may become a steep slope or a sheer dropoff. Off the coast of Florida, in the Gulf of Mexico, the continental shelf ends in a dropoff that runs for 800 kilometers (480 miles) and at some points has perpendicular drops of 1.6 kilometers (1 mile). By contrast, off the coast of Chile, the continental shelf ends in a steep incline that descends uninterrupted to a depth of over 6 kilometers (3.6 miles).

Worldwide, rivers annually carry almost 750 million tons of sediment to coastal seas. Some ends up deposited on the bottom of the continental shelves, and the rest is dissolved in the seawater. The sediments contain nutrients such as nitrogen, phosphorus, calcium, and silica that are essential for life in the ocean. Nutrients are also carried to the coastal seas by upwellings (see Chapter 4). The heavy concentration of nutrients and the large amount of sunlight this area receives combine to make continental shelves such productive areas.

Figure 16–1

Commercial Fishes. The large numbers of plankton in coastal seas support large schools of commercial fishes such as these herring. (Norbert Wu)

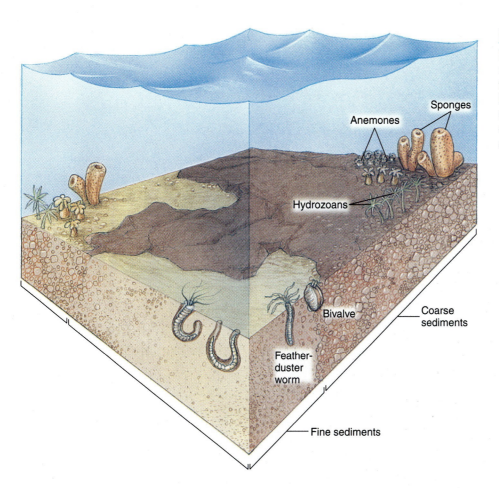

Figure 16–2

Patchiness. *Patchiness* refers to the uneven distribution of benthic organisms that results from the uneven distribution of bottom sediments. Areas of coarse sediments and fine sediments are randomly distributed on the continental shelf, and each supports different types of organisms.

Labels in figure: Anemones · Sponges · Hydrozoans · Bivalve · Coarse sediments · Feather-duster worm · Fine sediments

LIFE ON CONTINENTAL SHELVES

The Role of Sediments

The number and type of organisms that can live in and on the bottom of continental shelves are greatly influenced by the type and characteristics of the sediments that comprise the bottom. In areas of a continental shelf where currents flow, the bottom is composed of coarse sediments. The moving water tends to carry away the fine, light sediments such as silt, leaving behind larger sand and rock particles. The bottom in these areas is constantly shifting and is not a favorable habitat for burrowing and interstitial animals that cannot withstand abrasion or the constant shifting of the sediments. The moving water, however, does carry a large supply of food. Sedentary or sessile filter-feeders and suspension-feeders are well adapted to this type of bottom. The size of the sediment particles gives them a firm foundation to which they can attach, and the large supply of suspended food provides a constant source of nutrients. Sandy bottoms favor burrowing filter-feeders like worms, amphipods, and clams. Coarser sediments are ideal for sponges, anemones, and colonial cnidarians.

Where bottom currents are weak, the bottom is composed of fine sediments, like silt. The sediments here are more stable and support a variety of organisms that construct permanent burrows, such as polychaete worms and some crustaceans. Most of the organisms living in this habitat are deposit-feeders, feeding on the organic material that settles from above and becomes trapped in the soft sediments. These bottoms do not support many filter-feeders because of the scarcity of suspended food and the problems associated with the fine sediments interfering with the animal's filtering structures.

The different types of sediments are not evenly distributed along the bottom, and, as a result, there is an uneven distribution of benthic organisms. This characteristic distribution of organisms in a population or community is referred to as **patchiness** (Figure 16–2).

Bottom Dwellers

A very different animal community from that of the shallow subtidal zone exists on the seafloor of continental shelves (Figure 16–3). Food in the form of detritus rains down from the sunlit waters above. The detritus consists of dead plankton, dead and dying larger animals, and organic debris, including human sewage, washed into the sea from land by rivers and runoff. Sessile filter-feeders such as sponges, tu-

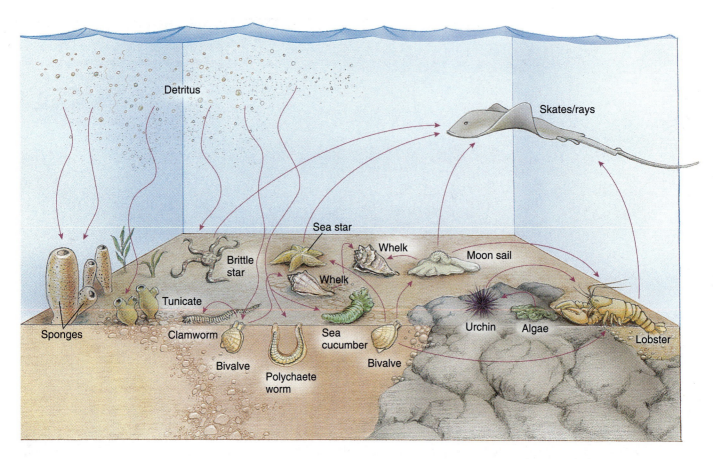

Figure 16–3

Benthic Food Web. Detritus forms the basis of the benthic food web in coastal seas.

nicates, and bivalves feed directly on the detritus by filtering large volumes of water containing detritus through their bodies. Several species of polychaete worm feed on trapped detritus as they consume bottom ooze in the process of making their burrows. Sea cucumbers use their sticky tentacles to gather food off the bottom, and their relatives the sea urchins move slowly over the bottom, chewing off bits of algae and detritus that have become attached to the firmer bottom sediments.

Burrowers, sedentary filter-feeders, and slow-moving grazers are preyed upon by more active animals. Polychaetes known as *clamworms* (*Nereis*), some as long as 0.5 meter (19.5 inches) feed on small bivalves that are burrowed in the soft bottom. Carnivorous snails, such as members of the family Buccinidae, feed on both bivalves and other snails, usually by boring a small hole through the shell so that the contents can be sucked out (Figure 16–4).

Sea stars move slowly along the bottom, feeding on mussels, oysters, and scallops. Brittle stars slither across the bottom powered by undulations of their serpentine rays. Some species burrow into the bottom, where they feed on decaying material, while others feed on detritus and small

Figure 16–4

Whelks. Whelks are carnivorous snails that feed on other molluscs. (Norbert Wu)

Figure 16–5

Lobster Trap. Lobsters are important commercial shellfish. They are caught in traps like these that are placed on the bottom overnight. (Ralph A. Reinhold/Earth Scenes/Animals Animals)

Figure 16–7

Commercial Fishes. This flounder represents just one of many important commercial fishes that are caught in coastal seas. (Trevor McDonald/Natural History Photographic Agency)

organisms that are passed along their tube feet toward their mouth.

Lobsters inhabit holes and crevices in the bottom by day and come out at night to feed on molluscs, crustaceans, and urchins as well as to scavenge. The American lobster (*Homarus americanus*) is fished commercially using pots or traps. The animals enter these seeking shelter or attracted by bait and are unable to get out (Figure 16–5). The majority of frozen lobster tails sold in supermarkets are species of spiny lobsters (*Panulirus*) and slipper lobsters (*Scyllarides*) from tropical and subtropical coastal waters.

Skates, rays, angel sharks (*Squatina*), and batfishes (Ogcocephalidae) are just a few of the fishes that forage along the bottom, feeding on molluscs and crustaceans that they crush with their powerful jaws (Figure 16–6). These animals are well adapted to dwelling and feeding on the bottom. They usually have flattened or compressed bodies, retractable fishing lures, and specialized sense organs. They also have well-camouflaged dorsums that allow them to blend with the sea bottom.

A large number of fishes spend their lives on or near the bottom of the continental shelf. Large schooling fishes such as haddock (*Melanogramus*), hake (Gadidae), pollock (*Pollachius virens*), cod, and ocean perch (*Sebastes marinus*) are important commercially (Figure 16–7). They are caught in large numbers in most of the coastal waters of the Northern Hemisphere (Table 16–1). These fishes feed on mol-

Figure 16–6

Cownose Ray. This cownose ray has powerful jaws to crush the hard shells of the molluscs and crustaceans on which it feeds. (Zig Leszczynski/Animals Animals)

Table 16–1
Size and Value of the Domestic Catch of Selected Fishes

Species	Quantity (in 1,000 lb)	Value (in $1,000)
Haddock	254,519	5,582
Atlantic pollock	15,843	10,543
Atlantic cod	95,881	52,013
Pacific cod	550,528	132,480
Ocean perch (Atlantic)	1,322	790
Ocean perch (Pacific)	60,972	13,561
Flounder	645,829	143,511
Halibut	70,454	53,773

From U.S. National Oceanic and Atmospheric Administration, National Marine Fisheries Service, Fisheries of the United States annual.

luscs, echinoderms, crabs, other fishes, each other, and even their own young. Living directly on the bottom is another important group of commercial fishes, the flatfishes (Pleuronectiformes). These include flounder, halibut, turbot, and various species of sole.

Life in the Water Column

Coastal seas receive freshwater runoff from the neighboring land. The runoff provides enough nutrients to support the growth of relatively large producers called **microphytoplankton** (20 to 200 microns) as well as smaller forms, the **nanophytoplankton** (less than 20 microns) (see Chapter 6). In the waters above the continental shelves, phytoplankton populations are frequently so dense as to make the water green.

In the colder waters around Antarctica and in the northern Pacific and Atlantic oceans, diatoms dominate the microphytoplankton. Planktonic organisms called *coccolithophores* (see Chapter 6) are sometimes so numerous in areas of the North Sea that the water turns white. At certain times of the year, they even outnumber the diatoms. In warmer coastal waters, dinoflagellates are more numerous. Unicellular green algae occur in both tropical and polar seas. The composition of the phytoplankton varies from one region to another, from season to season, and sometimes even within a single season. Grazing by zooplankton and other animals also affects the composition of the phytoplankton. When the number of one type of phytoplankton decreases

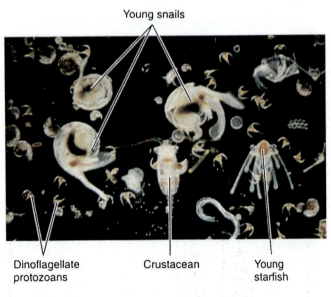

Young snails

Dinoflagellate protozoans   Crustacean   Young starfish

Figure 16–8

Zooplankton. Members of the zooplankton include protists, adult crustaceans, jellyfishes, and the larvae of both invertebrate and vertebrate animals. (D.P. Wilson/Science Source/Photo Researchers, Inc.)

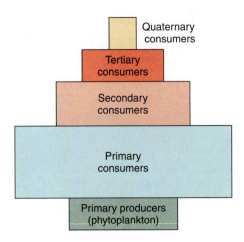

Quaternary consumers

Tertiary consumers

Secondary consumers

Primary consumers

Primary producers (phytoplankton)

Figure 16–9

Productivity of Coastal Seas. In this generalized pyramid, the size of each box indicates the relative amount of biomass at each trophic level. Because of the abundance of sunlight and nutrients, phytoplankton in coastal seas can reproduce at extremely rapid rates and can therefore support about five times their own biomass in primary consumers. These primary consumers, which include enormous amounts of zooplankton, are in turn a rich food supply for large populations of higher-order consumers. (D.P. Wilson/Science Source/Photo Researchers, Inc.)

as the result of heavy grazing, another species will usually proliferate to take its place.

Drifting animals that make up the zooplankton (Figure 16–8) feed on the millions of diatoms, dinoflagellates, and other phytoplankton. The most abundant members of marine zooplankton are crustaceans called *copepods.* In coastal waters, the concentration of copepods can be as high as 100,000 individuals per cubic meter of water. The primary reasons for such large numbers are the tremendous reproductive capacity of these animals and the rich food supply. After being fertilized, a female can produce as many as 100 eggs, and some species produce a new clutch of eggs every four or five days. Between one and two weeks after hatching, the new generation is mature and ready to reproduce. Copepods are the primary consumers of diatoms. A single copepod can consume as many as 120,000 diatoms per day.

Since tiny phytoplankton can be either eaten directly by small zooplankton or filtered from the water by benthic filter-feeders such as clams and worms, food chains in coastal seas are frequently two or more steps shorter than those of the open sea. Small zooplankton are the preferred food of large fishes like menhaden (*Brevoortia*) and alewives (*Alosa pseudoharengus*), while large bottom fishes, such as cod and haddock, prey on filter-feeders. In both cases there are only three trophic levels between primary producers and consumers of reasonable size (Figure 16–9). Even larger animals such as tuna, sharks, and humans are only four trophic levels from producers. The combination of higher productivity and shorter food chains sup-

ECOLOGY AND THE MARINE ENVIRONMENT

The Highly Productive Southern Ocean

At first, one might think that the frigid waters surrounding the continent of Antarctica wouldn't harbor many living organisms. In reality, though, the Southern Ocean, like other coastal seas, is an extremely productive area. The productivity of the Southern Ocean supports not only large numbers of aquatic organisms but also most of the animals that live in Antarctica. The basis of this rich food web are the numerous diatoms that dominate the phytoplankton. Dominant among the zooplankton that feed on these diatoms is krill (*Euphausia superba*). This herbivorous crustacean accounts for almost 50% of the zooplankton in the Southern Ocean. Although krill are found throughout the region, their distribution is not uniform. The greatest concentrations have been recorded in the Weddell Sea, where some swarms of krill have been estimated to be 2 million tons. It has been estimated that the annual biomass of krill in the Southern Ocean is between 750 million and 1,300 million tons.

During the summer, primary production in the Southern Ocean increases as the pack ice melts and the polar region receives more sunlight. At this time, krill move to the surface to graze on the in-creased numbers of diatoms. During the winter, less light, increased pack-ice cover, and increased turbulence lower primary production to near zero. At this time, the krill move to deeper water and apparently feed mostly on detritus.

As many as 20 species of squid, some relying heavily on krill as a food source, are found in the Southern Ocean. The squids provide food for numerous species of toothed whales, seals, and birds. It is estimated that these animals consume about 35 million tons of squid annually from this region. Many fish species are also found in the Southern Ocean. Some species are permanent residents, while others are seasonal visitors, arriving in summer to feed on the large numbers of krill. It is estimated that all of the fish in the Southern Ocean consume 100 million tons of krill annually.

Although Antarctica is home to few species of birds, those that do live there are usually represented by large populations. Their diet consists of crustaceans, mainly krill and copepods, squids, fish, and carrion. Krill accounts for almost 78% of all the food they eat. The total annual consumption of krill directly or indirectly by birds is approximately 115 million tons. Seven species of seal are found in the Southern Ocean. Of these, the crabeater seal's diet consists of 94% krill, the leopard seal's diet consists of 37% krill, and the fur seals of southern Georgia feed almost exclusively on krill. The other seal species feed on fish and squid that in turn rely on krill. Larger baleen whales, such as the blue whale, feed mainly on krill. They consume about 43 million tons of krill annually. As a result of the pressures of whaling, the number and size of baleen whales in the Southern Ocean has decreased. It is estimated that 150 million tons of krill that was formerly consumed by baleen whales is now available to other krill-feeding species. Smaller whales, seals, and penguins appear to be the main beneficiaries of this change, as indicated by increases in the size of their populations, presumably due to the increased availability of food. Krill are also being harvested for human consumption. The krill being netted by fishers is assumed to be available because of the decrease in the size of whale populations. There is still some doubt, however, whether the productivity of krill is sufficient to feed humans without endangering other animal populations in the area.

ports a larger number of higher-level consumers in coastal seas.

The most productive of the planktonic ecosystems are located on continental shelves, where the combination of winds, ocean currents, and shape of the seafloor interact to bring nutrients into the photic zone from the ocean floor (upwelling zones; see Chapter 4). The shallow banks along the coasts of the North Atlantic and the Pacific coast of Peru are examples of such productive areas. These upwelling areas yield almost half the world's supply of commercial fishes. In these regions, the almost continuous supply of nu-trients supports a phytoplankton community dominated by large, chain-forming diatoms. These chains, which consist of several cells linked together, are large enough to be eaten directly by large zooplankton like the shrimplike krill (*Euphausia*) of the Southern Ocean and small fishes like anchovies (Engraulidae) of the Pacific Ocean off the coast of South America. Krill and anchovies, in turn, are large enough to be worthwhile prey for large fishes, seabirds, seals, whales, and even humans. The high productivity and short food chains of upwelling areas support the greatest biomass of any planktonic system.

Biologists estimate that as much as one seventh of the anchovies caught worldwide by humans and seabirds come from Peru's coastal seas in years when upwelling currents are strong. Smaller upwellings along the Pacific coast of the United States and islands like Japan support locally productive fisheries. It is no wonder that so many animals, including humans, come to upwelling areas to feed. The primary production in these regions may be as much as six times higher than that of the open sea, and they produce biomass at a rate in excess of 36,000 times that of the open sea. These limited areas represent valuable commercial resources that are protected by laws and require conservation.

The ecological importance of plankton goes beyond their role of supplying food for animals. An overwhelming number of animal species spend at least a part of their lives as members of the plankton. Sedentary animals such as barnacles, mussels, and coral, which spend their adult lives fixed in one place, rely on their planktonic larval forms, such as the nauplius, trochophore, veliger, and planula, to colonize new territories. When spawning, these animals release hundreds to thousands of tiny eggs and sperm that unite in the water column and then scatter in the currents. Almost all crabs, shrimps, and lobsters have larval stages that feed and grow as part of the plankton before settling down to lead their adult existence. Even free-swimming fishes such as herring and eels have tiny planktonic larvae and juvenile stages almost as small that join the zooplankton for the first weeks or months of their lives. Only a small fraction of the larvae in plankton will eventually develop into adults. Odds are that only 1 in every 100,000 eggs found in the plankton will survive to adulthood. The health of plankton is important not only because of the role they play in supporting commercial fisheries but also because any disturbance to their community can cause unexpected changes in populations of animals whose adults are anything but planktonic.

NUTRIENT CYCLING

The diversity of life on the bottom and in the water column of continental shelves would not be possible without the activity of microorganisms like bacteria. Bacteria break down the dead remains of plants and animals, releasing the nutrients to be used again. These nutrients are then returned by upwellings to the surface, where they can be recycled by plankton.

Except in the tropics, shallow coastal seas exhibit very definite seasonal changes in productivity. During the spring in temperate waters, increased sunlight bathes water that is rich in nutrients that have been brought to the surface by the action of winter storms and the breakdown of the thermocline during winter. The combination of nutrients and sunlight supports a bloom of phytoplankton. This large increase in phytoplankton is followed by a proportional increase in the population of zooplankton. As the summer progresses, increased numbers of plankton deplete nutrients in the warm surface water. Colder bottom waters contain more nutrients, but, because of the density of the water, these nutrients do not rise (see Chapter 4). Eventually, phytoplankton and zooplankton populations also begin to decline due to the decreased nutrient concentration. Mixing occurs again in the fall as temperatures decline and surface water density increases to match the density of bottom water. In response to new nutrients present in the surface water, there is another bloom, although not as large as the one that occurs in spring. During the winter, more nutrients are brought to the surface waters, but there is usually not enough sunlight to support large phytoplankton populations. As spring returns, the cycle repeats itself.

KELP FORESTS

One of the most productive marine communities on some continental shelves is the kelp bed. Kelp is a type of brown algae (see Chapter 7) that requires a hard, rocky bottom, cold water, and a continuous supply of nutrients to support its high level of photosynthetic activity. Most kelp beds are found in water that is no more than 20 meters (66 feet) deep, but if the water is exceptionally clear, they may occur in water as deep as 30 meters (99 feet). Kelp are strictly cold-water organisms that rarely survive in areas where the average surface water temperature exceeds 20°C. Some kelp beds are like underwater forests, forming a canopy that shades smaller algal species and an understory that is home to many animal species. Giant kelps of the genus *Macrocystis* are the world's largest algae, reaching lengths of 20 to 40 meters (66 to 132 feet). Giant kelps also grow at amazing rates. A mature blade of *Macrocystis* can add as much as 50 centimeters (19.5 inches) of new tissue in a single day when conditions are favorable.

In the kelp zones of the Pacific and southern Atlantic oceans, *Macrocystis* dominates (Figure 16–10a). The tall, narrow blades of this kelp grow in tight clusters, and there is enough space between them for animals to move about with ease. Beneath the canopy of the giant kelp is a smaller species, the sea palm (*Eisenia*). The stipe of this species is thick and elastic and is able to bend with water currents. The sea palm depends on the minimal light that penetrates the kelp canopy to supply its needs for photosynthesis. The rocky bottom is carpeted by several species of red algae that can survive on the wavelengths of light that penetrate the canopy and reach the bottom.

The genus *Laminaria* is dominant in the North Atlantic Ocean. *Laminaria* grows closely packed to form thickets, and the dense nature of the growth makes this habitat more suitable for crawling rather than swimming animals (Figure 16–10b).

(a)

Figure 16–10

Kelp Beds. (a) The giant kelp *Macrocystis,* which forms underwater forests, dominates the kelp zones of the Pacific and southern Atlantic oceans. (b) The dominant kelp in the North Atlantic Ocean is *Laminaria,* which forms dense thickets. (a, Norbert Wu; b, Fred Bavendam/Peter Arnold, Inc.)

(b)

The life cycle of kelp begins when spores settle on rocky bottom that receives enough light to satisfy the organism's needs for energy. The spores germinate and develop into a microscopic form that is preyed upon by herbivores. The density of herbivores in an area is a primary factor in determining whether new algae can establish themselves. Once established, kelp grow quickly. Stipes bearing the flattened blades that carry out photosynthesis grow toward the light at the surface. At the surface, they spread out, forming a canopy similar to that of a terrestrial forest. As they grow, the kelp's appearance changes as each different species exhibits a characteristic growth form. *Macrocystis* has gas-filled floats at the base of each blade to help buoy up this photosynthetic structure. An entire individual, consisting of a stipe, floats, and as many as 200 blades, grows as a single unit and lives for about six months. After six months, these individuals are replaced by new ones that frequently grow in pairs at the astounding rate of 3 to 5 meters (10 to 16.5 feet) per week under good conditions. *Macrocystis* is a perennial, usually living three to seven years before it dies from any number of causes. *Laminaria,* on the other hand, grows like a conveyor belt, with new tissue added at the base as older tissue at the tip erodes away.

Large numbers of organisms use kelp for food, shelter, or both. The dense network of kelp blades slows currents and decreases the force of all but the most energetic storm waves, thus providing convenient shelter for many fishes and mammals such as otters and seals. The large, tree-sized algae greatly increase the amount of useful habitat in the area. In kelp forests of the West Coast of the United States, echinoderms, like the omnivorous bat star (*Patiria*), move slowly among the kelp, feeding on both animals and algae. Its relative, the many-rayed sunflower sea star (*Pycnopodia helianthoides*), feeds on urchins, other sea stars, and molluscs that it finds on the kelp forest floor. Fishes such as pipefish, blennies, eels, and the Garibaldi (*Hypsypops rubicundus*), as well as many larger fish species, find the kelp an ideal environment in which to live. Sheepshead (*Semicossyphus pulcher*) come to feed on many of the larger invertebrates, while several species of rockfish (*Sebastes*) feed on both invertebrates and other fishes. Without the kelp forests for habitat, these species would either disappear or be present in smaller numbers (Figure 16–11).

While kelps are constantly growing, they are also constantly eroding, producing an almost steady stream of detritus that plays a significant role in the kelp forest food web. A species of mussel that lives in kelp beds is one of the few multicellular organisms that produces an enzyme that can digest cellulose. This species can feed not only on bacteria that coat particles of kelp, as do other species, but they can also feed on the kelp fragments themselves. Other filter-

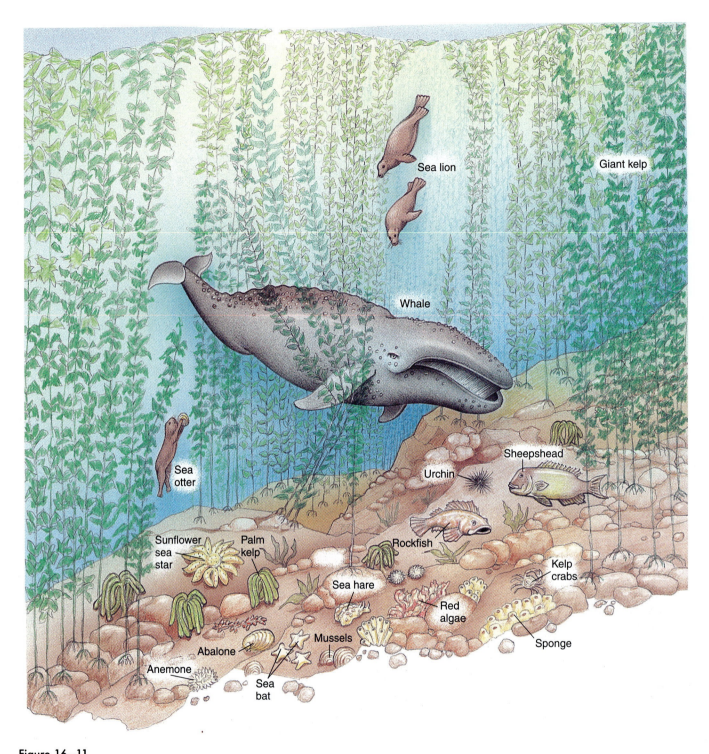

Figure 16–11

Kelp Forest Community. Kelps provide both habitat and food for a large number of marine organisms.

feeders attach directly to the surface of parts of the kelp, some growing over the surface of new blades almost as fast as they are produced. Sponges grow at the base of kelp and, together with the kelp's holdfasts, form a habitat that shelters tiny shrimp, crabs, lobsters, brittle stars, and small fishes. The commercially important molluscs known as

abalone (*Haliotis*) are common residents of kelp beds of the Pacific coast, while American lobsters are a common resident of North Atlantic kelp beds.

Although most of the organisms that live on kelp are filter-feeders, some are herbivores that feed directly on the kelp itself. Snails crawl along the stipes, using their radula to scrape the outer layer of cells from the alga. Burrowing through the matted holdfasts are hordes of termite-like crustaceans called *gribbles (Phycolimnoria)*. In some instances, the action of gribbles can be so destructive that the weakened holdfasts can no longer anchor the alga, and storm waves will uproot it and carry it away. It is thought that in a stable kelp forest, gribbles may be the primary cause of mortality among adult kelp.

Kelps are also a favorite food of sea urchins. Wave action and predatory fishes that feed on sea urchins usually prevent urchins from doing significant damage to the upper portions of the kelp and the canopy, but there are few predators that prevent these animals from devouring young algae and damaging the important holdfasts. A decline in the sea otter population along the West Coast and the lobster population along the North Atlantic coast has resulted in population explosions of urchins. Some researchers believe

that increased numbers of urchins, along with naturally occurring events such as disease and climate changes, may be primarily responsible for the decline in kelp beds in some areas over the last 50 years. Off the coast of Nova Scotia, the lobster population decreased by as much as 50%, the result of overfishing in the 1970s. During the same time period an estimated 70% of the kelp beds in this area disappeared.

In areas where kelp has been completely removed, enough sea urchins remain to make it nearly impossible for new kelps to reestablish themselves. Although urchins may ultimately decimate the kelp, the animals continue to survive. Since they are primarily generalist feeders, sea urchins simply switch to feeding on other species of algae when kelps are not available. In southern California, sea urchins feed on the amino acids from treated sewage. Other animal species that depend on kelp for food or shelter are not as fortunate and soon disappear along with the kelp.

CHAPTER SUMMARY

The coastal seas that lie above continental shelves contain enormous numbers of phytoplankton, which form the basis of an extremely productive food web. The high productivity of these regions is due to a combination of physical and biological factors. Rivers and runoff deposit large amounts of nutrients directly into coastal seas, and the geography of continental shelves favors upwellings that bring nutrients from the bottom to the photic zone, where they can be utilized by plankton.

The number and kinds of organisms that can live on the bottom of continental shelves is greatly influenced by sediment characteristics. Coarse sediments favor sedentary filter-feeders and suspension-feeders. Fine sediments provide a more stable bottom that favors burrowing organisms that are primarily deposit-feeders. Much of the food that supports the primary consumers of bottom communities drifts down from the sunlit waters above and consists mostly of detritus and organic debris.

In surface waters, phytoplankton support large numbers of zooplankton and some large animals as well. Those large predators that do not prey directly upon the plankton feed on smaller animals that are plankton-feeders. As plankton and larger organisms die, their remains settle to the bottom, where bacteria decompose the remains, releasing the nutrients for use by other organisms. Although these nutrients tend to accumulate on the bottom, seasonal changes in water density result in vertical turnover in temperate zones.

In areas of the continental shelf where the bottom is hard, the water is cold, and nutrients are readily available, kelps dominate. These algae form complex, three-dimensional habitats for large numbers of invertebrate and vertebrate animals. Kelps also serve as an important food source, directly or indirectly, for many of the species that live in and around kelp beds.

SELECTED KEY TERMS

patchiness, *p. 301*
microphytoplankton, *p. 304*

nanophytoplankton, *p. 304*

QUESTIONS FOR REVIEW

MULTIPLE CHOICE

1. Most of the world's harvest of commercial fish comes from
   a. estuaries
   b. bays
   c. coastal seas
   d. the open ocean
   e. benthic regions

2. The number and types of benthic organisms found on the continental shelf are most influenced by the

a. amount of sunlight
b. water temperature
c. characteristics of the sediments
d. salinity
e. water clarity

3. Sedentary and sessile filter-feeders and suspension-feeders are better adapted to
a. hard bottoms
b. soft bottoms

4. The uneven distribution of bottom organisms and sediment type on the continental shelf is referred to as
a. reticulation
b. patchiness
c. diversity
d. sediment selection
e. benthic orientation

5. In colder waters, _____ dominate the phytoplankton.
a. krill
b. copepods
c. diatoms
d. dinoflagellates
e. kelp

6. The most abundant members of the zooplankton are
a. krill
b. copepods
c. jellyfish
d. diatoms
e. dinoflagellates

7. Nutrients produced by bacteria in the bottom sediments are returned to the surface waters by
a. storms
b. upwellings
c. seasonal changes in water density
d. all of the above

8. In temperate seas during the summer months, populations of phytoplankton decline due to
a. increased temperature
b. increased sunlight
c. decreased predation
d. decreased nutrient availability
e. changes in salinity

9. The world's most productive upwelling area is located off the coast of
a. California
b. Peru
c. China
d. South Africa
e. Europe

10. The dominant alga in southern California kelp forests is
a. *Fucus*
b. *Laminaria*
c. *Eisenia*
d. *Sargassum*
e. *Macrocystis*

SHORT ANSWER

1. What are the main sources of nutrient input into coastal seas?

2. What factors affect the size of plankton populations?

3. Explain how the type of bottom sediments influences the diversity of life on the floor of the continental shelf.

4. Describe the seasonal changes that occur in the plankton of temperate coastal seas.

5. Explain how decreases in the size of American lobster populations have affected North Atlantic kelp beds.

6. Why are kelp beds frequently compared to terrestrial rainforests?

7. Why are coastal seas and continental shelves such productive areas?

8. Diagram a simple food web for the continental shelf.

THINKING CRITICALLY

1. Currently there is concern about the possibility of global warming due to the "greenhouse effect." How might this affect the productivity of coastal seas?

2. Why do so many species of small fish found in coastal seas travel in large schools?

SUGGESTIONS FOR FURTHER READING

Campbell, D. G. 1992. The Bottom of the Bottom of the World, *Natural History* 101(11):46–52.

Gore, R. 1990. Between Monterey Tides, *National Geographic* 177(2):2–43.

McPeak, R. H. and D. A. Glantz. 1984. Harvesting California's Kelp Forests, *Oceanus* 27(1):19–26.

Smith, W. O., Jr. and D. M. Nelson. 1986. Importance of Ice Edge Phytoplankton Production in the Southern Ocean, *Bioscience* 36:251–257.

Winston, J. E. 1990. Life in Antarctic Depths, *Natural History* 99(4):70–75.

Chapter 17

The Open Sea

KEY CONCEPTS

1. The open ocean is a pelagic ecosystem.

2. Phytoplankton are the primary producers in open-ocean food webs, and their productivity is limited by the scarcity of nutrients.

3. Limited primary production and food webs with several energy-wasting steps limit the number of large animals the open ocean can support.

4. Gelatinous plankton such as salps and ctenophores play a significant role in concentrating nutrients in open-ocean ecosystems.

5. The open ocean is a fairly stable environment.

6. Animals of the open sea display a number of interesting adaptations that help them avoid predation.

7. Several structural features and behaviors have evolved to help organisms that are not active swimmers from sinking.

8. Crustaceans, such as copepods, dominate the zooplankton in the open ocean.

9. Large zooplankton include jellyfishes and gastropod molluscs.

10. Fishes, molluscs, and mammals make up most of the nekton in the open sea.

In 1967, a swordfish 3 meters (10 feet) long attacked the research submarine *Alvin* off the coast of Georgia at a depth of about 600 meters (1980 feet). The entire length of the bill, which measured 1 meter (3 feet), penetrated the vessel's outer fiberglass hull, forcing it to return to the surface. No one was injured in the incident, and the crew ate the fish for dinner after extracting it from the submersible.

The swordfish is just one of several large animals that inhabit the open ocean, an immense area that contains 1.4 billion cubic kilometers (300 million cubic miles) of salt water and provides several hundred times the living space of all terrestrial habitats combined. Compared to the productive coastal seas, however, the open ocean seems quite barren. Limited amounts of nutrients restrict the numbers of primary producers, and, although the open ocean is immense, it supports relatively few large animals. The combination of low productivity and lack of hiding and resting places makes the open sea a quite different habitat from coastal seas.

CHARACTERISTICS OF THE OPEN SEA

Beyond the shallow coastal seas of the continental shelves lies the open ocean. When viewing the ocean from above, the open expanse of water appears quite the same and somewhat monotonous. Even though it lacks obvious boundaries, the open sea can still be divided into regions based upon the physical characteristics of the water and the life forms that inhabit them. The open ocean can be divided into vertical zones based on the depth to which light penetrates to support the process of photosynthesis. As you may recall from Chapter 2, the photic zone is the layer that receives enough sunlight to power photosynthesis. In the open ocean the photic zone extends down to a maximum depth of 200 meters (660 feet). Below this is the aphotic zone that extends to the ocean bottom. In this zone, light rapidly disappears until the environment is totally dark. In this chapter we will be dealing with life in the photic zone. Life in the aphotic zone will be covered in the next chapter.

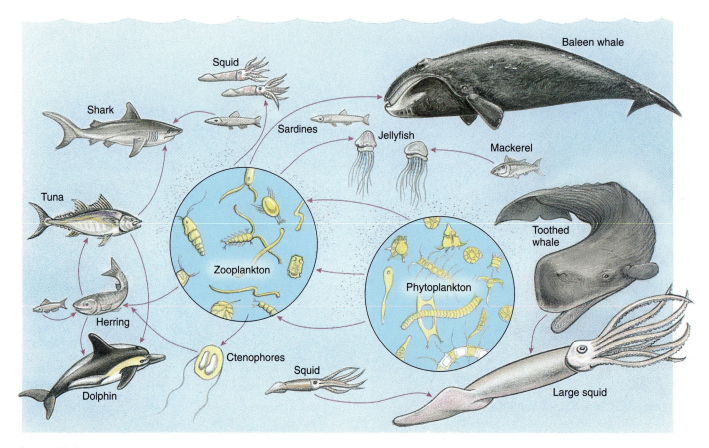

Figure 17–1

Open-Ocean Food Web. In open-ocean food webs, the major herbivores are zooplankton, which are the primary food of many large animals.

The open sea represents a pelagic ecosystem, one in which the inhabitants live in the water column. In a pelagic ecosystem, the basis of food chains is the many small species of phytoplankton. There are good reasons for the primary producers in this ecosystem to be so small. These organisms derive the nutrients that they need from the seawater that surrounds them. The smaller the organism, the greater the relative surface area that is exposed for the absorption of nutrients and sunlight. The major herbivores in the open ocean are zooplankton, and these supply food for the nekton—large, free-swimming animals, such as tuna and other pelagic fishes, and whales (Figure 17–1).

PRODUCTIVITY IN THE OPEN SEA

In the great expanse of the open ocean there are no plants or seaweeds that can serve as primary producers to support consumers. The surfaces to which plants and seaweeds might attach are deeply submerged and shrouded in darkness by the light-absorbing properties of the overlying water. In the open sea, all higher forms of life rely on the various species of plankton, the organisms floating and drifting in the water column, to supply food. Many plankton are single-celled and microscopic, while others are just large enough to see with the naked eye. Some members of the plankton, such as large jellyfishes, may be 50 or more centimeters across.

A dynamic, floating ecosystem composed of plankton stretches across the entire surface of the sea and down into the depths for hundreds of meters. Pelagic ecosystems such as these include thousands of algal and animal species, each with its own light, temperature, nutrient requirements, growth characteristics, and relationships with nutrients and predators. Studying such a complex system and its inhabitants is no easy task, and our knowledge about open-water plankton communities continues to grow and change rapidly.

Although surface waters of the open ocean receive large amounts of sunlight, they do not receive any of the nutrients that are washed to sea from the land or that are brought to the surface by upwellings. This results in water that contains low levels of the nutrients, such as nitrogen and phosphorus, that are necessary to support the life of phytoplankton. Waters just above the bottom contain higher concentrations of nutrients but do not receive enough sunlight to make them productive. The problem is compounded by the fact that there is very little mixing of deeper, high-nutrient water with the low-nutrient water at the surface.

The lack of nutrients in the open ocean is most pronounced in tropical waters. These waters have more or less permanent layers that are separated by a thermocline (see Chapter 4) with a lighter, less dense layer of water on top of colder, denser water. This arrangement prevents significant exchange of nutrients from the deep water to the surface. Thus, even though tropical seas are the warmest and offer plankton large quantities of intense sunlight, the limited amount of nutrients limits the size of phytoplankton populations. Phytoplankton production in the central South Pacific Ocean is only about half as much as the productivity off the California coast and only about one sixth as much as some coastal areas off Saudi Arabia or western Africa, where coastal upwellings bring nutrients to the surface. The small amount of phytoplankton in tropical seas supports even fewer numbers of the zooplankton that graze on the phytoplankton.

Lacking sufficient quantities of key nutrients, the crystalline, blue waters of the open ocean support phytoplankton whose growth is so nutrient-limited that they grow as slowly as possible to support life. Such nutrient-poor waters support only the tiny primary producers called **nanoplankton** (plankton less than 20 microns in size) (Figure 17–2a). These small organisms slip through the filtering structures of most large zooplankton. These tiny cells can be eaten efficiently only by **microzooplankton** (zooplankton 20 to 200 microns in size). At the next trophic level, microzooplankton fall prey to **macrozooplankton** (200 to 2,000 microns), which in turn provide food for **megazooplankton** (over 2,000 microns) such as openwater shrimp species. Shrimp are a primary source of food for small, open-water fishes, and these small fishes are preyed upon by salmon, tuna, and squid.

The number of large animals that the open sea can support is quite limited for two reasons. First, they have few nutrients available to support primary production, which places a low ceiling on the total biomass that the area can

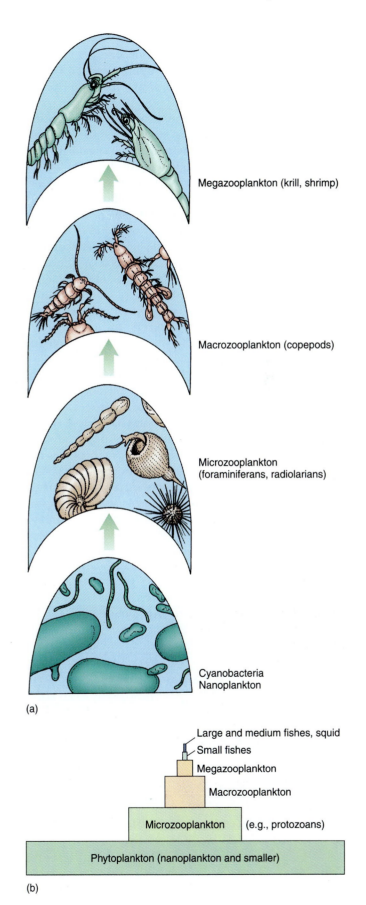

Megazooplankton (krill, shrimp)

Macrozooplankton (copepods)

Microzooplankton (foraminiferans, radiolarians)

Cyanobacteria
Nanoplankton

(a)

Large and medium fishes, squid
Small fishes
Megazooplankton
Macrozooplankton
Microzooplankton   (e.g., protozoans)
Phytoplankton (nanoplankton and smaller)

(b)

Figure 17–2

Planktonic Food Chain. (a) The nutrient-poor waters of the open ocean support food chains with several links composed of small planktonic organisms. (b) Because open-ocean food chains consist of so many energy-wasting steps, the open seas support relatively few large consumers, as indicated in this generalized pyramid of biomass.

sustain. Second, a food web involving several energy-wasting steps between primary producers and final consumers guarantees that the final trophic level will contain relatively few organisms, which means there can be few large predators (Figure 17–2b).

In order to learn more about these little-known oceanic ecosystems, in some recent studies researchers have donned SCUBA gear to observe and collect specimens firsthand. Several important components of the plankton slip through plankton nets, while others, such as the gelatinous ctenophores, are so fragile that net collecting damages them beyond recognition. SCUBA divers can locate ctenophores by the way they reflect light and by the rainbow of colors produced when sunlight is refracted by their rows of beating cilia. Though these animals do not normally survive even the gentlest of standard collecting techniques, a careful diver can scoop them into jars without injuring them. Other species of gelatinous zooplankton, foraminiferans; clusters of floating algal filaments; and collections of translucent, drifting, organic particles called **marine snow,** can be collected by hand in a similar fashion.

Collections and observations made by SCUBA-diving researchers have yielded a wealth of information about the gelatinous zooplankton known as *salps* (Thaliacea) and *larvaceans* (Larvacea; see Chapter 9). Salps have barrel-shaped bodies that are open at both ends. Some species produce and maintain a mucous net inside their oral openings and use it to filter water that is pumped through their bodies. When filled, the net and its contents are digested. Salps eat bacteria too small for most sizable plankton to capture and make that source of organic material available to their own predators. On the other hand, salps produce fecal pellets from planktonic food in abundance, increasing the flow of nutrients down and out of the photic zone.

Because their bodies are at least 95 percent water, these adaptable organisms can grow in size and reproduce very rapidly into swarms that can stretch over kilometers and contain up to 500 individuals in each cubic meter of water. There are reports of such colonies, known as **pyrosomes** (Figure 17–3a), that measure as long as 14 meters (46 feet), although the usual range is from a few centimeters (1 inch) to 3 meters (10 feet). Each pyrosome is composed of hundreds of individual animals that are joined together to form a hollow cylinder that appears to be a single organism. Each member of the colony faces outward and sucks in water and small plankton. The incurrent water enters the center of the colony, flowing through and out the back, moving the colony through the water by means of slow jet propulsion. Pyrosomes are found worldwide but most commonly in tropical and subtropical seas.

Sometimes, a salp will be inhabited by an amphipod crustacean of the genus *Phronima* (Figure 17–3b). This amphipod eats the internal organs of the salp and then uses the salp's exterior as a sort of mobile home in which it lives and broods its young. The salp's body continues to move through the water, propelled by the swimming action of the crustacean.

Larvaceans secrete structures composed of mucus called *houses*. Water is drawn through the house when the larvacean waves its tail. Inside the house, tiny plankton are trapped in a wing-shaped feeding filter and are digested. Since the houses have no opening for the elimination of feces, larvaceans must abandon their houses several times a day as they become clogged with feces and large plankton and then produce new ones. Interestingly, the discarded houses of larvaceans, which often contain up to 50,000 trapped, living, phytoplankton cells, immediately become homes for bacteria and end up as particles of marine snow.

(a)

(b)

Figure 17–3

Salps and Larvaceans. (a) Large colonies of salps such as this one are called *pyrosomes*. (b) Amphipods of the genus *Phronima* feed on salps. (a, Copyright 1995 Norbert Wu)

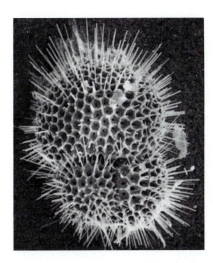

Figure 17—4

Foraminiferans. The spiny projections on this foraminiferan increase the organism's ability to capture food by increasing the surface area available for food-gathering rhizopodia. (From A.W.H. Be, 1968. *Science* 161:881—884. Copyright 1968 the American Association for the Advancement of Science)

The types of feeding exhibited by salps and larvaceans allow these gelatinous zooplankton to filter enormous volumes of water very efficiently.

SCUBA collections of oceanic plankton have also yielded important information about foraminiferans. Since these organisms are usually damaged beyond repair in collecting nets, it had not been possible to study the living organisms prior to collection by SCUBA divers. The spherical shells of these organisms are usually less than 1 millimeter in diameter, but some species are covered with a network of radiating spines that increases their diameters up to ten times (Figure 17—4). Along these spiny supports, foraminiferans spread dense, sticky nets of living cytoplasm called **rhizopodia**, which are used to trap food.

The ability of organisms that are essentially spherical to trap suspended food increases dramatically with increased size. This is because the surface area of a sphere increases as the square of its radius. A small increase in radius can produce a significant increase in the organism's surface area, the part of the organism that traps food. Because spines and rhizopodia increase a foraminiferan's radius, and thus its surface area, the organism's food-gathering ability increases hundreds of times.

Some foraminiferan species exhibit unexpected feeding techniques. Instead of being exclusively passive interceptors of detritus, bacteria, and algae, many tropical species actively prey on copepods nearly their own size. The rhizopodia act first as a spider's web, snaring the prey and dragging it toward the shell. Then, these flexible yet powerful strands penetrate the prey's body, rupturing the muscle tissue into fragments that are then wrenched out of place

and engulfed. These captured bits of tissue fragments are pulled into the foraminiferan's cell, where they are digested.

Many foraminiferans from nutrient-poor waters maintain ecologically efficient relationships with small dinoflagellates that live symbiotically within the foraminiferan's cytoplasm. By day, the dinoflagellates migrate out along the rhizopodia, gaining maximum exposure to sunlight. At night, they migrate deep into the central shell.

Since it is difficult to keep planktonic foraminiferans alive in the laboratory, marine biologists have not yet been able to study the chemical basis of this relationship thoroughly. Experiments have demonstrated, however, that foraminiferans with dinoflagellates, when kept in normal sunlight, grow and reproduce more rapidly than those kept in the dark or those that have had the dinoflagellates chemically eliminated.

Although the chemical evidence is lacking, the relationship seems to involve an exchange of materials between the foraminiferans and their symbionts similar to that which occurs between corals and their zooxanthellae. The foraminiferans generate waste products such as carbon dioxide and ammonia, which, although toxic to the foraminiferan, are perfect nutrients for the dinoflagellates. Since the dinoflagellates grow within the cell where the wastes are generated, they are able to absorb them with minimal effort. As the nutrients are not diluted by the surrounding seawater, the dinoflagellates do not have to expend energy to reconcentrate them. This self-contained fertilizing and waste-removal system is extremely efficient for supporting life in nutrient-poor water.

On a smaller scale, plankton biologists have found what they call **micropatchiness** throughout the photic zone. On bits of marine snow and on clusters of floating algal threads live aggregations of bacteria up to 10,000 times more concentrated than bacteria in the open water. Along with these bacteria, which include both primary producers and decomposers, drift hordes of their protozoan predators. Further study of these drifting particles has led to discoveries indicating that the microscopic communities they harbor may in fact be complete ecosystems in miniature. On each of these floating islands of life, bacteria grow, respire, and are eaten. Primary producers reabsorb any nutrients released in this process, and grazing zooplankton feed on the primary producers. The whole system cycles through intense microbial activity with little input from the outside other than sunlight.

These floating microenvironments are seen by some biologists as highly evolved, stable associations of primary producers and consumers. By remaining close together, microscopic producers and consumers can concentrate and store nutrients in their immediate vicinity, allowing them to survive in the nutrient-poor macroenvironment that surrounds them.

Chain-forming diatoms of several genera, for example, grow into clumps in which individual algal cells are sepa-

ECOLOGY AND THE MARINE ENVIRONMENT

Are Open Oceans Really as Barren as Deserts?

Recent research indicates that the large expanses of open ocean may not be as devoid of life as previously thought. Data recently collected in the northern Pacific Ocean show two to three times more organic matter produced by photosynthesis than had been reported previously. Some biologists think that the open ocean has not been sampled often enough to catch periods of high productivity, resulting in low productivity assumptions. Others think that the sampling techniques themselves may have been responsible for erroneous results.

Previous sampling methods and contaminated containers may be responsible for low estimates of productivity. Phytoplankton are delicate organisms that can easily be damaged by the collecting techniques. Contaminated or dirty containers may also have caused the death of relatively large numbers of photosynthetic organisms during experiments designed to measure production rates. In coastal and upwelling areas, where phytoplankton levels are very high, a small level of mortality during the experiments would not greatly affect the results nor the calculations based on those results. In open-ocean areas, however, where

populations of these organisms are sparse to begin with, the loss of substantial numbers could greatly affect the outcome of the experiments. The experiments that indicated high levels of production in the northern Pacific Ocean were performed under ultraclean conditions to avoid killing any of the phytoplankton. Yet, even with all the care that was taken, microscopic examination of the bottles used in the experiment showed dead and dying organisms.

Data from an effort called the Plankton Rate Process in Oligotrophic Oceans Research Project conducted in 1985 showed productivity rates two to four times greater than those previously published. Researchers were careful to avoid contaminating the phytoplankton with trace metals and causing unnecessary mortality. They also used finer filters in gathering samples that could trap even the smallest photosynthesizers such as cyanobacteria and coccolithophorids.

Another way of measuring productivity is to use large areas of ocean water separated by natural ocean processes, instead of water artificially contained in sample bottles. Researchers at Woods Hole Oceanographic Institution used

data that were collected from the Sargasso Sea near Bermuda over a period of 18 years. Their calculations indicated levels of production so high that there could not be enough nitrogen in the surface water to support it. Two of the researchers, William Jenkins and Joel Goldman, suggest that the nitrogen is coming from deeper water being brought to the surface at infrequent intervals by storms. They also propose that the ocean studied is a two-layer system. The top layer shows the low productivity that is associated with the open sea, whereas the level just below that layer and just above the nutrient supply shows much higher productivity. Thus samples taken from slightly different depths may yield dramatically different results. There is also some evidence for a similar double-layer system in the northern Pacific Ocean.

If indeed open-ocean productivity is greater than assumed, this very large area of the world is more ecologically important on a global scale than previously thought. Additional field studies are being conducted, and previous data are being reevaluated in an effort to better understand the productivity of this enormous area.

rate but held together by a mucous sheath. That sheath houses a complete community of bacteria, diatoms, single-celled animals of several species, and even small, shrimplike zooplankton. The larger animals in these associations feed on bacteria, diatoms, and even on each other, forming a complete ecosystem bundled up in one tiny, floating package. The cycle of primary production, grazing, release of nutrients, and assimilation of those nutrients goes around and around, either on those tiny floating particles or in water patches too small to be sampled by standard techniques.

Other researchers have uncovered evidence that floating bacteria and other tiny organisms are far more impor-

tant as a group in open ocean ecosystems than previously imagined. These organisms are so small that they are designated as **ultrananoplankton** (0.2 to 20 microns). These minute bacteria and protozoa slip through the mesh of standard plankton nets, leaving no way to estimate their numbers or ecological importance from plankton tows. Recent estimates show that microbial activity in the water column may in fact dominate the ocean's surface-layer food webs.

In their self-contained microenvironments, individual organisms show few, if any, indications of being nutrient-limited. The difference between these systems and those of upwelling areas, then, is not in the organisms' absolute growth rates, but in the structure of the food chain and in the difference in total biomass supported as a result of the input of new nutrients. The greater the nutrient input from outside the system, the more biomass can be supported.

SURVIVAL IN THE OPEN SEA

Remaining Afloat

Members of both the phytoplankton and zooplankton exhibit adaptations that allow them to live near the surface of the water without sinking into the ocean depths (Figure 17–5). Phytoplankton must remain in the photic zone in order to carry out photosynthesis, and the distribution of phytoplankton dictates the distribution of zooplankton that rely on these organisms for food. Many planktonic organisms have projections from their bodies, such as needle-like spines, fine hairs, or even wing-shaped appendages. These structural adaptations add little weight to the organisms but increase friction and produce drag, thus increasing resistance to sinking as they move through the water.

Adaptations that increase friction do not prevent an organism from sinking; they merely slow the process. Even the tiniest members of the plankton must display other adaptations that allow them to counteract gravity and remain in the sunlit surface waters. You may recall that diatoms produce droplets of oil that help them remain buoyant (see Chapter 6). Copepods also store food as oil droplets. Some species of marine bacteria store food in the form of oil droplets, while others produce vacuoles containing carbon dioxide or carbon monoxide, which increase buoyancy. Radiolarians have chambers containing water and carbon dioxide that help them to remain afloat.

Some phytoplankton and many zooplankton remain afloat by actively swimming. Dinoflagellates use their whip-like flagella to swim, and in a similar fashion the numerous copepods use their legs and antennae to actively keep afloat, moving them back and forth as fast as 600 times a second. The bell-shaped body of a jellyfish is more than 90% water and is very close to the density of seawater. Most species are able to undulate the bell, producing a slight upward push that prevents them from sinking. Sense organs that include photoreceptors and statocysts help the animal to orient itself properly in the water so that it doesn't swim downward.

Many open-ocean zooplankton make daily migrations from the surface of the sea to depths of nearly 1.6 kilometers (1 mile). Some marine biologists believe that these movements take advantage of the benefits of feeding on the phytoplankton that live only in the photic zone, while reducing to some extent the threat of predation by plankton-eating fishes, which are also more abundant in the upper regions. These migratory zooplankton are often so densely

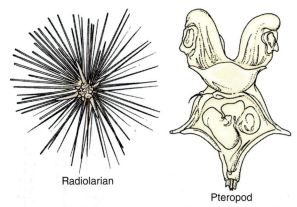

Radiolarian

Pteropod

(a) Projections like the needles of the radiolarian and the wings of the pteropod create friction and slow sinking.

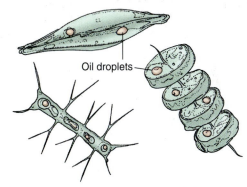

Oil droplets

(b) Diatoms store their food reserves as oil. The oil makes them more bouyant and offsets some of the weight of their shells.

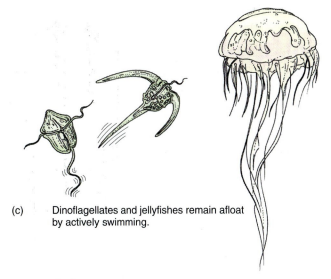

(c) Dinoflagellates and jellyfishes remain afloat by actively swimming.

Figure 17–5

How Open-Water Organisms Remain Afloat. Organisms that inhabit the water column of the open sea slow or prevent sinking by (a) increasing friction, (b) having adaptations that increase buoyancy, or (c) actively swimming.

packed that they form what is called a **deep scattering layer,** a mixed group of zooplankton and fishes that gives sonar systems a false image of a nearly solid surface hanging in midwater.

Avoiding Predation

Despite lower productivity, the open ocean provides an environment that is generally more stable than others we have examined. Sudden fluctuations in salinity and temperature do not occur in the open seas, and even the most violent storms have little effect a few meters below the surface. Organisms that live in the well lit surface waters face other problems of survival, though, such as a lack of places to hide to avoid predation. When threatened, animals of the open sea cannot quickly hide in a rock crevice, burrow into bottom sediments, or hide behind a stand of coral. There are no large seaweeds, like kelps, to provide a quick refuge.

In light of this problem, pelagic organisms display a variety of adaptations that help to increase their chances of survival. Larger members of the plankton, such as jellyfish, have stinging cells that help to protect them from potential enemies. Some animals elude predators through the speed with which they swim, while others, like flying fishes and some species of squid, take to the air when pursued. Many species of fish that make up the nekton find safety in numbers by swimming together in schools.

In an environment that lacks hiding places, camouflage plays a central role in survival. Larger animals from nudibranchs and pelagic snails to sharks and whales exhibit countershading (Figure 17–6). These animals have dorsal surfaces that are dark blue, grey, or green, and ventral surfaces that are silvery or white. This arrangement of colors makes them difficult to see from both above and below.

Many planktonic species are not conspicuous because they are nearly transparent. For instance, the most abundant members of the zooplankton, copepods, are often transpar-

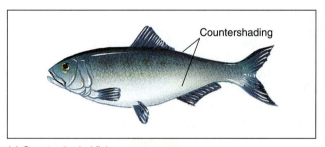

(a) Countershaded fish

Figure 17–6

Countershading. (a) In a countershaded animal, the body surface that is usually exposed to sunlight is a dark color, while the opposite surface is a paler color or white. In the ocean environment, a uniformly colored fish (b) is highlighted from above and easy to see, while a countershaded fish (c) is effectively camouflaged.

(b) Uniformly colored fish in natural lighting

(c) Countershaded fish in natural lighting

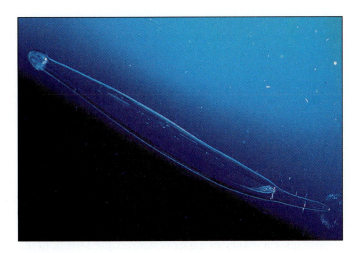

Figure 17–7

Arrowworms. Arrowworms are aggressive predators with transparent bodies. They are considered commercial pests because they consume large numbers of planktonic fish larvae. (Copyright 1995 Peter Parks/Mo Yung Productions/Norbert Wu)

ent, as are salps, larvaceans, ctenophores, and jellyfishes. Arrowworms (*Sagitta*) (Figure 17–7) are so transparent that they are invisible in the water and were not discovered until 1768, even though they are relatively large (up to 10 centimeters or 4 inches) and are distributed worldwide.

Arrowworms are formidable predators with eyes that see in all directions and sensory hairs that detect the slightest movement in the water. They hang nearly motionless in the water, where they wait for prey, like a copepod or a larval fish, to come within striking distance. When the arrowworm senses prey, it darts toward it with amazing speed, grasping the prey with the hard pincer-like structures around its mouth.

Some animals increase their chances of survival in the open sea by forming colonies. One of the most successful of these is the cnidarian group Siphonophora. Although siphonophore (sy-FAHN-uh-fohr) colonies look like a single individual, they are actually made up of thousands of individuals, none of which can live independently of the colony. Some members of the colony specialize in capturing prey, while others digest food, maintain the colony, or reproduce new individuals. Siphonophores are common in all oceans, where they drift on or just beneath the surface of the water, some species propelled by the gentle pulsations of several tiny bells.

One well-known siphonophore is the Portuguese man-of-war (*Physalia physalis*), which produces a gas-filled float that helps to propel the colony by catching the wind. Sometimes thousands of these colonies passively gather in drift lines caused by currents, forming a mass of siphonophores that will stretch for miles across the open sea. The gas sac of a large man-of-war may be as long as 40 centimeters (16 inches), and periodically the colony submerges the float to keep it moist. If the float drys out, it will crack, the gas will escape, and the colony will sink. The tentacles formed by feeding polyps in the colony can be as long as 8 meters (26 feet) in a large individual.

Although the stinging cells of the man-of-war are capable of killing a large mackerel, one of the colony's main food items, there is a small fish that lives symbiotically with the colony that is not harmed by the stings. The man-of-war fish (*Nomeus gronovii*) spends most of its time swimming among the stinging tentacles (Figure 17–8). The little fish averages 8 centimeters (1.5 inches) in length and is deep blue with vertical black stripes, allowing the fish to blend quite well with the Portuguese man-of-war's tentacles. How the man-of-war fish is able to avoid being killed by its host is not known. The fish has been observed to feed on some of the polyps, and possibly ingestion of the venom helps to immunize it against the stings. In laboratory experiments, the fish was found to be able to survive venom injections ten times that which would be needed to kill another fish species of its size.

Figure 17–8

Man-of-War Fish. The small man-of-war fish gains protection by spending most of its time among the tentacles of the Portuguese man-of-war. (Copyright 1995 Peter Parks/Mo Yung Productions/Norbert Wu)

Even with protective adaptations, the odds against survival in the open sea are tremendous. Many species compensate for the low survival rate by producing enormous numbers of offspring, thereby increasing the odds that a few will survive to propagate. The female ocean sunfish (*Mola mola*), for instance, produces as many as thirty million eggs in a single breeding season. Only two or three offspring need to survive to perpetuate the species.

LARGE PLANKTON

The largest members of the plankton are jellyfishes. One of the most common is the moon jellyfish (*Aurelia*). This animal's bell-shaped body measures from 15 to 45 centimeters (6 to 9 inches). A fringe of thin, hairlike tentacles surrounds the bottom of the bell, and within the clear body are four oval, pigmented gonads. In addition to the tentacles, the outer surface of the bell is covered by bands of sticky mucus that ensnare any small organism that comes into contact with it. Cilia on the body surface constantly move the mucus and any food that it has trapped to the edge of the bell. The large blobs of accumulated mucus and food are then removed by fleshy projections on the underside of the bell and brought into the mouth for digestion. The moon jellyfish feeds mainly on copepods and other small zooplankton.

Larger and more dangerous are the jellyfishes known as *lion's mane jellyfishes* (*Cyanea*) (Figure 17–9a). One enormous species found in the Arctic Ocean has a bell that averages 2.5 meters (7 feet) in diameter and has tentacles that extend downward for 30 meters (99 feet) or more, giving it the distinction of being the largest member of the zooplankton in the world. Species found in temperate regions of the Atlantic and Pacific oceans tend to be smaller. They feed on surface fishes, and their sting can easily kill a 30-centimeter (12-inch) fish.

One of the jellyfish species best adapted to life in the open ocean is *Pelagia noctiluca* (Figure 17–9b). This species has a beautiful, pastel body, and at night it is bioluminescent. Unlike most species of jellyfish that have an asexual polyp stage in their life cycle, *Pelagia* larvae mature in the open sea without going through a polyp stage.

Even some molluscs have become adapted to life in the open sea. Pteropods (TAYR-uh-pahdz) (*ptero* meaning "wing"; pod meaning "foot"), popularly known as *sea butterflies* (Figure 17–10a), are related to snails. The animal's foot has two large winglike projections, and the shell is much reduced or absent. The projections on the foot propel the animal through the water. The surface of these projections is covered with cilia that create a current that drives plankton, mainly diatoms, toward the mouth. These gastropods are abundant in the open ocean and sometimes form dense groups typically preyed upon by herring, cod, and haddock. In some areas of the south Atlantic Ocean, a species of pteropod (*Hyalostylus striatus*) that produces a light, cone-shaped shell is so numerous that large areas of the bottom are covered with the shells of these dead animals.

(a)

(b)

Figure 17–9

Jellyfishes of the Open Sea. (a) This species of the lion's mane jellyfish, *Cyanea capillata,* can achieve bell sizes of 3 meters (10 feet) in diameter and weigh 1 ton. (b) *Pelagia noctiluca* is bioluminescent and lacks an asexual polyp stage in its life cycle. (a, Norbert Wu; b, Richard Herrmann)

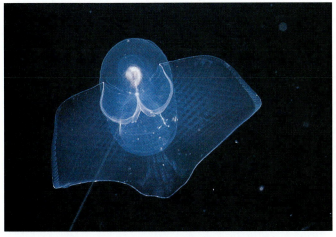

(a)

Bubble raft

(b)

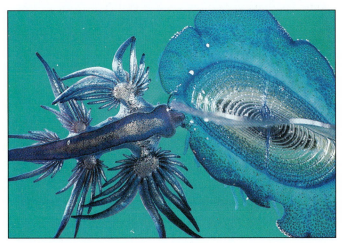

(c)

Figure 17–10

Molluscs of the Open Sea. (a) Pteropods, or sea butterflies, have a foot that is modified to form a pair of winglike structures that animals use to propel themselves through the water column. (b) The purple sea snail keeps itself afloat by clinging to a bubble raft that it produces. This specimen is extending its proboscis toward the nearby siphonophore *Vellela* (by-the-wind-sailor), one of its favorite foods. (c) The pelagic nudibranch (*Glaucus*) swims upside-down just beneath the surface of the water. This specimen is feeding on a by-the-wind-sailor. (a, Richard Herrmann; b, Copyright 1995 Peter Parks/Mo Yung Productions/Norbert Wu; c, Copyright Oxford Scientific Films/Animals Animals)

Another prominent mollusc in the open sea is the purple sea snail (*Janthina;* Figure 17–10b). Unlike pteropods, the purple sea snail has retained a relatively large, although light, shell. It keeps itself from sinking by producing a raft of bubbles surrounded by mucus. The snail clings upside down to the underside of this bubble raft. This upside-down lifestyle has led to a reversal of the typical coloration of open-water organisms. What would normally be the animal's upper surface is mainly white, while its under surface is dark purple. The purple sea snail is widely distributed in all tropical waters, where it feeds on larger members of the zooplankton such as copepods, jellyfishes, and by-the-wind sailors (*Vellela*), a siphonophore related to the Portuguese man-of-war.

Some species of nudibranch have also enjoyed success in the open sea. Some of the most numerous are species in the genus *Glaucus* (Figure 17–10c). These small, blue nudibranchs glide along upside-down beneath the surface of the water. Like the purple sea snail, *Glaucus* also displays a reverse color pattern with a dark ventral surface and a lighter-colored dorsum. *Glaucus* feeds on by-the-wind-sailors, and,

like some other species of nudibranch, it retains the stinging cells in its body for its own defense. *Glaucus* has been observed to leave clutches of eggs attached to the floats of by-the-wind-sailors after having devoured all of the colony's polyps.

NEKTON

The nekton is composed of all actively swimming organisms whose movements are not governed by currents or tides. Included in this group of animals are some larger invertebrates, fishes, and mammals.

The invertebrates that reign supreme in the open sea are the squids. Speed, keen eyesight, and intelligence make the squid a formidable predator. The animal's body is streamlined, and it has sucker-laden tentacles for grasping prey. By drawing water into the mantle cavity and forcefully expelling it through its siphon, squids are able to achieve, over short distances, bursts of speed that are faster than the fastest fishes. Although squids normally swim backward,

(a)

(b)

Figure 17–11

Fishes of the Open Sea. (a) The John Dory has jaws that protract. (b) The jaw of the sailfish is modified to form a bill. (a, Daniel Heuclin/Natural History Photographic Agency; b, Norbert Wu)

they can reverse their direction by adjusting the position of their siphon. This is especially useful when maneuvering within a school of fish.

The nekton includes many species of fish, and these species exhibit a variety of adaptations that enable them to survive in this niche. The majority of bony fishes that inhabit the open sea have mouths at the front of the body with lower jaws that protrude farther forward than the upper jaw. This arrangement allows the fish to grab prey from a variety of positions. Some species, like the John Dory (*Zenopsis;* Figure 17–11a), have jaws that protract. When the fish opens its mouth to seize its prey, the jaws project forward. They then retract quickly as the mouth closes.

The upper jaw of species like marlin and sailfishes (Istiophoridae) is greatly elongated, forming a bill (Figure 17–11b). These fishes lack the teeth characteristic of carnivores and use their bills to club their prey. The bill of a swordfish (*Xiphias gladius*) is broad and flat. As they swim into schools of mackerel, menhaden (*Brevoortia*), or squid, they flail their bill left and right, beating their prey and then swallowing their stunned victims whole.

Swordfishes will attack virtually anything, often without any apparent provocation. The broken bills of swordfishes have been found in the timbers of wooden ships, sometimes penetrating two layers of oak planks. Captured blue whales and fin whales have been found to have broken bills embedded in their sides and backs attesting to swordfish attacks on these large mammals.

The most wide-ranging fishes in the open ocean are tunas, which belong to the mackerel family (Figure 17–12). The largest species of tuna are the bluefin tuna (*Thunnus thynnus*), with individuals achieving lengths up to 4 meters

(13 feet), and the yellowfin tuna (*Thunnus albacares*), which achieves lengths of 2 meters (7 feet) or more. Smaller relatives include the albacore (*Thunnus alalunga,* slightly over 1 meter [3 feet] in length) and the ocean bonito (*Sarda,* not quite 1 meter).

Tunas must swim constantly, or they will sink. This level of activity requires a large amount of energy and a good supply of oxygen. To supply the needed oxygen, these animals must swim fast and move large volumes of water past their gills. One adaptation for fast swimming in some tuna species is a body temperature that averages 8°C to 10°C higher than the surrounding water (see Chapter 10 for details). The higher temperature allows the fish to metabolize faster. Digestion is more rapid, nerve impulses travel more quickly, and the large skeletal muscles used for swimming contract and relax about three times faster than in other fishes. These adaptations help to account for the tuna's great speed and strength.

Many species of tuna exhibit bursts of speed that can exceed 54 kilometers per hour (more than 33 miles per hour). The main propulsive force for swimming comes from the high-keeled, sickle-shaped tail. In order to decrease resistance in the water, their bodies are streamlined and their body surface is smooth. Their gill covers fit tightly against the sides of their body, and their pectoral fins lie retracted in grooves in their sides. Yellowfin tunas can reach speeds of 75 kilometers per hour (45 miles per hour), and one relative, the wahoo (*Acanthocybium solanderi*), has been clocked at 80 kilometers per hour (48 miles per hour). When they sense food, tunas quickly accelerate, reaching top speed in less than 1 second. Tunas must consume large amounts of food in order to supply the energy they need

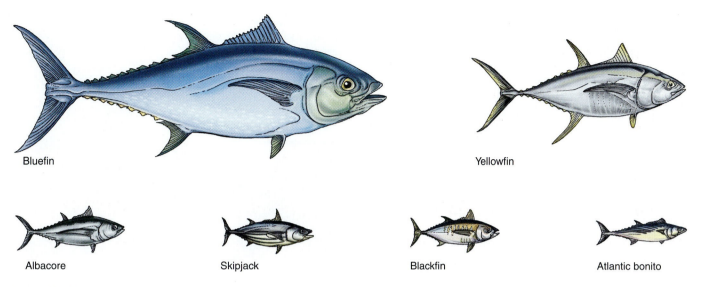

Bluefin

Yellowfin

Albacore

Skipjack

Blackfin

Atlantic bonito

Figure 17–12

Tuna. Tunas are fast swimmers that can be found throughout all seas. They range in size from the large bluefin tuna to the small Atlantic bonito.

for such vigorous swimming. Their diet is varied, consisting mainly of herring, anchovies, mackerel, sardines, flying fishes, and squids.

Quite different from the very active lifestyle of tunas is the more relaxed swimming of the ocean sunfish (*Mola mola;* Figure 17–13). Adults appear to spend most of their time lying on their sides at the surface of the water, apparently basking in the sun; this behavior gives them their common name. The ocean sunfish has a highly modified caudal fin and prominent dorsal and anal fins. These fish are quite large, reaching lengths of over 3 meters (10 feet) and weights in excess of 1 ton. They feed on larger zooplankton, especially jellyfish, and seem to have few natural predators.

The ocean sunfish is protected by a layer of cartilage 5 to 10 centimeters (2 to 4 inches) thick just beneath the surface of the skin. Although this thick layer offers protection against most predators, it is a haven for parasites. Almost every specimen that has been examined has been full of internal and external parasites, which may cause the death of many individuals. On the other hand, since parasite-infested animals are more easily caught, the data may show a bias.

Stomach analysis of some specimens reveals the remains of animals found only in deeper water, indicating that these large fish may not be as lazy as once perceived but may actually feed below the surface as well. Some biologists suggest that the normal habitat for this fish is actually deep water, and only diseased individuals are found at the surface, where they come to die and can be collected. In California, ocean sunfish regularly come near shore in summer to visit cleaning stations near kelp beds.

Some of the most efficient predators of the open ocean are sharks, whose streamlined bodies are well adapted to a pelagic lifestyle. Their reproductive behavior also reflects adaptations to life in the open sea. Unlike most bony fishes, which employ external fertilization, sharks copulate (see

Figure 17–13

Ocean Sunfish. Not much is known of the life history of this large, odd-looking fish. (Copyright 1994 Mark Conlin/Mo Yung Productions/ Norbert Wu)

Chapter 10), with the male transferring sperm to the female. This increases the likelihood that the maximal number of eggs will be fertilized, and thus fewer eggs are produced. In many species the eggs are retained in the female's reproductive tract where they hatch and for a time are nourished on a milky fluid. This strategy helps to ensure better survival of the young when they enter the environment. Hammerhead sharks (*Sphyrna*) and blue sharks (*Prionace glauca*) not only hold their young within the body but also have a connection that supplies nutrients to the young from the mother's blood in a manner similar to that of mammals.

Sharks produce fewer young each breeding season than other fishes because their reproductive strategies are more efficient, and the survival rate of their offspring is better. For instance, while most bony fishes produce thousands of offspring each breeding season, the blue shark bears only about 30 young each season. Each is 60 to 70 centimeters (24 to 28 inches) long and fully independent when born, with a full set of teeth for feeding and defense.

Another inhabitant of the nekton that is related to the shark is the manta ray or devilfish (Mobulidae) (Figure 17–14). Fully adult mantas may measure 6 meters (20 feet) from the tip of one fin to the tip of the other and weigh as much as 1.5 tons. Mariners of old believed that manta rays could grab ships by their anchor chains and drag them to the ocean depths. It was also believed that these creatures would envelope swimmers in their large fins and devour them. Neither of these beliefs is true of course, but injured or provoked manta rays have been known to reduce small, wooden fishing boats to masses of floating splinters. Manta

Labial flaps

Figure 17–14

Manta Ray. The manta ray, or devilfish, feeds on small fishes and plankton that it scoops up with its large mouth. (Norbert Wu)

rays feed primarily on small fishes and plankton that they channel into their mouths with the large fleshy extensions called *labial flaps.*

The largest members of the nekton are whales. Baleen whales filter krill, small plankton, and sometimes fish, from the seawater. Toothed whales, such as sperm whales, feed on squid and fish (see Chapter 12 for a discussion of cetacean lifestyles).

CHAPTER SUMMARY

Although the open ocean appears quite uniform, it can be divided into regions based on the physical characteristics of the water and the kinds of organisms inhabiting them. It can also be divided into vertical zones, based on the amount of sunlight available. The open ocean lacks well-defined communities and is termed a *pelagic ecosystem.*

The primary producers of the open ocean are phytoplankton. Even though the open sea receives large amounts of sunlight, it lacks the supply of nutrients that make coastal seas so productive. The lack of nutrients is most pronounced in tropical oceans, where the water forms more or less permanent layers that are separated by thermoclines. The open sea does not support many large animals because of limited primary production and food webs involving several energy-wasting steps between primary producers and final consumers.

Delicate planktonic organisms, such as ctenophores, salps, and foraminiferans, which are frequently damaged when collected by plankton nets, can be readily captured by SCUBA-diving biologists. Study of the living organisms

has allowed marine biologists to determine that these organisms play a significant role in concentrating nutrients in planktonic ecosystems.

Many organisms have evolved special adaptations that allow them to be successful and multiply in nutrient-poor waters. Plankton biologists have also discovered that bits of organic matter and clusters of floating algae support miniature ecosystems composed of bacteria and numerous single-celled organisms.

Although the level of productivity is low, the open ocean provides a fairly stable environment. Organisms that live in this region do not have to adapt to changing environmental conditions, but they do face other challenges. The open sea does not provide animals with places to hide, so they have evolved other ways of avoiding predation, including toxins, speed, and camouflage. Another problem faced by pelagic organisms is remaining afloat. Some remain in the surface water by storing nutrient reserves in the form of oil droplets. Others have chambers that trap gases, while still others have body adaptations that increase friction and decrease the rate at which they sink. Some organisms keep themselves afloat by actively swimming.

Members of the zooplankton feed on the phytoplankton. Among the small animals of the zooplankton, crustaceans known as copepods dominate. Larger members of the zooplankton include jellyfish and some gastropod molluscs. The nekton consists of animals that are free swimming and whose movements are not governed by currents or tides, such as squids, fishes, and mammals. Members of the nekton feed on both phytoplankton and zooplankton as well as each other.

SELECTED KEY TERMS

microzooplankton, *p. 313*

macrozooplankton, *p. 313*

megazooplankton, *p. 313*

marine snow, *p. 314*

pyrosome, *p. 314*

rhizopodia, *p. 315*

micropatchiness, *p. 315*

ultrananoplankton, *p. 316*

QUESTIONS FOR REVIEW

MULTIPLE CHOICE

1. The layer of ocean that receives enough sunlight to power photosynthesis is the
   a. photosynthetic zone
   b. photic zone
   c. aphotic zone
   d. abyss
   e. benthic zone

2. A collective term for animals whose distribution is not governed by waves or currents is
   a. zooplankton
   b. pelagic
   c. nekton

3. Productivity in the open ocean is limited in most cases by
   a. sunlight
   b. temperature
   c. space
   d. nutrient availability
   e. salinity

4. The type of coloration exhibited by many large animals of the open sea is
   a. countershading
   b. disruptive
   c. cryptic
   d. warning
   e. reticulated

5. The rate of sinking can be slowed by anatomical adaptations that increase
   a. mass
   b. volume
   c. friction
   d. streamlining
   e. density

6. The largest members of the zooplankton are
   a. copepods
   b. diatoms
   c. jellyfishes
   d. sharks
   e. whales

7. Pteropods are related to
   a. jellyfishes
   b. diatoms
   c. snails
   d. lobsters
   e. sea stars

8. The upper jaw of the marlin and swordfish has been modified to form a
   a. tusk
   b. bill
   c. tooth
   d. sucker
   e. beak

9. Pelagic sharks produce fewer offspring than bony fishes that live in the open sea because
   a. young sharks require more food
   b. most young sharks are not carnivorous
   c. it is more difficult for adult sharks to find mates
   d. shark reproduction is more efficient
   e. there are more predators that feed on small sharks

10. Manta rays feed on
    a. dolphins
    b. squid
    c. large fishes
    d. molluscs
    e. plankton

SHORT ANSWER

1. What are some adaptations found in open-ocean plankton that allow them to be productive despite the nutrient-poor environment?

2. What is marine snow and what is its importance?

3. What are some adaptations that help organisms in the open ocean remain afloat?

4. What aspects of reproduction in pelagic sharks indicate adaptations to life in the open sea?

5. Describe some of the ways that animals living in the open sea hide themselves from predators.

6. Describe some of the strategies animals use to increase their survival in the open sea.

7. Describe how the upside-down lifestyle of the purple sea snail and the nudibranch *Glaucus* has influenced their coloration.

8. Describe how a swordfish feeds.

9. Describe how tunas are adapted for fast swimming.

10. Why is it beneficial for the primary producers of a pelagic ecosystem to be small?

11. Why is the photic zone of the open ocean not as productive as the photic zone of coastal seas?

12. The open ocean does not support as many individual large animals as coastal seas even though it covers a much larger area. Why?

THINKING CRITICALLY

1. Why does it make good ecological sense for the largest sharks to be plankton-feeders rather than carnivores?

2. The water of the North Atlantic Ocean is colder than that of the Atlantic Ocean in tropical regions. Do you think this difference might affect the shapes of phytoplankton living in these two areas? Explain your answer.

3. Why is it difficult to keep large tunas alive in public aquariums?

SUGGESTIONS FOR FURTHER READING

Hardy, A. 1965. *The Open Sea: Its Natural History.* Houghton Mifflin, Boston.

Hardy, J. T. 1991. Where the Sea Meets the Sky, *Natural History* 100(5): 58–65.

Raymont, J. E. G. 1980. *Plankton and Productivity in the Oceans,* 2nd ed. Vol. 1: Phytoplankton. Pergamon Press, Elmsford, NY.

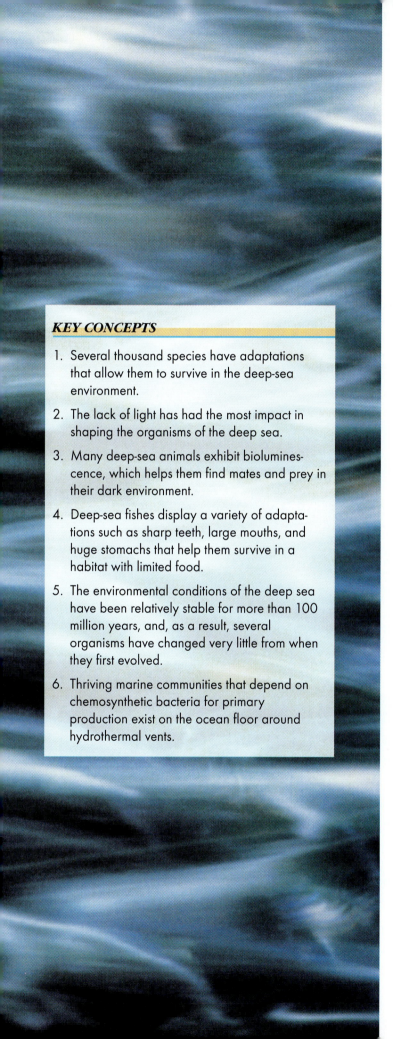

Chapter 18

Life in the Ocean's Depths

In the early years of marine biology research, scientists believed that no animal life could exist in the dark abyss of the sea. There was no sunlight to power photosynthesis, so there could not be any producers on which consumers would feed. The temperatures were too cold and the pressures too extreme to support life. As we noted in Chapter 1, retrieval of the transatlantic cable in the late 1800s demonstrated that animal life could exist beneath the photic zone. On July 21, 1951, the Royal Danish Research Vessel *Galathea* recovered a variety of animals, including sea anemones, sea cucumbers, bivalves, amphipods, and bristle worms from over 10,000 meters (33,000 feet) deep in the Philippine Trench east of the Philippine Islands. Eight years later, Jacques Piccard and Don Walsh, using the bathyscaphe *Trieste*, discovered animal life at even greater depths at the bottom of the ocean's deepest trench, the Challenger Deep. These and other discoveries proved that animal life not only could survive in the dark recesses of the ocean but could also thrive. In this chapter you will be introduced to the forbidding world of the ocean deep and the strange organisms that make it their home.

KEY CONCEPTS

1. Several thousand species have adaptations that allow them to survive in the deep-sea environment.

2. The lack of light has had the most impact in shaping the organisms of the deep sea.

3. Many deep-sea animals exhibit bioluminescence, which helps them find mates and prey in their dark environment.

4. Deep-sea fishes display a variety of adaptations such as sharp teeth, large mouths, and huge stomachs that help them survive in a habitat with limited food.

5. The environmental conditions of the deep sea have been relatively stable for more than 100 million years, and, as a result, several organisms have changed very little from when they first evolved.

6. Thriving marine communities that depend on chemosynthetic bacteria for primary production exist on the ocean floor around hydrothermal vents.

CHARACTERISTICS OF THE DEEP SEA

As late as 1843, the British naturalist Edward Forbes concluded that animal life could not exist in the sea below a depth of 55 meters (182 feet). At the time, his theory was completely plausible, since what was known of the conditions of the deep ocean would appear to make it a most inhospitable place for life. As we have noted previously, sunlight rapidly fades with increasing water depth. Even in the clearest ocean water, there is generally not enough light to drive the process of photosynthesis below 200 meters (660 feet). Heat energy is also absorbed quickly by surface waters. Temperatures in the deep remain frigid throughout the year.

Adding to these problems is the tremendous pressure of the ocean depths. The pressure at sea level is 1 atmosphere (760 mm Hg [millimeters of mercury] per square centimeter or about 15 pounds per square inch). About every 10 meters (33 feet) below the surface of the ocean, the pressure increases by another atmosphere. At this rate the pressure exerted on an organism at 1,000 meters (3,300 feet) would be 510 kilograms per square centimeter (1,500 pounds per square inch) or about 100 times greater than the pressure exerted at the surface. At a depth of 10,600 meters (35,000 feet), the pressure would be 5,100 kilograms per square centimeter (7.5 tons per square inch).

Even if living organisms could survive in such an environment, they would have to find their food and their mates in total darkness. As adverse as these conditions may

seem, they are at least stable, and, over millions of years of evolution, some organisms left the sunlit upper waters and adapted to life in the dark recesses of the sea (Figure 18–1).

Early marine biologists believed that the enormous pressures encountered in the ocean depths would literally crush an organism. Interestingly enough, several thousand species of animal, especially echinoderms, annelids, and molluscs, tolerate the pressures. The fluid pressure in these animals' tissues matches the pressure of the surrounding water. The tissue fluid pressure pushes against the surrounding pressure with an equal but opposite force, preventing the animal's body from being crushed.

Nearly all deep-sea animals have body temperatures that are close to the temperature of the surrounding water.

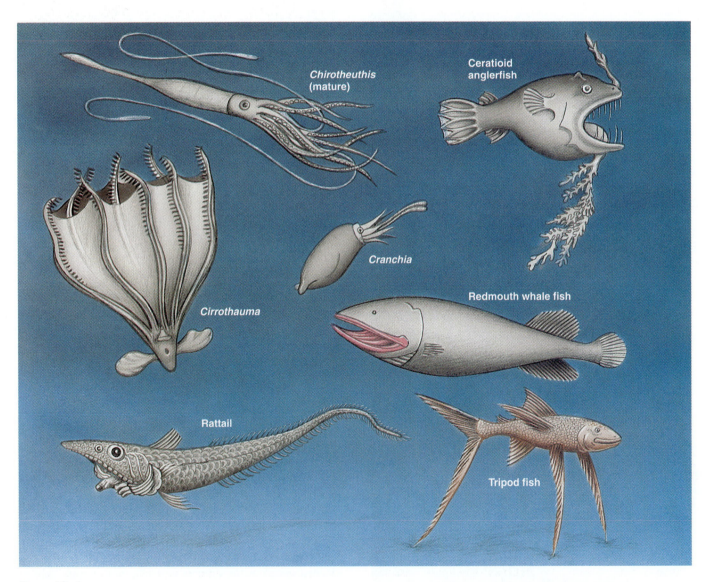

Figure 18–1

Deep-Sea Animals. Animals that live in the ocean's depths display a variety of intriguing adaptations that allow them to survive in a dark environment with cold temperatures and extreme pressures.

Their body temperatures are so low that their metabolism is quite slow. As a result they move more slowly, grow more slowly, reproduce less frequently and later in life, and generally live longer than similar species from warmer surface waters. A lower metabolic rate also means that these animals require less food, which in the ocean deep is scarce, to stay alive.

One advantage of the cold is the increased density of the water. Animals that live in the deep have body densities very close to the density of their environment and do not have to expend energy to keep from sinking, as do animals in the photic zone. (Recall from Chapter 10 that this is the reason deepwater fishes lack a swim bladder.)

LIFE IN THE DARK

Color in Deep-Sea Organisms

Of the three major factors that affect animals living in the ocean deep—light, temperature, and pressure—the lack of light has had the greatest evolutionary impact. As we noted in the previous chapter, fishes that inhabit the well-illuminated surface waters exhibit countershading: they have dark dorsal surfaces and lighter ventral surfaces. From 150 to 450 meters (500 to 1,500 feet) below the surface, a region referred to as the **mesopelagic zone** or **"twilight" zone,** there is still enough light to make countershading a useful means of camouflage. A common resident of these depths is the small hatchet fish (*Argyropelecus*; Figure 18–2). Its body is silvery or iridescent, with a dark dorsal surface and silvery underside. Like many deepwater fishes, hatchet fishes possess rows of light-producing organs called **photophores** along their bodies. These aid in species recognition, and the bioluminescence may make the ventral surface lighter, thus helping in camouflage.

Not all of the animals in this region, however, exhibit countershading. In the twilight zone, as well as the darker zones below, live animals that exhibit a variety of body colors. Black stomiatoid (STOH-mee-uh-toyd) fishes have an iridescent sheen. Fishes known as *gulper eels* and *swallowers* (Eurypharyngidae) are black or brown. These odd fish prowl the depths at 1,800 meters (6,000 feet), looking for prey. Whalefishes (Cetomimoidei) resemble whales in their general appearance but are only a few centimeters in length. Their fins and jaws are bright orange and red. Deep-sea species of squid are permanently colored deep red, purple, or brown, and many species are bioluminescent. Benthic species of squid are often white. Shrimp that are blood-red feed on red arrowworms and scarlet copepods. Recent studies by biologists at the Scripps Institution of Oceanography indicate the red colors of deep-sea crustaceans vary according to the state of the molt cycle and may actually be involved in energy transfer during molting.

Figure 18–2

Hatchet Fish. The silvery underside and dark dorsum of this hatchet fish is characteristic of the coloring of fishes that inhabit the twilight zone. (Norbert Wu)

The Roles of Bioluminescence

Even though sunlight cannot penetrate to the depths of the ocean, many deep-sea animals produce their own light in the form of bioluminescence. This characteristic is especially common in animals that are found between 300 and 2,400 meters (1,000 and 8,000 feet). Some, such as certain species of squid, crustacean, and fish, have their own luminescent organs, while others harbor bioluminescent bacteria in species-specific locations. This symbiosis between bacteria and their host species is an example of mutualism. The host receives light that can be used to locate and recognize a mate or find prey, and the bacteria are given a place to live and are supplied with food. Some species of fish are able to control the bioluminescence of their symbiotic bacteria by altering the flow of oxygenated blood to the regions

MARINE BIOLOGY AND THE HUMAN CONNECTION

Exploring the Ocean's Depths

For centuries humans have dreamed of being able to descend to the ocean's depths in order to explore them. Some even tried by descending in containers such as wooden barrels equipped with an air hose, but the depths they could reach were limited, as was their success in making observations. In this century, Dr. William Beebe pioneered underwater exploration. In 1934, he descended to a depth of 918 meters (3,028 feet) in a steel ball called a *diving bell* or *bathysphere*. The diving bell had portholes and lights and was supplied with air and power by a mother ship on the surface.

The first self-contained craft that could operate free from a mother ship was the bathyscaphe (BATH-eh-skayf), invented in 1948 by Auguste Piccard (Figure 18–A). The vessel was basically a sphere that could accommodate two people. It was suspended from a huge float that contained gasoline to power the craft's systems. Auguste's son Jacques and Don Walsh (a lieutenant in the U. S. Navy) used the bathyscaphe *Trieste* in 1960 to descend to the bottom of the Challenger Deep (10,800 meters or 35,800 feet).

Today's manned deep-sea submersibles are very sophisticated vessels equipped with the latest in photographic and robotic technology. Of the submersibles used by the United States, the best known is *Alvin* (see photo in Chapter 1), which can carry a crew of three. *Alvin* averages 150 dives per year and was used by Robert Ballard in 1977 to first view the animals of hydrothermal vent communities. Ballard used *Alvin* again in 1985 to view the sunken passenger ship *Titanic*.

The United States Navy operates a research-and-recovery vessel called the *Sea Cliff*. This vessel can operate at depths of 6,100 meters (20,000 feet), and in 1986 it was used by the United States Geological Survey to examine mineral deposits located off the coasts of California and Oregon. Two other government-owned submersibles are the *Johnson Sea Link 1* and the *Johnson Sea Link 2*. Each can carry a crew of four, and both participated in the efforts to recover pieces of the space shuttle *Chal-*

Figure 18–A

Early Manned Diving Devices. In 1960, the bathyscaphe *Trieste* was used by Jacques Piccard and Don Walsh to descend to the bottom of the Challenger Deep, the deepest point in the ocean. (UPI/Corbis-Bettman)

lenger following its disastrous explosion shortly after takeoff in January 1986.

Manned submersibles have been used to photograph sharks, survey coral reefs, and produce motion-picture footage for documentaries on the sea and its creatures. They are also used commercially in offshore petroleum developments and other engineering projects.

The drawbacks of using manned submersibles are that they are expensive to operate and pose risks to the passengers. To deal with these problems, engineers have developed a set of unmanned units called Remotely Operated Vehicles (ROVs; Figure 18–B). These carry photographic equipment, robotic arms, and other types of sampling equipment, and they can be controlled from the surface, eliminating the need for an on-board operator. The main disadvantage of ROVs is the need for a sizeable cable through which power is transmitted to the unit and data are transmitted back from the unit. However, this one disadvantage is far outweighed by lower construction and operating costs, not to mention the minimal risk to human life.

During the exploration of the *Titanic* in 1986, *Alvin* carried with it a small ROV called *Jason Jr.,* or JJ. JJ was attached to *Alvin* and was used to explore the inside of the *Titanic,* something the larger, more cumbersome *Alvin* could not do. The costs of ROVs are decreasing, and they are now frequently used in engineering projects such as offshore drilling work, bridge and dam construction, and laying pipeline. Military and law enforcement agencies are interested in the possible applications of ROVs, and there is a growing demand for them in the area of transportation, where they are used for maintenance and routine inspection of large ships. Although ROVs will probably not replace the manned submersibles, they do offer the ability to examine larger areas of the sea for longer periods of time at a greatly reduced cost and risk.

Figure 18–B

Remote Operating Vehicles, or ROVs. Jason, an ROV, is used by marine biologists to explore deep-sea ecosystems. (Robert Ballard/News Office, Woods Hole Oceanographic Institution, MA)

Figure 18–3

Bioluminescent Organs. Luminescent organs help species identify each other, locate prey, and avoid predation. They are frequently arranged in rows on the fish's body, as in this Pacific viperfish, or in depressions on the head. (Norbert Wu)

where the bacteria reside. When oxygen levels are high, the bacteria luminesce, and when the oxygen levels are low, they do not.

Bioluminescence occurs when a protein called *luciferin* is combined with oxygen in the presence of an enzyme called *luciferase* and ATP. During the series of reactions that follows, the chemical energy of ATP is converted into light energy. The process is efficient, producing almost 100% light and little heat. Most bioluminescence associated with marine organisms is blue-green, although red and yellow have also been reported.

Luminescent organs can be located at a variety of positions on an animal's body (Figure 18–3). Many deep-sea fishes display rows of photophores along their sides or on their bellies. Other species of fish may have depressions containing bioluminescent bacteria located on their heads or on fleshy growths attached to the head. Deep-sea squids exhibit bioluminescent spots on their tentacles and circling their eyes.

It would seem that such illuminated animals in a dark world would be easy marks for predators. In the twilight zone, where sunlight is fading into darkness, the presence of bioluminescence on the animal's ventral surface may take the place of countershading in camouflaging the animal. If the intensity of the bioluminescence matches the intensity of sunlight at that level, the fish will not produce any shadows or silhouettes when seen from below.

Bioluminescence can also play an important role in mating. During breeding times, the pattern of lights identifies an individual as being male or female. In lanternfishes (*Myctophum*), for example, males carry bright lights at the tops of their tails, while females have only weak lights on the underside of their tails (Figure 18–4). Like fireflies, some

species of deepwater fish signal their readiness to mate by a series of light flashes.

The pattern of lights also serves for species identification, especially among members of closely related species. For instance, one species of lantern fish has three rows of light spots along its ventral surface, while another closely related species displays only two. This would be of particular advantage in finding mates or in forming schools. The marine biologist William Beebe once remarked that the pattern of lights on deepwater fishes was so distinctive that he could recognize species in absolute darkness just by the light pattern that they presented.

Bioluminescence also functions to attract prey. Anglerfish (*Melanocetus johnsonii*) and stomiatoids attract prey with bioluminescent lures (Figure 18–5). Smaller fishes whose normal prey are luminescent are attracted by the light and, when close enough, become a meal for the larger predator. Dr. Beebe observed that a lanternfish apparently uses the light from its ventral surface to find prey. Copepods that were illuminated by the lights from the ventral surface of this fish were quickly eaten. Some species of stomiatoid fish have lights around their eyes that illuminate whatever the fish looks at. They have been observed to "fix" krill in the beam of light and then rapidly devour them. Deepwater krill also have illuminated spots around their eyes that may serve to help them find smaller prey.

Some animals use bioluminescence for defense. In the dark waters of the deep, the inky cloud produced by surface squids would have no effect in deterring a predator. Some deepwater squid species, however, release a bioluminescent fluid that clouds the water with light, confusing potential predators. This tactic is also used by some species of deepwater shrimp. The scarlet-colored opossum shrimp

Figure 18–4

A lanternfish. The bright, bioluminescent spots on the tail of this lanternfish identify it as a male. The pattern of bioluminescent spots helps the lanternfish recognize the opposite sex during the mating season. (Norbert Wu)

Figure 18–5

Anglerfish. The fleshy projection on the head of this anglerfish contains a bioluminescent "lure" at the end. Smaller fishes are attracted to the light and become an easy meal for the anglerfish. (Norbert Wu)

(Mysidae) gets its name because the female carries her eggs in a pouch under the thorax. When these shrimp are threatened, they release a substance that bursts into a cloud of miniature light particles that look like stars in the sky at night. The sudden burst of light frightens and confuses potential predators, allowing the shrimp to make an escape.

Seeing in the Dark

Do fishes in the aphotic zone of the sea perceive anything other than the light produced by other animals? To answer this question, the British biologist N. B. Marshall made an extensive study of the brains and eyes of deep-sea fishes, some species living as deep as 900 meters (3,000 feet) below the surface. He concluded that even though there is vir-

tually no light at these depths, these fishes can probably see their food, their own species, and their predators, utilizing what little light is available to stimulate vision.

Whereas the typical vertebrate eye is essentially spherical, the eyes of many deep-sea fishes are tubular and contain two retinas instead of one (Figure 18–6). One retina is used to fix images of distant objects, while the other is used to fix images of things that are closer. Not only do fishes with tubular eyes see better in dim light, but they also have better depth perception. This allows them to judge the distance to their prey quite accurately so that they are less likely to miss their catch. This is a definite advantage in an area where food is so scarce and where speed is not common.

The biologist G. L. Walls has noted that between 900 and 1,500 meters (3,000 and 5,000 feet), the eyes of fishes become smaller and less functional. The anglerfish has well-developed eyes during its early life, which is spent in well-lit surface waters. As it grows into an adult, it begins to sink and eventually resides at depths around 1,800 meters (6,000 feet). At this stage in its life, the eyes stop growing and sometimes begin to degenerate.

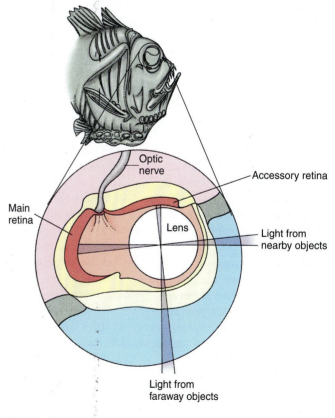

Figure 18–6

The Eye of Deepwater Fishes. Fishes that live in the ocean's depths have eyes that are adapted for seeing in dim light. They are tubular and contain two retinas. The shape of the eye gives them better depth perception so they are less likely to miss their prey when they strike.

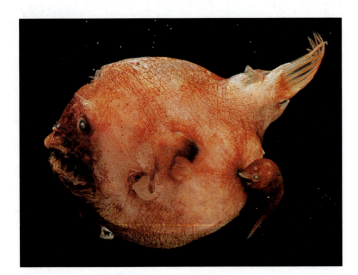

Figure 18–7

Mating in Anglerfishes. Male anglerfishes attach themselves to a female, and in some species the relationship is permanent, with the male becoming a lifelong parasite of the female. (Norbert Wu)

Deep-sea squids also have strange-looking eyes that can be barrel-shaped, stalked, or unequal in size. It is thought that these eyes are probably adapted to the amount of light available in their habitat and to their method of locating and capturing prey.

Some animal species, both vertebrate and invertebrate, that occupy the deepest recesses of the sea have tiny eyes that are only slightly functional, or they are totally blind. These organisms rely more on tactile and chemical stimuli to find their prey and mates and to avoid predation.

Some species of deepwater anglerfish have solved the problem of finding a mate in the dark in an unusual way. Since mates are hard to find, when a male encounters a female, he bites her and remains attached, sometimes for the rest of his life (Figure 18–7). In some species, the male becomes a lifelong parasite. The skin around the male's mouth and jaws fuses with the female's body, and only a small opening remains on either side of the mouth for gas exchange. The eyes and most of the male's internal organs degenerate, and his circulatory system becomes connected to the female's. The male, in effect, becomes an external, sperm-producing appendage.

It is believed that the female releases her eggs in deep water, and fertilization is external. The fertilized eggs quickly float upward to hatch in surface waters, where the young feed on copepods and other small plankton. As the young grow, females develop fishing lures, and males develop gripping teeth on both their snouts and chins. When the fish mature from larvae to adults, they descend and begin their deepwater existence.

Finding Food in the Dark

In the depths of the sea, food is quite scarce. There are no photosynthetic organisms to produce food, but organic wastes, scraps of food, and dead and dying organisms drift down slowly from above. Some of this food is consumed before it reaches the bottom, and the rest reaches the floor of the ocean, where it supports a variety of benthic organisms and active deep-sea scavengers. Detritus-feeders of the benthic community are, in turn, food for more active predators.

Not all of the deep-sea food chains, however, rely on the rain of detritus from the surface. Many small fishes and invertebrates rise at night to feed in the rich surface waters, returning during the day to the deeper recesses of the sea. These vertical migrations help to bring more food to the lower depths. Larger predators also make daily migrations to feed on smaller organisms.

Even at the best of times, food is scarce, and animals that survive in this environment exhibit some bizarre adaptations for feeding. Many deep-sea fishes appear to be nothing more than a huge, tooth-filled mouth that is attached to an expandable stomach with a small tail for swimming. Fish species called *gulper eels* (Figure 18–8) are common at depths below 1,500 meters (5,000 feet). They have jaws that are hinged like a trapdoor and stomachs that can expand several times their normal size to accommodate prey that is larger than the gulper eel itself. A research vessel captured a 15-centimeter (6-inch) gulper specimen with a 23-centimeter (9-inch) fish coiled in its stomach. Some species of gulper eel grow to 1.8 meters (6 feet) long. Most of this length is in their long, whiplike tail. The tip of the tail is lu-

Figure 18–8

Gulper Eel. Many deepwater fishes such as this gulper eel have huge mouths that can engulf prey as large as the predator. (Norbert Wu)

Figure 18–9

Stomiatoid Fishes. Stomiatoid fishes, like this black sea dragon, usually have a fleshy projection called a *barbel* dangling from their chin or throat. It is believed these structures function as lures to attract prey and as sensors for finding food in the bottom ooze. (Norbert Wu)

minescent and may play a role in luring potential prey close to the animal's mouth.

Stomiatoids (Figure 18–9) are also capable of ingesting prey larger than themselves. This gives them the ability to take advantage of the rare, large food items. Most species are 15 to 18 centimeters (6 to 7 inches) long. They have large heads with curved, fanglike teeth and elongated bodies that taper into a small tail. They are prominent residents of the deep from 180 to 600 meters (600 to 2,000 feet).

The typical stomiatoid has a fleshy projection called a **barbel** that dangles below its chin or throat. The precise role of these intriguing structures is not known, but it has been speculated that they may serve as lures, or they may be used to probe the bottom ooze for food. They may even serve in species recognition during mating.

The characteristics of the barbel vary from species to species. In some the barbel is a short hair, while in others it is a whiplike structure that may be ten times longer than the animal's body. One species that is only 4 centimeters (1.5 inches) long has a barbel that is 40 centimeters (15 inches) long, and a 21-centimeter (8.5-inch) species has a barbel nearly 1 meter (3 feet) long. The shape of the barbel is as variable as the length. Some are single strands, while others branch repeatedly; still others resemble a strange flower or even a cluster of grapes. Many are also bioluminescent.

Deep-sea anglerfishes that are found at depths of 1,970 meters (6,500 feet) have a bony projection with a luminescent lure at the tip that is a modification of one of the spines of the fish's dorsal fin. Through evolution, the first spine of the dorsal fin separated from the others, moved forward, and was modified into a fishing pole that, when not in use, lies in a groove on the top of the fish's head. Some species have short, stubby poles, while others have long, slender ones. Sets of muscles evolved that can move the pole forward, back toward the mouth, or out of the way when the animal is eating. At the end of the pole is a luminous "lure." The anglerfish's prey may mistake the lure for a worm or shrimp, and as it moves closer to investigate, the anglerfish moves the lure closer to its mouth. When the prey gets close enough, the anglerfish's lower jaw suddenly drops and the gill covers expand, creating an opening large enough to devour a prey as big as the predator. This rapid action creates a suction that sweeps the unwary victim into the anglerfish's waiting mouth. The jaws snap shut, impaling the victim on long, curved teeth on the roof of the mouth. The prey is then swallowed whole, pushed into the stomach by teeth in the predator's throat (pharyngeal teeth).

GIANTS OF THE DEEP

Most deep-sea animals are small compared to those living in surface waters, but a few are giants compared to their shallow-water relatives. One possible reason for the large size of these organisms is that they live longer than their shallow-water relatives and thus have time to grow larger. Sea urchins that live on the ocean floor have bodies that measure 30 centimeters (1 foot) in diameter, compared to shallow-water species that measure only a few centimeters across. In the deep sea off the coast of Japan live hydroids that reach the amazing height of 2.5 meters (8 feet). Sea pens that are ordinarily 0.5 meter high also grow as large as 2.5 meters in the ocean depths, and isopod crustaceans (related to terrestrial pill bugs) as large as 20 centimeters (8 inches) are found there. Shrimp species that are bright red

Figure 18—10

Giant Squid. The giant squid is the largest of all invertebrates. It feeds on fishes and is preyed upon by sperm whales. This specimen was found in the waters off New Zealand. (New Zealand Herald, Auckland)

or bioluminescent reach lengths nearly 30 centimeters (1 foot) long. Some of these species have antennae that are twice their body length, which may be used to capture small prey.

Perhaps the most spectacular of the deep-sea giants is a mollusc, the giant squid (*Architeuthis*; Figure 18–10). The giant squid is the largest of all invertebrates, averaging from 9 to 16 meters (about 30 to 53 feet) in length, including the

tentacles. Its arms are as thick as a human thigh and are covered with thousands of suckers. The two tentacles can be more than 12 meters (40 feet) long, and their flattened ends bear 100 or more suckers with serrated edges. As in other species of squid, the arms and tentacles capture prey and carry it to the animal's beak, where it is shredded into small pieces. The largest giant squid recorded was washed up on a New Zealand beach in 1888. It measured 18 meters (59 feet) in total length, with the tentacles accounting for 15 meters (49 feet) of the length. Although no giant squid has ever been accurately weighed, estimates run as high as 1 ton for a large specimen.

Little is known about the natural history of giant squids. They are found in all oceans, spending most of their lives at depths of 180 meters (600 feet) or more. They use the same method of swimming as other squids, a type of jet propulsion. Although there are reports of giant squids passing ships traveling 23 kilometers per hour (14 miles per hour), their anatomy suggests that they are weak swimmers that probably cannot capture active prey. No one knows for sure what these animals eat. It is assumed that they feed on a variety of small invertebrates and fishes. A major predator of giant squids is the sperm whale. Specimens of sperm whales have been found with scars of the serrated suckers on their bodies. Analysis of the stomach contents of captured specimens have revealed the remains of giant squids.

Living Fossils from the Deep

The process of evolution by natural selection is stimulated by changes in the environment (see Chapter 5). Since environmental conditions of the deep sea are believed to have remained nearly stable for over 100 million years, biologists have theorized that the sea's depths may hold many organisms that have undergone very little change from their early ancestors. Marine biologists were encouraged in this pursuit of "living fossils" by the discovery in 1864 by Norwegian oceanographers of a large sea lily that was dredged from 540 meters (1,800 feet). Similar species were known from fossils 120 million years old, but until this discovery no living specimens had ever been discovered. In 1870 another expedition discovered a large, red sea urchin from the depths of the North Atlantic. This genus was previously known only from 100 million-year-old fossils found in the white chalk cliffs of Dover, England.

The *Challenger* expedition of 1872 to 1876 netted a living specimen of *Spirula*, a mollusc that is named for its spiral-shaped internal shell. *Spirula* individuals are small molluscs that resemble squids and octopuses. They average 7.5 centimeters (3 inches) in length and have a barrel-shaped body with short, thick arms. The animal swims with its head pointing downward and contains an internal shell that is divided into gas-filled chambers. Similar molluscs called

Belemnites were common in the seas 100 million years ago but disappeared around 50 million years ago, with the exception of the ancestors of the *Spirula.*

In 1903 another strange animal called the *vampire squid* (Vampyromorpha; Figure 18–11) was discovered. The squid received its name because the webbing between the arms and dark color suggested the sinister figure of a mythical vampire to Carl Chun, who named it. The vampire squid appears to be the descendant of a group of molluscs that were intermediate between octopuses and squids. These animals, some with webbing between the arms and others with paddle-like fins, disappeared from the fossil record about 100 million years ago. It appears the vampire squid's ancestors may have avoided extinction by retreating to the ocean's depths around 900 to 2,700 meters (3,000 to 9,000 feet).

The vampire squid's muscles are soft and poorly developed, implying that the animal is not a very good swimmer. It probably drifts in the depths or moves feebly with its head down and the arms hanging limp. The arms are connected by a tissue that forms a webbing, and the combination forms a loose bag around the animal's mouth. Originally, the vampire squid was thought to be an octopus, but then researchers found two additional arms, bringing the total to ten, which is characteristic of squids. These two additional arms are long and lack suckers and are quite different from the tentacles of other squids. They are thought to act as feelers, and, when not in use, they are coiled into special pockets in the animal's web. Because of the animal's differences from squids and octopuses, they are placed in a different order (Vampyromorpha).

Vampire squids apparently spend their entire lives in deep water. Like other cephalopods, they have good eyesight, and they possess several bioluminescent organs, including two particularly large ones that can be covered by flaps of skin. They are thought to feed on small animals that are not particularly active, possibly using their long feelers to find their prey.

Vampire squids mate in a fashion similar to the octopuses and squids, with the male using one of his arms to insert a packet of sperm into the female genital opening. Spawning and hatching occur at depths of 2,000 to 2,500 meters (6,500 to 8,500 feet). When the young first hatch, they have eight arms. They develop the other two arms, the webbing, and light organs when they are about 2 centimeters (0.66 inch) in length.

Vampire squids have been collected in several areas of the world, including off the coasts of South Africa, India, Indonesia, and New Zealand. One was taken from a submarine canyon north of New Zealand at a depth of 3,000 meters (9,850 feet). The largest vampire squid ever taken measured 20 centimeters (8.5 inches) and was a female. No

Figure 18–11

Vampire Squid. This deepwater animal is not really a squid but an intermediate between squids and octopods. It received its name because of the webbing between the arms.

Figure 18–12

Neopilina. This limpet-like mollusc was known only from fossils until the discovery of a live specimen in 1957. (William C. Jorgensen/Visuals Unlimited)

males larger than 12.5 centimeters (5 inches) have ever been recorded.

In 1938, fishermen caught an unusual fish about 1.8 meters (6 feet) in length off the coast of South Africa. The fish had large, thick scales and fleshy bundles between its body and its fins. It was identified as the coelacanth (*Latimeria chalumnae*; see Chapter 10 for photo), a species thought to be extinct for 70 million years. This first known specimen was taken in relatively shallow water, 73 meters (240 feet). Subsequently, specimens have been caught at depths of 180 to 260 meters (600 to 1,200 feet). The discovery of the coelacanth gave biologists essential insights into the evolution of tetrapods.

In 1952, researchers aboard the vessel *Galathea* discovered an unusual, limpet-like mollusc called *Neopilina* (Figure 18–12) at 3,600 meters (12,000 feet) in the muddy clay off the Pacific coast of Costa Rica. This primitive mollusc belongs to the class Monoplacophora and was thought to be extinct for 350 million years. It measured 4 centimeters (1.5 inches) in length, and had a lightweight, conical shell. The animal had a large pink-and-blue foot, probably used to crawl along the bottom mud in search of food that it gathers with the short, fleshy tentacles behind the mouth. Since this first discovery, many other species of monoplacophorans have been found.

Discoveries of such living fossils from the deep sea are the exception rather than the rule, but marine biologists still believe that there are many more strange creatures and important discoveries waiting to be made in the ocean's depths.

VENT COMMUNITIES

Life exists not only in the deepest waters of the oceans, but also on the ocean floor. In 1977, oceanographers discovered a thriving community off the Galapagos Islands along volcanic ridges in the ocean floor. Mineral-rich water, heated by the earth's core, raises the temperature of the water in these regions (normally 2°C to 8°C) to 16°C. Since this initial discovery, marine scientists have discovered several other **vent communities** in other parts of the world including off the coast of Oregon, the west coast of Florida, the Gulf of Mexico, and in the central Gulf of California.

Despite the extremes of temperature and pressure, these self-contained communities are some of the most productive in the sea and demonstrate that communities can exist without solar energy. Not all vent communities surround hydrothermal vents. For instance, the Florida and Oregon vent communities are associated with cold-water seepage areas.

You will recall from our discussion of geology in Chapter 3 that vents form at spreading centers when cold seawater seeps down through cracks and fissures in the ocean floor. The water comes into contact with hot, basaltic lava, where it loses some minerals to the lava and picks up others, notably sulfur, iron, copper, and zinc. The superheated water returns to the sea through chimney-like structures formed by minerals that have precipitated from the hot water. Some of these rise as high as 13 meters (43 feet) off the floor of the sea. The chimneys are divided into two groups:

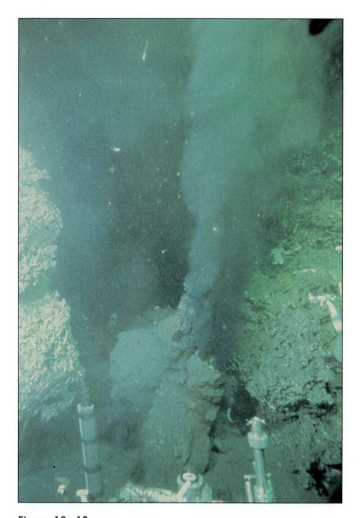

Figure 18—13

Black Smoker. The superheated water emitted from these chimneys is rich in sulfides. When the hot, sulfide-laden water mixes with the cold ocean water the sulfides form black precipitates that give the appearance of smoke rising from the chimneys. (Photo Researchers, Inc.)

white smokers and black smokers. White smokers produce a stream of milky fluid rich in zinc sulfide. The temperature of this water is normally less than 300°C. Black smokers are narrow chimneys that initially emit a clear water with temperatures of 300° to 450°C that is rich in copper sulfides. As the clear hot water encounters the cold ocean water, sulfide deposits begin to precipitate, producing the black color (Figure 18–13).

Surrounding deep-sea vents for a few meters in all directions are rich marine communities (Figure 18–14). The residents of these communities include large clams, mussels, anemones, barnacles, limpets, crabs, worms, and fishes. The clams (*Calyptogena*) are very large and exhibit the fastest known growth rate of any deep-sea animal (4 centimeters or 1.5 inches per year). Hydrothermal vent

communities also contain large worms called **vestimentiferan worms** (phylum Pogonophora) because of the covering produced to protect their bodies. Some of these worms may be as much as 3 meters (10 feet) in length and several centimeters in diameter.

The primary producers in hydrothermal vent communities are chemosynthetic bacteria (see Chapter 6) that oxidize compounds like hydrogen sulfide (H_2S) to provide the energy for producing organic compounds from carbon dioxide. Small animals, such as crustaceans, may feed directly on bacteria, while soft-bodied animals may be able to absorb organic molecules that are released by the bacteria when they die.

The clams, mussels, and worms are primary consumers, filtering the bacteria from the water or grazing on

Figure 18—14

Vent Community. Vestimentiferan worms and crabs are some of the larger animals found in this vent community in the Galapagos rift. (Science Visuals Unlimited/WHOI/D. Foster)

MARINE ADAPTATION

Light from Hydrothermal Vent Communities

Cindy Lee Van Dover, a biologist from the Woods Hole Oceanographic Institution, is interested in energy transfer and predator–prey relationships in hydrothermal vent communities. While engaged in a study of shrimp populations associated with Atlantic Ocean vents, she made an intriguing discovery. She noticed that, even though these animals lack eyes, they have a reflective spot located just behind the head. The spot consists of two lobes, each connected by large nerves to the brain, and it contains a light-sensitive pigment that is similar to the visual pigment found in the eyes of other animals. The evidence suggested that the spots might be modifications of the compound eyes found in other species of crustacean. Since the organ lacks a lens, the animal cannot see images, but it can sense the presence of light or similar radiation. Analysis of the pigment suggests that it is sensitive to the longer wavelengths of visible light (red) and some of the infrared spectrum. Since these are the first wavelengths to be absorbed by water, they are the least available in the ocean's depths. Why does this deep-sea shrimp have an organ that is sensitive to this portion of the light spectrum? One hypothesis is that it helps the shrimp locate prey by sensing the source of particular wavelengths of radiation emitted by the prey.

In order to confirm or deny this hypothesis, it was first necessary to determine if wavelengths of light in this range were actually present in hydrothermal vent communities. Answers to the question were provided by using the submersible *Alvin* to examine the vent community associated with the Juan de Fuca ridge system off the northwestern coast of the United States. The investigations involved taking digitized pictures of the vent with an electronic still camera, using different types of illumination. The vent was photographed without illumination, illuminated by a halogen lamp mounted on *Alvin*, and finally with illumination from a flashlight shining through *Alvin*'s port. When these photographs were analyzed, the data indicated that radiation coming from the vents was in the long wavelength range of the visible spectrum (800 to 900 nm). From these data, researchers concluded that either the waters of the vent and/or the substances they contain give off this radiation, or the internal wall of the vent chimney does, and the surrounding waters scatter and diffuse it.

The radiation that is associated with hot vents is an energy source that may be used in some way by bacteria that inhabit the area. The shrimp may use this energy to orient themselves relative to the vent or to locate their food. This discovery of light at the hot vents poses many new questions, and researchers are currently investigating how this energy may be related to the vent's biological communities.

---

the bacterial film that covers the rocks. The clams *Calyptogena,* mussels *Bathymodiolus,* and vestimentiferan worms *Riftia* have symbiotic chemosynthetic bacteria located in their tissues. *Riftia* has no mouth or digestive system; instead, the soft tissues of its body cavity are filled with symbiotic bacteria. The mussels have only a rudimentary digestive system, and both the clams and mussels harbor large numbers of bacteria in their gills.

The bivalves and worms have red flesh and red blood. The red color is due to the presence of the oxygen-carrying protein hemoglobin. Hemoglobin supplies tissues with the large amounts of oxygen that are necessary for oxidation of molecules like hydrogen sulfide. It also supplies the oxygen needs of the animal's tissues for growth and maintenance.

Sulfides that the bacteria require for chemosynthesis are provided by the host's circulatory system. *Riftia,* for instance, has a sulfide-binding protein in its blood that allows the animal to concentrate sulfides from the environment and transport them to the bacteria. Most animals would be poisoned by the high concentrations of sulfide circulating in the worm's body, but the sulfide-binding protein has such a high affinity for free sulfide ions that it effectively binds them up, thus preventing them from accumulating in the blood and entering and poisoning the worm's cells.

It is generally believed that deepwater vents are formed and ultimately become dormant on a regular cycle. Shortly after the vents form, they are colonized by living organisms. These organisms thrive until geological changes cause the vent to become inactive. Without a steady source of nutrients for chemosynthetic bacteria, the vent community dies, and the process is repeated at new vent sites. At some inactive vent sites, researchers have discovered pieces of clam shells. Since they know how long it takes the clams to grow and that it takes approximately 15 years for the shells to completely dissolve under the conditions around the dormant vents, they estimate that active vents must last about 20 years. Radiometric dating of materials from vent sites supports this estimate.

For the organisms of vent communities to continue to survive, then, they must colonize new vents as these vents develop. It has been demonstrated that, with their high growth rate, large clams can become mature in four to six years. It is believed that their large body size not only allows them to harbor large numbers of symbiotic bacteria but also to produce large numbers of larvae. If the larvae remain suspended in the water for periods of several weeks to several months, the deepwater currents could carry them over hundreds of kilometers, allowing them to disperse to other vent sites. Some biologists hypothesize that the larvae may rise to the surface to be distributed more rapidly by surface currents before sinking back to the seafloor. Another hypothesis suggests that the larvae of vent animals use whale carcasses as an intermediate habitat where they can grow, and in some cases asexually reproduce, before colonizing new vents. More research still needs to be done, however, before any of these hypotheses, or others that may be proposed, will ultimately be confirmed.

CHAPTER SUMMARY

Until the mid-1800s, biologists believed that life could not exist in the deepest parts of the ocean. There was no sunlight to support photosynthesis, and the cold temperatures and extreme pressures at great depths were thought to limit, if not exclude, animal life. Although the conditions of the deep are adverse, they are stable. This has allowed several thousand species to adapt to this habitat, an exciting discovery since the deep sea was expected to be a species-poor region.

Deep-sea animals metabolize more slowly, move more slowly, grow more slowly, reproduce later in life, and generally live longer compared to their counterparts in warmer surface water. Since their metabolic rate is low, deep-sea animals do not have to eat as much or as often to survive. Cold water is denser than warm water, so deep-sea animals do not have the problem of remaining buoyant that their surface relatives have.

Of all the factors that have shaped life in the ocean's depths, the lack of light has had the most impact. In the dark, camouflage is not a problem, since it is difficult for one animal to see another. Animals that inhabit the twilight zone exhibit countershading that helps to camouflage them in this region of minimal light.

Many deep-sea animals exhibit bioluminescence. Some have their own bioluminescent organs, while others harbor symbiotic bacteria that produce light. The bioluminescence helps animals to find prey and mates in a world of almost total darkness. It also plays a role in species identification and, in some species, acts to attract prey. Bioluminescence may also be used as a means of defense.

Food in the ocean's depths is scarce. Detritus that falls from above is consumed by some organisms before it reaches the floor, but most supports benthic organisms that are in turn food for more active predators. The nightly migrations of some fish and invertebrates to feed in the surface waters brings more food to the depths. The shortage of food has led to many interesting adaptations in deep-sea fishes.

Since the conditions of the deep have changed very little over the past 100 million years, biologists have speculated that some animals living in the deep may have remained unchanged since they first evolved. Over the years several examples of "living fossils" have been recovered from the ocean's depths, including invertebrates and fishes.

In 1977 oceanographers discovered thriving animal communities on the ocean floor. These communities are located around deep-sea hydrothermal vents that bring superheated water and nutrients from beneath the earth's crust. The basis of the food chain in these communities is chemosynthetic bacteria. These communities are found in both the Pacific and Atlantic oceans and are the focus of much current research.

SELECTED KEY TERMS

mesopelagic zone, *p. 329*

"twilight" zone, *p. 329*

photophore, *p. 329*

vent community, *p. 338*

QUESTIONS FOR REVIEW

MULTIPLE CHOICE

1. Compared to animals with higher body temperatures, animals with lower body temperatures tend to
   a. consume more food
   b. move more slowly
   c. grow more quickly
   d. reproduce earlier in their life cycle
   e. all of the above

2. The environmental factor that has had the greatest evolutionary impact on animal life in the deep is
   a. temperature
   b. pressure
   c. light
   d. salinity
   e. oxygen levels

3. The eyes of many deep-sea fishes are
   a. round
   b. spherical
   c. compound
   d. tubular
   e. identical to land vertebrates

4. Many deepwater fishes use _____ to attract prey.
   a. color
   b. bioluminescence
   c. sound
   d. odor
   e. tactile sensors

5. In deepwater animals, bioluminescence functions in
   a. species identification
   b. locating mates
   c. locating prey
   d. defense
   e. all of the above

6. Worms known as *vestimentiferans* lack a _____ system.
   a. respiratory
   b. circulatory
   c. nervous
   d. digestive
   e. reproductive

7. The worm *Riftia* has a sulfur-binding protein in its blood that allows it to
   a. produce its own food
   b. photosynthesize
   c. feed on chemosynthetic bacteria
   d. concentrate high levels of sulfide in its blood
   e. carry oxygen for bacterial metabolism

8. The coelacanth is thought to be related to the ancestor of
   a. molluscs
   b. sea stars
   c. octopuses and squids
   d. tetrapods
   e. vestimentiferan worms

SHORT ANSWER

1. What role does bioluminescence play in life in the deep?

2. What adaptations have evolved in deepwater fishes to help them find and capture prey?

3. What role do vertical migrations play in bringing nutrients into deep water?

4. Why did early biologists think that animal life could not survive in the ocean's depths?

5. Why are some deep-water animals bright red?

6. Why do some biologists believe that the sea's depths may harbor organisms that have changed very little from their early ancestors?

THINKING CRITICALLY

1. How might you determine whether an animal in a hydrothermal vent community derived its nutrition from symbionts or from feeding on other vent organisms?

2. The anglerfish spends the early part of its life in the upper regions of the ocean and its adult life in the abyss. What kinds of anatomical changes would you expect to observe as the animal matures?

3. Assuming deepwater fishes could survive the physical environment of surface waters, do you think that they would be able to compete effectively with fishes already adapted to that environment? Explain.

SUGGESTIONS FOR FURTHER READING

Childress, J., H. Felbeck, and G. Somero. 1987. Symbiosis in the Deep Sea, *Scientific American* 256(5):114–120.

Conniff, R. 1996. Clyde Roper Can't Wait to Be Attacked by the Giant Squid, *Smithsonian* 27(2):126–137.

Grassle, J. F. 1987/88. A Plethora of Unexpected Life, *Oceanus* 31(4):41–46.

Lutz, R. A. and R. M. Haymon. 1994. Rebirth of a Deep Sea Vent, *National Geographic* 186(5):114–126.

MacDonald, I. R. and C. Fisher. 1996. Life Without Light. *National Geographic* 190(4):86–97.

Van Dover, C. L. 1987/88. Do "Eyeless" Shrimp See the Light of the Glowing Deep Sea Vents? *Oceanus* 31(4):47–52.

Wu, N. 1990. Fangtooth, Viperfish, and Black Swallower: At 3,000 feet, the Light Goes Out and Life Depends on Strange Adaptations, *Sea Frontiers* 36(5):32–39.

Sport and commercial fishing are two of the many activities in which humans interact with the ocean and its inhabitants. Pollution, as in the exhaust from this outboard motor, is an example of human impact on the marine environment.
(Richard Gibson/Hi–Seas, Inc.)

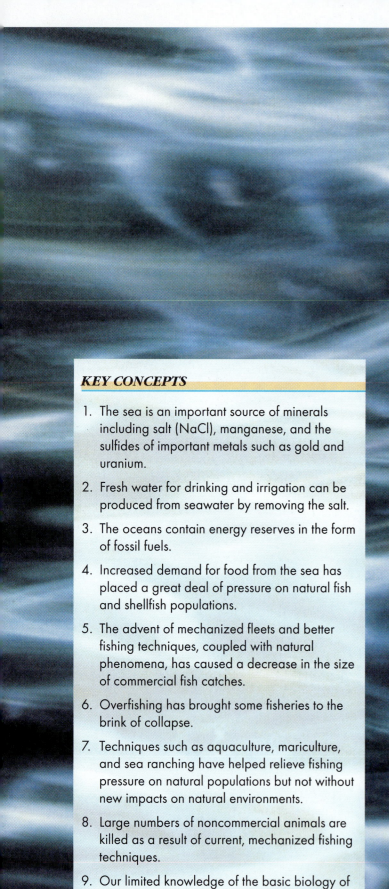

Harvesting the Ocean's Resources

The sea has traditionally been a major source of food for human populations. At one time, it was believed that the oceans were an infinite source of food, but, as we move toward the end of this century, we now realize that we must conserve precious fishery resources if they are to last for our generation and that of our children.

The sea is also seen as a potential new source of a variety of minerals and materials. The demand for natural resources has increased dramatically in the last hundred years, and the ability of onshore sources to meet the demand is being taxed. As terrestrial mineral resources are depleted, society will look with increasing interest toward the sea as a possible source of these minerals. Whether these mineral resources will be tapped to the fullest extent depends on a number of factors, and many ecologists are concerned that the development of these resources could have devastating effects on an already stressed ecosystem.

SALT AND WATER

The most obvious products of the sea are its major constituents: salt and water. About 30% of the world's supply of salt (NaCl) comes from seawater. The remaining 70% comes from terrestrial salt deposits that were formed when water from ancient seas evaporated. In order to keep the cost of production low, as little industrial energy as possible is used in the extraction process. In the south of France, Puerto Rico, central California, the Bahamas, Hawaii, and the Netherland Antilles, sea salt is extracted and refined to produce table salt. The process begins by allowing seawater to enter shallow ponds (Figure 19–1). Evaporation of the water produces a concentrated salt solution to which more seawater is added. Finally, the water is allowed to totally evaporate, leaving a thick salt layer behind that can be processed commercially. In colder regions where evaporation of the water is not feasible, seawater is allowed to enter shallow ponds and freeze. The ice layer that forms on top is nearly pure water, concentrating the salt in the water beneath. This highly concentrated saltwater is then removed and heated to drive off the remaining water.

In addition to salt, the sea provides a vast reservoir of water. Unfortunately, the salts must be removed to make the water useful for irrigation or drinking—a process known as **desalination.** The greatest problem with producing fresh water from seawater is the cost. The energy required to power desalination processes is expensive. In general, it costs more to produce fresh water by desalination than to obtain it from groundwater or surface water supplies. Whether desalination is used depends on how badly the water is needed, what other sources of water are available, the uses for the water produced, and the local cost of desalination.

In many areas of the world, the amount of fresh water is the major limiting factor for human populations and industrial expansion. These areas include Israel, Saudi Arabia, Morocco, Malta, Kuwait, and many Caribbean islands. In each of these areas, thousands of cubic meters of fresh water are produced daily by desalination. In Texas and California, desalination plants produce fresh water but at twice the cost of fresh water from other sources. The water produced by desalination in these states is primarily used for drinking and some light industry. It is too expensive, however, to be used for agriculture, which requires extremely large volumes of water.

MINERAL RESOURCES

The oceans contain large amounts of minerals, but not all of them can be easily obtained for commercial use. About 60% of the world's supply of magnesium and 70% of the world's supply of bromine come from seawater. Magnesium is combined with other metals to form alloys used in making portable tools and business machines and in the aerospace industry. Bromine is used as a disinfecting and bleaching agent and a reagent in numerous commercial chemical applications.

It is estimated that the seas contain as much as 10 million tons of dissolved gold and 4 billion tons of uranium. The concentrations, however, are on the order of 1 part per billion or less, and no one has yet devised a method of extracting these minerals from seawater where the costs of extraction did not greatly exceed the value of the product.

Sulfides

During the 1970s and 1980s, deposits of sulfides (mineral compounds containing sulfur) were found at several sites in the oceans. These deposits form when hot molten material beneath the earth's crust rises along rift valleys, heating the rocks and causing them to fracture, forming mineral-rich solutions. When the solutions come into contact with the colder seawater, the minerals precipitate to the seafloor and can form massive deposits more than 10 meters (33 feet) high and hundreds of meters long. Minerals found in the form of sulfide ores in various regions of the world include zinc, iron, copper, gold, silver, platinum, molybdenum, lead, and chromium. At this time, however, the technology does not exist for selective sampling of these deposits or for mining them.

During the 1960s, muds containing metallic sulfides of iron, copper, zinc, and small amounts of gold and silver were discovered in the Red Sea. These deposits were 100 meters (330 feet) thick in small basins at depths of 1,900 to 2,200 meters (6,200 to 7,200 feet). The salt water that lies

Figure 19-1

Salt Ponds. A salt pond on the island of Bonaire in the Caribbean. Seawater is trapped in ponds such as this. When the water evaporates, it leaves behind deposits of salt, Bonaire's major export. The high-salinity water teems with brine shrimp and larval brine flies, the favorite foods of West Indian flamingos like this one. (Copyright 1990 C. Allan Morgan/Peter Arnold, Inc.)

over these deposits contains hundreds of times the concentration of these minerals compared to normal seawater. The estimated value of these mineral deposits is in the billions of dollars.

Manganese

Manganese is an element used in industry as a component of several alloys. In combination with iron it adds strength to steel and makes it easier to mold. In combination with copper it forms alloys that are quite sensitive to temperature changes, and it is used in temperature-activated switches such as thermostats. This very important mineral is found in many areas of the deep ocean along the floor in the form of manganese nodules. It is estimated that 16 million tons of these nodules accumulate on the ocean floor each year. Although first discovered in the 19th century by the *Challenger* expedition (see Chapter 1), only during the last 25 years has industry concentrated on ways of mining and extracting these nodules. Since the 1960s, several multinational groups have spent over $600 million dollars to locate the nodules and develop methods to collect them. Progress has been slow, and by the 1980s many groups had terminated or suspended their activities. The depressed market for metals as well as legal problems with ownership are two reasons for the decreased activity.

SAND AND GRAVEL

The most widespread seafloor mining operations extract sand and gravel that is used for making cement and concrete and for building artificial beaches. The process of extracting the material is essentially the same as in land-based operations, and the major cost is based on the distance the material must be transported. Approximately 112 billion tons are extracted worldwide annually. The size of the ocean's reserves is estimated at 600 billion tons. The removal of sand and gravel from the seabed is the only major seabed mining that is done by the United States at this time. The United States has an estimated 450 billion tons of sand reserves off the northeastern coast and large gravel deposits off the coast of New England in the area of the Georges Bank as well as off the coast of New York City.

Deposits of calcium carbonate are found along the coastal areas of Texas, Louisiana, and Florida. The calcium carbonate from these deposits is used for lime and cement, as a source of calcium oxide for the removal of magnesium from seawater, and as a gravel substitute in road construction. In the Bahamas, sands are mined as a source of calcium carbonate, and reef sands are mined in Florida, Hawaii, and Fiji.

Coastal sands in some areas of the world contain deposits of iron, tin, uranium, platinum, gold, and even diamonds. For hundreds of years tin has been extracted from sand dredged from the coastal regions of Southeast Asia from Thailand, along Malaysia to Indonesia. Today 1% of the world's tin is obtained from coastal sand in this region. In Thailand, tin mining has resulted in heavy deposits of silt in the intertidal and subtidal regions—all but destroying the productivity of these coastal habitats. It is estimated that sands in the shallow coastal waters of Japan hold a reserve of 36 million tons of iron.

Since 1972, the (former) Soviet Union has extracted uranium from bottom sediments of the Black Sea; the United States, Australia, and South Africa extract platinum from some coastal sands. In all of these instances, there is widespread concern about the effects of pollution and habitat destruction that accompany the extraction of minerals from the seabed.

ENERGY SOURCES: COAL, OIL, AND NATURAL GAS

Coal, oil, and natural gas are collectively referred to as **fossil fuels** because they are formed from the remains of plants and microorganisms that lived millions of years ago. Coal was formed from the remains of plants, such as ferns, that lived in prehistoric swamps. When these plants died, they generally fell into the swamp and were covered with water. The fungi that play a major role in the decomposition of woody plant material could not survive in the anaerobic swamp water, and other decomposers, such as anaerobic bacteria, don't decompose wood very rapidly. As a result, very little decomposition took place, and, over time, the plant material became buried under layers of sediment. The sediment compressed the organic material, and, over millions of years, the heat generated from the minimal decomposition and the pressure of being buried in the earth's crust converted the plant material to coal. Some of these deposits were eventually submerged under the sea and offer a source of coal when the quantity and accessibility make it feasible. In Japan coal is mined from under the sea using shafts that originate on land or that descend from artificial islands.

Ninety percent of the mineral value taken from the sea at this time is in the form of oil and natural gas. Oil and natural gas were formed from the remains of microorganisms, such as diatoms. When these organisms died, their remains settled to the bottom of the sea and became covered with sediments. The pressure and heat of being buried under tons of sediments converted the remains of these organisms to oil and natural gas over millions of years. Major offshore oil deposits are located in the Persian Gulf, North Sea, Gulf

Figure 19–2

Oil Platform. Pumping equipment used to remove oil from the seabed is set up on large platforms such as this one in the Gulf of Mexico. (C. Lockwood/Earth Scenes/Animals Animals)

of Mexico, the northern coast of Australia, the southern coast of California, and the coastlines that border the Arctic Ocean. Many areas are still unexplored, such as the continental shelves of Asia, Africa, parts of South America, and Antarctica. Recently, oil has been discovered off the mouth of the Amazon river in South America and near the Philippine Islands in the Pacific Ocean. In order to extract the oil and gas, industry has had to develop huge drilling platforms and specialized equipment that can be used for drilling and developing wells at great depths (Figure 19–2). This was a particularly challenging task for engineers in the North Sea, where heavy seas and frequent storms are common.

Presently, the annual revenue worldwide from offshore oil and gas production is $80 billion dollars. The offshore reserves represent about one third of the world's estimated total reserves. Although the cost of drilling and extracting offshore oil is about three to four times greater than the same development on land, the size of the deposits make the process financially worthwhile in some regions. Little is known about the oil reserves in the deep sea, but the deeper the reserves, the more expensive it will be to recover. Methods and equipment that have been developed for oceanographic research and deep-sea drilling have provided industry with models for developing the next generation of oil- and gas-drilling equipment. The development of various offshore sites proceeds slowly because of environmental concerns, legal restraints, and the uncertainty associated with global oil supplies. It appears that petroleum exploration and development will continue to be the main focus of ocean mining for the future.

COMMERCIAL FISHING

During the past 50 years there has been a dramatic increase in the amount of fish taken from the sea. In 1950 the total world catch of marine fish was approximately 21 million tons. By 1994, it reached 90 million tons. The increase in commercial fishing is the result of increased demand brought about by the growth in the number of humans (Figure 19–3). The increased demand was followed by more intense fishing and the development of more sophisticated fishing techniques. Technological advances played a major role in increasing both the yield and the catch of commercial fisheries. However, recently the world catch has not been increasing proportionately to the expanded fishing effort.

During the same period of time, there has been a change in the use of the commercial catch. In 1950, 90% of the world's catch was used for human consumption, and the remaining 10% was used for fish meal products, mainly to feed the livestock of more developed countries. By 1980, 60% of the catch went to feeding people and 40% to feeding domestic animals, a trend that continues today. Recall the 10% rule (see Chapter 2) that states only about 10% of the energy available at one trophic level is passed on to the next. Based on this observation, it becomes apparent that feeding livestock with commercial catch is not a very

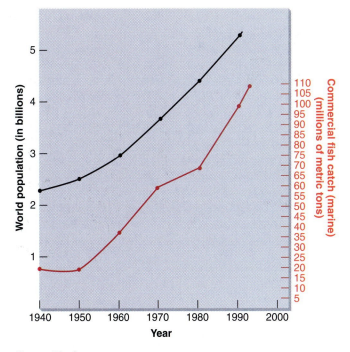

Figure 19–3

World Population and Commercial Fishing. Since 1940, as the population of the world has increased, so has the demand for fishery products. (Data from the United Nations Organization)

MARINE BIOLOGY AND THE HUMAN CONNECTION

Using Genetics to Improve Ocean Fisheries

Fisheries biologists in several countries are beginning to use the modern techniques of gene and chromosome manipulation to improve ocean fisheries. One technique involves the production of triploid individuals (animals that have an extra set of chromosomes in each of their cells). Triploid salmon can be produced by exposing salmon eggs to particular stimuli such as changes in temperature or chemical shock. Such stimuli interfere with the ability of chromosomes to separate properly during cell division, resulting in triploid individuals that fail to mature sexually. Since the triploid fish cannot spawn, they feed longer and grow to greater size. The energy that they would normally invest in reproduction is used for increased growth. Triploid coho salmon (*Oncorhynchus kisutch*) are currently produced in fish farms in Washington state.

Triploid salmon have also been hybridized with triploid trout in vitro. Hybrids between salmon and trout with normal numbers of chromosomes do not survive, but hybrids between triploids do. Hybrids of rainbow trout (*Salmo gairdneri*) and coho salmon are resistant to a viral disease that kills many trout and causes great economic losses for trout farmers. Currently these hybrids are being raised on fish farms in Idaho.

A technique called *gynogenesis* uses ultraviolet light to alter the chromosomes in fish sperm. Eggs that are fertilized by this sperm will produce only females. In many commercial fish species, females live longer and grow larger. Female flounders, for instance, grow to twice the size of males. Increasing the number of females in a population would then increase the yield. It is possible to reverse the sex of a female fish that is produced through gynogenesis by treating it with male hormones. The fish is still genetically female but develops testes and produces viable sperm. Eggs fertilized by such "males" will yield only female offspring.

Researchers are currently using the techniques of genetic engineering and recombinant DNA to remove genes for a particular trait from one fish and insert them into another. These techniques are especially useful for transferring genes for disease resistance or growth factors. The techniques of bioengineering are not restricted to fish. Biologists are experimenting with commercially important shellfish as well. For instance, at the University of Maine, triploid oysters (*Crassostrea*) known as "four seasons" oysters are being produced. Oyster fishing is generally a seasonal activity since mortality tends to be high in the summer. During the summer reproductive phase, oyster meat is of poor quality and not marketable. Since triploid oysters are sterile, they continue to grow instead of reproducing, producing more meat of better quality during the usually poor summer months. Researchers are now experimenting with a triploid form of the blue mussel (*Mytilus edulis*).

The bulk of genetically engineered fish are raised in mariculture and not released into the ocean. Before large numbers of these sterile or genetically altered animals are released into the environment, biologists must answer several questions. Will the inbreeding of genetically altered species produce animals that have less resistance to disease or environmental variables? Do the sterile fish grow at the same rate as their natural counterparts? Will the introduction of too many sterile fish drastically reduce the breeding potential of natural populations? Will the introduced animals compete with natural populations for vital resources, and, if so, what will be the effects of the competition?

For the time being, these genetically altered animals are producing important yields and helping to relieve the pressures on natural populations, allowing them to recover. It is possible that in the future genetically altered commercial catch may help to preserve important fisheries.

efficient use of this resource. For each ton of commercial fish fed to animals, only about 0.1 ton of animal protein would be produced. Most of the commercial fish catch consists of fishes that are in higher trophic levels, such as tunas. This is a greater waste of energy than harvesting fishes at lower trophic levels.

The pressures of demand for fish have caused some significant decreases in traditional fisheries such as tuna, while stimulating the development of new fishery-related products. One such new product is surimi, which is made from Alaskan pollock (*Theragra chalcogramma*), a bottom fish (Figure 19–4). The flesh is processed to remove the fats and

Figure 19–4

Surimi Products. Many processed seafood products are made from surimi, a refined protein made from a bottom fish, the Alaskan pollock. (George Semple)

oils that give the fish its characteristic flavor, and a highly re-fined fish protein, surimi, is produced. The surimi can then be flavored to produce artificial crab, shrimp, lobster, or scallops. Surimi is a major fish product in the Japanese mar-ket and is the product with the fastest-growing demand in the U.S. fish market.

Anchovies

Anchovies (Engraulidae; Figure 19–5a) are small, fast-grow-ing fishes that feed on the bountiful phytoplankton in up-welling areas along the coast of Peru. Since anchovies travel in large, dense schools, it is possible to catch large numbers with nets. Anchovies represent the world's greatest fish catch for any single species. Most of the anchovies that are caught are not used for human consumption but end up as fish meal used to feed domestic animals. The fishing of an-chovies commercially for fish meal began in 1950, with an annual harvest of 7,000 tons (Figure 19–5b). By 1962, world demand for fish meal had so increased that the an-chovy fishery yielded 6.5 million tons of fish. By 1970, this amount had increased to 12.3 million tons. As the tonnage taken increased over the years, the size of the fish being caught decreased, a common effect of overfishing. This means it took more and more fish to make up one ton.

(a)

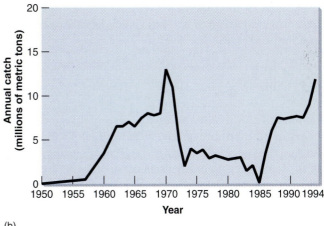

(b)

Figure 19–5

Anchovies. (a) Anchovies swim in dense schools. (b) The Peruvian anchovy catch since 1950 shows the combined impact of overfishing and natural phenomena such as El Niños. (a, Norbert Wu; b, Data from the United Nations Food and Agriculture Organization)

In 1953, 1957 to 1958, 1965, and 1972 to 1973, the Peruvian coast was hit by major El Niños (see the box on Ecology and the Marine Environment: The El Niño Southern Oscillation in Chapter 4). During an El Niño, the change in wind patterns produces changes in surface currents and upwellings. This leads to a decrease in nutrients in the surface water, a lack of production by phytoplankton, and a massive die-off of zooplankton. The decomposing organic matter robs the surface water of oxygen and adds to the death toll. Seabirds and other animals that depend on anchovies for food die of starvation or migrate.

The intense fishing that produced the record catch of 1972 coupled with an El Niño event later that year caused a dramatic decrease in the 1973 harvest to only 2 million tons. Alarmed by the situation, the Peruvian government set restrictions and placed quotas on the anchovy catch. As a result of these measures, the fishery began a slow recovery in 1974 and 1975. Then in 1976 another El Niño caused another major decrease in the catch. From 1977 to 1982 catches again began to show a small increase, reaching 2 million tons again by the end of 1982. An El Niño in 1982 and 1983 caused another decline, and between 1983 to 1985, the catch was below 150,000 tons. By 1986, the catch was back to 3.4 million tons, but El Niños continue to appear periodically, the last in 1993, and this will continue to cause fluctuations in the anchovy catch. As the anchovy population declines, there is some evidence that sardines may be entering their niche, placing more pressure on the anchovy population.

There is a general lesson to be learned from these fluctuations in the anchovy fishery. Overfishing, especially when combined with unpredictable natural phenomena, can result in disastrous changes in key marine resources. Overfishing results in fewer reproductive adults in the population, which yields fewer young. Over time, the population continues to get smaller. Without proper safeguards and conservation methods, important fisheries may collapse forever.

Figure 19–6

Red Drum. Red drum are popular sport fish as well as an important commercial catch. (Marc Epstein/Visuals Unlimited)

Red Drum

The red drum or redfish (*Sciaenops ocellatus;* Figure 19–6) occurs in both inshore and offshore waters of the Gulf of Mexico. The larger fish, which are believed to be the spawning stock, congregate in large schools in offshore waters. In 1983, the commercial redfish catch was 1,500 tons. Of the commercial catch, 7% came from offshore waters. By 1986, the commercial catch had doubled, and nearly all of the fish were taken from offshore stocks. The dramatic increase in the commercial catch was attributed to the popularity of a Louisiana style of cooking known as *Cajun*, and of one dish in particular, blackened redfish.

In order to meet the high demand for redfish, commercial fleets set out with state-of-the-art equipment, including purse seines. Purse seines are huge nets with bottoms that can be closed off by pulling on a line, similar to the way in which a purse or bag of marbles can be closed by pulling on the drawstrings (Figure 19–7). A purse seine can catch as much as 68,000 kilograms (150,000 pounds) of fish each time the net is set. When a school of fish is located, a powerful skiff is launched from a large vessel called the *purse seiner*. The skiff hauls the purse seine in a circle around the school and back to the purse seiner. The bottom of the net is then closed off, and the closed seine containing the fish is hauled on board the purse seiner. This method of fishing is preferred because it yields a much larger catch with less effort than techniques such as pole fishing.

Alarmed by the possible consequences of overfishing, the U.S. Federal government placed a 500-ton-per-season limit on redfish caught in offshore waters. The coastal states of Texas, Louisiana, Mississippi, Alabama, and Florida passed legislation that prevented the unloading of any redfish caught in purse seines. As a result of these protective measures, the commercial catch decreased to 2,400 tons in 1987. Continuing analysis of the situation led to prohibitions on commercial fishing in some areas, while limiting the recreational catch. Both were banned in other areas. In 1988, the U.S. government closed federal waters to commercial fishing of redfish because the quota of 850 tons (combined recreational and commercial catch) had been exceeded. The following year the commercial catch of redfish was 144 tons. Recreational fishing of the redfish was also curtailed by most states in 1988 to allow existing stocks to recover, while protective legislative measures were determined. It is imperative that this important commercial species continue to be protected before it is completely overfished.

Tunas

Tuna fishing is big business for many countries including the United States. The fishing of tunas is a sophisticated and expensive operation, with a modern tuna boat costing in the neighborhood of $5 million. Tunas are caught with huge purse seines that can measure 1,100 meters (3,600 feet) long and 180 meters (595 feet) deep. These large nets take advantage of the schooling behavior of tuna. A single setting of the net can harvest as much as 150 tons of tuna. Although this method of tuna fishing is highly efficient, it does have one drawback. Dolphins are frequently trapped in the nets with the fish and drowned. For unknown reasons, dolphins follow schools of tuna on their ocean migrations, swimming in the surface water just above the school. Tuna fishers actually watch for dolphins to locate schools of tuna. The total number of dolphins killed in tuna fishing is unknown, but it is estimated that the largest kill occurred in 1970, when more than 500,000 dolphins died.

In 1972, the Marine Mammal Protection Act was passed. This act, among other things, made efforts to reduce the number of dolphins killed in tuna fishing. The act allows the National Marine Fisheries Service (a division of the National Oceanic and Atmospheric Administration, or NOAA) to set a limit on the number of dolphins killed and to recall the fishing fleet if the limit is exceeded. The limit was set at 20,500 dolphins, and the only time that the fleet was recalled for exceeding the limit was in 1977.

In addition to setting limits, the law mandates that an observer from the Marine Mammal Protection Agency of the

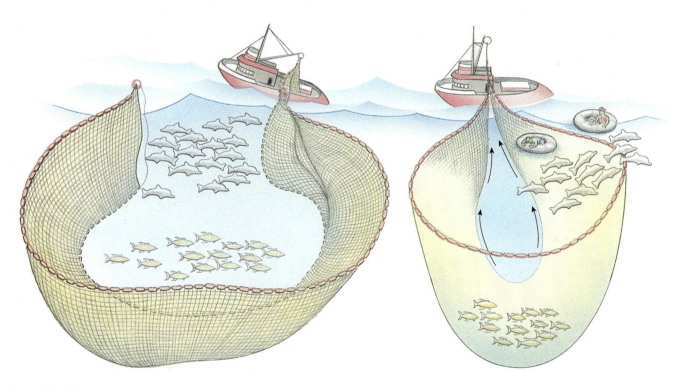

Figure 19—7

Purse Seine Net. The purse seine net is set in a large circle around a school of fish. The ends of the net are then drawn together and pulled closed, much like pulling on a drawstring bag. Unfortunately, when this technique is employed in tuna fishing, dolphins are also trapped in the net and drowned.

(a)

Figure 19—8

Salmon. (a) Salmon are popular with commercial fishers because the fish are large and bring a high price per pound. (b) Many salmon are now reared in hatcheries and released to the sea. This practice relieves some of the pressure on natural populations. (a, Shane Moore/Animals Animals)

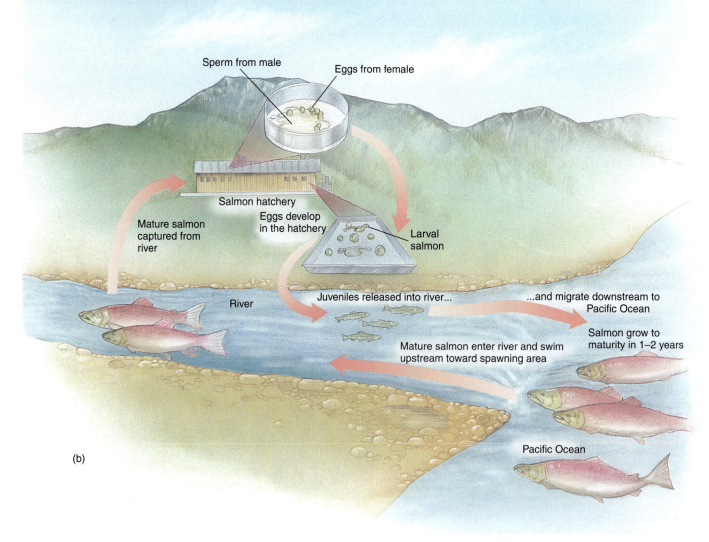

Sperm from male

Eggs from female

Salmon hatchery

Eggs develop in the hatchery

Larval salmon

Mature salmon captured from river

River

Juveniles released into river...

...and migrate downstream to Pacific Ocean

Salmon grow to maturity in 1–2 years

Mature salmon enter river and swim upstream toward spawning area

Pacific Ocean

(b)

National Oceanographic and Atmospheric Administration (NOAA) be present for at least part of the fishing season on the tuna boats. These observers not only count the number of dolphins that are killed but also monitor fishing techniques and methods used to free the dolphins. Based on these observations, the observers attempt to develop new strategies and equipment for protecting the dolphins. Some new strategies that have been developed have not been used. For instance, not setting the nets at night will reduce dolphin kills, but since this will also reduce the size of the tuna catch, it is not practiced. Most U.S. tuna boats are now based in foreign countries, where they are free from these restrictions.

By 1990 the National Marine Fisheries Service estimated that the number of dolphins killed in tuna fishing by the American fleet was zero. However, dolphins continued to be killed by foreign fishers who continued to use older fishing methods. Also in 1990, the major tuna packers in the United States agreed that they would only buy tuna from fishers that did not harm or kill dolphins in the fishing process. To indicate this, their products carry a seal indicating that the tuna is "dolphin free."

Salmon

Another large fishery in which the United States plays a prominent role is the salmon fishery. Pacific salmon (*Oncorhynchus*; Figure 19–8a) are fished in the coastal waters of the Pacific Northwest and Alaska. Salmon breed in fresh water (see Chapter 10) but spend most of their adult lives in the ocean. Most of the commercial and sport salmon catch is made when the adults begin to migrate back to inland streams to reproduce. Because there is a large demand for salmon, the fish bring a high price per pound. As with other highly exploited fishes, salmon are becoming scarce, and their fishing is tightly regulated. Current regulations have shortened the fishing season and decreased the allowable catch, but fishers from several Asian countries still take these fish illegally on the open ocean.

The problem of maintaining salmon populations does not involve just overfishing but also protecting their spawning grounds. Salmon require shaded, clear, cool, fast-running, unpolluted streams for reproduction. Damming of rivers for flood control and hydroelectric power has prevented salmon access to many streams. Harvesting of trees by the lumber industry removes shade and allows runoff of silt that clouds the water. Without shade, the water temperature rises. Pollutants such as human wastes have injured young fish and decreased water quality. Although these problems are being dealt with, they are not being dealt with effectively or quickly enough, and the spawning population is quite small.

The cost of salmon continues to increase not only because there are fewer fish but because of the cost of cleaning up their native streams and paying for the construction of hatcheries to supplement natural populations. About half of the salmon now caught at sea are from hatchery-reared juveniles (Figure 19–8b).

Shellfish

Another important component of marine fisheries are the many types of commercially important shellfish. Crustaceans such as crabs, shrimp, prawns, and lobsters; and molluscs such as clams, mussels, oysters, scallops, octopus, and squid are the most important. Estimates of the world's shellfish catch are over 6 million tons of molluscs and 4 million tons of crustaceans. Although the numbers, compared to fish, seem rather low, the high demand for shellfish places a high dollar value on each pound of catch. In the United States, oysters are harvested from New England, the Gulf of Mexico, Puget Sound and Washington coastal estuaries, and Chesapeake Bay (Figure 19–9). Lobsters are fished in New England (*Homarus americanus*) and Florida (*Panulirus* and *Scyllarides*). Shrimp are fished in the Gulf of Mexico, the Atlantic Ocean off the coasts of North and South Carolina and Georgia, and along the West Coast. Scallops (*Argopecten*) are fished in the Gulf of Mexico and along the southern Atlantic states. Crabs and clams are caught along virtually all of the coastal areas.

Shellfish fisheries face the same problems as the rest of the fishing industry. In addition, they are hard-hit by pollution that contaminates estuaries and nearshore waters and contributes to toxic algal blooms (red tides), which render shellfish poisonous.

In the late 1970s the demand for Alaskan king crab (*Paralithodes camtschatica*) increased dramatically. This animal is mainly fished in the Bering Sea, and as a result of the increased demand, there was a tremendous increase in fishing pressure. Between 1979 and 1980 more than 45,000 tons of crab were harvested. The following year the catch declined to 21,000 tons, less than half, and by 1984 the catch had declined to 4,000 tons. A combination of overfishing and lack of knowledge of the crab's natural history are blamed for the drop in commercial catch. King crabs migrate across the floor of the Bering Sea, but not enough is known about the stage in their life cycle during which they migrate or how many migrate in any given direction to implement appropriate protective measures. In 1986, the fishery came under tight government control, and by 1987 the commercial catch rose to 6,000 tons. Continued regulation of this fishery should help ensure the future catch, although a major problem, as with other fisheries, will still be the lack of enforcement of protective laws.

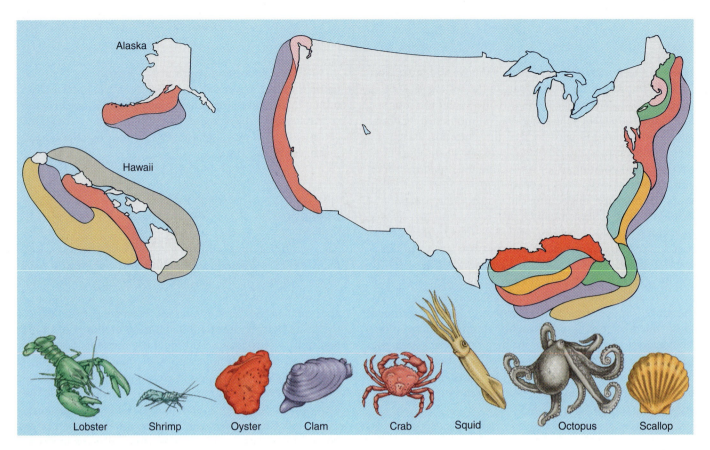

Figure 19–9

Commercial Shellfishing. This map shows the location of some of the more important shellfishery areas in the United States.

MARICULTURE

Many fisheries' biologists are looking into the possibility of increasing the commercial yield from the sea through **mariculture,** the use of agricultural techniques to breed and raise marine organisms. (Sometimes the term **aquaculture,** the farming of freshwater or marine organisms, is also used.) If only one species is raised, the process is called **monoculture;** if several species are raised together, it is called **polyculture.** The benefit of raising several species together in the same pond is that it makes more efficient use of all the available resources. The Chinese were engaged in the mariculture of oysters as far back as 1000 B.C., and the peoples of Japan and Southeast Asia have been engaged in the practice for centuries. In these cultures, the methods have traditionally been labor-intensive, and most operations are small, family-owned farms.

A good example of mariculture today can be found in Israel, where there is the problem of feeding a large population that is concentrated in a small area. The Israelis have acres of fish ponds that produce an annual yield of 6,500 tons of carp and mullet in polyculture. In the United States, fish farming produces only 2% of the fisheries' total product, and most of that is catfish and trout, not marine

species. Recently, there has been a worldwide increase in the farming of salmon. In 1988, the world mariculture of salmon produced 140,000 tons, the greatest amounts being produced in Norway and Scotland. The Canadian government has been encouraging salmon farming, with increased numbers of farms opening along the coast of British Columbia. The States of Washington, Oregon, California, and Maine allow salmon farming, but the procedure to obtain the proper permits is long and complicated. Texas and Florida have several farms that raise redfish (Figure 19–10). The initial costs, however, are quite high (approximately $400,000), which limits the number of individuals that can engage in this fishery.

Shellfish are also being raised in mariculture. Clams, mussels, and oysters are raised worldwide. In Europe and Asia a method of mariculture known as **raft culture** (Figure 19–10a) is popular. In raft culture, juveniles of commercially important molluscs are collected from natural populations and attached to ropes that are suspended from rafts. This allows the farmer to keep the juveniles in a portion of the ocean where food supplies are high, while minimizing their exposure to natural predators. Raft culture of mussels in Spain produced 120,000 tons in 1986, and in the same year raft culture of oysters in Japan yielded 251,000 tons.

(a)

(b)

Figure 19–10

Mariculture. (a) These oysters are being raised in raft cultures in Ireland. (b) This farm in Borneo raises large shrimp known as *prawns*. Although the prawns produced by this farm are an important food resource, the operation disrupts the ecology of the mangrove swamps where the farm is located. (a, Dale Sarver/Animals Animals; b, Copyright 1993 Frans Lanting/Minden Pictures)

The Japanese also culture scallops in net cages that hang from rafts as well as the more conventional bottom methods. This has resulted in an annual scallop harvest of 200,000 tons. The United States, on the other hand, is just beginning to use the techniques of raft culture.

Shrimp is another commercial shellfish that is produced in large quantities by mariculture (Figure 19–10b). In 1988, the world harvest of shrimp from mariculture was 450,000 tons. In Ecuador, shrimp farming is the second largest industry (oil production is the first). Some U.S. companies have shrimp farms in several Latin American countries, where costs are lower. In Thailand, the growing number of shrimp farms along coastal areas is beginning to have a negative impact on local mangrove communities.

At the Woods Hole Oceanographic Institution, biologist John Ryther has developed a model for using sewage waste in the mariculture of oysters. Ryther used nutrients in the sewage from a small town of 50,000 to produce algae that can be used to feed oysters. His pilot plant produced an annual yield of 800 tons of oyster meat using this method. Since oysters produce large amounts of their own solid waste, Ryther introduced polychaete worms that would feed on the oyster's waste and keep that problem under control. The worms could then be harvested and sold for fish bait.

In Long Island Sound, warm water from industrial cooling systems is used to increase oyster production. The average water temperature is 30°C (86°F), and it warms the animals enough to increase their metabolism and production. The success of the Long Island project has prompted some biologists to suggest that it would be possible to produce large amounts of oysters by mariculture in the zones around the equator, where the water is warmer. Nutrients could be supplied from sewage or by pumping nutrient-rich deep water to the surface, producing an artificial upwelling.

Another technique that is used to increase the commercial catch is **ocean ranching** or **sea ranching.** This involves raising young fish in hatcheries and returning them to the sea, where they can develop into adults and increase the size of the population. In the United States, both federal and state hatcheries produce millions of fish that are used to supplement both commercial and sport fishing stocks. Salmon is one commercial fish that is produced this way (Figure 19–11). In a program in Oregon, the Weyerhauser Company has built hatcheries next to their paper plant.

Figure 19–11

Salmon Hatcheries. Salmon raised in hatcheries such as this one in Newfoundland, Canada, supplement the commercial salmon catch. (Francis Lepine/Earth Scenes/Animals Animals)

Hawaiian Mariculture

Although Hawaii is better known for its rather warm coastal water, the deep waters (600 meters or 2,000 ft) off the islands are cold and nutrient-rich, providing a basis for a large and varied mariculture. Engineers at the Natural Energy Laboratory of Hawaii (NELH) located at Kailua-Kona on the island of Hawaii have developed ways of using a combination of cold water from the deep and warm surface water to generate electricity, grow health foods, raise seafood and vegetables, and produce fresh drinking water without pollution.

In addition to being cold, the deep-ocean water is rich in nutrients like nitrates, phosphates, and silicates that are usually scarce in surface waters. NELH uses nine large plastic pipes to bring this water to the surface, creating an artificial upwelling. This technology can potentially enrich areas that are not being served by natural upwellings. Another benefit of the cold seawater is its purity. It provides an essentially disease-free environment for the species being raised.

Phil and Joe Wilson, who run Aquaculture Enterprises, use the artificial upwellings to raise Maine lobsters in Hawaii without the need for boats or the worry of bad weather. The cold water is actually a little too cold for Maine lobsters, but the Wilsons mix the cold water with warmer surface water to produce a 21°C (70°F) bath that allows the lobsters to thrive. Lobsters in the wild usually require seven years to grow to a market weight of 0.5 kilogram (a little more than 1 pound). The Wilson brothers can accelerate growth by providing summertime water temperatures year-round, thus avoiding the three-month winter hibernation of animals in the wild. Under these conditions, lobsters reach market size in about three years.

Lobster fishers in Maine, however, do not have to worry about the Hawaiian competition. It is still cheaper to catch lobsters than to raise them in mariculture, and the Wilsons do not produce enough lobsters to make any noticeable impact on the northeastern market. The benefits of the Wilson's operation is that they can deliver lobsters year-round and at a fixed price. Recently, the Wilsons have pioneered a lobster mariculture that utilizes waste products of commercial fishing as a source of food for the lobsters.

Another group, Royal Hawaiian Sea Farms, grows a variety of seaweeds, such as nori (*Porphyra*), ogo (*Gracilaria*), and limu 'ele 'ele (*Enteromorpha*). These "sea vegetables" are considered common foods by many Asians and Pacific islanders. Royal Hawaiian's products are highly nutritious and are quite popular with supermarkets and health food stores in Hawaii and California. The seaweeds are grown from spores attached to rope nets that sway in large water tanks or from vegetative fragments that grow unattached in large rotating drums (tumble culture). Nourished by the rich seawater and the abundant sunlight, the seaweed can be harvested once a week. Although many of these organisms would grow without the cold nutrient-rich water, the water contributes to more vigorous growth.

Another operation, Cyanotech, grows a microalga called *Spirulina*. Spirulina is highly prized by health food enthusiasts as a source of B vitamins and beta carotene, and it contains up to 70% protein. Cyanotech produces 9 tons of spirulina per month, which is sold to wholesalers for pills, dips, seasonings, and food additives. Cyanotech also grows algae for the production of phycobiloprotein. The protein is a pigment that fluoresces blue or red and is used in many areas of biomedical research for marking molecules, making them easier to locate and identify. Highly purified forms of this pigment sell for $10,000 per gram. Researchers at Cyanotech are investigating the production of a lower-grade phycobiloprotein that can be sold more cheaply and be used as a coloring agent for cosmetics and food.

Ocean Farms of Hawaii raises thousands of salmon in large ponds, where they are fed on krill. The cold-water zooplankton thrive in the cold water of the artificial upwelling. Ocean Farms also uses water from the artificial upwelling to raise abalone, sea urchins, and oysters. The facility can produce 1.8 million kilograms (4 million pounds) of salmon, 15 million oysters, and 1 million abalone and sea urchins per year. Similar operations are now functioning in St. Croix, U.S. Virgin Islands and Muroto, Japan.

Warm water from the plant is used to accelerate the early development of the fish. The young are then moved to pens near the sea, imprinted with a chemical substance in the water, and then released. The imprinted adults return to the holding pens later to spawn, and, thus far, the project has been successful. Commercial fishers, however, are not all in favor of ocean ranching as they believe it will lead to control of the fishing industry by a few large companies that will be able to depress the price of fish on the market by increasing the availability of the fish.

PRACTICAL PROBLEMS OF FISHING AND FISHERIES

In 1977, the United States, along with most other coastal nations, increased the zone of ocean that it controls to 200 miles off the coast. These zones are called **exclusive economic zones (EEZ).** Other countries can fish in these coastal zones only by specific agreements with the controlling government. What areas can be fished, what fish can be caught, and the limits are all negotiated with the government of the country in whose water the operator is fishing. In the United States, it is the job of the Coast Guard to enforce the agreements, and fisheries observers are placed on fishing vessels to monitor and record the catch. Unfortunately, coastal limits are often not enforced, especially in poor, developing countries. Even with such attempts to carefully manage fishery resources, the catches are declining while costs rise.

The cost of commercial fishing accelerates as the price of boats, gear, fuel, and crew's wages increase. In the United States, most commercial fishers are independents who sell their catch directly to processors. Since the biggest demand in the U.S. market is for fish fillets and fish steaks, commercial fishers preferentially fish for the higher priced fishes such as swordfish, halibut, and salmon, since the filleting process uses only 20% to 50% of the fish's body weight. Compare this to the government-subsidized commercial fishing of some foreign countries, which processes the catch at sea at a lesser cost and uses substantially more of the animal's flesh. The net result is that it is cheaper to import some marine fisheries' products than to catch them domestically.

Biologists worldwide are concerned at the number of noncommercial marine animals that are killed each year because they are caught while fishing for commercial species. These animals are referred to as **incidental catch, bycatch,** or in the case of fish, "trash fish," and represent a terrible waste of marine resources. Dolphins killed in the tuna fishery are only a small part of the waste.

Fishing with drift nets produces especially large numbers of by-catch. **Drift nets** are large nets composed of sections called tans (Figure 19–12). Each tan averages 40 to 50 meters (132 to 165 feet) in length and 7 meters (23 feet) in height. A catcher boat sets as much as 60 kilometers (36 miles) of drift net in the evening and retrieves the net in the morning. The commercial catch, usually squid or salmon, is stored and taken to factory ships for processing. The by-catch is dumped overboard. In the North Pacific, drift-net fisheries result in the annual death of as many as 15,000 Dall porpoises and 700,000 seabirds, as well as other species such as turtles, fur seals, and sharks. These animals become

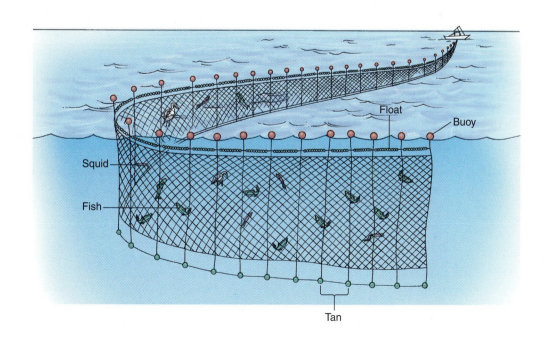

Float

Buoy

Squid

Fish

Tan

Figure 19–12

Drift Net Fishing. Drift nets are made up of sections called *tans*. As much as 60 kilometers of net are set in the evening, and fish and squid get tangled as they swim into the net. The nets are then retrieved in the morning, and the catch is removed and refrigerated. Seabirds and marine mammals may also get trapped in the net as they try to feed on the fish.

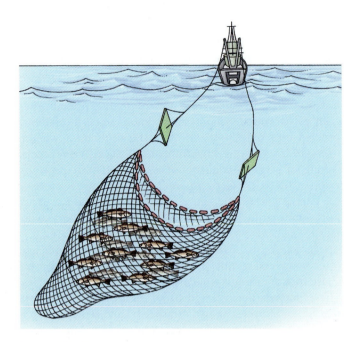

Figure 19–13

Trawl Fishing. Trawls are large nets that are either dragged along the bottom or in midwater. The trawl captures virtually everything that enters it.

entangled in the drift nets when they try to feed on the catch.

Many commercial fishes and shellfish are captured by trawls. **Trawls** are large nets that are dragged along the bottom or in midwater, depending on the type of catch, by ves-

sels called *trawlers* (Figure 19–13). Although trawling is an efficient means of fishing, it produces a large by-catch. It has been estimated that shrimp trawlers catch as many as 45,000 sea turtles annually, with a mortality of 12,000 or more.

Of the approximately 1.5 million tons of shrimp and 21 million tons of fish that are caught annually in U.S. waters, 3 to 5 million tons are discarded. The amount of the catch that is discarded varies from place to place. If the by-catch does not fetch a high enough price, or if processors are not available for the by-catch, it is discarded. In the past, shrimp fishing in the Gulf of Mexico usually produced 15 pounds of by-catch for each pound of shrimp. The figures are similar for the scallop industry. Almost all of the by-catch is returned to the ocean dead or carted off to landfills. Nets now used by shrimpers have a set of deflecting bars that push most of the by-catch below the net and not into it. This change has reduced the by-catch in the net to less than 5 pounds per pound of shrimp caught. New federal regulations require that these new nets be phased in over the next several years. By comparison, in Southeast Asia, the demand for protein is so high that almost all of the incidental catch finds a local market (Figure 19–14) and very little is wasted, but overfishing is still a problem.

A major reason for the difficulty in managing fishery resources is that we know so little about the basic biology of these commercially important organisms. The case of the Alaskan king crab cited earlier in this chapter is a good example. Until biologists better understand the life cycles, behavior, physiology, and ecology of commercial species, they will not be able to develop and implement the plans and conservation measures needed to properly manage these resources now and for future generations.

Figure 19–14

By-Catch. In Asia the demand for protein is so high that most of the catch is used, as can be seen by the variety of fish and shellfish in this fish market. (Naomi Duguid/Asia Access)

CHAPTER SUMMARY

Almost one third of the world's supply of salt (NaCl) is produced from seawater by the process of evaporation. Fresh water for drinking and irrigation can be obtained from seawater by removing the salt, a process called *desalination*.

The oceans contain vast amounts of minerals, but most of them are difficult to reach or expensive to extract. In some areas of the ocean, sulfides of a variety of important minerals accumulate in large amounts. Currently the technology does not exist for sampling or mining these deposits. Manganese in the form of manganese nodules is abundant on the seafloor. The largest seafloor mining operations extract sand and gravel from the seafloor for use in making cement, concrete, roads, and artificial beaches.

The oceans also contain important reserves of energy in the form of coal, natural gas, and oil. Most of what is removed from the sea at this time is natural gas and oil. Offshore reserves of oil and natural gas represent about one third of the world's total reserves.

In the past 50 years, the demand for food from the sea has increased, and so has commercial fishing. Natural populations of fish and shellfish were not able to withstand the pressure of increased fishing, and, as a result, the size of commercial catches decreased, and some fisheries have been brought to the brink of collapse.

Mariculture or aquaculture represents a way of increasing the commercial catch. Salmon farming is practiced in the United States and Canada, and some southeastern states are involved in redfish mariculture. A variety of mariculture techniques have been introduced for raising shellfish. Raft culture has been used to produce mussels and oysters, and in Japan net cages are used to culture scallops. Shrimp is another important species that is raised in mariculture. Sea ranching, or sea farming, a technique that involves raising fish to a certain stage and then releasing them into their natural habitat, is being used to increase natural populations of salmon.

The cost of commercial fishing is increasing, and since most of the commercial fishers in the United States are small, independent operators, they tend to restrict themselves to the large and more expensive fish that will yield a better profit at market. It is hard for them to compete with government-subsidized foreign operators for some catch, and, in many instances, it is cheaper for U.S. processors to buy the catch abroad rather than from local fishers.

Biologists are concerned about the number of noncommercial animals that are killed as a result of large-scale mechanized fishing. The incidental catch, or by-catch, represents a large waste of marine resources. Current changes in fishing regulations have resulted in equipment changes that decrease the by-catch and are improving the situation.

In addition to overfishing and pollution, our limited knowledge of the basic biology of many commercial species hinders our ability to properly regulate and conserve these resources.

SELECTED KEY TERMS

desalination, *p. 347*

fossil fuels; *p. 348*

purse seine, *p. 352*

mariculture, *p. 356*

aquaculture, *p. 356*

monoculture, *p. 356*

polyculture, *p. 356*

raft culture, *p. 356*

ocean or sea ranching, *p. 357*

exclusive economic zone, *p. 359*

by-catch, *p. 359*

drift net, *p. 359*

trawl, *p. 360*

QUESTIONS FOR REVIEW

MULTIPLE CHOICE

1. The largest seafloor mining operations extract
   a. salt
   b. manganese
   c. sulfides
   d. sand and gravel
   e. calcium

2. The most valuable substance (in dollar amount) that is removed from the sea is
   a. coal
   b. salt
   c. water
   d. oil
   e. sulfide

3. Many _____ are killed as a result of the methods of fishing for tuna.
   a. turtles
   b. seabirds
   c. dolphins
   d. noncommercial fishes
   e. molluscs

4. In the United States, most of the by-catch is

a. used for feeding livestock
b. consumed by humans
c. dumped into landfills
d. returned to the sea
e. exported to other countries

5. A major problem associated with preferentially fishing for large fish such as tuna is that
a. they are harder to catch in nets
b. they feed at higher trophic levels
c. fewer can be caught at one time
d. they yield more waste when processed
e. they do not swim in schools

SHORT ANSWER

1. What are some benefits of mariculture?

2. What is the relationship between the size of the world's population, fishing effort, and the size of the commercial catch of fish and shellfish?

3. How does the cost of fishing influence the kinds of fish caught and the methods used to process them?

4. Considering the vast mineral resources of the sea, why aren't more minerals mined from this rich area?

5. Why is it important for the fisheries industry to be regulated?

6. Why is it not ecologically sound to use anchovies for livestock feed?

7. Why are some commercial fishers opposed to ocean ranching?

THINKING CRITICALLY

1. From an ecological standpoint, would it make more sense for humans to eat anchovies or tuna? Explain.

2. When the size of the commercial catch increases, the size of individual fishes decreases. Why does this occur?

3. Suggest some ways that commercial fishers could decrease the size of their catch without becoming unemployed or going bankrupt.

4. In addition to the obvious problems associated with overfishing of commercial species, what is another way in which commercial fishing operations significantly disrupt ecosystems?

SUGGESTIONS FOR FURTHER READING

Bardach, J. 1987. Aquaculture, *Bioscience* 37(5):318–319.

Caddy, J. F., ed. 1988. *Marine Invertebrate Fisheries: Their Assessment and Management.* Wiley Interscience, New York.

Hapgood, F. 1989. The Quest for Oil, *National Geographic* 176(2): 226–259.

MacLeish, W. H. 1985. New England Fishermen Battle the Winter Ocean on Georges Bank, *Smithsonian* 16(2):40–51.

Parfit, M. 1995. Diminishing Returns: Exploiting the Ocean's Bounty, *National Geographic* 188(5):56–73.

Rudloe, J. and A. Rudloe. 1989. Shrimpers and Sea Turtles: A Conservation Impasse, *Smithsonian* 29(9):45–55.

Ryther, J. H. 1981. Mariculture, Ocean Ranching, and Other Culture-Based Fisheries, *Bioscience* 31:223–230.

Van Dyk, J. 1990. Long Journey of the Pacific Salmon, *National Geographic* 178(1):2–37.

Chapter 20

Oceans in Jeopardy

I n 1968, a group led by Thor Heyerdahl crossed the southern Atlantic Ocean on a papyrus raft. After completing the trip, Heyerdahl reported that the ocean was polluted. His navigator, Norman Baker, noted that even though for days they would see no land, no ships, and no other humans, they would still see garbage and gobs of oil. In the 30 years since this expedition, the condition of the ocean has not improved. If anything, it has gotten worse. For years humans have thought of the ocean as a huge waste receptacle, but now there is concern about what this waste is doing to the environment and what will become of the waste already placed in the ocean. Once-magnificent recreational beaches are being rendered dangerous. Medical wastes, such as syringes and other biological contaminants, are brought in by the tide. Tar residues clutter the sands. Development of oceanfront property has resulted in erosion and damage to offshore habitat that may never be reversed. The United States is not the only country experiencing such problems. Virtually every nation on earth is feeling the impact of a polluted sea. Without changes in federal and international policy, the situation will only become more serious.

POLLUTION

Ocean Dumping

Almost from the beginning of human history, coastal countries around the world have used the sea as a dumping site for trash and garbage (Figure 20–1a). Over the years, domestic wastes, industrial wastes, chemicals, and, more recently, radioactive wastes have literally been poured and dumped into the ocean. In September of 1994, the ninth annual Coastal Cleanup coordinated by the Center for Marine Conservation cleared 2.8 million pounds of trash and debris

---

KEY CONCEPTS

1. The dumping of all sorts of wastes into coastal seas has resulted in a decrease in the economic and recreational value of this area and in some instances has even posed a health hazard.

2. Pollutants enter coastal seas by way of agricultural and urban runoff as well as by direct dumping.

3. Some pollutants accumulate and magnify in food chains, posing serious problems for higher-order consumers.

4. Plastic trash is a significant hazard for large marine animals.

5. Oil spills damage significant amounts of habitat and injure and kill marine life.

6. Development of coastal areas leads to loss of habitat and diminished numbers of marine life.

7. Destruction of wetlands results in decreased ocean productivity.

8. It is not too late to become involved with conserving the oceans and their inhabitants.

(a)

(b)

Figure 20–1

Ocean Dumping. (a) Much of the trash on this recreational beach is the result of littering by thoughtless individuals. (b) After military conflicts, such as World War II, wastes are frequently dumped into the ocean. (a, Mark Conlin/Mo Yung Productions/Norbert Wu; b, Hal Beral/Visuals Unlimited)

from U.S. coastlines in a 3-hour period. The cleanup effort drew 139,746 volunteers nationwide and cleaned 5,148 miles of the nation's beaches and waterways. The trash included a car hood in Palm Beach, Florida, a bag of undelivered mail on the coast of North Carolina, the remnants of refugee rafts in the Florida Keys, and two kitchen sinks! Florida volunteers collected 211,632 cigarette butts, 37,143 glass bottles, and 38,126 metal cans. Medical wastes including 4,400 syringes were recovered mostly on beaches of the Northeast and the Gulf Coast. The effort in Florida is estimated to have saved the state $500,000 in cleanup costs. Sixty-one percent of the trash collected along the beaches was plastic. This is an increase from previous years and is of special concern because plastics do not degrade quickly and they are hazardous to wildlife (see section on plastics later in this chapter). A total of 82 animals were found trapped in the debris. A project report released ten months after the cleanup concluded that U.S. beaches are still being polluted by a wide variety of materials, especially plastic wastes. If more concerned citizens will get involved with projects such as Coastal Cleanup, perhaps the state of our beaches will improve.

Following the major wars of this century, the ocean has been the dumping ground for discarded military hardware and munitions (Figure 20–1b). Toxic gases and chemicals removed from Germany following World War II were dumped in the North Atlantic Ocean; in the Pacific Ocean bays and lagoons were filled with jeeps, tanks, trucks, and munitions. After the Vietnam War, vehicles and munitions were unloaded into the coastal waters of Southeast Asia. Some of the discarded material either entangles or poisons marine birds and mammals. On the positive side, some of

this metallic junk has been converted into artificial reefs populated by fishes and invertebrates (Figure 20–2).

Today's electronic, chemical, and defense industries and nuclear facilities produce large amounts of highly toxic or radioactive wastes that have been stored in landfills and in above-ground repositories. As concern for their effects on human health increases, it has been suggested that they be deposited on the ocean floor, for instance, in the deep-sea trenches far from continental land masses. Supporters of this proposal point out that since the trenches are sub-

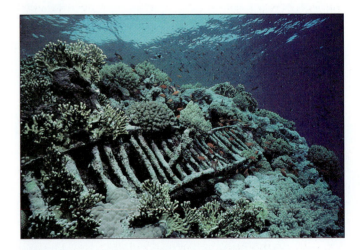

Figure 20–2

Artificial Reefs. Sunken vessels like this one become artificial reefs, providing habitat for a large variety of marine organisms. (Jeffrey Rotman/Peter Arnold, Inc.)

duction zones (see Chapter 3) the wastes would ultimately be taken into the mantle of the earth and this would eliminate the problem of long-term storage. Opponents of this plan point out that if the containers holding the waste decompose before subduction occurs, their contents could be released into the surrounding seawater, causing considerable habitat damage. Considering our general ignorance of the deep-sea benthos and deep-sea food webs, no one really knows the long-range effects or dangers of such plans. This plan is currently restricted by the Ocean Dumping Act of 1972, which requires any legal dumping of radioactive waste in the ocean be preceded by an environmental impact statement and approval of both houses of Congress.

Throughout the world, major population centers have developed around estuaries and along the rivers that drain into them. In the United States, over 50% of the population lives within 50 miles of a coastline (including the Great Lakes). Worldwide, the numbers are similar and sometimes higher. This dense population with its needs for energy, industry, and waste treatment has placed a large burden on coastal seas and coastal habitats. It was once thought that coastal waters had an infinite capacity to absorb and remove wastes from human populations. Unfortunately, too much waste has entered too small an area in too short a time. This has resulted in a decrease in the economic and recreational value of coastal areas and, in some cases, has endangered public health and safety. During the 1960s and 1970s, environmental legislation was passed, creating agencies to control and monitor resources, including the ocean. Controlling the problem now, however, is not sufficient, since we must also contend with problems created in the past.

The dumping of human and industrial wastes into coastal waters was and is an inexpensive but short-sighted solution to eliminating these wastes. Since 1890, a variety of garbage, sewage, and toxic chemicals have been dumped into the New York Bight, the area that lies off the mouth of the Hudson River. Floating debris from the dumping ends up on beaches. By 1934, the problem had reached such large proportions that legislation was passed prohibiting the dumping of any floating waste. The legislation did not, however, limit the dumping of building wastes, storm and sewage wastes, or industrial wastes. It has been estimated that between 1890 and 1971, 1.4 million cubic meters of solid waste was dumped into the New York Bight—enough to cover all of Manhattan Island with a layer of waste as high as a six-story building. Needless to say, this amount of dumping has greatly decreased water quality and the quality of bottom sediments. The materials dumped were both toxic and oxygen-demanding (decomposition of these materials by microorganisms tends to deplete the water of oxygen). Occasionally, the contaminated water would upwell along the coast, causing the death of shallow-water organisms. The offshore dump was finally closed in 1987 by the Environmental Protection Agency. The agency did, how-

Figure 20–3

Medical Wastes. Toxic and medical wastes, such as this syringe, make recreational beaches unfit for visitors and kill marine organisms. (Peter K. Ziminski/Visuals Unlimited)

ever, allow New York City, as well as several New Jersey communities, to continue dumping sewage sludge, a product of sewage treatment, at a site 171 kilometers (106 miles) out to sea at the edge of the continental shelf. An excess of 8 million tons of sludge was dumped at this site annually.

In the summer of 1988, many northeastern beaches had to be closed because of the polluted waters and the amount of dangerous wastes that were washing up on the beaches. In addition to gummy balls of sewage as much as two inches in diameter, medical wastes such as syringes, vials of blood, and medicine bottles were washing up on shore (Figure 20–3). On a New Jersey beach just south of Long Island, New York, over 100 vials of blood washed up. When the blood in the vials was tested, 5 vials were found to contain antibodies for AIDS. The public scare caused by the medical wastes washing up caused attendance at some New York and New Jersey beaches to drop by 85%.

In response to this situation, Congress passed legislation that prevented the dumping of sewage sludge or industrial wastes in the ocean after January 1, 1992. This legislation does not solve all of the problems, however. Much of the material that contaminated beaches in 1988 continues to contaminate beaches today. It is the product of overflowing storm sewers that empty into the ocean, as well as improperly dumped wastes.

The bigger threat to coastal water quality is not the dumping of wastes but the increased size of coastal populations and improperly controlled residential and commercial development, discussed later in the chapter. In theory we could have large populations and clean water if we spend the money to treat the sewage. Whether we decide to spend the money depends on ethical, economic, and political considerations.

Pollution via Environmental Routes

Not all pollutants enter the ocean through indiscriminate dumping. Pesticides, including DDT (dichloro-diphenyl-thrichloro-ethane); toxic organic compounds, such as PCBs (polychlorinated biphenyls); and heavy metals such as mercury, lead, zinc, and chromium continue to enter the sea through various environmental routes (Figure 20–4) even though they are no longer used and closely monitored. The use of DDT was suspended in the United States in the 1960s, and the United States ceased production of PCBs in the late 1970s. Both DDT and PCBs are persistent toxic substances that remain in the environment for long periods, allowing them to continuously pollute the ocean. DDT, a pesticide, is carried to the sea by runoff from the land. The year

that DDT was banned, another pesticide, DICOFOL (made from DDT) was approved for use. Hundreds of thousands of tons of this pesticide were used around the nation until 1988, when researchers found levels of 1% to 7% DDT contaminants per bag of DICOFOL.

PCBs were used in a number of consumer articles including plastics, solvents, and electrical insulators. They are still present in many electrical devices that were manufactured before 1979. When these devices are burned, the PCBs enter the atmosphere and re-enter the water supply with precipitation.

Toxic materials like DDT, PCBs, and heavy metals are always found in higher quantities in sediments than they are in the overlying water. This allows them to continue seeping into the water for years to come.

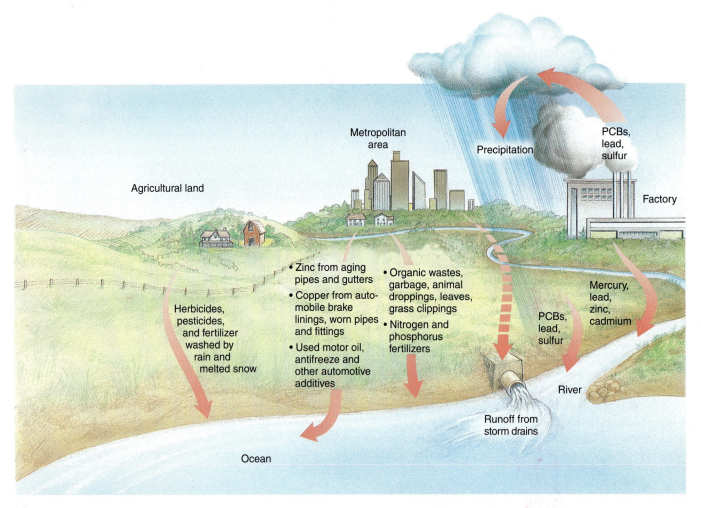

Figure 20–4

Pollution from Land. Pollutants, including toxic organic compounds, pesticides, sewage, petroleum products, and heavy metals, are carried into the sea by rivers and washed into the sea by rainwater and melting snow.

Another source of marine pollution is surface runoff from agricultural lands. Surface runoff containing pesticides and nutrients can be carried by streams and rivers to the sea, where it can poison and overfertilize the water. An increase in nutrients can cause algal blooms. The increase in algae outstrips the environment's ability to support it, and the algae die. The decomposition of the dead algae removes excessive amounts of oxygen from the water, and this causes more organisms to die, resulting in even lower levels of oxygen.

Surface runoff from metropolitan areas contains pesticides, fertilizers from residential areas, chemicals from oil and gasoline, residues from industry, and bacteria from sewage. On the East Coast, water from storm drains is mixed with sewage. This causes an overflow at sewage treatment facilities when there are heavy rains and in the winter months when snow melts. The overflow enters the ocean. This type of pollution damages recreational areas as well as coastal fisheries. Metropolitan pollution in Long Island Sound has killed many species of wildlife, and the stench of the sewage and dead organisms has rendered many of its beaches useless for recreation.

Cottages built along the shore use septic tanks to process sewage and drain wastes into the sandy soils. Commercial developments intentionally and unintentionally drain a variety of wastes into the ground, and coastal cities and small towns dump raw sewage into the shallow water off the coasts. Incoming tides bring sewage-contaminated water onto beaches. This contaminated water, added to the runoff from septic tanks, makes beaches unsafe for humans and contaminates the invertebrates. Filter-feeders such as clams and oysters concentrate large amounts of bacteria from sewage in their tissues, resulting in their meat being unfit for human consumption. In many parts of the United States, bacteria from cities and individual septic systems have forced the closure of clamming beaches and oyster beds. In some cases, the result has been economic hardship for clam and oyster fishers.

In addition to the problems of untreated runoff, coastal cities discharge treated sewage that adds its own array of contaminants to the ocean water. Chlorine that is added to drinking water and then later to sewage during treatment may form chlorinated organic compounds in seawater that are toxic to some marine organisms.

Efforts to reduce the discharge of toxic materials into the marine environment are gradually succeeding. Although there has been an increased input of nutrients from fertilizers used in agricultural areas, better sewage treatment has decreased the nutrient content of discharged sewage, so in some areas there has been no net increase in the level of nutrients reaching the sea. Unfortunately, this is not true everywhere.

Since 1980, the increased use of unleaded fuels has significantly decreased the amount of organic lead entering the marine environment. The concentration of lead in surface waters of the Sargasso Sea dropped by 30% between 1980 and 1984, and in the last ten years, the amount of lead entering the Gulf of Mexico from the Mississippi River has decreased 40%.

POLLUTANTS IN MARINE FOOD CHAINS

Not all pollutants that enter the marine environment dissolve in the water. Some are absorbed by particles of clay and detritus that are suspended in the water. These, in turn, clump together, forming denser masses that sink to the bottom. The tainted detritus as well as contaminated phytoplankton are used by various marine organisms as a source of food. Since many of these organisms have no means of breaking down the toxins or excreting them, the toxins become concentrated in the body tissues of the organisms that feed on them. These toxins can be passed along to predators, where they are concentrated further, a process known as **biological magnification** (Figure 20–5). At some level, the concentration of toxins can be so high that the next predator to feed on the contaminated flesh will become seriously ill or die. During the 1970s, this is exactly what occurred when mercury from industrial pollution accumulated in the food chain and contaminated tuna. Humans who consumed the contaminated tuna suffered from mercury toxicity, which caused a variety of neurological problems and death.

Between 1953 and 1960 an industrial plant in Minamata, Japan, dumped industrial wastes containing high levels of mercury into a local embayment. The mercury readily formed organic complexes that entered the food chain and were concentrated in the shellfish and fish that the people of the town used as a major food source. Many cases of severe mercury poisoning and even death resulted from this concentration of a pollutant in a marine food chain. The damage to the nervous system caused by the mercury was particularly noticeable in children whose mothers had consumed contaminated shellfish or fish while they were pregnant. Bottom fish in all of the estuaries of the United States show some level of toxic materials. Shellfish can concentrate certain toxic materials, such as mercury, lead, and cadmium, at several thousand times the concentration that these materials are found in the surrounding water. Scallops can concentrate cadmium to levels 2 million times that of its concentration in the water, and oysters can concentrate DDT over 90,000 times that of its concentration in water.

Phytoplankton are particularly sensitive to chronic pollution by toxic substances, as are many forms of zooplankton and larger crustaceans. Toxic pollutants inhibit photosynthesis, growth, and cell division in marine phytoplankton. They also affect the growth and development of filter-feeding zooplankton, as well as the early developmental stages of other organisms, by interfering with the process of cell division. The impact of marine pollution on

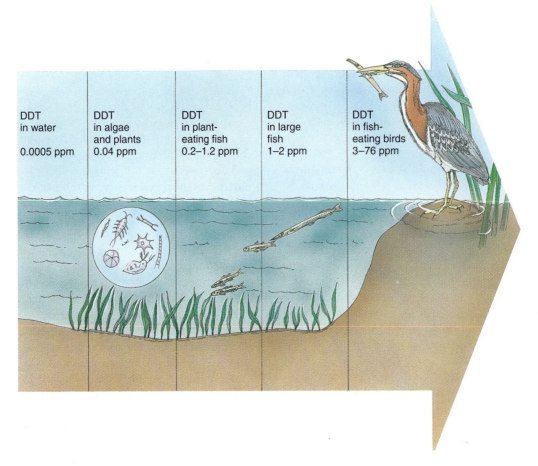

DDT
in water

0.0005 ppm

DDT
in algae
and plants

0.04 ppm

DDT
in plant-
eating fish

0.2–1.2 ppm

DDT
in large
fish

1–2 ppm

DDT
in fish-
eating birds

3–76 ppm

Figure 20–5

Biological Magnification. Toxic substances can accumulate in the tissues of consumers as they feed on organisms that contain low levels of toxic material. Notice how the level of DDT in this example becomes more concentrated in the tissues of organisms as you move up the food chain. Ultimately, the toxic substance will reach a lethal level and cause the death of a higher-order consumer.

plankton has barely been studied, but it is known that pollution in the North Atlantic Ocean has resulted in a decrease in the number of species of both phytoplankton and zooplankton. This should serve as a warning. In the long term, pollution causes an overall reduction in marine primary productivity, altering interactions at every trophic level throughout the food web.

PLASTIC TRASH

It is estimated that naval and merchant ships legally dump 77 tons of plastic into the ocean annually. In addition to this, the National Academy of Science estimates that the fishing industry annually discards or loses 149,000 tons of fishing gear (nets, ropes, traps, buoys) mainly composed of plastic and then dumps another 2,600 tons of plastic packaging material into the sea. To this, add the discarded plastic from pleasure boats, ocean liners, oil platforms, and visitors to the beach. In total, about 1 million tons of plastic waste enters the ocean annually. Carried by the currents, this refuse can appear on even the most remote beaches of the world.

The very characteristics that make plastic such an important product of manufacturing—its strength, durability, and cost—also make it one of the most hazardous materials dumped into the sea. Many marine biologists consider plastic trash to be as great a killer of marine life as oil spills and toxic chemicals. No one knows for sure how long plastic remains in the marine environment, but it is thought that a plastic six-pack ring could last as long as 450 years.

The greatest danger of plastic wastes is to larger marine animals and seabirds. As many as 30,000 fur seals are killed annually when they become trapped in discarded fishing nets and cargo straps (Figure 20–6). Discarded netting also traps fish, turtles, and other marine life. Traps used to catch shellfish such as lobsters and crabs are frequently made in part or entirely of plastic. When these are lost, they continue to catch animals. The animals are unable to escape the traps and ultimately die, frequently of starvation. On the western coast of Florida, more than 100,000 of these traps are set out each year, and the loss rate is about 25%. Many seabirds are killed when they become entangled in six-pack rings and fishing line. Whales and dolphins have been found suffocated by plastic bags or sheets of plastic. Seabirds, sea turtles, and mammals all have a tendency to eat plastics. The plastics form an undigestible mass that blocks the digestive tract and causes the animals to die of starvation.

Currently, the Marine Plastic Pollution Research and Control Act, which was passed in 1987 and became effective in 1988, prohibits the dumping of plastic in the ocean and requires ports and terminals to provide facilities for the disposal of this waste. The U.S. Navy, however, can still legally dump trash and plastic into the ocean. Similar laws have been passed by other nations, and there exists an international agreement to prohibit the dumping of plastic waste in the ocean. In the United States, the Coast Guard is mandated to enforce the law, but with all of its other duties and small resources, violations still occur. Other nations are even less successful in enforcing dumping restrictions, and plastic continues to be a problem. Some manufacturers have started to produce plastics that are more biodegradable, with a shorter lifespan in the environment. Unfortunately, this will probably not do much to reduce the problem in the ocean. Only education and responsible action by those who use the sea for commerce and recreation will reduce this problem.

Figure 20–6

Plastic Trash. Plastic trash poses the single greatest threat to large marine animals, especially turtles, birds, and mammals. Animals like this seal get tangled in the plastic and lose appendages or drown. Some consume the plastic, possibly mistaking it for a jellyfish, and die of intestinal blockage. (Norbert Wu)

OIL SPILLS

The demand for oil to supply the needs of industry and motorists has exposed the ocean to one of its greatest threats. Not only is petroleum pumped from the seafloor by offshore drilling operations, it is transported from sites of production to processors and industrial areas with the aid of tankers and supertankers that travel the seas. At each step of the process there is the potential for accidents, well blow-outs, and leakage, not to mention the entry into the sea of petroleum and petroleum products that are released into the environment by runoff, industrial discharge, and other processes. The transport of oil and petroleum products by tankers provides the greatest risk of large spills.

Oil is a mixture of hundreds of substances that can react with the environment. When it is released into water it spreads a film across the surface. The lighter components of the oil either evaporate or are absorbed by particles of clay and other sediments in the water and sink to the bottom. Some inorganic components become dissolved in the seawater, while some of the organic components are degraded by the action of bacteria and fungi. Many bacteria can digest the simpler organic molecules in oil, but they have a more difficult time digesting the more complex molecules. These persist in the sea for long periods of time as tarry chunks that float on the surface or lie on the bottom. Little is known about the effect of oil on bottom-dwelling organisms of the deep. We do know that oil can suffocate organisms at the bottom of bays and harbors, and that some of the soluble components of oil are extremely toxic to many forms of marine life, especially eggs and young.

Ecological Effects of Large Oil Spills

When a large oil spill occurs near a coastal area, the damage to the environment and marine life can be substantial. Millions of seabirds, such as cormorants and diving ducks, and thousands of marine mammals, especially seals and sea otters, have fallen victim to oil spills. A heavy coating of oil mats a bird's feathers, impairing its ability to fly and swim. Oil also reduces the ability of large outer feathers to repel water and destroys the insulating effects of down feathers. During cold weather, a spot of oil no bigger than a dime in a vital area can cause death by hypothermia (lowering the body temperature). As birds preen to remove oil from their feathers, they may ingest fatal amounts of oil. In otters, oil destroys the ability of the fur to insulate the animal, causing them to die of hypothermia (Figure 20–7). In other mammals, including otters, the oil clogs the ears and nostrils and irritates the eyes. Oil has also been shown to cause cancer in sea otters.

Less visible, but just as deadly, are the effects on invertebrate, algal, and plant life along the shore. On sandy shores, sand crabs and other organisms that live in spaces between sand particles are killed by the toxic components of oil or are smothered by a coating of oil. Even after tides have washed the beaches clean, a residual layer of oil may still persist several meters below the surface. Although some intertidal invertebrates, such as bivalve molluscs, seem to be resistant to oil pollution, their flesh becomes tainted with petroleum chemicals, which are passed along the food chain. On rocky shores, barnacles are fairly resistant to pollution, although in some instances they are smothered by the oil, but grazing molluscs like limpets, periwinkles, and the carnivorous whelks are vulnerable. Toxic components of the oil act as a narcotic, causing molluscs to lose their hold on the rocks and be washed away by the tides.

The elimination of these grazers allows seaweeds to colonize the vacated area. Eventually, the seaweeds become encrusted with oil and are torn away by the waves. The ultimate outcome of oil pollution in the intertidal zone is a decrease in species diversity, a simplification of the food web, and a disproportionate increase in the populations of resistant species (Figure 20–8).

The immediate damage from large oil spills is quite obvious. The damage to the environment from gasoline, diesel fuel, and other more toxic petroleum products is more insidious. Because many of these petroleum products evaporate and disperse quickly, it is difficult to assess their immediate effects, but since they are constantly released into the sea, they cause chronic problems. Since most petroleum that is transported by sea is offloaded or transferred at ma-

Figure 20–7

Oil-Related Injury. This otter was a casualty of the *Exxon Valdez* oil spill. Even a small spot of oil on the feathers of a bird or the fur of a marine mammal can cause the animal to lose body heat and die. (AP/Wide World Photos)

(a) Rocky shore before an oil spill

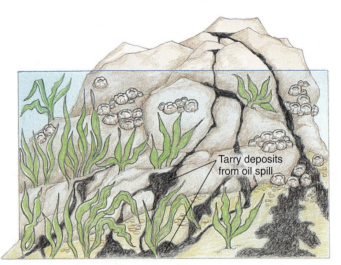

Tarry deposits from oil spill

(b) Rocky shore after an oil spill

Figure 20–8

Generalized Changes in Habitat That Result from Oil Pollution. (a) A generalized view of a temperate zone rocky shore before an oil spill. (b) The same rocky shore after an oil spill. Oil kills molluscs, allowing seaweeds to overgrow the intertidal rocks. Wave action will eventually remove much of the algae, resulting in a decrease in species diversity.

(a)

(b)

Figure 20–9

Oil Spills. (a) Oil booms are used in an attempt to limit the spread of oil leaking from the damaged tanker *Exxon Valdez*. (b) Oil from the *Exxon Valdez* covers some of the rocky coast of Prince William Sound, Alaska, killing intertidal organisms. (a, Courtesy of the U.S. Coast Guard; b, Joel Bennett/Peter Arnold, Inc.)

jor ports and most major ports are located in the world's largest bays and estuaries, these delicate, productive, and valuable areas remain at risk of damage from oil and petroleum products. When oil spills occur in the open sea, the effects are difficult to assess. Because they occur far from land, they are difficult to monitor, and wind, waves, and currents dissipate the oil quickly. Presumably, the damage to marine life is the same as in coastal spills.

The largest oil spill in the United States occurred in March of 1989 when the tanker *Exxon Valdez* ran onto a rocky reef 25 miles out of its home port of Valdez, Alaska. The accident tore five huge holes in the hull, and the ship spilled more than 240,000 barrels (10 million gallons) of oil into Alaska's pristine Prince William Sound (Figure 20–9a).

Several factors intensified the problem and hampered cleanup. Local plans for dealing with oil spills were not designed to handle a spill of this magnitude. Cleanup equipment and personnel were not actively in service, and there was a delay while these were assembled. The rugged terrain where the spill occurred combined with the tidal currents of the enclosed area also aggravated the problem. The spill quickly spread over 2,610 square kilometers (900 square miles) of Alaska's most picturesque and biologically productive coastal areas. As the oil moved out of the sound, currents carried it down the Alaskan coast, where it continued to coat rocky shores and damage marine and terrestrial habitats (Figure 20–9b). Thousands of birds, fish, and marine mammals died as a result of this oil spill. The cold temperatures of this subarctic region slow the natural decomposition of oil by photochemical reactions and bacteria. The marine life of the area also live longer and reproduce more slowly, so although the region is expected to recover

from this accident, it will probably take a longer time than if it occurred in a temperate or tropical zone.

Not all large oil spills are the result of shipping accidents. The largest and longest-lasting oil spill in the world occurred in June of 1979 when an offshore oil well in the Gulf of Mexico, the *Ixtoc I*, owned and operated by a Mexican company, blew out and caught fire. It required almost a year to cap the well, during which time over 137 million gallons of petroleum poured into the Gulf of Mexico. Similar events have occurred in the North Sea, Persian Gulf, and off the coasts of the United States.

During the Iran–Iraq war, the bombing of oil facilities resulted in large amounts of oil entering the Persian Gulf. One bombed oil rig pumped 50,000 gallons per day into the Gulf for almost three months. Miles of beaches on the west side of the Gulf are still covered with a hardened surface of sun-baked oil. Current reports indicate that 40% of the coral reefs off the coast of Qatar and 30% of the reefs off the Bahrain coast are dead as the result of the oil spills. During the Persian Gulf War, oil facilities sabotaged when the Iraqi army retreated from Kuwait released thousands of gallons of oil into the Persian Gulf around Kuwait. The damage from this ecological disaster is still being assessed.

Oil Spill Cleanup

Current technology for dealing with ocean oil spills includes the use of oil booms and oil skimmers (Figure 20–10) that help to confine the spill to a smaller area and recover some of the oil. These techniques are most efficient when the spill occurs in an area that is easily accessible and

Figure 20–10

Oil Spill Cleanup. Absorbent booms (red) are used to contain and clean up oil spills. (Doug Wechsler/Earth Scenes/Animals Animals)

where equipment can be rapidly mobilized. In a spill like the one involving the *Exxon Valdez,* where weather, sea conditions, rugged coastline, and distance from cleanup equipment are major factors, little of the oil is actually recovered. In some oil spills along an accessible coast, straw is used to soak up the oil. The oil-soaked straw is then collected by hand and disposed of by burning. Although incineration of the oil-soaked straw removes the pollutant from the marine environment, it creates another ecological problem in the form of air pollution. Microbiologists have developed a genetically engineered bacterium that is capable of degrading most of the organic compounds that are present in crude oil. Current legislation prevents the use of such organisms in nature, and it is still being tested to see what other environmental impact it may have before it is actively used against oil spills. Naturally occurring bacteria from oil seeps on the ocean floor are currently being used with some success to clean up oil spills.

HABITAT DESTRUCTION

Wetlands

Wetlands, such as salt marshes, provide nutrients, shelter, and spawning areas for a variety of marine organisms including crabs, shrimp, oysters, and many commercial fishes. In the past, these areas have been drained, filled, and dredged to provide more ground for industry, channels for large vessels to enter ports and harbors, and beachfront real estate for the growing number of individuals who desire to live and vacation near the sea (Figure 20–11). It is estimated that in 1776, the United States originally had between 125

and 215 million acres of wetlands. By 1975 the amount of wetlands had decreased to 99 million acres. It is estimated that 18,000 acres of wetlands were lost annually between 1950 and 1970. The loss of estuarine wetlands has been greatest in the states of Florida, California, Texas, New Jersey, and Louisiana. At this time, over half of the coastal wetlands in the contiguous 48 states has been destroyed, and the other half is still under pressure as the demand for residential and recreational development along the coasts increases.

We now recognize that wetlands, including estuaries, are among the most productive of marine environments and that they play a major role as a nursery for a number of important commercial fish and shellfish (see Chapter 13). In recognition of this fact, both federal and state governments have passed environmental legislation that closely regulates any development or modification of these areas. Whenever possible, these valuable areas are being carefully managed and restored to their natural state. If estuarine or wetland areas must be altered, the current policies demand that other areas that have been damaged be restored to replace the wetlands that are used. Although this practice sounds good in principle, in the few instances where it has actually been tried it has generally been unsuccessful because of loopholes in the legislation. One exception is the current restoration of wetlands in the San Francisco Bay area. Environmentalists are cooperating with federal and local government as well as local industries to make their plan for restoring wetlands work. To further complicate matters, the federal government, in response to pressure from special interest groups, continues to change its definition of what constitutes a protected wetland and what it means to re-

Figure 20–11

Wetland Destruction. This shoreline development at Barnegat Bay, New Jersey, was once a productive wetland that was drained and filled to allow for real estate development. (R. Ballou/Earth Scenes/ Animals Animals)

store an area. Changes such as these severely weaken the laws that were meant to protect these resources.

In the 1970s and 1980s, increased public awareness led to the passage of new laws protecting wetlands in many states and enforcing of federal regulations more strictly. Wetlands along the West Coast are still under heavy pressure for industrial and residential development, and net loss of these valuable areas is still continuing.

Beaches

Coastal areas are popular vacation and retirement destinations. Travel brochures display pictures of beautiful, immaculate, sandy beaches backed by vigorous shoreline vegetation (Figure 20–12a). Few show the reality—overcrowded beaches backed by hotels, resorts, and shops (Figure 20–12b). Development of the seashore for recreational and commercial use, combined with intensive seasonal use, has had a severe impact on intertidal areas. The heavy usage has had serious effects on intertidal wildlife, especially those of the sandy shores. Beach-nesting birds like the piping plover (*Charadrius melodus*) and the least tern (*Sterna albifrons*) are in danger of extinction along the U.S. Gulf Coast because of disturbances created by bathers, pets, and dune buggies. Other shorebirds are subjected to competition for nesting sites and egg predation from the growing populations of several species of large gulls. Gulls (see Chapter 11) are highly tolerant of humans and can thrive on the garbage humans generate. Populations of sea turtles and horseshoe crabs are also declining due to a loss of nesting sites on sandy beaches and predation by domestic animals. The construction of beachfront houses, docks, and seawalls disrupts habitats and removes areas that species need for nesting sites, reproduction, and feeding. Some coastal communities are attempting to control the problems by setting aside stretches of beach for bird and turtle nests, levying fines for littering beaches, and participating in projects like the coastal cleanup mentioned at the beginning of this chapter.

Development of beaches for recreational use frequently destroys sand dunes and beach vegetation that play an important role in preventing beach erosion. The loss of vegetation and human traffic cause significant beach erosion to occur, and ironically, vacation areas often spend hundreds of thousands of dollars each year to refortify beaches that were destroyed by tourism in the previous season.

Coral Reefs

As the popularity of the sport of SCUBA diving increases, more and more humans are invading reefs. Although the mass-marketing of SCUBA training dates back a mere 30 years, today there are more than 4 million certified divers in the United States alone. This number is expected to increase by another 10 million in the next ten years. SCUBA diving may be the world's fastest growing recreational sport, and it has spawned a multibillion dollar industry that involves not only diver training and equipment sales but also travel and resorts catering solely to divers. With modern air service and well-equipped dive boats, divers have access to even the most remote dive destinations in the world.

(a)

(b)

Figure 20–12

Beach Development. (a) Travel brochures entice vacationers with pictures of beautiful sandy beaches and aquamarine water. (b) In reality, the beach is really a small strip of sand crowded with tourists and surrounded by hotels, condominiums, shopping areas, and restaurants. (a, C. Dani/I. Jeske/Earth Scenes/ Animals Animals; b, J. R. Williams/Earth Scenes/Animals Animals)

The Effects of Artificial Processes on Beach Formation

Human interference with natural processes can have a pronounced effect on beaches. When rivers are dammed to control floods or to produce power, sand and gravel that once moved down the river to be deposited on the coast become deposited in the lake that forms behind the dam. This activity removes an important source of new material that is needed to replace sediments that are removed by natural erosion processes.

Engineering projects such as breakwaters and jetties in coastal zones also produce changes in beaches. Breakwaters and jetties are built to protect coastal areas from wave action. The areas behind these structures are quiet, and sediments that settle out there are not available for longshore currents to carry them down to other beaches. Longshore currents are generated by waves that break at an angle to the beach. These currents move parallel to the shore in the surf zone and carry sediments along as they do, in what is known as the *longshore transport process.*

Historically, coastal engineering projects have triggered a series of chain reactions. Facilities that have been constructed in one area create ecological problems that must be solved by other engineering projects, which create problems of their own, and so on. According to the U.S. Coastal Survey, 43% of our national shorelines, excluding Alaska, are losing more sediments than they receive. In other words, our beaches are disappearing into the sea. In some areas, eroding beaches are maintained by bringing in sand and gravel from inland areas or dredging it from offshore sandbars. These programs are expensive, but necessary, if the beaches are to be preserved.

An example of interfering with the natural processes of the coastal zone can be seen in the Santa Barbara harbor project in California. In this section of the California coast, the longshore current transports sediment down the coast of California from north to south (see figure). To supply the needs of recreational boaters, a jetty and breakwater were constructed at Santa Barbara to form a boat harbor. The jetty is on the north side of the harbor and proceeds out to sea before turning south to form a breakwater. This structure not only produces a wave-sheltered area, it also blocks the longshore currents. To the north of the structure, more sand was deposited, and the beach began to grow, while beaches south of the structure began to disappear. Eventually, the beach to the north grew until it reached the seaward limit of the jetty and the longshore currents could again move sand to the south. Instead of carrying sand to more southerly beaches, however, the current began to deposit sand at the end of the breakwater, forming a sand spit that began to fill in the harbor, while the southern beaches continued to erode and disappear. In order to deal with the problem, a dredge pumps the sand deposited in the harbor back into the longshore current south of the harbor, where it can be carried to southern beaches. In this case, interference with a natural process resulted in the expenditure of large amounts of time, money, and energy to recreate a process that nature did for free.

Figure 20–A

California Longshore Currents. (a) The action of longshore currents along the coast of southern California before the Santa Barbara harbor project. (b) Changes that resulted after the boat harbor at Santa Barbara was built. (c) New problems occur as a result of alterations in the longshore current.

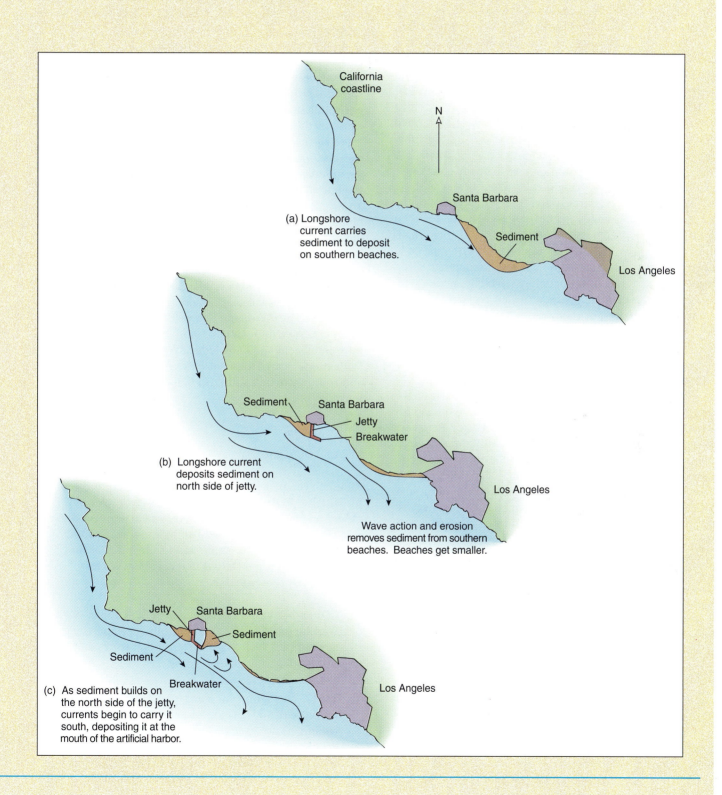

California
coastline

N

Santa Barbara

(a) Longshore
current carries
sediment to deposit
on southern beaches.

Sediment

Los Angeles

Sediment

Santa Barbara

Jetty

Breakwater

(b) Longshore current
deposits sediment on
north side of jetty.

Los Angeles

Wave action and erosion
removes sediment from southern
beaches. Beaches get smaller.

Jetty

Santa Barbara

Sediment

Sediment

Breakwater

Los Angeles

(c) As sediment builds on
the north side of the jetty,
currents begin to carry it
south, depositing it at the
mouth of the artificial harbor.

This increase in human visitors to reefs has caused a great deal of concern among marine biologists and dive operators. The world's coral reefs already suffer damage from coral bleaching (see Chapter 15), pollution, oil spills, boat groundings, overfishing, overcollecting, and dynamite fishing. The multitudes of flipper-clad visitors only add more pressure to the already beleaguered reefs. Careless anchoring of pleasure craft and dive boats damages large amounts of coral. As divers descend onto a reef they kick and grab corals, breaking off pieces, sometimes by accident and sometimes on purpose. They stir up bottom sediments that coat and suffocate corals, and they beat against them with a variety of devices that hang from their bodies. Uninformed divers injure and even kill many organisms by touching, squeezing, bumping, and inadvertently crushing them.

Improper buoyancy control seems to be the chief cause of physical damage to living corals. When divers wearing too much weight enter the water, they rapidly sink to the bottom, smashing corals in the process. The divers then push off from corals to raise themselves or hold corals in order to stay down. All of this interaction with the coral causes extensive damage. Although some concerned diver-naturalists who run dive operations carefully monitor the activities of the people they escort to the reefs, most divers are not escorted by a concerned or knowledgeable individual. As a result, boatloads of divers and snorkelers constantly lay siege to the reef. The problem is compounded by the fact that many divers do not realize that corals are alive and thus are not aware of the damage they are doing.

During the summer of 1989, Helen Talge, a marine biologist from the University of South Florida, anonymously studied divers visiting Looe Key National Marine Sanctuary in the Florida Keys. The purpose of the study was to determine what type of reef visitor does the most damage to coral. During the course of the study, Talge observed 206 divers and snorkelers as they swam around the reef. Talge monitored how often a diver bumped into, scraped, or pushed off of coral, and she also monitored how much sediment each diver stirred up. The results were recorded in terms of negative interactions per minute. Her results indicated that the type of diver that did the most damage was a young, male SCUBA diver wearing gloves. Divers having the highest negative interactions (1 per minute in the course of a 30-minute dive) were those with a specific objective such as photography, shell collecting, or lobstering. The profile of the most careful diver was a snorkeler who dove without gloves.

The average diver visiting the sanctuary had at least six negative interactions with the reef. Since more than 1 million people visit these reefs per year, it becomes obvious that even the most casual contact can have a dramatic, cumulative effect on the reef. Unless more is done to educate divers, snorkelers, and dive operators about the reef and its inhabitants, there will be fewer reefs for SCUBA enthusiasts to visit and admire in the future.

EPILOGUE

The basic mechanisms of evolution and ecology are inextricably interrelated, and human activity changes those interactions. The scope and pace of the changes caused by humans are very different from those of nature. Nature works by means of small changes over long periods of time. Pollution, oil spills, habitat destruction, and other human changes frequently involve entire biological communities and sometimes occur instantaneously. The result is widespread disruption and damage of marine ecosystems. The environmental damage that results is significant not only because of the damage done this month or this year but because of the long-term changes in ecological relationships and disappearing niches that lead to long-term changes in evolutionary patterns and ultimately species extinction. Although all the various regions of the sea are unique, the knowledge that we gain from studying one area helps us to better understand others. An understanding of the underlying patterns and processes in the sea will help us to judge the impact of our actions and help us determine just how much an area can be modified without jeopardizing its environmental or economic value. Humans will continue to interact with the sea, especially along coastal areas, and problems caused by this interaction will continue to occur in the future. We owe it to ourselves and our children to make the most intelligent and knowledgeable use of these resources. With proper care, these resources can renew themselves and continue to be a source of delight, amazement, and sustenance.

What You Can Do

Although the picture of our oceans presented in this chapter is not a pretty one, the situation is not hopeless. Many groups and individuals are working hard to preserve our marine resources. If you would like to know more about current efforts to save the ocean and its creatures or how you as an individual can make a difference, here is a list of some organizations you can contact.

Center for Marine Conservation, 1725 DeSales NW, Washington, DC 20009. This group educates the public on topics such as marine pollution issues and marine wildlife sanctuaries.

Cousteau Society, 930 West 21st St., Norfolk, VA 23517. This group encourages protection and preservation of the oceans for future generations. Student volunteers are welcome to work at their Norfolk office. Students might wish to participate in Project Ocean Search, in which people from a variety of backgrounds spend two weeks studying an island ecosystem.

Greenpeace USA, Inc., 1432 U Street NW, Suite 201-A, Washington, DC 20009. This international organization focuses on ocean ecology, hazardous wastes, disarmament, and atmospheric pollution issues.

League of Conservation Voters, 1150 Connecticut Ave. NW, Suite 201, Washington, DC 20036. This is a nonpartisan political action group that works to elect proenvironmental candidates to Congress. They also publish a scorecard that rates current members of Congress on their environmental votes.

National Wildlife Federation, 1400 16th St. NW, Washington, DC 20036. Conservation education is the primary mission of this group.

The Nature Conservancy, 1815 North Lynn St., Arlington, VA 22209. This group fosters global preservation of natural diversity by finding, protecting, and maintaining the best examples of communities, ecosystems, and endangered species in the natural world.

Resources for the Future, 1616 P St. NW, Washington, DC 20036. This group is concerned with the quality of the environment and the conservation of natural resources.

World Wildlife Fund, 1250 24th St. NW, Washington, DC 20036. This group strives to save endangered species and acquire wildlife habitats.

CHAPTER SUMMARY

Almost from the beginning of human history, people have dumped wastes into the ocean. These include garbage, sewage, and industrial wastes, to name a few. Currently there is a move to dump radioactive and toxic wastes from industry into the sea in the region of subduction zones. As populations in coastal areas have increased, so has the amount of wastes dumped into coastal seas. This indiscriminate dumping has resulted in a decrease in the economic and recreational value of some beaches and in some instances has even posed a health hazard.

Indiscriminate dumping is not the only way in which pollutants enter the marine environment. Agricultural and urban runoff contains a variety of pesticides as well as human and industrial wastes. Some substances when incinerated produce gases that can dissolve in precipitation and enter the ocean when it rains. Some toxic materials can combine with bottom sediments, increasing the length of time that they remain in the environment and contaminate the water. Legislation and the use of less toxic materials, such as unleaded fuel, is gradually helping decrease the level of pollution in some regions.

Not all pollutants that enter the marine environment dissolve in the water. Some are absorbed by particles of matter that are consumed by animals and channeled through food chains. As one animal feeds upon another, the level of toxic material in a food chain increases to levels that can cause disease or death in higher-order consumers.

Plastic trash constitutes another pollutant. Because of its durability, it persists in the environment for long periods. Large marine animals like birds, reptiles, and mammals are most affected by plastic trash. Lost or discarded plastic traps that are used to catch fish, crabs, and lobsters continue to trap and kill animals for long periods.

Oil spills are another environmental problem. Crude oil contains a mixture of substances, many of which are toxic and can damage marine communities. Although there exist protective measures to prevent spills and to clean them up when they occur, they do not always work, and there have been numerous examples of severe environmental damage due to oil spills.

The heavy development of coastal regions has severely damaged habitats, especially wetlands and sandy beaches. Development physically disrupts the habitat, and it increases siltation and the amount of sewage that will enter the ocean. The large numbers of humans, along with their pets, that visit sandy beaches annually destroy habitats, kill marine organisms, and pollute the environment. Destruction of wetlands results in decreases in ocean productivity and in the populations of many important commercial species that rely on these areas as nursery grounds. Increased numbers of sport divers visiting coral reefs is causing substantial damage to a habitat already stressed by pollution and other human-related problems.

SELECTED KEY TERM

biological magnification,
p. 367

QUESTIONS FOR REVIEW

MULTIPLE CHOICE

1. An example of a persistent toxin would be
   a. fertilizer
   b. human waste
   c. DDT
   d. red tide toxin
   e. lead

2. *Biological magnification* refers to
   a. the increase in size of organisms as trophic levels increase
   b. the increase in populations of algae in response to nutrients
   c. the change in population size from generation to generation
   d. the accumulation of toxins in the flesh of animals at higher trophic levels
   e. the accumulation of pollutants in the marine environment

3. Birds that are tolerant of humans and thrive on human garbage are
   a. pelicans
   b. lesser terns
   c. piper plovers
   d. gulls
   e. sandpipers

4. Sewage from beachfront cottages usually enters
   a. a sewer system
   b. septic tanks
   c. the sea directly
   d. drainage ditches
   e. compost heaps

5. Damage to _____ has the greatest impact on many commercial species of fish and shellfish.
   a. sand beaches
   b. rocky beaches
   c. wetlands
   d. the benthos
   e. the open sea

SHORT ANSWER

1. What environmental problems are associated with plastic trash?
2. What are the major sources of oil pollution?
3. What effect does an oil spill have on the ecology of a rocky shore?
4. What activities are most damaging to wetlands?
5. Describe how toxins and other pollutants can enter ocean food chains.
6. List some of the major problems that are associated with the agricultural runoff that enters the ocean.
7. Describe how an oil spill causes injury to birds and mammals.
8. Describe how recreational and commercial use of beaches affects beach ecology.

THINKING CRITICALLY

1. Seashell collectors are frequently blamed for decreases in local mollusc populations in Florida. The collectors state that development is more to blame than overcollecting. Do you think that the collector's argument is valid? Explain.
2. What portion of the East Coast of the United States do you think would be more likely to experience algal blooms resulting from agricultural runoff?
3. An industrial waste is being dumped into the ocean off the coast of California, and you are asked to determine if it is accumulating in the aquatic food chains. How might you determine this?

SUGGESTIONS FOR FURTHER READING

Brower, K. 1989. State of the Reef, *Audubon* 91(2):56–81.

Bruemmer, F. 1995. La Arribada, *Natural History* 104(8):37–42.

Dolan, R. and H. Lins. 1987. Beaches and Barrier Islands, *Scientific American* 257(1):68–77.

Farrington, J. 1985. Oil Pollution: A Decade of Monitoring, *Oceanus* 29(3):69–77.

Hodgson, B. 1990. Alaska's Big Spill: Can the Wilderness Heal? *National Geographic* 177(1):4–43.

Ragotzkie, R., ed. 1983. *Man and the Marine Environment.* CRC Press, Boca Raton, FL.

Ward, F. 1990. Florida's Coral Reefs are Imperiled, *National Geographic* 178(1):114–132.

Weisskopf, M. 1988. Plastics Reap a Grim Harvest in the Oceans of the World, *Smithsonian* 18(12):58–67.

Answers to Multiple Choice Questions

Chapter 1

1. d **2.** b **3.** b **4.** c **5.** c

Chapter 2

1. b **2.** c **3.** c **4.** e **5.** a **6.** d **7.** b
8. c **9.** d **10.** d

Chapter 3

1. d **2.** c **3.** d **4.** c **5.** b **6.** a **7.** e
8. d **9.** e **10.** a

Chapter 4

1. b **2.** a **3.** c **4.** d **5.** c **6.** b **7.** e
8. d **9.** d **10.** c

Chapter 5

1. d **2.** d **3.** e **4.** d **5.** c

Chapter 6

1. c **2.** b **3.** d **4.** b **5.** a **6.** c **7.** d
8. d **9.** c **10.** d

Chapter 7

1. b **2.** c **3.** a **4.** d **5.** c **6.** e **7.** b
8. d **9.** d **10.** a

Chapter 8

1. c **2.** c **3.** a **4.** d **5.** e **6.** c **7.** d
8. c **9.** d **10.** e

Chapter 9

1. e **2.** b **3.** e **4.** b **5.** d **6.** c **7.** b
8. b **9.** c **10.** d

Chapter 10

1. c **2.** b **3.** d **4.** a **5.** c **6.** a **7.** d
8. d **9.** e **10.** a

Chapter 11

1. c **2.** b **3.** d **4.** e **5.** c **6.** b **7.** e
8. c **9.** d **10.** c

Chapter 12

1. c **2.** d **3.** c **4.** a **5.** c **6.** d **7.** d
8. c **9.** e **10.** b

Chapter 13

1. a **2.** b **3.** a **4.** b **5.** c **6.** b **7.** c
8. c **9.** c **10.** b

Chapter 14

1. c **2.** d **3.** e **4.** a **5.** d **6.** d **7.** c
8. b

Chapter 15

1. a **2.** b **3.** b **4.** e **5.** d **6.** c **7.** e
8. b **9.** c **10.** d

Chapter 16

1. c   **2.** c   **3.** a   **4.** b   **5.** c   **6.** b   **7.** d
8. d   **9.** b   **10.** e

Chapter 17

1. b   **2.** c   **3.** d   **4.** a   **5.** c   **6.** c   **7.** c
8. b   **9.** d   **10.** e

Chapter 18

1. b   **2.** c   **3.** d   **4.** b   **5.** e   **6.** d   **7.** d
8. d

Chapter 19

1. d   **2.** d   **3.** c   **4.** c   **5.** b

Chapter 20

1. c   **2.** d   **3.** d   **4.** b   **5.** c

abiotic environment (ay-by-AH-tik) The physical or nonliving environment in which an organism lives.

abiotic factors (ay-by-AH-tik) Physical factors—such as temperature, salinity, pH, sunlight, currents, waves, and the type and size of sediment particles—that influence the life and distribution of living organisms.

absorption The process of taking up or taking in material from the surrounding environment.

abyssal plain (uh-BIS-suhl) A flat expanse at the bottom of an ocean basin.

abyssal zone (uh-BIS-suhl) The ocean bottom that extends from a depth of 4,000 to 6,000 meters.

adductor muscle (a-DUHK-tir) A large muscle that functions in the opening and closing of a bivalve's shell.

adhesion A property of a liquid by which the liquid is attracted to the surface of objects that carry an electrical charge.

adsorption The process by which ions adhere to the surface of an object.

aerobic (ay-ROH-bik) An adjective meaning in the presence of oxygen.

aerobic respiration (ay-ROH-bik) A metabolic process requiring oxygen in which organic food molecules are broken down and energy is released.

agar (AH-gar) An algal polysaccharide that forms thick gels at very low concentrations. It is resistant to degradation by most microorganisms and is used to culture bacteria in microbiology laboratories.

allantois (eh-LAN-toys) An embryonic support membrane found in some vertebrates that functions in the elimination of wastes.

allopatric speciation (al-oh-PAT-rik) The formation of new species that occurs when two or more populations or parts of a large population become geographically isolated from each other.

ambergris (am-BER-gris) A digestive by-product of sperm whales that is used in making perfume.

ambulaccral groove (am-byu-LAK-ruhl) A groove through which the tube feet of an echinoderm extend.

ammocoete (AM-oh-seet) The larval stage of a lamprey.

amnion (AM-nee-uhn) A liquid-filled sac that contains the developing embryo of some vertebrate animals.

amniotic egg (am-nee-AH-tik) An egg covered by a protective shell and containing a liquid-filled sac called the *amnion*.

amoeba (uh-MEE-buh) A single-celled heterotrophic protozoan that moves and captures food by means of pseudopods.

ampulla (am-POOL-uh) A saclike structure that attaches to the tube foot of an echinoderm and acts as a water reservoir.

ampullae of Lorenzini (am-POOL-ee of loh-ren-ZEE-nee) Organs found in sharks and other cartilaginous fishes that can detect electrical signals in the water.

anadromous (uh-NAD-ruh-muhs) A term applied to fishes that reproduce in fresh water and spend their adult lives in the marine environment.

anaerobe (AN-uh-rohb) An organism that lives in an environment that lacks oxygen.

anaerobic (an-uh-ROH-bik) Lacking oxygen.

Animalia The kingdom that contains the animals—heterotrophic, eukaryotic organisms whose cells lack cell walls.

annelid (AN-eh-lid) A segmented worm belonging to the phylum Annelida.

anthozoan (an-thuh-ZOH-uhn) An animal belonging to the cnidarian class Anthozoa, which includes corals and sea anemones.

aphotic zone (ay-FOH-tik) The portion of the pelagic division where sunlight is absent.

aquaculture The farming of freshwater or marine organisms.

Aristotle's lantern A chewing structure composed of five teeth found in the mouths of sea urchins.

arthropod (AR-thruh-pahd) An animal belonging to the phylum Arthropoda. Arthropods are characterized by jointed appendages and an hard exterior covering (exoskeleton).

artificial selection The process by which farmers and animal breeders select only animals and plants with certain desirable traits for breeding in an effort to produce more organisms with the same desirable traits.

asconoid sponge (AS-kuh-noyd) A type of sponge whose body has only a single spongocoel that does not contain invaginations.

asexual reproduction The process by which offspring are produced from a single parent without the fusion of sex cells.

asthenosphere (as-THEN-uh-sfeer) The region of mantle that lies below the earth's crust.

atoll (a-TOHL) A coral reef that is somewhat circular in shape with a centrally located lagoon.

ATP (*abbreviation for* adenosine triphosphate) The major energy-carrying molecule in cells.

autotroph (AW-toh-trohf)   An organism that is capable of producing its own food; also known as a *producer*.

back reef   The area opposite the reef front.

bacteria   Microscopic organisms that belong to the kingdom Prokaryotae.

bacterial rhodopsin   A purple pigment very similar to the visual pigment in vertebrate eyes that is used by purple bacteria to capture light energy for the synthesis of ATP.

baleen (buh-LEEN)   A proteinaceous structure used to strain food from the water; it takes the place of teeth in baleen whales.

barbel (BAHR-behl)   A fleshy projection on the head of some fishes that may fulfill a sensory role or function as a lure.

barnacle   A sessile crustacean whose body is usually covered by plates composed of calcium carbonate.

barrier reef   A reef separated by a lagoon from the land mass with which it is associated.

bathyal zone (BATH-ee-uhl)   The portion of the ocean bottom that extends from the edge of the continental shelf to a depth of 4,000 meters.

bathygraphic features (bath-eh-GRAF-ik)   The physical features of the ocean bottom.

bathymetric chart (bath-eh-MET-rik)   Chart of the ocean that shows lines connecting points of similar depth.

bathyscaphe (BATH-eh-skayf)   A self-contained deep-sea research craft that can operate free of a mother ship.

benthic division (BEN-thik)   The division of the ocean environment composed of the ocean bottom.

benthic organism (BEN-thik)   An organism that lives primarily in or on bottom sediments.

benthos (BEN-thohs)   A collective term for the organisms living in or on the sea bottom.

bilateral symmetry   A type of symmetry in which body parts are arranged in such a way that only one plane through the midline of the central axis (the midsagittal plane) divides the organism into similar right and left halves.

binary fission   A type of asexual reproduction in which one cell splits into two after the original cell has duplicated its genetic material.

binomial nomenclature (by-NOH-mee-uhl NOH-men-klay-chur)   The system of using two words (that is, a genus-and-species epithet) to name an organism.

biogenous sediments (by-AH-gen-is)   Sediments formed from the remains of living organisms.

biogeochemical cycles   A combination of biological, chemical, and physical processes that act to recycle nutrients within the biosphere.

biological magnification   The concentration of pollutants or toxins in higher trophic levels of a food chain.

biosphere (BY-OH-sfeer)   The collection of all of the earth's ecosystems together.

biotic environment (by-AH-tik)   The living portion of an organism's environment.

biotic factors (by-AH-tik)   Interactions among living organisms, such as predator–prey relationships, competition, and symbioses.

blade   The large, flat, leaflike structures of brown algae.

blowhole   A hole on the top of the head of a cetacean that serves as the opening to the animals respiratory system.

blubber   A thick layer of fat found beneath the skin of some marine mammals.

brachiopod (BRAK-ee-uh-pawd)   A type of generally benthic lophophorate with a bivalved shell belonging to the phylum Brachiopoda; it is commoly known as a *lamp shell*.

breaker   A type of wave, the lower part of which is slowed by friction but whose crest continues to move toward the shore at a speed greater than that of the rest of the wave.

breeching   A behavior thought to be a mating ritual of male humpback whales in which the animal jumps out of the water and comes crashing back down, creating a loud noise.

brown algae   Algae that belong to the division Phaeophyta. In addition to chlorophyll, they contain the accessory pigment fucoxanthin, which gives them their characteristic olive-brown color.

bryozoan (bry-oh-ZOH-uhn)   Tiny lophophorates belonging to the phylum Bryozoa. They are commonly called *moss animals* and form colonies on a wide variety of solid surfaces.

bubble net   A ring of bubbles produced by feeding humpback whales that is used to capture food.

bycatch   The noncommercial animals killed during fishing for commercial species. Also known as *incidental catch*.

byssal threads (BIS-suhl)   Strong protein fibers secreted by mussels and used to fasten the animal to rocks or other solid surface.

capillary action   The ability of water to rise in narrow spaces.

capillary wave   A wave for which the restoring force is the surface tension of the water.

carnivore   An animal that feeds on other animals.

carotenoids (kuh-RAHT-in-noydz)   Yellow and orange accessory pigments that absorb green light and function in photosynthesis.

carrageenan (kar-uh-JEE-nuhn)   An algal polysaccharide from red algae that is used commercially as a thickening and binding agent.

catadromous (kuh-TAD-ruh-muhs)   A term applied to fishes that reproduce in the marine environment and spend their adult lives in fresh water.

caudal fin (KAW-duhl)   The tailfin of a fish.

cephalization (sef-uh-luh-ZAY-shuhn)   The concentration of sensory organs in the head region of an animal.

cephalopod (SEF-uh-loh-pahd)   An animal belonging to the molluscan class Cephalopoda, which includes squids, octopods, cuttlefishes, and nautiloids.

cephalothorax (sef-uh-loh-THOR-aks)   In an animal, the combination of two body regions: the head and the thorax.

cerata (sir-AH-tuh)   Projections on the body of a nudibranch that function in gas exchange.

cetacean (seh-TAY-shen)   An animal belonging to the mammalian order Cetacean, which includes whales, dolphins, and porpoises.

chelicera (keh-LI-suh-ruh)   An appendage found in chelicerates that is modified for the purpose of feeding and takes the place of mouthparts.

chelicerate (keh-LI-suh-reht)   An arthropod belonging to the subphylum Chelicerata. Chelicerates are characterized by the presence of feeding structures called *chelicerae*, which take the place of mouthparts.

chelipeds (KEE-leh-pehd)   The first pair of legs of many decapods; these two are modified to form a claw that is used for capturing prey and for defense.

chemoreceptor (kee-moh-ree-SEP-tuhr)   A sense organ capable of detecting changes in the chemical composition of an organism's environment.

chemosynthesis (kee-moh-SIN-theh-sis)   The process of producing organic food molecules through the use of the energy supplied by chemical reactions.

chemosynthetic (kee-moh-sin-THEH-tik)   Having the ability to use the energy from chemical reactions to construct organic food molecules.

chimaera (ky-MEER-uh)   Fishes that belong to the subclass Holocephali. These fishes are related to sharks and generally have an unusual body shape.

chitin (KY-tin)   A polysaccharide found in the cell walls of fungi and in the exoskeletons of many arthropods.

chiton (KY-tuhn)   A mollusc belonging to the class Polyplacophora. Chitons are characterized by a shell composed of eight separate plates bound together by a leathery girdle.

chlorophyll a   A green pigment that functions in photosynthesis. It absorbs primarily violet and orange light.

chlorophyll b   A green pigment that functions in photosynthesis. It absorbs primarily blue and orange light.

chloroplast   An organelle found in eukaryotic photosynthetic organisms that contains the pigments necessary for photosynthesis.

choanocyte (koh-AN-oh-syt)   A flagellated cell found lining cavities within the body of a sponge. Choanocytes are responsible for moving water through the body of a sponge. Also known as a *collar cell*.

chorion (KOR-ee-uhn)   An embryonic support membrane found in some vertebrates that functions in gas exchange.

chromatophore (kroh-MAT-uh-fohr)   A special cell in an animal's skin that contains pigment molecules.

chromosome (KROH-muh-sohm)   Structures consisting of DNA and protein that are located in the nucleus of eukaryotic cells.

chronometer (kroh-NAHM-i-ter)   A seagoing clock that keeps accurate time and that is used to determine longitude.

cilia (SIL-ee-uh)   Hairlike structures found on some cells that function in movement and/or feeding.

cirriped (SER-uh-ped)   Feathery appendages used by barnacles to filter food from the surrounding water.

claspers   Modifications of the pelvic fins of male sharks that function to transfer sperm to the genital opening of the female.

cleaning station   A territory established by one of several cleaner organisms where fishes will visit at regular intervals to have parasites and dead tissue removed.

cnidocil (NYD-uh-sil)   A bristle-like structure that extends from one end of a cnidocyte and functions as a trigger.

cnidocyte (NYD-uh-syt)   The stinging cell found in members of the phylum Cnidaria.

coccolithophore (kahk-oh-LITH-oh-for)   Photosynthetic protist that belongs to the phylum Chrysophyta. Coccolithophores are covered by calcareous plates.

coenocytic (see-nuh-SIT-ik)   The condition of having a body consisting of one giant cell or a few large cells containing more than one nucleus.

collar cell   A flagellated cell found lining the cavities within sponges. They are responsible for producing the water current that flows through a sponge's body. Also known as a *choanocyte*.

comb jelly   An animal belonging to the phylum Ctenophora. These animals resemble jellyfishes and move by means of cilia arranged in rows called *comb plates* or *ctenes*.

commensalism   A symbiotic relationship in which one organism benefits while the other is neither harmed nor benefited.

community   An assembly of populations of different species that occupy the same habitat at the same time.

competitive exclusion   The process by which the less successful competitor for a limited resource is driven to extinction.

conchiolin (kahn-KY-uh-lin; kahn-CHEE-uh-lin)   The protein that makes up the periostracum of a molluscan shell.

consumer   An organism that relies on another organism for food. Also known as a *heterotroph*.

continental drift   The movement of continental masses as the result of seafloor spreading.

continental rise   A gentle slope at the base of a steep continental slope.

continental shelf   The edge of a continental land mass.

control set   The trial set in an experiment that does not contain the experimental variable.

coral bleaching   The process by which corals, for reasons unknown, expel their zooxanthellae.

coralline algae   Species of red algae that have a coating of calcium carbonate and that resemble coral.

Coriolis effect (kohr-ee-OH-lis)   The apparent deflection of the path of winds and ocean currents that results from the rotation of the earth.

crinoid (KRY-noyd)   An animal belonging to the echinoderm class Crinoidea. These generally sessile animals have bodies that resemble flowers and are commonly referred to as *sea lilies* and *feather stars*.

crop   A digestive organ found in some animals that stores food before it is processed.

crust   The outermost, thinnest, and coolest layer of the earth.

crustacean An animal belonging to the arthropod class Crustacea, which includes crabs, lobsters, shrimp, and barnacles.

ctenoid scales (TEE-noyd) Thin overlapping scales found in the more advanced bony fishes. Their exposed posterior edges have small spines.

ctenophore (TEEN-uh-fohr) An animal belonging to the phylum Ctenophora. Also known as a *comb jelly*.

cuticle A thick, multilayer protective covering found on the exposed parts of some algae.

cyanobacteria (sy-AN-oh-bak-TEER-ee-uh) A group of photosynthetic prokaryotes that contain chlorophyll a and that release oxygen as a by-product of their photosynthesis. Cyanobacteria are sometimes referred to as *blue-green bacteria*.

cycloid scales (SY-kloyd) Thin, overlapping scales with a smooth posterior edge that are found in some of the more primitive bony fishes.

cyprid larva (SIP-rid) A pelagic larval stage in the life cycle of a barnacle. A cyprid larva has compound eyes and a carapace composed of two shell plates.

decapod (DEK-uh-pahd) An animal with five pairs of walking legs that belongs to the arthropod order Decapoda, which includes crabs, lobsters, and shrimp.

decomposer An organism that breaks down the tissue of dead plants and animals, as well as animal wastes.

deep scattering layer A mixed group of zooplankton and fishes that causes sonar systems to generate a false image of a nearly solid surface hanging in midwater.

deepwater wave A wave that occurs in water that is deeper than one half of the wave's wavelength.

delphinid (del-FI-nid) A collective term referring to dolphins and porpoises.

density The mass of a substance in a given volume of that substance.

deposit feeder An animal that feeds on bottom sediments.

desalination (dee-sal-uh-NAY-shun) The process of removing salt from seawater.

desiccation (dehs-ik-KAY-shun) The process of drying out or losing moisture.

detritivore (deh-TRY-ti-vor) An organism that feeds on detritus.

detritus (deh-TRY-tuhs) Organic matter such as animal wastes and bits of decaying tissue.

diatoms (DY-uh-tahmz) Photosynthetic, unicellular protists that belong to the phylum Chrysophyta. Diatoms have a glassy covering composed of silica.

dinoflagellates (dy-noh-FLA)-eh-laytz) Photosynthetic, usually single-celled protists that possess two flagella. Dinoflagellates belong to the phylum Pyrrophyta.

disruptive coloration A type of coloration found in some animals in which the background color of the body is broken up by lines that frequently run in the vertical direction.

diurnal tide (dy-YUR-nuhl) The condition of having only one high tide and one low tide each day.

DNA (*abbreviation for* deoxyribonucleic acid) A double-stranded molecule with the shape of a helix that contains an organism's genetic information (genes).

dorsal fin A fin found on the dorsal surface of a fish.

downwelling zone An area where surface water is sinking and displacing deeper water to the surface.

drift net Large nets composed of sections called *tans* that may stretch for as much as 60 kilometers. Drift nets entangle fish, squid, and other marine animals that swim into them.

ebb tide A falling tide.

echinoderm (eh-KY-noh-derm) An animal belonging to the phylum Echinodermata, which includes sea stars, brittle stars, sea urchins, sea cucumbers, and crinoids.

echolocation A process that allows some cetaceans to use sound waves to distinguish and home in on objects from distances of several meters.

ecological efficiency The percentage of energy that is taken in as food by one trophic level and then passed on as food to the next highest trophic level.

ecological equivalents Two different groups of animal that have evolved independently along the same lines in similar habitats and which therefore display similar adaptations.

ecosystem A system that is composed of living organisms and their nonliving environment.

ectotherm An animal that obtains most of its body heat from its surroundings.

embayment An area of coastline where portions of the ocean are cut off from the rest of the sea.

encephalization quotient (EQ) (en-sef-uh-ly-ZAY-shun) The ratio of actual brain mass to expected brain mass for a defined body size.

endoskeleton (EN-doh-SKEL-eh-tuhn) An internal skeleton.

endotherm (EN-doh-therm) An animal that maintains a constant body temperature by generating heat internally.

epidermis (ehp-i-DER-mis) An outer layer of cells or the cellular covering of an organism.

epifauna (EP-i-faw-nuh) Benthic organisms that live on the ocean's bottom.

epiphyte (EP-i-fyt) A plant or alga that grows on another plant or alga.

epitoky (EP-i-toh-kee) A reproductive strategy in some polychaetes that involves the formation of a pelagic reproductive individual, or epitoke, that is different from the nonreproducing form of the worm.

equator A circle drawn around the center of the earth that is perpendicular to its axis of rotation.

errant polychaete (ER-ent PAHL-eh-keet) An actively mobile polychaete that has a mouth equipped with jaws or teeth.

estuary A region where fresh water is mixed with salt water.

eukaryote (yoo-KAR-ee-oht) An organism whose cells contain a nucleus and membrane-bound organelles.

evolution The process by which populations of organisms change over time.

Exclusive Economic Zones (EEZ) The zones of ocean exclusively controlled by a coastal nation.

exoskeleton (EK-soh-SKEL-eh-tuhn) A hard protective exterior skeleton such as that found in arthropods.

experimental set The trial set in an experiment that contains the experimental variable.

fault An area where the earth's plates move past each other.

feeding polyp A type of polyp found in hydrozoan colonies that functions in capturing food and feeding the colony. Also known as a *gastrozooid*.

fermentation A metabolic process in which organic nutrients are broken down and energy is released in the absence of oxygen.

filter-feeder An organism that filters its food from the water.

fitness An organism's biological success as measured by the number of its own genes that are present in the next generation of a population.

fjord (fyord) A deep valley cut into the coastline by glaciers and filled with a mixture of fresh water and salt water.

flood tide A rising tide.

fluke A type of parasitic flatworm belonging to the class Trematoda.

food chain A sequence of feeding relationships among a group of organisms that begins with producers and proceeds in a linear fashion to higher-level consumers.

food web A representation of the complex feeding networks that exist in an ecosystem.

foraminiferan (for-uh-muh-NIF-uh-ran) An amoeba-like protozoan that produces an elaborate shell of calcium carbonate and uses pseudopods to capture food.

fore reef The seaward side of a coral reef.

fossil fuel Fuel—including coal, oil, and natural gas—formed from the remains of plants and microorganisms that lived millions of years ago.

fragmentation A type of asexual reproduction in which a part of the parent organism breaks off and forms a new individual.

fringing reef A reef that is found close to and surrounding newer volcanic islands or that borders continental land masses.

fruit A structure composed of several layers of tissue that surround a seed, protecting it and aiding in its dispersal.

frustule (FRUHS-tyool) The glassy structure composed of silica that covers diatom cells.

fucoxanthin (few-koh-ZAN-thin) A xanthophyll pigment that is brown to golden brown in color.

fungus (*plural*: fungi) An organism that belongs to the kingdom Fungi. Fungi are eukaryotic, have cell walls containing chitin, and are not photosynthetic.

gamete (GAM-eet) A sex cell, such as a sperm, egg, or pollen grain.

gametophyte (ga-MEE-toh-fyt) The stage in the life cycle of an alga or plant during which gametes are produced.

ganoid scale (GAN-oyd) Thick bony scales that do not overlap. Ganoid scales are found in some primitive bony fishes.

gas bladder A gas-filled structure that helps to buoy the large blades of an alga.

gastropod (GAS-troh-pahd) A member of the molluscan class Gastropoda, which includes snails, limpets, abalones, and nudibranchs.

gastrovascular cavity (gas-troh-VAS-kyoo-ler) A central cavity found in the body of cnidarians that functions in digestion and in the movement of materials within in the animal.

gene A unit of hereditary information that is located on an organism's DNA.

gene pool All of the genes in a given population that exist at a given time.

generating force A force that disturbs the surface of water, producing a wave.

genus (JEE-nuhs) The first name in the two-part Latin name for an organism. It is the taxonomic group that is one step above the species level.

gill filaments Thin, rodlike structures that make up the gills found in some animals.

gill rakers Finger-like projections on the gill arches of a fish that function in trapping material and preventing it from fouling the gill.

gravity wave A wave for which the restoring force is gravity.

green algae Algae that belong to the division Chlorophyta. Green algae contain chlorophyll and the accessory pigments called *carotenoids*.

guano (GWAH-noh) A phosphate-rich manure produced by birds.

gular pouch (GYOO-ler; GUH-ler) A sac of skin that hangs between the flexible bones of a pelican's lower mandible.

habitat The specific place in the environment where an organism lives.

hadal zone (HAYD-uhl) The portion of the ocean bottom that lies at depths greater than 6,000 meters.

halocline (HAL-oh-klyn) A zone in the ocean characterized by a rapid change in salinity with increasing depth.

halophyte (HAL-oh-fyt) A salt-tolerant flowering plant.

hard corals Corals that produce hard skeletons composed of calcium carbonate and which are primarily responsible for building coral reefs. Also known as *scleractinian corals*.

herbivore An animal that eats only plants and algae.

heterocercal tail (het-uh-roh-SIR-kuhl) The type of tail found in some fishes in which the upper lobe is larger than the lower lobe and the vertebral column bends slightly upward into the upper lobe.

heterocysts (HET-uh-roh-sists) Thick-walled chambers that contain enzymes for fixing nitrogen. Heterocysts are associated with colonies of cyanobacteria.

heterotroph (HET-uh-roh-trohf)   An organism that relies on other organisms for food. Also known as a *consumer*.

high marsh   The region of a salt marsh closest to shore that is covered briefly by salt water each day.

holdfasts   A branching system of fibers that resembles roots. Holdfasts function to anchor algae to the bottom and to hard surfaces.

homeostasis (HOH-mee-oh-STAY-sis)   The internal balance that living organisms must maintain in order to survive.

homocercal tail (hoh-moh-SIR-kuhl)   A tail in which the upper and lower lobes are generally equal and into which the vertebral column does not extend.

hydrophyte (HYD-roh-fyt)   A flowering plant that generally lives submerged under water.

hydrostatic skeleton (hyd-roh-STAT-ik)   Body support that results from maintaining fluid under relatively high pressure in an animal's internal body cavity.

hydrozoan (hyd-roh-ZOH-en)   A generally colonial organism that belongs to the cnidarian class Hydrozoa.

hypothesis   An explanation for observed events that can be tested by experiments.

incidental catch   Noncommercial animals that are killed each year during fishing for commercial species. Also known as *bycatch*.

infauna (IN-faw-nuh)   Benthic organisms that live in the bottom sediments.

interspecific competition   Competition between similar species for a limited resource.

intertidal zone   The region of a shore that is covered by high tide and exposed at low tide.

intraspecific competition   Competition between members of a single species for a limited resource.

invertebrate   An animal that lacks a vertebral column (backbone).

ion   A particle that carries an electrical charge.

isolating mechanisms   Mechanisms that prevent members of different species from reproducing in nature.

isopycnal (eye-soh-PIK-nuhl)   Having the same density at the top as at the bottom.

isosmotic (eye-sahz-MAHT-ik)   Having the same concentration of solutes as the fluid environment.

keystone predator   An animal whose presence in a community makes it possible for many other species to live there. Also known as a *keystone species*.

keystone species   An animal whose presence in a community makes it possible for many other species to live there. Also known as a *keystone predator*.

krill   Pelagic, shrimplike creatures that belong to the arthropod order Euphausiacea.

larvacean   A free-swimming tunicate that resembles a tadpole.

lateral line system   A system of canals running the length of a fish's body and over its head that functions in detecting movement in the water.

latitude   The angular distance north and south of the equator measured in degrees.

leuconoid sponge (LOO-keh-noyd)   A sponge with a complex body containing many spongocoels and chambers leading to them.

lithosphere (LITH-oh-sfeer)   The part of the earth comprising the crust and upper mantle.

longitude   The angular distance east and west of the prime (or Greenwich) meridian measured in degrees. Also known as a *meridian*.

lophophorate (lohf-uh-FOHR-ayt)   A sessile animal that lacks a distinct head and possesses a feeding device called a *lophophore*.

lophophore (LOHF-uh-fohr)   An arrangement of ciliated tentacles that surrounds the mouth of animals known as *lophophorates*. It functions in feeding and in respiration.

low marsh   A salt marsh found in the lower intertidal zone that is covered by tidal water much of the day.

macrozooplankton (maK-roh-ZOH-oh-plank-tuhn)   Zooplankton that range in size from 200 to 2,000 micrometers.

madreporite (mad-ruh-POHR-ryt)   The structure at which water enters the water vascular system of an echinoderm.

magma   Molten material located deep in the earth's mantle.

mammary glands   Special glands found in female mammals that produce milk.

mangal (MAN-guhl)   A community dominated by plants called *mangroves*.

mantle (earth's)   The thickest layer of the earth and the one that contains the greatest mass of material.

mantle   The part of a mollusc's body that secretes its shell.

mantle cavity   The space between a mollusc's mantle and its body.

mariculture (MAR-i-kuhl-chuhr)   The use of agricultural techniques to breed and raise marine organisms. Also known as *aquaculture*.

marine biology   The study of the living organisms that inhabit the seas and of their interactions with each other and their environment.

marine snow   Translucent, drifting organic particles and aggregates found in the water column.

medusa   The free-floating form of cnidarian that resembles an umbrella or a bell.

megazooplankton (meg-uh-ZOH-oh-plank-tuhn)   Zooplankton that measure over 2,000 micrometers in size.

meiofauna (MY-oh-fawn-uh)   The tiny organisms that are adapted to living in the spaces between sediment particles.

melon   An oval mass of fatty, waxy material that is located between the blowhole and the end of the head in cetaceans capable of echolocation. The melon serves to direct and focus the sound waves produced by an animal.

meridian   A line of longitude.

mesopelagic zone (MEZ-oh-peh-LAJ-ik)   The region between 150 meters and 450 meters below the ocean's surface where there is not enough light to power photosynthesis. Also known as the *twilight zone.*

metabolism   The sum of all of the chemical reactions that occur within living cells.

microhabitats   The smaller subdivisions of a habitat.

micropatchiness   The uneven distribution of microorganisms on bits of marine snow.

microphytoplankton (my-kroh-FYT-oh-plank-tuhn)   Phytoplankton that range in size from 20 to 200 micrometers.

microplankton   Plankton that range in size from 20 to 200 micrometers.

microzooplankton   Zooplankton that range in size from 20 to 200 micrometers.

midocean ridge   A long mountain range that forms along cracks on the ocean floor where erupting magma breaks through the earth's crust.

mixed semidiurnal tide   A tide in which the high tide and the low tide are of different levels.

mollusc   A soft-bodied animal whose body is frequently covered by a shell. Molluscs are members of the phylum Mollusca.

molting   In arthropods, the process in which an old exoskeleton is shed and a new one is formed.

monoculture   The process of raising only one species in mariculture.

morphology (mohr-FAHL-uh-jee)   The structure and/or appearance of an organism.

mucilage (MYOO-suh-lij)   The slimy, gelatinous covering found on algal cells.

mutualism   A symbiotic relationship in which both organisms benefit.

nacreous layer (NAY-kree-uhs)   The innermost layer of a molluscan shell. In oysters, it also is known as the *mother-of-pearl layer.*

nanophytoplankton (NAN-oh-FYT-oh-plank-tuhn)   Phytoplankton that range in size from 5 to 20 micrometers.

nanoplankton (NAN-oh-plank-tuhn)   Plankton that range in size from 5 to 20 micrometers.

natural selection   The mechanism that explains why organisms that possess variations best-suited to their particular environments exhibit a better survival rate and reproductive capacity than do less well-suited organisms.

nauplius larva (NAW-plee-uhs)   A larval stage in the life cycle of many crustaceans. Nauplius larvae are characterized by three pairs of appendages and a median eye.

nautical mile   A unit of distance equal to 1 minute of latitude, 1.85 kilometers, or 1.15 land miles.

neap tide   The tide that exhibits the smallest change between the high- and low-tide marks. Neap tide occurs when the sun and the moon are at right angles to each other.

nekton   All actively swimming organisms whose movements are not governed by currents or tides.

nematocyst (neh-MAT-uh-sist)   The stinging organelle found within the stinging cell of cnidarians.

nematode (NEM-uh-tohd)   A round, wormlike animal that belongs to the phylum Nematoda.

neritic zone (neh-RIT-ik)   The zone of water that lies over the continental shelves.

net plankton   Plankton that can be captured with standard-sized plankton nets.

neuromast (NOO-roh-mast)   A sense organ that can detect vibrations in the fluid that fills the canals of the lateral line system.

niche (nish; neesh)   An organism's role in its environment.

nictitating membrane (NICK-ti-tay-ting)   A clear membrane that covers and protects the eye of some vertebrates.

nitrification (ny-truh-fi-KAY-shun)   The process by which ammonia from animal wastes and dead tissue are converted into nitrate ions.

nitrifying bacteria (NY-truh-fy-ing)   Bacteria that are capable of converting the ammonia from animal wastes and dead tissue into nitrate ions.

nitrogen fixation   The process by which some microorganisms are able to convert atmospheric nitrogen into a form that is useable by producer organisms.

nucleus   A structure found in eukaryotic cells that contains the cells' chromosomes and acts as the cellular control center.

nudibranch (NOO-di-brangk)   A gastropod mollusc that does not have a shell and that has many projections from its body called *cerata.*

nutrient   Any organic or inorganic material that an organism needs to metabolize, grow, and reproduce.

obliterative countershading   A type of coloration found in some animals in which the body surface exposed to the sun (usually the dorsal surface) is of a dark color and the opposite surface (usually the ventral surface) is light in color or white.

ocean ranching   The process of raising young fish in hatcheries and returning them to the sea when they have reached adulthood. Also known as *sea ranching.*

oceanic zone   The zone of ocean composed of the water that covers the deep ocean basins.

oceanography   The study of the oceans and their phenomena, such as waves, currents, and tides.

omnivore   An animal that feeds on both producers and consumers.

ooze   Sediment composed of fine biogenous particles.

operculum (oh-PER-kyoo-luhm)   A hard or tough covering found in some molluscan gastropods that closes the opening to the shell when the animal retracts. In fishes, the protective covering of the animal's gills.

osculum (AHS-kyuh-luhm; *plural*: oscula)   The opening through which water exits the body of a sponge.

osmoconformer (ahz-moh-kuhn-FOHR-mer)   An animal whose tissues and cells can tolerate dilution.

osmoregulator (ahz-moh-REG-yoo-lay-tir)   An animal that can maintain an optimal salt concentration in its tissues regardless of the salt content of its environment.

osmosis (ahz-MOH-sis)   The movement of water across a semipermeable barrier in response to differences in solute concentration on either side of the barrier.

ostium (AHS-tee-uhm; *plural*: ostia)   A tiny hole or pore that allows water to enter the body of a sponge.

oviparous reproduction (oh-VIP-uh-ruhs)   A type of reproduction in which the young develop in eggs that are laid outside of the mother's body.

ovoviviparous reproduction (oh-voh-vy-VIP-uh-ruhs)   A type of reproduction in which the young develop in eggs retained within the mother's body and are born live.

paralytic shellfish poisoning (PSP)   A toxic condition that occurs following the ingestion of shellfish contaminated by a dinoflagellate toxin.

parasitism   A symbiotic relationship in which one organism benefits and the other is harmed.

parthenogenesis (pahr-theh-noh-JEN-eh-sis)   The process by which an egg develops without being fertilized by a sperm.

partially mixed estuary   An estuary that has a strong surface flow of fresh water and a strong influx of seawater.

patch reef   Small patch of reef located in a lagoon associated with an atoll or a barrier reef.

patchiness   The uneven distribution of organisms in a population or a community.

pectoral fins   The anterior paired fins of a fish.

pedicellariae (ped-uh-suh-LEHR-ee-eh)   Tiny pincerlike structures found in some echinoderms that function to keep the surface of the body clean and free of parasites and the settling larvae of fouling species. In some echinoderm species, they may also aid in obtaining food.

pedicle (PED-i-kuhl)   A fleshy stalk found in some lamp shells that is used to fasten the animal to a solid surface.

pelagic division (pe-LAJ-ik)   The division of the marine environment composed of the ocean's water.

pelvic fins   The posterior pair of fins of a fish.

pen   An internal strip of hard protein that helps support the mantle of a squid.

periostracum (per-ee-AHS-treh-kuhm)   The outermost layer of a molluscan shell. The periostracum is composed of the protein *conchiolin*.

pH scale   A scale that indicates the acidity or basicity of a solution. The pH scale runs from 0 to 14, with 7 being the neutral point. The pH number reflects the concentration of hydrogen ions in a solution. Acids have a pH between 0 and 7, whereas bases have a pH between 7 and 14.

pharynx (FA-ringks)   A muscular tube that forms part of the digestive tract of an animal.

pheromone (FAYR-eh-mohn)   A hormone released into the environment by one individual that controls the behavior and/or development of other individuals of the same species.

phoronid (FOHR-oh-nid)   A small, wormlike lophophorate that secretes a tube of leathery protein or chitin.

photic zone (FOH-tik)   The portion of the pelagic division of the ocean that receives enough sunlight to support photosynthesis.

photophore (FOH-toh-fohr)   A specialized organ found in some organisms that functions in producing bioluminescence.

photosynthesis   The process by which some organisms use the energy of sunlight to produce organic molecules, usually from carbon dioxide and water.

phycocyanin (fy-koh-SY-uh-nin)   A blue pigment that absorbs green light. In some organisms, phycocyanin functions as an accessory pigment in photosynthesis.

phycoerythrin (fy-koh-e-RITH-rin)   A red pigment that absorbs blue and green light. In some organisms, it functions as an accessory pigment in photosynthesis.

physiographic chart (fiz-ee-oh-GRAF-ik)   A chart of the ocean that uses perspective drawing, coloring, or shading to show the various depths.

phytoplankton (FY-toh-plank-tuhn)   Tiny photosynthetic organisms that float in the oceans currents.

picoplankton (PY-koh-plank-tuhn)   Plankton that measure less than 2 micrometers in size.

pinniped (PIN-i-ped)   An animal belonging to the mammalian order Pinnipedia, which includes seals, sea lions, elephant seals, and walruses.

placenta (plah-SEN-tuh)   An organ present only during pregnancy in some female mammals that functions in maintaining the fetus until birth.

placoid scale (PLAK-oyd)   A type of scale found in cartilaginous fishes that has a structure resembling the teeth of other vertebrates.

plankton   Organisms that float or drift in the sea's currents.

Plantae (PLAN-ty)   The kingdom that contains plants—multicellular, eukaryotic organisms that have cells with cellulose-containing walls and which are capable of photosynthesis.

pneumatophore (noo-MAT-uh-fohr)   An erect aerial root of the black mangrove.

pod   A group of related cetaceans.

polar   Having a positively charged end and a negatively charged end.

polychaete (PAHL-eh-keet)   A type of annelid worm belonging to the class Polychaeta.

polyculture (PAHL-ee-kuhl-chur)   The raising of more than one species in mariculture.

polygynous (pah-LIJ-eh-nehs)   Relating to a male having more than one female mate.

polyp (PAHL-uhp)   A generally benthic form of cnidarian characterized by a cylindrical body that has an opening at one end that is usually surrounded by tentacles.

population   A group of individuals of the same species that occupies a specified area.

prismatic layer (priz-MAT-ik)   The middle layer of a molluscan shell and the layer that contains most of the mass of the shell.

producer   An organism that can produce its own food. Also known as an *autotroph*.

progressive wave   A wave that moves in a particular direction, as opposed to one that moves randomly.

Prokaryotae (proh-kar-ee-OH-tee)   The kingdom that contains all of the unicellular organisms that lack a nucleus and membrane-bound organelles.

prokaryote (proh-KAR-ee-oht)   A unicellular organism without a nucleus; a member of the kingdom Prokaryotae.

prop root   An above-ground root that helps support a plant.

Protista (proh-TIS-tuh)   The kingdom that contains any eukaryotic organism that is not a plant, fungus, or animal.

protozoa   Protists that exhibit animal-like characteristics such as heterotrophy and the ability to move.

pseudofeces (SOO-doh-fee-sees)   Large, semisolid particles produced by bivalves and consisting of phytoplankton and detritus that they filter but do not consume.

pseudopod (SOO-doh-pahd)   Finger-like projections of cytoplasm and membrane that function in both locomotion and feeding in amoebae and their relatives.

purse seine   Huge nets with bottoms that can be closed off by pulling on a line in a manner similar to the pulling of the drawstring of a purse.

pycnocline (PIK-noh-klyn)   A zone in the ocean that is characterized by a rapid change in density with depth.

pyrosome (PY-roh-sohm)   A large colony of salps that may contain as many as 500 individuals per cubic meter of water and stretch over several kilometers.

radial symmetry   The symmetrical organization of body parts around a central axis that is similar to the arrangement of spokes on a wheel.

radiolarian (ray-dee-oh-LAYR-ee-uhn)   A protozoan that has an intricate shell made of silica and that uses pseudopods to capture prey.

radula (RAJ-oo-luh; RAD-yoo-luh)   A ribbon of tissue that contains teeth. This structure is unique to molluscs.

raft culture   A process in which juvenile commercial molluscs are attached to ropes that are suspended from rafts floating in regions of the ocean where food is plentiful and exposure to natural predators is minimized.

red algae   Algae that belong to the division Rhodophyta. They contain the accessory pigments phycoerythrin and phycocyanin.

red tide   A condition that occurs as the result of a population explosion of certain dinoflagellates that sometimes imparts a reddish color to the water.

reef crest   The highest point on a coral reef.

reef flat   The area opposite the reef front. Also known as the *back reef*.

reproductive isolation   The absence of interbreeding between two populations.

reproductive polyp   A type of polyp found in hydrozoan colonies that is specialized for the process of asexual reproduction. Also known as a *gonangium*.

resource partitioning   The process by which organisms share a resource.

restoring force   The force that causes the water in a wave to return to its undisturbed level.

rhizoids (RY-zoydz)   Rootlike structures that develop from the germinating spores of some brown algae and that serve to fasten them to the bottom. Holdfasts eventually develop from these structures.

rhizome (RY-zohm)   Major stems of a flowering plant that grow horizontally, usually just below the surface of bottom sediments.

rhizopodia (ry-zoh-POH-dee-uh)   Dense, sticky nets of living cytoplasm that are used by foraminiferans to capture prey.

rift zone   A region where the lithosphere splits, separates, and moves apart as new crust is formed.

rostrum (RAHS-truhm)   The noselike structure that projects over the mouth of a shark.

salinity   A measure of the concentration of dissolved inorganic salts in water.

salt glands   Glands found in some vertebrates that are specialized for salt secretion.

salt wedge   The angled boundary between salt water and fresh water in an estuary that occurs when the rapid flow of river water prevents the salt water from mixing with the fresh water.

salt wedge estuary   An estuary in which the seawater moves in and out along an angled boundary known as a *salt wedge*.

scaphopod (SKA-foh-pahd)   A mollusc belonging to the class Scaphopoda. Scaphopods are also known as *tusk shells* because their shells are shaped like the tusks of elephants.

scyphozoan (sy-fuh-ZOH-uhn)   An animal that belongs to the cnidarian class Scyphozoa. These are commonly called *jellyfish*.

sea anemone   Large, heavy, complex polyps that belong to the cnidarian class Anthozoa.

sea ranching   The process of raising young fish in hatcheries and returning the adults to the sea. Also known as *ocean ranching*.

seafloor spreading   The process by which magma driven by convection currents is turned back by the lithosphere, moves laterally, and then descends, causing lateral movement of the earth's crust.

seamount   A steep-sided formation that rises sharply from the ocean bottom.

sedentary polychaete (PAHL-eh-keet)   A sessile polychaete that usually forms some sort of tube to cover its body.

seed   A specialized structure found in some species of plants. The seed contains the plant embryo and a supply of nutrients surrounded by a protective outer layer.

selective forces   The physical and biological characteristics of the environment, such as temperature, salinity, predation, and food availability, that favor the survival of one species over another.

semidiurnal tide   The condition of having two high tides and two low tides each day.

sepia (SEE-pee-uh)   A dark fluid produced in the ink glands of cephalopods.

setae (SEE-tee)   Bristles that are found on the bodies of annelid worms.

sextant   A device used to measure the angle of a star with respect to the horizon.

sexual dimorphism (dy-MOR-fiz-uhm)   The condition in that males and females look different.

sexual reproduction   The process by which two parent organisms produce an offspring by the fusion of sex cells produced by each parent.

shelf zone   The part of the ocean bottom that extends from the line of lowest tide to the edge of the continental shelf.

shipworm   A bivalve mollusc that resembles a worm and burrows into wood, causing extensive damage of wooden structures.

silicoflagellate (sil-i-koh-FLAJ-uh-layt)   A tiny photosynthetic protist that belongs to the phylum Chrysophyta. Silicoflagellates possess internal shells composed of silica and one or two flagella.

siphon   A tubular structure formed from the mantle of a bivalve that directs the flow of water in and out of the animal's body.

siphonophore (sy-FAHN-uh-fohr)   An animal belonging to the cnidarian order Siphonophora. These pelagic colonies resemble single individuals and are represented by the Portuguese man-of-war and by-the-wind sailors.

siphuncle (SY-fuhn-kuhl)   A cord of tissue that runs through the chambers of a nautilus shell and functions in the removal of seawater from the chambers and in the regulation of the gas content of the chambers.

sirenian (sy-REE-nee-uhn)   An animal belonging to the mammalian order Sirenia, which includes manatees and dugongs.

slack water   The period during a change in tide when tidal currents slow down and then reverse.

soft corals   Corals that do not produce stony exterior skeletons.

solute   A dissolved substance.

speciation (spee-see-AY-shun)   The process by which new species are formed.

species   One or more populations of potentially interbreeding organisms that are reproductively isolated from other such groups.

species epithet (EHP-i-thet)   The second part of the two-part Latin name for an organism.

specific heat   The amount of heat energy required to increase the temperature of 1 gram of a substance by 1°C. Also known as the *thermal capacity*.

spermaceti (spur-meh-SEH-tee)   A thin, colorless, transparent oil that forms a waxy material when it comes into contact with air. Spermaceti is found in a cavity in the head of sperm whales.

spicule (SPIK-yool)   A support structure found in sponges. A spicule can be composed of calcium carbonate, silica, or the protein spongin.

spiracle (SPEER-uh-kuhl)   One of a pair of openings on the head of some cartilaginous fishes that functions in the passing of water to the gills.

spiral valve   A valve shaped like a spiral staircase found in the intestine of sharks.

splash zone   The uppermost area of a rocky shore that is covered by only the highest tides and usually is just dampened by the spray of crashing waves. Also known as the *supratidal zone*.

sponge   The common name for animals belonging to the phylum Porifera. These animals are characterized by their lack of symmetry (that is, they are asymmetrical) and by bodies that contain numerous holes.

spongin (SPUN-jin)   A structural protein found in some sponge spicules.

spongocoel (SPUN-joh-seel)   A spacious, water-filled cavity found within the body of a sponge.

sporophyte (SPOHR-oh-fyt)   The stage in the life cycle of an alga or a plant in which diploid spores are produced.

spring tide   A tide that exhibits the greatest change between the high and low tide marks. Spring tide occurs when the sun and the moon are in line with each other.

spur-and-groove formation   The usual shape of a reef front, characterized by finger-like projections of coral that protrude seaward.

squalene (SKWAY-leen)   An oily material produced by the liver of sharks.

stinging cell   A cell found in animals belonging to the phylum Cnidaria that functions in feeding and defense. Also known as a *cnidocyte*.

stipe   A stemlike structure found in brown algae.

stomata (stoh-MAH-tuh)   Openings in a plant leaf that allow water to escape and gases to enter.

subduction zone   Region of the ocean floor where old crust is removed and recycled back to the earth's core.

subtidal zone   The region of the shore that is covered by water, even during low tide.

supratidal zone   The uppermost area of a rocky shore; it is covered only by the highest tides and is usually just dampened by the spray of crashing waves. Also known as the *splash zone*.

surf zone   The area of a coast where waves slow down before striking the shore.

surface tension   A property of liquids characterized by the clinging together of molecules at the exposed surface as the result of cohesive forces.

suspension feeder   An organism that feeds on material suspended in the water.

swells   Long-period, uniform waves that appear as a regular pattern of wave crests on the ocean's surface.

swim bladder   A gas-filled sac found in some bony fishes that allows them to maintain neutral buoyancy.

syconoid sponge (SY-kuh-noyd)   A sponge with a single spongocoel that has many invaginations.

symbiosis (sim-by-OH-sis)   An intimate living arrangement between two different kinds of organisms.

table reef   A small reef found in the open ocean that has no central island or lagoon.

tapeworm   A parasitic flatworm belonging to the class Cestoda. Tapeworms are found in the intestines of vertebrate animals.

taxonomy (tak-SAHN-uh-mee)   The science of naming and classifying organisms.

test   The hard endoskeleton of a sea urchin.

thaliacean (tha-lee-AY-shun)   Barrel-shaped, free-swimming tunicates belonging to the urochordate class Thaliacea.

thallus (THAL-uhs)   The body of an alga.

theory   A body of observations and their experimental supports that have stood the test of time.

theory of plate tectonics   The theory that states that the movement of continental masses is the result of the movement of the rigid slabs, or plates, on which they rest.

thermal capacity   The amount of heat energy required to increase the temperature of 1 gram of a substance by 1°C. Also known as the *specific heat*.

thermocline (THER-moh-klyn)   A zone in the ocean that is characterized by a rapid change in temperature with increasing depth.

tidal currents   Currents that occur as the result of changing tides.

tidal flat   An area of estuary that is exposed at low tide and covered at high tide.

tide   The changes in sea level that occur as the result of the gravitational pull of the moon and the sun on the water in the oceans.

tidepools   Depressions in intertidal rocks or in the intertidal zone of sandy beaches that continue to hold water during a low tide.

tintinnid (tin-TIN-id)   Planktonic protozoans that have a body covering called a *lorica* that is formed from foreign particles and that use cilia to move and capture food.

trace elements   Elements dissolved in seawater that are present in concentrations less than one part per million.

trawl   Large nets that are dragged along the bottom or in midwater, depending on the catch, by vessels called *trawlers*.

trochophore larva (TROHK-eh-fohr)   A free-swimming ciliated larva found in the life cycles of many marine molluscs as well as some other marine organisms.

trophic level (TROH-fik)   An energy-storing level in a food chain.

tropical zone   The area of the earth that lies between the Tropic of Cancer (23.5 degrees N) and the Tropic of Capricorn (23.5 degrees S).

tsunami (soo-NAH-mee)   A large seismic sea wave that increases in amplitude as it approaches a shore. Sometimes referred to as a *tidal wave*.

tube feet   Tubular structures found in echinoderms that function in locomotion and feeding.

tubules of Cuvier   Tubules expelled from the anus of some species of sea cucumber when they are disturbed. Upon contact with seawater, the tubules become sticky.

tunicate (TOO-ni-kayt)   Animals belonging to the subphylum Urochordata. Tunicates are named for their body covering: a tunic composed of a substance similar to cellulose.

turbidity current   A swift avalanche of sediment and water that erodes a slope as it sweeps down it and picks up speed.

twilight zone   The region of ocean between 150 and 450 meters where there is not enough light to power photosynthesis. Also known as the *mesopelagic zone*.

ultrananoplankton (uhl-truh-NAN-oh-plank-tuhn)   Plankton that range in size from 0.2 to 5 micrometers.

ultraplankton (UHL-truh-plank-tuhn)   Plankton that range in size from 0.2 to 5 micrometers.

umbo   The area around the hinge of a bivalve's shell. This area represents the oldest part of the shell.

upwelling   The process by which a combination of wind and ocean currents bring nutrient-laden material from the ocean bottom into the photic zone.

upwelling zone   An area where bottom water rises to the surface and displaces surface water to the bottom.

valves   The two jointed halves of a shell that surround the bodies of bivalves and lamp shells.

vascular plants   Plants that contain vascular tissue.

vascular tissue   Tissue composed of specialized vessels that carry fluids and provide structural support for some species of plant.

veliger larva (VEL-uh-jer)   A free-swimming larval stage that develops from the trochophore larva of some molluscs.

vent community   The populations of organisms that live around a deep-sea vent.

vertebrate   An animal that possesses an internal skeletal rod (commonly called a *backbone*) composed of units known as *vertebrae*.

vestimentiferan worms (ves-ti-men-TIF-eh-ruhn)   Large worms of the phylum Pogonophora that are found in deep-sea vent communities.

viviparous reproduction (vy-VIP-uh-ruhs)   A type of reproduction in which the young develop from embryos that are attached internally to the mother and that are born live.

water vascular system   A hydraulic system unique to echinoderms that functions in locomotion, feeding, respiration, and excretion.

wave shock   The force of waves as they crash against the rocks and the organisms that live on the rocks.

well-mixed estuary   An estuary in which there is a seaward flow of water with uniform salinity at all depths.

xanthophylls (ZAN-thuh-filz)   A group of accessory pigments that function in the process of photosynthesis in some organisms.

zonation   The separation of organisms in a habitat into definite zones or bands.

zone of dry sand   The uppermost vertical zone in the high in-tertidal region of a sandy beach that contains no moisture.

zone of drying sand   The vertical zone in the high intertidal region of a sandy beach characterized by the presence of mois-ture only during the highest tides.

zones of intolerance   Regions that are so far removed from an organism's optimal range for an environmental variable that the organism cannot survive.

zone of resurgence   A vertical zone of the mid- and low inter-tidal regions of a sandy beach that retains water at low tide.

zone of retention   A vertical zone in the high intertidal region of a sandy beach that retains water during low tide as the result of capillary action.

zone of saturation   The vertical zone of the sandy intertidal re-gion that is constantly moist.

zones of stress   Regions above or below an organism's optimal range for an environmental variable in which the organism must expend more energy than normal to maintain homeostasis.

Zooxanthellae (ZOH-oh-zan-thel-ee)   Dinoflagellates that are important symbionts of jellyfishes, corals, and molluscs.

zygote (ZY-goht)   The diploid cell that is formed when a male gamete fuses with a female gamete.

Note: Page numbers in *italics* indicate figures, "t" following a page number indicates a table.